ASTROPHYSICS I

STARS

ASTROPHYSICS I

STARS

Richard L. Bowers
LOS ALAMOS NATIONAL LABORATORY

Terry Deeming
DIGICON GEOPHYSICAL CORPORATION

JONES AND BARTLETT PUBLISHERS, INC.
Boston Portola Valley

Editorial offices: Jones and Bartlett Publishers, Inc., 30 Granada Court, Portola Valley, CA 94025.

Sales and customer service offices: Jones and Bartlett Publishers, Inc., 20 Park Plaza, Boston, MA 02116.

Library of Congress Cataloging in Publication Data

Bowers, Richard, 1941–
Astrophysics.
Bibliography
Includes index.
Contents: 1. Stars— 2. Interstellar matter and galaxies.
1. Astrophysics. I. Deeming, Terry. II. Title.
QB461.B64 1984 523.01 83–17234
ISBN 0-86720-018-9

Publisher: Arthur C. Bartlett
Production: Bookman Productions
Book and cover design: Hal Lockwood
Copyeditor: Aidan Kelly
Illustrator: Nancy Warner
Composition: Science Press
Printing and binding: Halliday Lithograph

Printed in the United States of America

Printing number (last digit)
10 9 8 7 6 5 4 3 2 1

The authors gratefully acknowledge permission to use the following figures: **Figure 3.2:** G. O. Abell, *Exploration of the Universe,* 2nd ed. (New York: Holt, Rinehart & Winston, 1969), Fig. 23.3. **Figure 3.3:** W. Becker, "Applications of Multicolor Photometry," in *Basic Astronomical Data,* Vol. 3 of *Stars and Stellar Systems,* edited by K. Aa. Strand (Chicago: University of Chicago Press, 1963), Fig. 27. **Figure 3.4(a):** H. A. Arp, *Ap. J.* 135 (1962):311, Fig. 3. **Figure 3.4(b):** R. L. Wildey, *Ap. J. Supp. 8* (1964):439, Fig. 17. **Figure 3.5:** H. C. Arp, "The Hertzsprung–Russell Diagram," in *Handbuch der Physik,* edited by S. Flügge (Berlin: Springer-Verlag, 1958), Fig. 22. **Figure 3.6:** G. O. Abell, *Exploration of the Universe,* 2nd ed. (New York: Holt, Rinehart & Winston, 1969), Fig. 21.1. **Figure 3.7:** Z. Kopal, *Close Binary Systems* (New York: Wiley, 1959), Fig. 6.1. **Figure 4.3:** S. Chandrasekhar, *Principles of Stellar Dynamics* (New York: Dover Publications, 1960), Fig. 25. **Figure 6.13:** C. Hayashi, R. Hoshi, and D. Sugimoto, *Prog. Theor. Phys. Supp. 22* (Kyoto), 1962, Fig. 3.1. **Figure 8.1:** I. Iben, Jr., "Stellar Evolution Within and Off the Main Sequence," in *Ann. Rev. Ast. Ap.,* 5 (1967):571, Fig. 1. **Figures 8.2 and 8.3:** I. Iben, Jr., *Ap. J.* 141 (1965): 993, Figs. 17 and 2. **Figure 8.4:** M. F. Walker, *Ap. J. Supp. 2* (1956):365, Fig. 4. **Figure 9.1:** E. Novotny, *Introduction to Stellar Atmospheres and Interiors* (London: Oxford University Press, 1973), Fig. 7.25. **Figures 9.2 and 9.3:** Adapted from I. Iben, Jr., *Ap. J.* 147 (1967):624, Figs. 8 and 9. **Figure 9.4:** Adapted from I. Iben, Jr., *Ap. J.* 143 (1966):483, Fig. 4. **Figures 10.7 and 10.8:** I. Iben, Jr., *Ap. J.* 147 (1967), Figs. 10 and 11. **Figure 10.9:** Adapted from R. F. Stein, "Stellar Evolution: A Survey with Analytical Models," in *Stellar Evolution,* edited by R. F. Stein and A. G. W. Cameron (New York: Plenum Press, 1966), Fig. 16. **Figures 10.10, 10.11, 10.12, and 10.13:** I. Iben, Jr., *Ap. J.* 143 (1966), Figs, 5, 8, 10, and 12 and 13. **Figure 10.14:** R. Kippenhahn, H. C. Thomas, and A. Weingert, *Zeitung für Astrophysik* 61 (1965):241, Fig. 2. **Figure 10.16:** Adapted from E. L. Hallgren and J. P. Cox, *Ap. J.* 162 (1970):933, Fig. 1. **Figure 11.2:** R. F. Christy, *Ap. J.* 144 (1966):108, Fig. 21. **Figure 11.3:** R. F. Christy, *Rev. Mod. Phys.* 36 (1964):555, Fig. 9. **Figure 11.4:** R. F. Christy, *Ap. J.* 144 (1966), Fig. 1. **Figures 12.4 and 12.5:** D. D. Clayton, *Principles of Stellar Evolution and Nucleosynthesis* (New York: McGraw-Hill, 1968), Fig. 7.4. **Figure 13.4:** G. Beaudet, V. Petrosian, and E. E. Salpeter, *Ap. J.* 150 (1967): 979, Fig. 3. **Figure 13.5:** H. Y. Chiu, *Stellar Physics,* Vol. I (Waltham, Mass.: Blaisdell, 1968), Fig. 6.23. **Figure 13.7:** D. Ezer and A. G. W. Cameron, *Canad. J. Phys.* 43 (1965):1497, Fig. 6. **Figure 15.1:** R. P. Kirshner, "Supernovas in Other Galaxies," *Sci. Amer.,* Dec. 1976, p. 92. **Figure 15.2:** Adapted from F. Zwicky, "Supernovae," in *Stellar Structures,* edited by L. H. Aller and D. B. McLaughlin (Chicago: University of Chicago Press, 1965), Fig. 16. **Figures 15.4 and 15.5:** R. P. Kirshner, "Supernovas in Other Galaxies," *Sci. Amer.,* Dec. 1976, pp. 97, 94. **Figure 15.7:** R. P. Kirshner and J. Kwan, *Ap. J.* 193 (1974):27, Fig. 2. **Figure 16.10:** R. N. Manchester and J. H. Taylor, *Pulsars* (San Francisco: W. H. Freeman, 1977), Figs. 1.3 and 2.1. **Figure 16.11:** R. N. Manchester and J. H. Taylor, *Pulsars* (San Francisco: W. H. Freeman, 1977), Fig. 1.4. **Figure 16.14:** Adapted from P. Goldreich and W. H. Julian, *Ap. J.* 157 (1969):869, Fig. 1. **Figures 16.15 and 16.16:** R. N. Manchester and J. H. Taylor, *Pulsars* (San Francisco: W. H. Freeman, 1977), Figs. 4.6 and 5.1, and 4.2. **Figure 17.5:** I. D. Novikov and K. S. Thorne, "Black Hole Astrophysics," in *Black Holes,* edited by C. DeWitt and B. DeWitt (New York: Gordon and Breach, 1973), Fig. 5.2.1. **Figures 17.6 and 17.7:** B. Paczynski, "Close Binaries," in *Stellar Evolution,* edited by H. Y. Chiu and A. Murriel (Cambridge, Mass.: MIT Press, 1972), Figs. 1 and 2. **Figure 17.8:** I. D. Novikov and K. S. Thorne, "Black Hole Astrophysics," in *Black Holes,* edited by C. DeWitt and B. DeWitt (New York: Gordon and Breach, 1973), Fig. 5.2.2. **Figure 17.9:** B. Paczynski, "Close Binaries," in *Stellar Evolution,* edited by H. Y. Chiu and A. Murriel (Cambridge, Mass.: MIT Press, 1972), Fig. 3. **Figure 17.10:** Adapted from B. Paczynski, "Close Binaries," in *Stellar Evolution,* edited by H. Y. Chiu and A. Murriel (Cambridge, Mass.: MIT Press, 1972), Fig. 3. **Figure 17.11:** P. Gorenstein and W. H. Tucker, "Supernova Remnants," in *New Frontiers in Astronomy,* edited by O. Gingerlich (San Francisco: W. H. Freeman, 1970), p. 276. **Figure 17.14:** R. N. Manchester and J. H. Taylor, *Pulsars* (San Francisco: W. H. Freeman, 1977), Fig. 5.4. **Figure 17.15:** H. Tannenbaum and W. H. Tucker, "Compact X-Ray Sources," in *X-Ray Astronomy,* edited by R. Giaconi and H. Gursky (Dordrecht, Holland: Reidel Publishing Co., 1974), Fig. 6.8. **Figure 17.16:** H. Gursky and E. Schreier, "The Galactic X-Ray Sources," in *Neutron Stars, Black Holes and Binary X-Ray Sources,* edited by H. Gursky and R. Ruffini (Dordrecht, Holland: Reidel Publishing Co., 1975), Fig. 13. **Figure 17.17:** J. H. Taylor and P. M. McCulloch, *Ann. N.Y. Acad. Phys.* 336 (1980):445, Fig. 2. **Figure 17.18:** D. B. McLaughlin, "The Spectra of Novae," in *Stellar Atmospheres,* edited by J. L. Greenstein (Chicago: University of Chicago Press, 1960), Fig. 1.

Contents

VOLUME I

Preface to Volume I

Student interest in astronomy and astrophysics has grown dramatically during the past decade, and from it has sprung a need for modern texts reflecting the advances in theoretical and observational astronomy of the sixties and seventies. This need has previously been met largely at the introductory (descriptive) level, and at the advanced level. *Astrophysics* is intended to fill the intermediate range. Much of the material that follows developed from course material and lectures presented over a five-year period to upper-level undergraduate and graduate students in the Department of Astronomy and the Department of Physics at the University of Texas at Austin and from a one-semester course presented in the Department of Physics and Astronomy at Texas A & M University.

The text is intended for a senior-level or first-year graduate-level course in astrophysics. Volume I covers a wide range of subjects in stellar astrophysics, and Volume II treats nonstellar astrophysics. We attempt to present a relatively self-contained discussion of core topics that are established and basic to advanced work in the field. More speculative topics that have not been fully explored theoretically, or that do not rest on unambiguous observational data, have been excluded. We did so not because these topics are uninteresting but simply because of length limitations.

We have chosen to emphasize the theoretical approach to astrophysics without attempting to cover the extensive field of observational methods. Nevertheless, ample reference is made to the results of observations and to specific astrophysical systems that confirm or restrict the theory. Detailed models of most astrophysical systems involve a variety of underlying physical processes and principles, and their analysis must often be carried out on high-speed computers. Because an extensive literature on numerical methods already exists, and because their use often involves problems specific to the individual model, we do not discuss computational methods. Instead, we emphasize simple analytic models wherever possible. The results of numerical analysis are, however, incorporated as illustrations.

The subject matter is presented at a level for senior students in physics, astronomy, or physical science who have completed undergraduate course work in mechanics, modern physics, electromagnetic theory, and calculus through differential equations, or for first-year graduate students. A knowledge of descriptive astronomy as developed in any standard introductory astronomy text is also assumed.

The subject matter is organized along more or less conventional lines. The first volume, which deals with stellar astrophysics, develops many topics that are used in the parts on nonstellar astrophysics in the second volume. Part 1, particularly Chapter 1, presents a relatively nonmathematical overview of topics to be covered in more detail in the remaining parts. It serves two primary purposes. The first is to establish a framework within which the reader can structure his knowledge of astronomy and astrophysics. The second is to establish an appreciation at the order-of-magnitude level of basic astrophysical parameters, which serve as benchmarks for more detailed discussions that follow. The first chapter emphasizes order-of-magnitude arguments that, though they may often be mathematically naive, illustrate how basic physical concepts can be used to extract qualitative estimates of astrophysical parameters without recourse to elaborate numerical or analytical methods.

The remaining parts of Volume I develop in a more mathematical way the astrophysics of stellar structure and stellar evolution. Volume II introduces the astrophysics of interstellar matter, the structure and evolution of galaxies and stellar systems, and cosmology.

January 1984

Problems have been included in the text to serve three purposes: (1) to supply order-of-magnitude values obtained from models developed in the text; (2) to extend the discussion once basic concepts have been presented; and (3) to supply details of derivations whose results are used in subsequent discussions. In the last case, sufficient guidance is given so that these derivations are straightforward.

We acknowledge support by the Department of Astronomy and the Department of Physics at the University of Texas at Austin. We are indebted to the faculty and our colleagues for their criticisms and suggestions. In particular, we are grateful to Dimitri Mihalas and Austin Gleeson for valuable discussions and suggestions at the time the manuscript was being developed. We also thank Margaret Burbidge and William Kauffman, who read portions of the manuscript. We acknowledge Digicon Geophysical Corporation and Los Alamos National Laboratory for support during the final stages of manuscript preparation and the Correspondence Center at Lawrence Livermore National Laboratory for typing assistance. Finally, we thank the many students who read and patiently endured preliminary sets of notes.

Richard L. Bowers
Terry Deeming

Part 1

INTRODUCTION

Chapter 1

AN OVERVIEW OF STELLAR STRUCTURE AND EVOLUTION

1.1. Stars

A star is a self-gravitating ball of gas, radiating energy into space. This energy is produced mainly by thermonuclear reactions taking place in the deep interior, but may also be released during contraction or collapse of the stellar core. The star must produce energy in order to maintain enough internal pressure to support itself against its own gravitational field. Stellar structure and evolution are controlled by these two opposing effects: gravity, which tends to collapse the star, and pressure, which tends to expand it. Sometimes one or the other wins slightly, and the star may expand or contract, but there is no doubt which will win in the end. Gravitational collapse must ultimately turn the star into a cold, dead object, like the white-dwarf companion to Sirius, or perhaps into a neutron star, like the pulsar in the Crab nebula, or possibly into a black hole. It may also set off the explosive event we call a supernova. Gravitation is also responsible for the initial formation of stars out of protostellar material. As soon as a large-enough mass of this material becomes detached from its surroundings, it begins to contract, and one or more stars may form from it. Gravitation is the dominant creative force in the universe, and, because of gravitational collapse, it is the dominant destructive force as well.

Stellar Atmospheres

The Sun is a typical star, although we see far more detail on it than on any other star. When we look at a typical star, we see only the outermost layers, because stellar surface material is relatively opaque. These outer regions (the atmosphere) are the only parts generally accessible to direct observation.

Our primary source of information about a star is the light emitted from its surface. For example, from a star's visible light we can measure its total luminosity, and therefore its total energy output. We can examine the star's spectrum, which is its energy output as a function of wavelength or frequency. The analysis of stellar spectra, which makes possible the study of stellar atmospheres, is fundamental to stellar astrophysics. Such analysis gives us information on the physical structure of the atmosphere, for example, the temperatures, pressures, and densities encountered, and may indicate the presence of turbulence, convection, or magnetic fields. It enables us to deduce the chemical composition of the atmosphere. Finally, some large-scale properties of a star, such as its surface gravity and rotational velocity, can often be calculated. Needless to say, it is for the Sun that these quantities are known most accurately. However, the Sun is not an especially convenient object on which to test our understanding of stellar atmospheres, because its outer layers are turbulent, and current theories and mathematical techniques are not quite capable of handling turbulence in atmospheres.

Stellar interiors are in some ways simpler than stellar atmospheres, in other ways more complicated. The most important complication is that they are generally unobservable, although detection of neutrinos escaping directly from the core of the Sun and other stars could offer indirect observational information on their interior structure. In general, therefore, we rely heavily on theory for conclusions about stellar interiors. The theory of stellar interiors attempts to relate the intrinsic properties of the star (total mass, chemical composition, and possibly rotation) to its gross observable properties (luminosity, surface gravity, surface temperature).

We would also like to know how these quantities are likely to change with time; for example, is the Sun likely to explode as a supernova tomorrow, in a few billion years, or at all? Astronomers refer to the way that the properties of a single star change with time as *stellar evolution.* The word evolution is used differently here than it is in biology, where it refers to changes in the general characteristics of a species after several generations of individual births and deaths. This kind of change from generation to generation is also encountered in astrophysics. For example, most stars probably lose mass during some stage of their evolution. The process may be gradual (stellar winds, planetary nebula formation), more dramatic (novae, mass exchange in binary systems), or cataclysmic (supernovae). In any case, some of the ejected material may have undergone significant changes in composition through nucleosynthesis. In this way the interstellar gas is enriched, and the evolution of subsequent generations of stars will be modified. The two types of evolution are therefore related.

The time-scale on which stellar evolution takes place is obviously very long. How can we hope to verify any theories about stellar evolution when most changes require many lifetimes? The answer lies in the multitude of stars in the sky.

Although we can never hope to observe significant changes in more than a very few stars (such as variable stars, novae, or supernovae), we can say that star A is probably what star B looked like a few billion years ago, or will look like a few billion years hence. That is, we can try to relate our predictions about stellar evolution to the various types of stars we see around us, and thus be able to say that such and such a star is an object of mass M, and some (initial) chemical composition, in some stage of evolution. This line of attack is aided by the fortunate existence of star clusters whose members are so obviously physically associated that we can be confident that they all began with nearly the same initial composition at essentially the same time, and are therefore all of the same age. The major differences between them should result from the differences in their masses (and possibly from their rotational velocities and magnetic fields).

1.2. Energy Transport and Generation in Stars

One way in which stellar interiors are simpler than stellar atmospheres is in the way energy is transported from point to point. This is called *energy transfer.* Most of the time energy is transported by electromagnetic radiation. In the deep interior, the photon's mean free path is extremely small compared with the dimensions over which other stellar variables change, and this radiation has an almost isotropic distribution. Consequently, energy transport can be adequately described by the diffusion approximation. This fact greatly simplifies the analysis of radiative energy transport.

In the atmosphere, on the other hand, the radiation field is strongly anisotropic (or else no radiation would escape the star), the opacity is strongly wavelength-dependent (otherwise we would see no characteristic line spectra), and the photon mean free path is long relative to the scale height. The problem of radiative transfer in a stellar atmosphere is therefore complicated and difficult. At some stages of evolution, if the opacity of the stellar material becomes too high, the temperature gradient too large, or both, the star becomes unstable, and convection sets in. This represents an additional means of transferring energy from point to point and may cause chemical mixing. In this way, some of the chemical inhomogeneities that develop in a star may be ironed out. If convection occurs near the surface of a star (as it does in the Sun), it may extend into the visible part of the stellar atmosphere and produce additional complications, such as solar granulation, sunspots, or flares.

Stellar Energy Generation

Long before nuclear and thermonuclear energy became of interest on the Earth, astronomers realized that the source of stellar energy must be subatomic. The principal subatomic source of stellar energy is the thermonuclear conversion of hydrogen to helium in the deep interior; this is true for most stars for most of their lifetimes. Sometimes other sources of thermonuclear energy become important; sometimes gravitational energy released during collapse, or even neutrino energy release, is important. Nevertheless, it is the conversion of hydrogen to helium that establishes the time-scale of a star's life. The nuclear parameters for this reaction determine the time-scale on which the stars evolve, and thereby help to determine the time-scale of the universe.

Energy Loss (Luminosity)

The rate of energy loss from a star is important for its structure. It can be calculated for the Sun by measuring the amount of solar energy falling on a unit area above the Earth's absorbing atmosphere. From this quantity, known as the solar constant (1.374×10^6 erg $\text{cm}^{-2}\text{sec}^{-1}$), and the mean distance of the Sun from the Earth, which is defined as one astronomical unit or A.U. (one A.U. $= 1.496 \times 10^{13}$ cm), it is easily shown that the energy output for the sun is $L_\odot = 3.86 \times 10^{33}$ erg sec^{-1}. The solar luminosity $L_\odot$ represents a standard unit of energy-loss rates in all branches of astrophysics.

Problem 1.1. What collecting area for solar radiation is required to light a 100-watt light bulb, if solar energy can be converted to electrical energy with a 100 percent efficiency?

Problem 1.2. By what percentage does the solar "constant" vary during the year because of the eccentricity ($e = 0.017$) of the Earth's orbit around the Sun?

The *absolute luminosity* of a star refers to the total electromagnetic energy output at all wavelengths, ignoring the possibility that energy is being output in other forms (e.g., particles, neutrinos, gravitons). This luminosity is called the *bolometric luminosity*. In practice, we can not observe all wavelengths of the spectrum. Instead the absolute luminosity is obtained by observing the object in a restricted part of the electromagnetic spectrum, and then applying a "bolometric correction" to allow for the unobserved wavelengths.

Related to the absolute luminosity of a star is the star's effective temperature, which is defined in the following way. A unit area of a perfect radiator or *black body* at an absolute temperature T radiates energy at a rate

$$\sigma T^4 \text{ erg cm}^{-2} \text{ sec}^{-1}, \tag{1.1}$$

where σ is the Stefan-Boltzmann constant. A star does not have a well-defined surface or a well-defined surface temperature, but if it did, and if the stellar surface were a black body, the total luminosity would be

$$L = 4\pi\sigma R^2 T^4. \tag{1.2}$$

We can use this relation to define an effective temperature, T_{eff} such that, for a real star with absolute luminosity L and radius R,

$$L = 4\pi\sigma R^2 T_{\text{eff}}^4. \tag{1.3}$$

Real stars, not being black bodies, will have effective temperatures that are not the same as the atmospheric temperatures deduced by spectral analysis. Nevertheless, the differences are generally not great—perhaps a few hundred degrees—so effective temperatures can be taken as close to atmospheric temperatures for many purposes.

Problem 1.3. What is the effective temperature of the Sun? (Answer: about 6,000 K)

1.3. STELLAR TIME-SCALES

We said above that self-gravitation in effect determines the state and evolution of a star. How do we know this? To put the question more specifically, suppose there were no internal support for the Sun. How long would it take for the Sun's self-gravity to cause a significant change in its radius? We can arrive at an order-of-magnitude estimate of the answer in two ways; both are simple, instructive, and typical of many arguments we will make in later sections.

Free-fall Time-scale

First, we ask what physical quantities the answer will depend on. Since we are talking about gravitational free fall of the star, with no restraining forces, the answer depends only on the strength of the gravitational field and on the physical dimensions involved. The time-scale must therefore be a function only of the mass M and radius R of the star, and of the Newtonian gravitational constant G, which, given the mass and the radius, determines the strength of the field. Now, there is only one expression with the physical dimensions of time that can be constructed from G, M, and R. It is

$$t = (R^3/GM)^{1/2}. \tag{1.4}$$

Problem 1.4. Prove this by showing that the dimensions of $(R^3/MG)^{1/2}$ are time. The force law $F = -M^2G/R^2$ gives the units of G.

The answer must consist of some dimensionless constant multiplied by expression (1.4). It is a result of experience (and to some extent a tenet of faith) that dimensionless multiplying factors are usually of the order of magnitude of unity (i.e., about one); so if we were to take expression (1.4) by itself as the answer to the problem, we would usually not be wrong by more than an order of magnitude.

The other approach is possible when we can write down a mathematical model of what is happening physically. Since we assume that there is no internal pressure in the star, a small mass m at the surface will fall unhindered along a path $r(t)$ that satisfies the differential equation

$$\frac{d^2r}{dt^2} = -\frac{GM}{r^2}, \tag{1.5}$$

where $r = R$ at $t = 0$. We can solve this differential equation directly, or with more insight say that during the fall of the particle over a distance R in a time t, the average value of d^2r/dt^2 must be $\approx -R/t^2$. Therefore we expect

$$-\frac{R}{t^2} \approx -\frac{MG}{R^2} \tag{1.6}$$

and hence

$$t = (R^3/GM)^{1/2}, \tag{1.7}$$

which is the same result as (1.4).

Problem 1.5. Imagine that equation (1.5) is the correct equation of motion for this problem, and that the surface begins with $r = R$ at $t = 0$. How long does it take for the star to collapse to half its initial size? How does this compare to the simpler estimate?

With mass and radius measured in solar units, $M_\odot = 1.99 \times 10^{33}$ g and $R_\odot = 6.96 \times 10^{10}$ cm, this time is easily shown to be

$$t_{ff} = 1.59 \times 10^3 \, (M/M_\odot)^{-1/2} \, (R/R_\odot)^{3/2} \text{ sec.} \tag{1.8}$$

For the Sun this is about 0.44 hrs, or 26.6 min. We have written t_{ff} to identify this as the free-fall time-scale.

Notice that the same result applies roughly to any situation in which gravitational fields arise from a mass M moving a distance scale R in a time t. For instance, the orbital period of the Earth about the Sun is given roughly by the same expression where the distance of the Earth from the Sun is used for R (Kepler's third law).

Problem 1.6. Think of some other astronomical situations in which this analysis can be applied, and verify that it gives the correct result.

Notice also that, since M/R^3 is proportional to the mean density, result (1.7) can be written

$$t \sqrt{\bar{\rho}} \simeq 1/\sqrt{G}. \tag{1.9}$$

(For the record, the mean density of the Sun is $\bar{\rho} = 3M/4\pi R^3 = 1.41$ gm cm^{-3}.) This result is important for the theory of pulsating variable stars, where the balance between pressure and gravity is only maintained on average. The characteristic period of oscillation P is related to the mean density of the star by the simple relation $P\bar{\rho}^{1/2} \simeq G^{-1/2}$.

Problem 1.7. Show from the analysis above that $P\bar{\rho}^{1/2} \simeq G^{-1/2}$.

Generally, numerical estimates obtained in this way are reasonable. In any event, the result reveals the way observable properties (in this case, the collapse time) depend on characteristics of the system. For example, result (1.7) indicates that collapse time decreases slowly with increasing mass, but increases more rapidly with increasing initial radius.

Problem 1.8. Suppose that two pulsating stars have equal masses but different radii. How would you expect their periods to compare?

Clearly the free-fall time for the Sun is very fast by astronomical standards. It is much less than the evolutionary time-scale t_{ev}, which we know (from the existence of objects that old on the Earth) is at least many billions of years:

$$t_{ff} \ll t_{ev}. \tag{1.10}$$

Apparently the Sun can adjust very quickly to an imbalance in the pressure-gravity equilibrium. A star that can do so is, to a high degree of accuracy, in a state of hydrostatic equilibrium. During stages of pulsational instability, equilibrium is maintained in terms of the average over several pulsation periods. When $t_{ff} \sim t_{ev}$, the evolution is rapid, and the concept of hydrostatic equilibrium becomes inapplicable. The hydrostatic-equilibrium condition is one of the basic physical principles we will use to construct a conceptual model of a stellar interior and a stellar atmosphere.

Kelvin-Helmholtz Time-scale

The rate of stellar evolution is determined primarily by the available energy. A major energy reserve is the star's gravitational potential energy. Two masses m and m' a distance r apart have a gravitational potential energy given by the classical expression

$$\Omega = -\frac{Gmm'}{r}. \tag{1.11}$$

The total gravitational potential energy is the integral of this expression over the star. Dimensionally, it must have the form $\alpha M^2G/R$, where M is the total mass, R the radius, and α a factor usually of order unity. If this estimate is applied (with $\alpha = 1$) to the Sun, one finds

$$\Omega_\odot = -M^2G/R$$
$$\simeq 4 \times 10^{48} \text{ ergs}. \tag{1.12}$$

Another important time-scale may now be constructed. Suppose the past energy output of the Sun could all have come from the release of gravitational potential energy during contraction from an initially very large (quasi-infinite) radius to its present size. This would release about 4×10^{48} ergs. The present luminosity of the Sun is 4×10^{33} erg sec^{-1}, and it does not seem to be changing much. At this constant rate, assuming 100 percent efficiency for the conversion of gravitational to radiative energy (which is theoretically not possible, but sets an upper limit), the Sun could have been shining at more or less its present brightness for about

$$t_K = 4 \times 10^{48}/4 \times 10^{33}$$
$$= 10^{15} \text{ sec} = 30 \text{ million yrs}. \tag{1.13}$$

The subscript K identifies this as the Kelvin (or Kelvin-Helmholtz) time-scale. Although it is obvious that $t_{ev} \gg t_{ff}$ for the Sun, it is perhaps not so obvious that $t_{ev} > t_K$. Indeed, until early in this century, it was believed that the energy of the Sun did come from its gravitational contraction, and that the Sun, and presumably the Earth, was a few million years old. Because of this rather entrenched theory, Charles Darwin and Kelvin had several public disagreements when the former's *Origin of the Species* was published, because the time-scale required for biological evolution was considerably greater than t_K. At about the same time, the geologist Sir Charles Lyell maintained that the ocean floor had to be at least a billion years old, if the behavior of sediments were to be understood. With the application of radioactive dating methods to rocks by (among others) Rutherford, it became clear that the Earth itself is several billion years old, and the Sun must be at least as old as the Earth.

Problem 1.9. Review the evidence that the Earth is a few billion years old, rather than a few million.

Einstein Time-scale

In 1929, Eddington took up the crucial question of how to deal with this discrepancy, and gave the historically interesting arguments leading to the conclusion that the Kelvin-Helmholtz theory is wrong and that there must be a subatomic source of energy. Eddington also gave another interesting quantity, the total energy equivalent of the mass of the Sun (recall that Einstein published his work on special relativity in 1905–1906.) This energy equivalent is, for the Sun,

$$E = M_{\odot} c^2 = 1.79 \times 10^{54} \text{ ergs.}$$

If all this energy could be converted to radiation, the Sun could continue shining at its present rate for as long as

$$t_E = 4.6 \times 10^{20} \text{ sec} = 1.4 \times 10^{13} \text{ yrs}$$

(E stands for Einstein). Thermonuclear fusion, which is responsible for most of the energy release, occurs only within the inner 10 percent of the Sun's mass. If all of this matter were converted into energy, the Sun could radiate at its current rate for approximately 1.4×10^{12} years. Anticipating the result that, for the Sun, t_{ev} is about nine billion years, we can say that the efficiency of conversion of mass to energy in the Sun is about 0.007.

The evolutionary time for stars like the Sun on the main sequence is determined by a series of nuclear processes that convert hydrogen into helium. The rate for the process is set by the slowest of its steps, which is the formation of deuterium:

$$\mathrm{p + p \rightarrow p + n + e^+ + \nu_e \rightarrow d + e^+ + \nu_e}.$$

This reaction involves two processes: (1) the weak interactions, which convert a proton into a neutron; and (2) the strong interactions, which bind the neutron and proton to form deuterium. The essential characteristic of the first step is its extreme weakness. For example, the lifetime of a free neutron, which is governed by the weak interactions, is about 15.3 minutes, whereas the time-scale characteristic of the strong interactions is typically about 10^{-23} seconds.

We can estimate the main-sequence lifetime of the Sun from the assumption that deuterium formation sets the overall rate for helium formation. By the argument detailed in Chapter 7, the evolutionary time-scale for the Sun turns out to be

$$t_{ev} \approx 10^{10} \text{ yrs.} \tag{1.14}$$

1.4. Static Configurations (Hydrostatic Equilibrium)

We have thus far emphasized the role of gravitation in the pressure-gravity equilibrium. Now consider the pressures required to support a star of mass M and radius R. A simple estimate is possible. The sphere of radius $r = 2^{-1/3}R = 0.79R$ splits a star of uniform density into two equal mass portions, each of $M/2$. The surface area of the inner sphere is $4\pi 2^{-2/3} R^2$. The total downward gravitational force at a point on the surface of this sphere is $G\ (M/2)^2/(R/2^{1/3})^2$. The pressure required for support at this point is therefore the gravitational force divided by the area, or

$$P = \frac{(M^2G/4)(2^{2/3}/R^2)}{4\pi\, 2^{-2/3}\, R^2} = \frac{2^{4/3}}{16\pi}\frac{M^2G}{R^4}. \tag{1.15}$$

This result could also be obtained by dimensional arguments. We expect that stars are not uniform spheres, but we may say that pressures of the order of GM^2/R^4 will be encountered in their interiors. (Incidentally, the analysis is not restricted to stars; any self-gravitating object of mass M and radius R supported by internal pressure, such as a planet, will need pressures of this order.) For typical pressures inside the Sun,

$$P \simeq M_{\odot}^2 G/R_{\odot}^4 = 1.12 \times 10^{16} \text{ dynes cm}^{-2}. \tag{1.16}$$

In a more exact model for the Sun, the central pressure is 1.3×10^{17} dynes cm^{-2}, and the pressure at $r = R/2$ is 5.9×10^{14} dynes cm^{-2}. At $r = 0.79R$, $P = 5.6 \times 10^{12}$ dynes cm^{-2}. A word or two of warning can be drawn from these comparisons. First, our rough estimates yield the right general order of magnitude. Second, numerical coefficients are not always of the order of magnitude of unity. For example, the factor $2^{4/3}/16\pi$

is about 0.05; so even if the uniform-density model were correct, we have "lost" almost two orders of magnitude by neglecting that factor. This is one reason why our value of 10^{16} dynes cm^{-2} is close to the central pressure of the Sun but is rather high for an "average" internal pressure. Finally, the uniform-density model is clearly too naive for the Sun, and this "geometrical" effect can contribute to the error. Nevertheless, rough estimates of physical quantities are extremely useful, and often enable us to decide what is important and what is not. Frequently they can be very good if obtained with some care.

Equations of State

Suppose we now want to know the temperature in the interior of the Sun. This involves something new, because only pressure is determined by the gravitational field. Temperature must be related to pressure by an equation of state, which can be written $P = P(\rho, T, C)$, where ρ is the density and C the chemical composition. A typical equation of state is the perfect gas law

$$P = nkT = \frac{\rho kT}{m} = \frac{\rho kT}{\mu m_H}, \tag{1.17}$$

where n is the particle density, m is the mass per particle, $m_H = 1.67 \times 10^{-23}$ g, and μ is the mean molecular weight. The perfect gas law is a good approximation to the state of matter in most regions of most stars, but it is not always applicable. In particular, highly collapsed stars (white dwarfs or neutron stars) may have degenerate equations of state in which the pressure is much less sensitive to the temperature than in the perfect gas law. Pressure may even be independent of temperature. For example, in a highly condensed white dwarf, the equation of state is approximately

$$P = K_1\rho^{5/3}; \tag{1.18}$$

in even more compact objects it can be approximated by

$$P = K_2\rho^{4/3}. \tag{1.19}$$

Furthermore, in hot stars, although the matter may behave like a perfect gas, the radiation field is so strong that a significant part of the support for the star comes from radiation pressure. The radiation pressure in a thermodynamic enclosure at temperature T is given by

$$P = \frac{1}{3}aT^4. \tag{1.20}$$

Even in nondegenerate stellar material (i.e., in ordinary stars), the total equation of state contains both gas pressure and radiation pressure, and has the form

$$P = \frac{\rho kT}{\mu m_H} + \frac{1}{3}aT^4. \tag{1.21}$$

Stellar Temperatures

Suppose, then, that we want to estimate the interior temperature of the Sun. We will try first pure gas pressure, then pure radiation pressure, and then compare the two. For gas-pressure support, the temperature is given by (1.17):

$$T = \mu m_H P/\rho k. \tag{1.22}$$

If we take $\rho = 3M_\odot/4\pi R_\odot^3$, and use (1.19) for the pressure $P = \alpha\, GM^2/R^4$, where α is a numerical factor, we find, after eliminating ρ and P, the gas temperature

$$T_g = \frac{4\pi}{3}\mu\frac{m_H G}{k}\alpha\frac{M}{R}. \tag{1.23}$$

We must next decide what to take for the molecular weight. This is an annoying detail that is easy to guess at, but hard to work out exactly. Suppose we already know that the stellar material is mostly hydrogen and that it is almost completely ionized. There are thus two particles for every m_H, and the mean molecular weight is ½. Taking $\mu = ½$, and inserting the physical constants ($m_H = 1.67 \times 10^{-24}$ g, $k = 1.38 \times 10^{-16}$ erg deg^{-1}), we find

$$\begin{aligned} T &= 4.81\,\alpha \times 10^7\text{K} \\ &= 2.4 \times 10^7\text{K} \end{aligned}$$

if $\alpha \simeq 0.5$. To support the Sun on gas pressure alone would require internal temperatures of the order of 20 million degrees.

Now consider the temperature required if radiation pressure were the sole support. Here (1.20) and (1.15) may be combined to give

$$\frac{1}{3}aT^4 = \alpha\frac{M^2G}{R^4}. \tag{1.24}$$

Solving for the temperature, we find

$$T_R = \left(\frac{3\alpha G}{a}\right)^{1/4}\frac{M^{1/2}}{R}. \tag{1.25}$$

Putting in the constants, we get

$$\begin{aligned} T_R &= 4.6 \times 10^7\alpha^{1/4}\text{K} \\ &= 3.9 \times 10^7\text{K} \end{aligned}$$

if $\alpha \simeq 0.5$. Thus a slightly higher (but not much higher) temperature is required to support the sun using pure radiation pressure instead of pure gas pressure. (The comparison is also sensitive to α). Although gas pressure is most important in the Sun, radiation pressure thus cannot be neglected. Notice that both expressions for T depend on the same power of the radius (R^{-1}), whereas T_R increases much more slowly with mass than does T_g. Therefore, the higher the mass of the star, the more easily radiation pressure can support it, because it can do so at a lower temperature. In principle, we can estimate the mass of the star at which radiation pressure provides more support than gas pressure by equating the two expressions for temperature and solving for M:

$$\left(\frac{3\alpha G}{a}\right)^{1/4}\frac{M^{1/2}}{R} = \frac{4\pi}{3}\frac{\mu G m_H}{k}\alpha\frac{M}{R}, \tag{1.26}$$

$$M_{\text{crit}} = \alpha^{-3/2}\frac{3\sqrt{3}}{16\pi^2}\frac{1}{\mu^2}\frac{k^2}{G^{3/2}m_H{}^2a^{1/2}}. \tag{1.27}$$

Before we go further with a calculation that produces a result like this, it will be wise to make sure that the factor containing physical constants, i.e., $k^2/G^{3/2} \times M^2a^{1/2}$, really has the dimension of a mass. We find

$$\frac{k^2}{G^{3/2}m_H{}^2a^{1/2}} = 4.61 \times 10^{33}g = 2.3\ M_\odot,$$

which yields

$$M_{\text{crit}} = \frac{0.23}{\mu^2}\alpha^{-3/2}\ M_\odot. \tag{1.28}$$

For $\mu = ½$ and $\alpha = 0.5$, $M_{\text{crit}} \simeq 2.5\ M_\odot$.

Most stars have a mass for which both radiation pressure and gas pressure have to be taken into account. Detailed calculations show that gas pressure is most important for stars of less than about two solar masses, whereas radiation pressure becomes increasingly important for stars of larger mass (at least on the main sequence).

1.5. The Virial Theorem

A star is composed of many particles, about 10^{57} in the Sun. The total energy of this system is made up of (a) the mutual gravitational energy of the particles, Ω, and (b) the energies (mostly kinetic) of the particles themselves (including the photons), which is known as the internal energy, U. The total energy of the system is then $E = U + \Omega$. Just as is true for a planet in orbit around the Sun, if the system is to be bound, its total energy must be negative. Furthermore, we shall discover later that in a gravitational system where the average particle energies are nonrelativistic, but which is in equilibrium, the internal kinetic energy is equal to exactly minus one-half the gravitational potential energy. That is,

$$\begin{aligned} U_k &= -\frac{1}{2}\Omega, \\ E &= U_k + \Omega = \frac{1}{2}\Omega < 0. \end{aligned} \tag{1.29}$$

This result is known as the virial theorem. For a star supported by particles whose energies are relativistic (such as photons), the virial theorem is

$$\begin{aligned} U_k &= -\Omega, \\ E &= U_k + \Omega = 0. \end{aligned} \tag{1.30}$$

A star described by (1.30) can be in equilibrium only if its total energy is exactly zero. This remarkable result, which we will examine in more detail later, explains why there is an upper limit to the mass that can be supported by photons (giving upper mass limit on the main sequence of about 50 $M_\odot$), or by electrons (white dwarfs, about 1.2 $M_\odot$) or by neutrons (neutron stars, about 2 $M_\odot$).

We may use the virial theorem to illustrate how energy changes occur in a star that is maintaining equilibrium. Specifically, suppose the star contracts slightly, reducing R and hence making Ω more negative. In order to restore equilibrium, the star must reduce its total energy by radiating away an amount $\frac{1}{2}|\Delta\Omega|$. In the process, the internal energy has increased by $\frac{1}{2}|\Delta\Omega|$. That is, half the released gravitational potential energy goes into heating the star, and half is radiated away. The star therefore gets hotter by losing energy. The specific heat (in equilibrium) is negative. This is an important general characteristic of systems in gravitational equilibrium. A more familiar example is that as an orbiting Earth satellite loses energy by friction with the upper atmosphere of the Earth, it moves faster in its orbit.

1.6. Relativistic Effects

Relativistic effects (both special and general) will be important in stellar structure when

(1) the velocities of the particles involved are comparable with the speed of light,
(2) energies are comparable with rest masses (through $E = mc^2$), or
(3) gravitational fields are strong.

These conditions are not all independent. We have already remarked that photons are relativistic particles. Furthermore, the electrons in white dwarfs and neutrons in neutron stars may, because of quantum-mechanical effects, acquire relativistic velocities.

Under what circumstances might we have to take account of the other possibilities? For example, in what kind of star would the gravitational potential energy be comparable with the rest mass of the star? We have already seen that the gravitational potential energy of the Sun is about 10^{48} ergs, and its rest-mass energy is about 10^{54} ergs. The mass-equivalence of the potential energy would therefore result in a correction of only about one part in 10^6. However, in more compact stars the correction may be significant. In general, such effects will be important when the gravitational potential energy is comparable to the rest-mass energy, $|\Omega| \sim Mc^2$. This happens when

$$R \sim \frac{MG}{c^2}. \tag{1.31}$$

For convenience we will define $R_g \equiv 2\,MG/c^2$ as the gravitational radius by an object of mass $M_\odot$ if $R \simeq R_g$. It is then no longer possible to describe gravitational fields using Newton's law. Instead Einstein's theory of gravitation (general relativity) must be used. When $MG/c^2 \ll R$, the distinction is usually unimportant. General relativity is important primarily for the structure of neutron stars, which have a mass near $M_\odot$ confined within a radius of a few tens of km.

Problem 1.10. What is the factor GM/Rc^2 for a neutron star?

Some care is required even where $MG/c^2 \ll R$. In normal stars, there is equilibrium between internal kinetic energy and gravitational potential energy, especially in supermassive stars, supported by photon pressure, where the total energy must be exactly zero. Under such conditions, a general-relativistic correction may be significant in deciding the equilibrium of the star, even though the correction is a relatively small fraction of the total energy.

An object whose radius $R \simeq R_g$ is referred to as relativistic, not only because its gravitational field is strong, but also because it can induce relativistic speeds in nearby objects. For example, a particle dropped from rest onto a relativistic star acquires a kinetic energy comparable to its own rest mass.

Problem 1.11. Show this for a relativistic star of arbitrary mass.

Black Holes

A gravitationally collapsed mass (black hole) is an extreme example of a relativistic object. To illustrate the properties of black holes, consider the ratio of a test particle's gravitational potential energy Ω to its total energy E,

$$\frac{|\Omega|}{E} \sim \frac{mMG/R}{mc^2} = \frac{MG}{Rc^2}, \tag{1.32}$$

which is independent of the test particle's mass m. Special relativity associates a mass equivalent, E/c^2, with any object whose energy is E; general relativity demands that all forms of mass act as sources of gravitational fields and respond to external gravitational fields. A photon of energy $h\nu$ must therefore lose energy as it moves away from a massive body; so (1.32) must apply to photons as well as to material particles.

Associating the reduction in photon energy $\Delta E = h\Delta\nu$ with the change in the photon's gravitational potential energy as it escapes from the mass M, $\Delta E = \Omega$, we find

$$\frac{|\Omega|}{E} \sim \frac{h\Delta\nu}{h\nu} = \frac{\Delta\nu}{\nu} \sim \frac{MG}{Rc^2}, \qquad (1.33)$$

where R is the radius of M. The reduction in photon energy

$$\Delta\nu \simeq -\nu\Delta\lambda/\lambda \qquad (1.34)$$

is referred to as a *gravitational redshift*. Since $R = R_g$, a black hole has the property that $\Delta\nu \simeq \nu$. In other words, the photon's energy is redshifted to zero (infinite wavelength). We conclude that the photon is in effect trapped within the black hole. Although (1.33) applies only to a photon assumed to reach arbitrarily great distances from M, general relativity shows that the same result holds for escape of a photon out to any point $r > R_g$.*

For a spherically symmetric mass M, the gravitational (or Schwarzschild) radius is

$$R_g = 2MG/c^2 = 2.96(M/M_\odot)\text{km}. \qquad (1.35)$$

The surface $4\pi R_g^{\,2}$ is called the black hole's event horizon, and bounds the region about which an external observer can obtain no information. We emphasize that R_g does not define a surface in the usual sense. The mass of the black hole is located at $r = 0$.

Problem 1.12. Use arguments similar to those in the preceding and in (1.33) to show that material particles cannot escape from a black hole.

Once a mass M is compacted within a region of extent R_g, it can no longer remain in hydrostatic equilibrium. Instead, it must dynamically collapse to arbitrarily high densities and zero volume in a time given approximately by (1.7). For an object of one solar mass and $R = R_g$, this time is $\sim 10^{-5}$ seconds. In collapsing to a black hole, the object may emit a significant amount of gravitational radiation. In fact, if a star collapsing to a black hole could liberate as much as 1 percent of the released potential energy in the form of gravitational radiation, the net energy release could equal that from all previous thermonuclear processes in a star of comparable mass. Such large energy processes may be important in understanding the massive energy losses observed in galactic nuclei and quasars.

Because of the intense gravitational fields near relativistic objects, unusual physical processes may be observable. The accretion of hydrogen onto either a neutron star or a black hole can release electromagnetic radiation in the gamma-ray or hard x-ray region. If the infalling matter is sufficiently dense, it will be optically thick, and the radiation that finally escapes may peak in the x-ray, the ultraviolet, or even the optical region. Theory predicts that x-ray luminosities as high as 10^{38} ergs/sec are possible. The recent detection of x-ray emission from compact galactic sources with $L_x \simeq 10^{38}$ ergs/sec supports the conjecture that some stars evolve into neutron stars or black holes and then accrete matter.

A significant fraction of a galaxy's mass could ultimately be locked up in black holes. In our Galaxy, roughly 40 percent of the mass in the disk is in stars more massive than about 1.4 $M_\odot$ (the mass limit for stable white dwarfs) and about 20 percent is in stars more massive than 5 $M_\odot$. Unless these objects can reduce their mass below 1.4 to 2.5 $M_\odot$ before ending their evolution, they will likely become black holes. Therefore 4 to 8 $\times$ 10^{10} $M_\odot$ (roughly 3 $\times$ 10^9 to 10^{10} objects) may undergo gravitational collapse in our Galaxy.

1.7. Star Formation

The estimated ages of stars in our Galaxy range from 10^{10} years (the Sun) to less than 10^6 years (the bright stars of the Orion nebula). The oldest stars may have formed out of primordial hydrogen and helium when the Galaxy formed, but the presence of young hot stars in regions of relatively dense interstellar matter that could not have been in existence for more than 10^6 to 10^7 years implies that star formation must be an ongoing process, at least in the disk of spiral galaxies. The following simple argument suggests how stars are born in regions of interstellar matter whose density ρ_0 is greater than average.

*Gravitational redshift from objects with $R > R_g$ in principle offers a useful method for measuring a system's mass. To date, gravitational redshift mass estimates have been made for white dwarfs and neutron stars.

Protostars

A hydrogen atom in an interstellar gas cloud of mass M and radius R will have a total energy of order

$$E_{\text{tot}} \simeq \frac{1}{2} m_H v^2 - \frac{m_H MG}{R}.$$

The atom will be gravitationally bound to the cloud if $E_{\text{tot}} < 0$ or, equivalently (if the cloud's density ρ_0 is assumed constant and its temperature $T \approx m_H v^2/k$), if the cloud's radius is greater than the Jeans radius,

$$R_J \simeq \sqrt{\frac{\pi kT}{m_H G \rho_0}}. \qquad (1.36)$$

Problem 1.13. Derive (1.36) for R_J, and explain why bound clouds must have $R > R_J$.

The mass of interstellar gas that becomes self-bound is therefore of order

$$\begin{aligned} M_J &\simeq 4\pi R_J^3 \rho_0/3 \\ &= (4\pi/3)(\pi k/m_H G)^{3/2}(T^3/\rho_0)^{1/2} \\ &= 5.1 \times 10^{-10}\,(T^3/\rho_0)^{1/2}\, M_\odot. \end{aligned} \qquad (1.37)$$

The bulk of interstellar matter consists of neutral hydrogen and is characterized by $T \simeq 100$ K and $\rho \lesssim 10^{-22}$ g/cm^3, so that $M_J \gtrsim 5 \times 10^4\, M_\odot$ or more. As the cloud contracts, it may fragment into 10^3 to 10^4 protostars having typical stellar masses, which is called a young stellar association. The details of such a scenario are complicated, because they are sensitive to the pressure and the properties of dust grains, to large-scale magnetic fields, and to rotation and turbulence.

The Zero-Age Main Sequence

A star is said to be born when its core commences hydrogen burning. The new star is then on the zero age main sequence, and it depends on thermonuclear burning for support. The less the initial stellar mass, the lower the central temperature that can be attained by gravitational contraction, as suggested by (1.23). If the total mass is small, gas pressure becomes ineffective for internal support, gravitational contraction occurs, and the density rises. At sufficiently high density, matter can support itself against gravity even at zero temperature, where the gas pressure vanishes. The source of pressure at high density is the electrons, which are released when atoms are pressure-ionized, and which obey the Pauli exclusion principle. The Pauli principle in effect states that no two electrons (or, more generally, two fermions) confined within a given region (such as a star) can have the same momentum and intrinsic angular momentum (spin). Most of the electrons in dense matter must be in a state of continual motion, and this continual motion results in a pressure that increases as the matter density increases. When this occurs, the matter is said to be degenerate. Consequently, a star must be massive enough to attain a core temperature of about 10^7 K before degeneracy halts gravitational contraction. This sets a lower limit of order

$$M_{\text{min}} \simeq 0.05\, M_\odot \qquad (1.38)$$

for an object able to reach the hydrogen-burning main sequence. We note that Jupiter has a mass not far below this limit.

An upper limit to main-sequence stellar masses may be set by two considerations. First, the luminosity increases with increasing mass; so massive stars will be bright, and will, on the main sequence, have high surface temperatures and high luminosities. Radiation is scattered by free electrons, and the continual bombardment will exert an outward force on them that increases with the luminosity. The electrons are electrostatically coupled to the positive ions (mostly H^+ in the outer stellar regions), and in order to maintain charge neutrality, the force is transmitted to these as well. We can estimate the magnitude of this force in the following way.

The flux of momentum carried by the radiation past a unit area a distance r from the star's center is $L/4\pi r^2 c$. The cross section for electron-photon scattering is roughly $\sigma_T = (8\pi/3)(e^2/m_e c^2)^2 = 6.7 \times 10^{-25}$ cm^2; so the effective outward force exerted by the radiation on an electron (and transmitted to the ions) is $\sigma_T L/4\pi r^2 c$. This outward force is opposed by the gravitational force $m_H MG/r^2$ of the star. Equating these two forces defines the Eddington limit,

$$\frac{m_H MG}{r^2} \approx \frac{\sigma_T L}{4\pi r^2 c}$$

or

$$L \simeq \frac{4\pi\, c m_H G}{\sigma_T} M.$$

If the luminosity exceeds the Eddington limit,

$$L \simeq 1.3 \times 10^{38}\, (M/M_\odot)\ \text{ergs/sec}, \qquad (1.39)$$

the radiation pressure acting on the envelope will exceed the envelope's weight, and matter will be expelled. Furthermore, for sufficiently massive main-sequence stars, which burn hydrogen by the CNO cycle and whose opacity is primarily electron-scattering, the envelope will become pulsationally unstable, and mass ejection will result. Both arguments lead to an upper mass limit of about

$$M_{\text{max}} \simeq 20\, M_\odot/\mu^2, \qquad (1.40)$$

where μ is the mean molecular weight ($1/\mu \approx 1.63$ for population I stars). Observations are consistent with this theoretical prediction that isolated stable main-sequence stars have masses in the range $0.05 \lesssim M/M_\odot \lesssim 60$.

1.8. Stellar Evolution

A star's lifetime is determined essentially by the amount of hydrogen that can be burned in its core, and this amount is proportional to the star's mass. However, the more massive a star is, the more rapidly it burns hydrogen to helium. Stellar-structure theory predicts that the main-sequence luminosity increases with stellar mass, as indicated by observations. The rate of increase depends on the mode of nuclear burning and is approximately $L \sim M^\alpha$. The value of α depends weakly on stellar mass and lies in the range $4.5 \gtrsim \alpha \gtrsim 3$.

Stellar Lifetimes

We estimate a star's main-sequence lifetime τ_{MS} as follows. Since only the core becomes hot enough to ignite hydrogen, about 10 percent of the star's mass represents main-sequence fuel, and the conversion of H to He liberates about 0.7 percent of the rest-mass energy per gram. Thus the total energy released by core H burning in a star of mass M is roughly $\eta M_c c^2$, where $\eta = 0.007$ and $M_c/M \simeq 0.1$. The rate of energy loss is given by the luminosity; so an estimate of τ_{MS} is

$$\tau_{\text{MS}} \simeq \frac{\eta M_c c^2}{L} \simeq 10^{10}\, (M/M_\odot)(L_\odot/L)\ \text{yrs}. \qquad (1.41)$$

When approximately 13 percent of a star's mass has been converted to helium, the evolutionary rate increases rapidly, and the star moves away from the main sequence. Theoretical studies indicate that more than 90 percent of a star's life is spent on the main sequence.

Thermonuclear Reactions

Nuclear evolution in a stellar core consists of successive stages of thermonuclear burning, interrupted by stages of gravitational contraction. Each nuclear-burning stage releases energy, which temporarily halts the previous stage of gravitational contraction. In low-mass stars, the central temperature never becomes high enough to ignite the ashes of He burning (C^{12} and O^{16}). Instead, electron degeneracy supplies the pressure necessary to halt collapse. The pressure-density equation of state is then given by (1.18) for low densities, where the electrons are nonrelativistic, or by (1.19) at higher density, where they are extremely relativistic. In either case we have $P \sim \rho^\gamma$, where γ is a constant. Noting that $M/R^3 \sim \rho$, we may rewrite the condition that the star be in hydrostatic equilibrium (1.16) in the approximate form

$$\rho^\gamma \sim M^2 G/R^4 \sim M^{2/3} \rho^{4/3} G \qquad (1.42)$$

or

$$M \sim \rho^{(\gamma - 4/3)3/2}. \qquad (1.43)$$

The average density of low-mass stars is low enough that $\gamma \simeq 5/3$, and $M \sim \rho^{1/2}$. For larger mass, $\gamma \longrightarrow 4/3$, and M becomes constant. This constant sets an upper limit to the mass of a cold star in stable hydrostatic equilibrium. If the mass exceeds this limit, the star will contract, usually on a free-fall time-scale. When the pressure is due to electrons, the star is a white dwarf, and the constant is the Chandrasekhar mass, M_{Ch}. Numerical studies show that $M_{\text{Ch}} \simeq 1.0$ to $1.4\ M_\odot$ depending on composition.

White Dwarfs

A star whose initial mass $M \lesssim M_{\mathrm{Ch}}$ will evolve through the stages of hydrogen burning and, if M is not too small, through He burning, ending its life as a white dwarf. What can we deduce about stars whose initial mass exceeds M_{Ch}? If the initial stellar mass is less than about 4 $M_\odot$, theory and observations suggest that the final state will be a white dwarf of mass M_{Ch}, the difference $M - M_{\mathrm{Ch}}$ being ejected, probably in the form of a planetary nebula. The gravitational potential energy of the Sun is given by (1.12). White dwarfs have radii of order $10^{-2} R_\odot$; so their potential energies are of order

$$\Omega_{\mathrm{W.D.}} \simeq 10^2\, \Omega_\odot. \tag{1.44}$$

When a 1 $M_\odot$ star evolves into a white dwarf, it loses about 4×10^{50} ergs of thermonuclear energy. The amount of energy needed to eject a 3.6 $M_\odot$ envelope from a 5 $M_\odot$ star prior to the white-dwarf stage is roughly 3×10^{48} ergs, which could come from either gravitational potential or thermonuclear energy. White dwarfs represent one possible end point of stellar evolution. Their masses are observed to be $\lesssim M_\odot$, and their luminosities L are usually 10^{-2} to 10^{-3} $L_\odot$. Spectral analysis yields surface temperatures of about 10^4 K, and (1.3) implies radii of about $10^{-2} R_\odot$.

Evolution of Massive Stars

Electron degeneracy will not halt contraction in stars more massive than about 5 $M_\odot$. Instead, the atomic weight increases with each new burning stage until the core consists predominately of Fe^{56} and iron-peak elements. Each fusion reaction leading to iron-group elements is exothermic and releases energy, but reactions using Fe^{56} or more massive nuclei as a fuel are endothermic and absorb energy from the star. The most important of these for stellar-core evolution is the photodisintegration of Fe^{56}, which occurs at high temperatures ($T \gtrsim 10^9$ K):

$$\mathrm{Fe}^{56} \longrightarrow 13\, \mathrm{He}^4 + 4\mathrm{n}. \tag{1.45}$$

Each photodisintegration reaction absorbs about 124 MeV from the radiation field (photons). A 1 $M_\odot$ core can contain approximately $1.2 \times 10^{57}/56$ iron nuclei; so the net energy absorbed is roughly $2 \times 10^{55} \times 124$ MeV, or 4.3×10^{51} ergs$/M_\odot$, which is comparable to the total energy radiated by the Sun in its main-sequence lifetime. The process is rapid, and the core becomes unstable and collapses. Collapse is accompanied by further heating, and at $T \gtrsim 10^9$ K the helium nuclei photodisintegrate,

$$\mathrm{He}^4 \longrightarrow 2\mathrm{n} + 2\mathrm{p}, \tag{1.46}$$

with the absorption of 28 MeV per reaction, or about 10^{52} ergs$/M_\odot$.

Problem 1.14. Verify that 1 $M_\odot$ consisting of 76 percent He and 24 percent neutrons releases about 10^{52} ergs/$M_\odot$ when helium photodissociates. Assume that 1 $M_\odot$ corresponds to about 1.2×10^{57} atomic mass units.

The dissociation of He^4 further accelerates the core's collapse, having absorbed 10^{52} ergs$/M_\odot$.

Supernova and Neutron-star Formation

The subsequent evolution of the star is quite complicated. In stars more massive than 4 to 6 $M_\odot$ a combination of neutrino, gravitational, and nuclear processes accompanying the collapse of the stellar core may trigger a supernova explosion, ejecting some or all of the stellar mass into interstellar space. If only the stellar envelope is ejected, the core will continue to collapse if its mass is not too great; pressure from degenerate neutrons will halt the collapse, and a neutron star will be formed. An analysis similar to (1.42) to (1.43) shows that there is a mass limit (similar to the Chandrasekhar mass) for neutron stars, in the range of 2 to 5 $M_\odot$. The gravitational potential energy of such an object is

$$|\Omega_{\mathrm{N.S.}}| \simeq \frac{M^2 G}{R} \approx 2 \times 10^{53}\ \mathrm{ergs} \tag{1.47}$$

for $M = M_\odot$ and $R \simeq 10^6$ cm; this exceeds the energy absorbed by (1.45) to (1.46) by nearly an order of magnitude. For a $M_\odot$ neutron star to form from a stellar core, roughly 10^{57} protons must be absorbed by the process (1.45) and (1.46) into neutrons. This conversion, taking place by weak interactions such as $\mathrm{p} + \mathrm{e}^- \longrightarrow \nu_e + \mathrm{n}$, produces a ν_e with a typical energy of the order of 15 to 20 MeV. In the conversion of approximately 10^{57} protons, approximately 2×10^{52}

ergs/$M_\odot$ goes into neutrino luminosity. If the supernova explosion does not disrupt the star, a neutron-star remnant may be left. Such a scenario would account for the pulsar observed in the Crab nebula (a 900-year-old supernova remnant), or for compact x-ray sources, such as Her X-1, observed in close binary systems. Should the supernova remnant's mass exceed the maximum mass of a stable neutron star (~2 or 3 $M_\odot$), the result could be a black hole (Cygnus X-1).

The matter ejected by the supernova may attain speeds of order 10^4 km/sec, and the accompanying energy output in the visible part of the spectrum may exceed 10^{45} ergs/sec, comparable to the luminosity of a medium-sized galaxy ($L \simeq 10^{11} L_\odot$). If the explosion is strong enough, the expanding material establishes a shock front that compresses the interstellar gas and may possibly trigger protostar formation. The shock front may also excite the gas enough to produce a nebula, such as the Veil nebula.

Stellar Nucleosynthesis

The gas out of which the first stars formed probably consisted of H and He (with abundances by weight $X \simeq 0.6$ and $Y \simeq 0.4$, respectively), and little if any heavy elements. The formation of stars and their subsequent evolution implies a concurrent evolution for the interstellar medium. Mass loss from individual stars, stellar winds, novae, and supernovae all return matter to the interstellar medium, whereas star formation removes it. Furthermore, some of the gas that condenses into stars must remain bound in the remnant (white dwarf, neutron star, or black hole) when the star's evolution has terminated. Each successive generation of star formation therefore returns a little less matter to the interstellar medium than it removed. As the gas density gradually decreases, so does the stellar birth rate. Eventually, the interstellar medium may become so dilute that star formation ceases altogether. The amount of gas available, and the time required for it to become bound up in dead stars, probably depends on the galactic type. In elliptical galaxies there appears to be little gas and no evidence of recent star formation; whereas in spiral and irregular galaxies the gas density is high, and the presence of bright, young stars argues that star formation is ongoing.* The evolutionary rate for interstellar matter is difficult to establish observationally, but we can make a simple estimate by assuming that the rate of decrease in the gas density ρ_g is proportional to the product of ρ_g and the fraction $f_R = M_R/M$ of an average star's mass that remains in the evolved remnant mass M_R:

$$\frac{d\rho_g}{dt} \simeq -\frac{f_R \rho_g}{\tau}. \tag{1.48}$$

Assuming both f_R and τ to be constant, we find

$$\rho_g(t) \simeq \rho_g(0) \exp(-f_R t/\tau).$$

The time-scale τ_{MS} can be taken to be the main-sequence lifetime of a typical lower-main-sequence star like the Sun, since these stars contain most of the mass in the Galaxy. Suppose that $M \simeq 1.5\ M_\odot$ and $M_R \simeq 1\ M_\odot$, and take $\tau_{\mathrm{MS}} \approx 5 \times 10^{10}$ years. Then $\rho_g(t_0)$, where t_0 is the present age of the universe, would be about 40 percent of its value when our Galaxy formed. Although this model is crude, it does suggest that the evolutionary rate of the interstellar medium may be comparable to the age of the galaxies.

*We don't know yet whether the evolution of interstellar matter plays a significant role in galactic evolution or not.

Chapter 2

PROPERTIES OF MATTER

2.1. Equations of State

Astrophysical systems generally are not in a state of exact thermal equilibrium. This is obvious for a star like the Sun, whose temperature ranges from near 10^7 K in the center to near 10^4 K at the surface. In most astrophysical systems, however, the change in such variables as the temperature over microscopically large volumes of matter is quite small. In the Sun, for example, the average change in temperature from the center to the surface is about $T_0/R_\odot \approx 10^7 \text{ K}/7 \times 10^{10} \text{ cm} \approx 10^{-4}$ K/cm.

The local properties of matter in an astrophysical system will also change with time. For example, as the material in a star expands or contracts, its composition or the degree of ionization of a given element may change. Furthermore, the time-scales that govern the system's rate of change are usually long relative to the time-scales that govern the microscopic processes determining the state of the material. The rates for microscopic changes are generally set by the time-scales for atomic processes, which are typically 10^{-8} sec or less, and are thus many orders of magnitude smaller than the time-scales for the fastest astrophysical processes. For these reasons, changes of state may be assumed to occur instantaneously, and a typical element of matter may be assumed to be in thermal equilibrium, and to be describable by the usual thermodynamic variables, including the internal energy U, the pressure P, and the temperature T. The material properties may then be expressed in terms of the equations of state, which relate, for example, the pressure or the energy density to the independent variables T and the mass density ρ.

The rest of this chapter presents the more important equations of state and thermodynamic relations that will be used in subsequent chapters. Much of the material in Sections 2.1 to 2.4 will be review, but it will also establish notation. The discussion of degeneracy in Section 2.5 is used in Chapters 13–16, and could be read at that time. Section 2.6 discusses formation and annihilation of particle-antiparticle pairs, and may be read in conjunction with Chapters 13, 15, and 26. Finally, Section 2.7 discusses regions in the ρ, T plane where the ideal-gas equations of state must be abandoned or modified, and may be read in conjunction with Chapters 15 and 16.

2.2. Ideal Gas

In a gas at high temperature and low density, there is very little interaction between particles, and the system pressure, number density, and temperature are adequately related by

$$P = \frac{N}{V} kT = nkT = \rho \frac{kT}{m}. \qquad (2.1)$$

The number of particles in the volume V is denoted by $N = nV$.

This is the most commonly encountered equation of state in astrophysics, since it may be used to describe: stellar matter during most of a star's lifetime, interstellar matter, nebulae, and the material content of the universe shortly after the Big Bang. It is also used to describe the "gas" of stars in simple models of globular clusters, galactic nuclei, and ellipsoidal galaxies; and in cosmology to describe the "gas" of galaxies filling the universe. A fluid satisfying (2.1) is called a perfect or ideal fluid. Several alternative forms may be obtained by replacing the particle number by

$$N = N_0 \bar{\mu} = \mathcal{R} \bar{\mu}/k, \qquad (2.2)$$

where N_0 is Avagadro's number, $\mathcal{R}$ is the universal gas constant, and $\bar{\mu}$ the number of molecules of gas per mole.

The internal energy* of an ideal gas is given by

$$U = \int_0^T c_V \, dT, \qquad (2.3)$$

where the heat capacity at constant volume is defined by

$$c_V = \left(\frac{\partial U}{\partial T}\right)_V = \left(\frac{\partial Q}{\partial T}\right)_V.$$

The last form follows from the first law of thermodynamics,

$$dU = dQ - P \, dV.$$

For an ideal gas, $c_V = c_V(T)$ is a function of temperature only. For a monatomic nonrelativistic ideal gas, $c_V = (3/2)Nk$ and $U = c_V T$. The specific heat (heat capacity per gram) of an ideal monatomic gas of atomic weight A is $1.247 \text{ A}^{-1} \times 10^8$ erg/gK.

The first law of thermodynamics yields a convenient relation between the heat capacity at constant pressure,

$$c_P = \left(\frac{\partial Q}{\partial T}\right)_P,$$

and c_V. Starting with the first law, and using $dU = c_V(T) \, dT$,

$$\left(\frac{\partial Q}{\partial T}\right)_P = c_V \, dT + P \left(\frac{\partial V}{\partial T}\right)_P = c_V + Nk. \qquad (2.4)$$

Since the left-hand side defines c_P, it follows that

$$c_P = c_V + Nk \qquad (2.5)$$

for any ideal gas. In general, c_P and c_V may be functions of T. The internal energy is therefore

$$dU = c_V \, dT = c_P \frac{d(PV)}{Nk} = \frac{d(PV)}{\gamma - 1},$$

where $\gamma \equiv c_P/c_V$. The internal energy density is often written in terms of an effective (and constant) γ' as

$$u = \frac{P}{\gamma' - 1}. \qquad (2.6)$$

In general the γ' appearing in (2.6) does not equal c_P/c_V. This relation is useful within a restricted range of pressures for which the ratio u/P is essentially constant. For a monatomic ideal gas, $\gamma' = \gamma = 5/3$.

We also introduce polytropic processes, which are defined by

$$\frac{dQ}{dT} = c = \text{constant}. \qquad (2.7)$$

The first law, the relation between c_P and c_V, and the ideal-gas equation of state may be combined to show that

$$c \, dT - c_V \, dT = P \, dV = (c_P - c_V) T \frac{dV}{V}. \qquad (2.8)$$

If we define the polytropic gamma by

$$\gamma = \frac{c_P - c}{c_V - c}, \qquad (2.9)$$

*We also define the energy density $u = U/V$, and the specific energy $\epsilon = u/\rho$.

equation (2.8) becomes $dT/T + (\gamma - 1)\,dV/V = 0$. Assuming that γ is a constant, and using the ideal-gas equation of state, we obtain the polytropic equations of state,

$$PV^{\gamma} = \text{const.},$$
$$P^{1-\gamma}\,T^{\gamma} = \text{const.}, \tag{2.10}$$
$$TV^{\gamma-1} = \text{const.}$$

Two important limits of (2.9) should be noted. Since $dQ = T\,dS$, where S is the entropy, a polytropic change at constant S corresponds to $c = 0$, and $\gamma = c_P/c_V$ is the adiabatic gamma. For isothermal changes, $c \longrightarrow \infty$, and the polytropic gamma is unity: $\gamma_{iso} = 1$.

Problem 2.1. Carry out the derivation of the polytropic equations of state (2.10).

Polytropic equations of state are extremely useful in analyzing the behavior of gravitating gaseous configurations, and will be used later to obtain several analytic solutions to one of the stellar-structure equations (the equation of hydrostatic equilibrium). In fact, simple models of homogeneous stars, white dwarfs, and neutron stars may be constructed by assuming that the matter is polytropic.

Although the preceding results yield a simple form of adiabatic processes as a special case, it will be convenient to introduce a more general form for later use. Therefore we define the adiabatic indices by the relations

$$PV^{\Gamma_1} = \text{const.},$$
$$P^{1-\Gamma_2}\,T^{\Gamma_2} = \text{const.}, \tag{2.11}$$
$$TV^{\Gamma_3-1} = \text{const.},$$

which apply to arbitrary gases.

Problem 2.2. Show that the relations in (2.11) hold for an adiabatic process for an ideal gas. Establish for a monatomic ideal gas the relations $\Gamma_1 = \Gamma_2 = \Gamma_3 = c_P/c_V$.

For ideal gases containing molecules, or for atoms at temperatures high enough to excite or ionize bound electrons, the Γ_i appearing in (2.11) will be unequal. For example, at temperatures above or below a narrow range around the ionization temperature of hydrogen, equations (2.11) are valid with $\Gamma_i = 5/3$; in the ionization zone they are less than $5/3$, and approach unity when the number of atoms and ions are comparable.

We will see that Γ_1 determines the dynamic stability of a star; Γ_2 governs convective instability; and Γ_3 delimits the regime of pulsational instability.

The entropy S of an ideal gas may be obtained from the equation of state and the first law of thermodynamics,

$$T\,dS = dU + P\,dV. \tag{2.12}$$

Using (2.3), (2.1), and the definition of γ,

$$S = Nk \ln\left(VT^{1/(\gamma-1)}\right) + \text{const.} \tag{2.13}$$

Problem 2.3. Carry out the derivation leading to (2.13).

The fundamental processes of astrophysics are usually irreversible, in the sense that locally defined entropy increases with time. Often (2.13) may be used to reach qualitative conclusions about a star or galaxy when the changes in P, ρ, or T are complex. The heat-loss rate $T\,ds/dt$, where s is the entropy per gram of matter, must also be included in models of stellar evolution.

The chemical potential of a specified type of particle in a thermodynamic system is defined as the change in the internal energy of the entire system when another particle of the same type is added:

$$\mu_i = \left(\frac{\partial U}{\partial N_i}\right)_{S,V,N_j} = \left(\frac{\partial u}{\partial n_i}\right)_{s,n_j}, \tag{2.14}$$

where the number N_j of all other particle types is held constant, as is the entropy and volume. Here $n_i = N_i/V$ and $s = S/V$. For an ideal gas

$$\mu_i = -kT \ln\left[\frac{g_i}{n_i}\left(\frac{m_i kT}{2\pi\hbar^2}\right)^{3/2}\right] + I_0, \tag{2.15}$$

where I_0 is the rest energy of the particle, $m_i c^2 = I_0$.

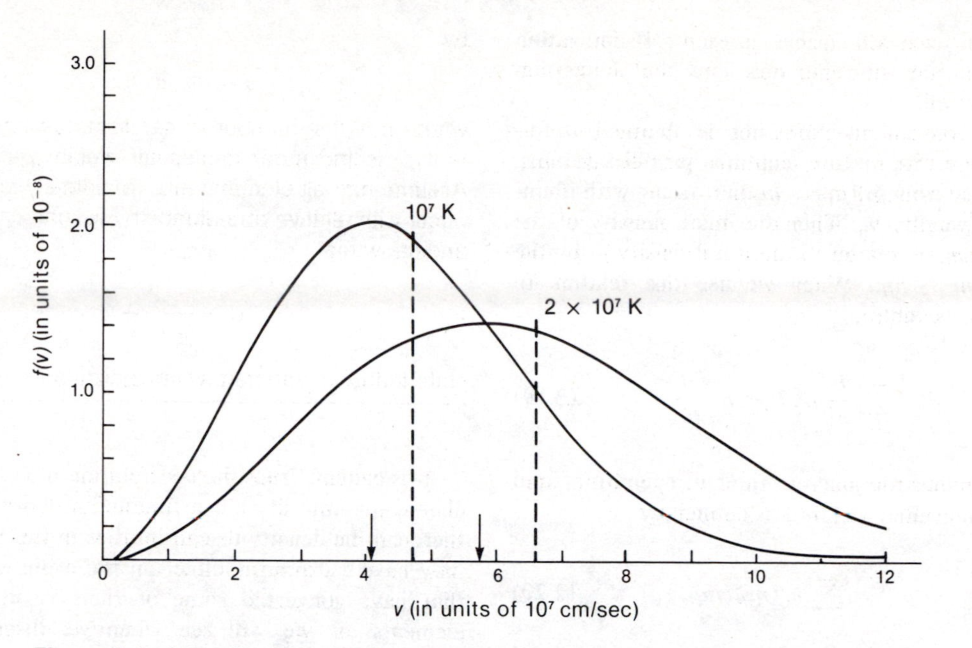

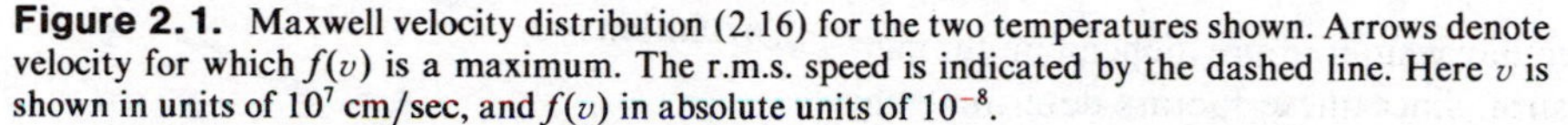

Figure 2.1. Maxwell velocity distribution (2.16) for the two temperatures shown. Arrows denote velocity for which $f(v)$ is a maximum. The r.m.s. speed is indicated by the dashed line. Here v is shown in units of 10^7 cm/sec, and $f(v)$ in absolute units of 10^{-8}.

The factor g_i is the multiplicity of the states, and will usually be taken to be $g_i = (2s_i + 1)$, where s_i is the quantum-mechanical spin of the particles. For electrons, neutrons, and protons, $g = 2$. For atomic nuclei, or atoms, g is replaced by the partition function, which takes into account internal degrees of freedom, which may be thermally excited.

The chemical potential is useful in relating the abundances of the reactants and products of chemical, thermonuclear, or particle reactions.

Particles of mass m in an ideal gas at temperature T have average velocities of the order of $(3kT/m)^{1/2}$. Their actual velocity distribution is given by the Maxwell distribution function,

$$f(v) = 4\pi \left(\frac{m}{2\pi kT}\right)^{3/2} \exp\left(-\frac{mv^2}{2kT}\right) v^2; \quad (2.16)$$

$f(v)\, dv$ gives the probability of finding a particle in the velocity range v to $v + dv$ at temperature T. Note that $\int_0^\infty f(v)\, dv = 1$. Figure 2.1 shows $f(v)$ for H atoms at temperatures of 10^7 K and 2×10^7 K (typical of hydrogen burning in the Sun). The most probable speed of a particle of mass m is $(2kT/m)^{1/2}$, and the root-mean-square speed is

$$\langle v^2 \rangle^{1/2} = \left(\frac{8kT}{\pi m}\right)^{1/2} \quad (2.17)$$

For hydrogen, with T in K,

$$\langle v^2 \rangle^{1/2} = 2.1 \times 10^4\, T^{1/2} \text{ cm/sec.}$$

The Maxwellian velocity distribution is also applicable to stars in the central regions of globular clusters and galactic nuclei, and to galaxies in rich clusters of galaxies.

2.3. Mixtures of Ideal Gases: Mean Molecular Weight

When a system contains a mixture of ideal gases, each of which obeys a perfect gas law $P_i = n_i kT$, with partial pressure P_i and number density n_i, the total pressure is given by

$$P = \sum_i P_i = \sum_i n_i kT. \quad (2.18)$$

The sum is over all species present. If ionization occurs, then the sum includes ions and ionization electrons as well.

A more convenient expression is obtained as follows. Suppose the mixture contains particles (atoms, ions, and electrons) of mass m_i that occur with abundance (by weight) $\bar{n}_i$. Then the mass density of the species i, $m_i n_i$, is related to the total density ρ by the relation $n_i m_i = \bar{n}\rho$. When we use this relation to eliminate n_i, we obtain

$$P = \sum_i \frac{\bar{n}_i}{m_i} \rho k T \equiv \rho \frac{kT}{\mu m_H}. \tag{2.19}$$

Here m_H denotes the mass of the hydrogen atom, and the mean molecular weight μ is defined by

$$\mu^{-1} \equiv \sum_i \bar{n}_i \, (m_H/m_i). \tag{2.20}$$

The mean molecular weight μ may depend on temperature and pressure, since these factors determine the degree of ionization, for example, which in turn determines the $\bar{n}_i$. A relatively simple and useful expression may be given when the constituents are assumed to be fully ionized hydrogen, helium, and metals. If we denote their atomic numbers and weights by (Z, A) and their mass fractions by X_A, with $X_1 \equiv X$, $X_2 \equiv Y$, and $X_3 \equiv Z$ for all $A > 4$ (heavy metals), it follows that

$$\bar{n}_i \rightarrow (Z + 1)\, X_A,$$

$$m_i \rightarrow A m_H,$$

and the mean molecular weight becomes, assuming $A = 2Z$ for heavy metals,

$$\mu^{-1} = \Sigma \frac{Z + 1}{A} X_A = 2X + (3/4)\, Y + (1/2)\, Z. \tag{2.21}$$

This relation is independent of T and P as long as ionization is complete. Note that the normalization condition $X + Y + Z = 1$ must hold.

Problem 2.4. For discussing degenerate matter in white dwarfs or highly evolved stars, it is convenient to define the matter density (due essentially to the ions) by

$$\rho = m_H \, \mu_e \, n_e, \tag{2.22}$$

where n_e is the number of electrons per unit volume, and μ_e is the mean molecular weight per electron. Assume that all elements are completely ionized and occur with relative abundance by weight X_A as above, and show that

$$\mu_e = \frac{2}{1 + X}. \tag{2.23}$$

Note that μ_e is constant when ionization is complete.

It is evident from the last relation in (2.19) that if chemical composition is a function of depth in a star, there can be density discontinuities in the star. These may have a dramatic effect on the evolution of stars that have converted some of their H into heavier elements, as we will see when we discuss stellar evolution.

Problem 2.5. Define the interface between a stellar core and its envelope as the surface across which the chemical composition changes. In equilibrium the pressure and temperature will be constant across this surface. Show that a density discontinuity must result. If the core is pure He and the envelope pure H, by what fraction must the density change?

It should be noted that for complete ionization μ ranges from about 1/2 to 2 and μ_e ranges from 1 to 2.

2.4. Radiation and Matter

The energy density and pressure of a photon gas (black-body radiation) are given by

$$u_R = aT^4 = \frac{4\sigma T^4}{c}, \tag{2.24}$$

$$P_R = \frac{1}{3} aT^4 = \frac{4\sigma T^4}{3}, \tag{2.25}$$

where $\sigma = ac/4$ is the Stefan-Boltzmann constant. The entropy is easily shown to be $S = (4/3)aT^3V$. Radiation behaves adiabatically like a gas with $\gamma = 4/3$, as follows directly from (2.11), (2.24)–(2.25), and the definition of adiabatic processes.

Radiation and gas may both contribute to the energy and pressure of the system. When the gas is ideal, the total pressure is

$$P = P_R + P_g = \frac{NkT}{V} + \frac{1}{3}aT^4, \qquad (2.26)$$

and the total energy is

$$U = uV = c_V T + aT^4 V. \qquad (2.27)$$

When both sources contribute, it is often convenient to define the quantity

$$\beta = P_g/P, \qquad 1 - \beta = P_R/P. \qquad (2.28)$$

Then the gas pressure may be written as $P_g = NkT/V = \beta P$, and the total pressure is

$$P = \frac{nkT}{\beta} = \frac{\rho kT}{\beta \mu m_H}. \qquad (2.29)$$

The advantage of this approach is that whenever β is constant, or the product $\beta\mu$ nearly constant, then the effect of radiation is formally equivalent to modifying the mean molecular weight. The results of calculations based on gas pressure alone may then be applied to matter and radiation simply by the replacement $\mu \longrightarrow \beta\mu$. In general, however, β will depend on position in the system.

Next consider the adiabatic behavior of a mixture of gas and radiation; in general, adiabatic processes satisfy

$$dQ = dU + P\,dV = 0. \qquad (2.30)$$

The change in U can be found from (2.27), and the definition of β may then be used to simplify the result. Thus

$$dU + P\,dV = c_V dT + 4aT^3 V\,dT + aT^4\,dV + P\,dV = 0,$$

$$c_V\,dT = \frac{c_V \beta PV}{Nk}\frac{dT}{T} = \frac{c_V}{c_P - c_V}\beta PV\,d\ln T = \frac{\beta PV}{\gamma - 1}\,d\ln T,$$

$$4aT^3 V\,dT = 12\frac{aT^4 V}{3}\,d\ln T = 12(1 - \beta)PV\,d\ln T,$$

$$aT^4\,dV = 3(1 - \beta)P\,dV.$$

Substituting these into the first law and dividing by PV gives, upon rearrangement,

$$[\beta + 12(1 - \beta)(\gamma - 1)]\,d\ln T + (4 - 3\beta)(\gamma - 1)\,d\ln V = 0. \qquad (2.31)$$

Equation (2.31) leads immediately to the relation between Γ_3, the ratio of specific heats $\gamma = c_P/c_V$ for the ideal gas, and the fraction $\beta = P_g/P$:

$$\Gamma_3 - 1 \equiv \left(\frac{\partial \ln T}{\partial \ln V}\right)_s = \frac{(4 - 3)(\gamma - 1)}{\beta + 12(1 - \beta)(\gamma - 1)}. \qquad (2.32)$$

In a similar way it is straightforward to show that

$$\Gamma_1 = \beta + \frac{(4 - 3\beta)^2(\gamma - 1)}{\beta + 12(1 - \beta)(\gamma - 1)}, \qquad (2.33)$$

$$\Gamma_2 = 1 + \frac{(4 - 3\beta)(\gamma - 1)}{\beta^2 + 3(\gamma - 1)(1 - \beta)(4 + \beta)}. \qquad (2.34)$$

For $\gamma = 5/3$ (monatomic, ideal gas) and no radiation ($\beta = 1$), equations (2.32)–(2.34) reduce to $\Gamma_1 = \Gamma_2 = \Gamma_3 = 5/3$ as expected; for pure radiation ($\beta = 0$) it follows that $\Gamma_1 = \Gamma_2 = \Gamma_3 = 4/3$. For intermediate values of β the adiabatic exponents are unequal. With increasing radiation content (decreasing β) they decrease steadily, as shown in Figure 2.2.

The preceding analysis demonstrates that, although an admixture of radiation can drive the Γ_i toward 4/3, it cannot reduce them below this limit.

The spectrum of electromagnetic radiation in thermal equilibrium with matter at temperature T is Planckian. Its energy spectrum, which gives the energy per unit volume in the frequency range ν to $\nu + d\nu$, is

$$u(\nu, T)\,d\nu = \frac{8\pi h\nu^3}{c^3}\frac{d\nu}{e^{h\nu/kT} - 1}. \qquad (2.35)$$

It is often convenient to describe the radiation in terms of photons, which are the quanta of the electromagnetic field. A photon corresponding to radiation of

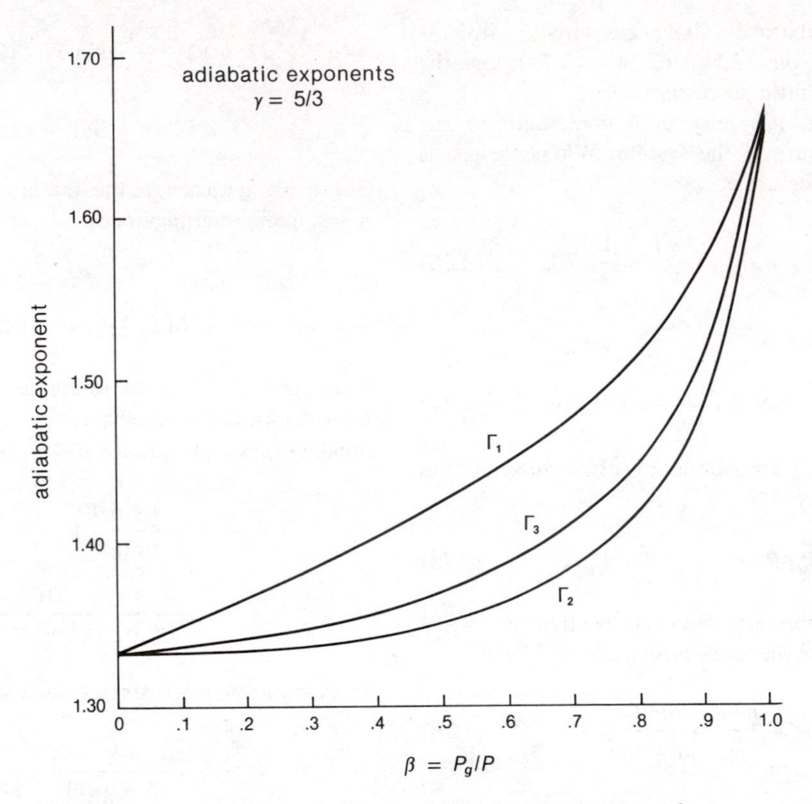

Figure 2.2. Adiabatic exponents versus ratio of gas pressure to total pressure.

wavelength $\lambda = c/\nu$ has energy $h\nu$, and radiation of energy density $u(\nu, T)$ may be considered to consist of

$$n = \int_0^\infty \frac{u(\nu, T)}{h\nu}\, d\nu \simeq 20\, T^3 \text{ photons per cm}^3, \quad (2.36)$$

where T is in K. The total energy density is given by the integral of (2.35), and is

$$u(T) = \int_0^\infty u(\nu, T)\, d\nu = aT^4. \quad (2.37)$$

2.5. Degenerate Matter

The particles of a gas that obeys the ideal-gas equation of state described in Section 2.4 have a probability of being in any given momentum state, $f(p) \ll 1$. As the gas temperature is lowered, $f(p)$ for particles of half-integer spin (fermions) approaches unity for particle energies $\epsilon_p \lesssim \mu$, and rapidly approaches zero for $\epsilon_p > \mu$, where μ is the chemical potential defined as in (2.14). For $T \simeq 0$, $f(\epsilon_p)$ is essentially a step function; at $T = 0$, $\mu \equiv \epsilon_F$, the Fermi energy, and

$$f(\epsilon_p) = \theta(\epsilon_F - \epsilon_p) = \begin{cases} 1 \text{ if } \epsilon_F \geq \epsilon_p, \\ 0 \text{ if } \epsilon_p > \epsilon_F. \end{cases}$$

When $T \approx 0$, $\mu \approx \epsilon_F$, and the width of the transition region $\epsilon_p < \epsilon_F$ to $\epsilon_p > \epsilon_F$ is of order kT/ϵ_F.

We can use these observations and the rudiments of kinetic theory to examine the thermodynamics of degenerate matter. The analysis is applicable to any fermion, in particular to electrons, nucleons (neutron or proton), neutrinos, and their antiparticles.

We need to understand degenerate matter in order to understand (a) white dwarfs, the cores of highly evolved stars whose masses are more than a few times $M_\odot$, and neutron stars; (b) why there are no stars less massive than a few hundredths $M_\odot$; and (c) much of planetary structure.

Under most astrophysical conditions, neutrinos have such large mean free paths in matter that they need not be described by an equation of state; that is, they exert no pressure. However, during the short interval of stellar core collapse, which may trigger a supernova explosion, conditions may lead to neutrino degeneracy. In this case an equation of state for the neutrinos will be needed.

Consider a gas containing N fermions; the probability that a particle is in a quantum state having momentum $\mathbf{p}$ and energy ϵ_p at temperature T is given by the Fermi-Dirac function,

$$f(p) = \frac{1}{e^{(\epsilon_p - \mu)/kT} + 1}. \tag{2.38}$$

The fermion chemical potential is denoted by μ. The fermion number density is

$$n = g_s \int f(p) \frac{d\mathbf{p}}{h^3}$$

$$= \frac{4\pi g_s}{(2\pi)^3 \hbar^3} \int_0^\infty \frac{p^2\, dp}{e^{(\epsilon_p - \mu)/kT} + 1}$$

$$= \frac{g_s}{2\pi^2 \hbar^3} \int_0^\infty \frac{p^2\, dp}{e^{(\epsilon_p - \mu)/kT} + 1}. \tag{2.39}$$

The factor g_s is the multiplicity factor that gives the number of quantum states corresponding to a given momentum $\mathbf{p}$. For electrons and nucleons, $g_s = 2s + 1$. Neutrinos have only one spin state; so $g_s = 1$ for them.

To find the energy density of the gas, multiply the energy of a single particle, ϵ_p, by the probability that the state is occupied, $f(p)$, and integrate over phase space:

$$u = g_s \int \epsilon_p f(p) \frac{d\mathbf{p}}{h^3}. \tag{2.40}$$

The pressure in a gas is the result of momentum transfer to the container walls by particles. If collisions with the walls are elastic, then the magnitude of the initial and final momenta are equal, and the momentum transfer is $2p \cos \theta$, where θ is the angle of incidence measured from the surface normal. The force exerted per particle in time Δt is therefore $2p \cos \theta / \Delta t$. The collision rate per unit area is $v_p\, \Delta t \cos \theta$ if v_p is the velocity of the particle of momentum $\mathbf{p}$. Combining these factors, multiplying by $f(p)$, and integrating over phase space yields

$$P = \int \frac{d\mathbf{p}}{h^3} \frac{2p\cos\theta}{\Delta t} v_p\, \Delta t \cos \theta\, f(p)$$

$$= \frac{1}{2\pi^2 \hbar^3} \int_0^\pi \cos^2 \theta \sin \theta\, d\theta \int_0^\infty v_p f(p) p^3\, dp$$

$$= \frac{1}{3\pi^2 \hbar^3} \int_0^\infty v_p f(p)\, p^3\, dp. \tag{2.41}$$

The results in (2.39) to (2.41) are valid for arbitrary temperatures (ignoring pair production), but can be evaluated analytically only at $T = 0$, where (2.39) shows that the Fermi energy $\epsilon_F = \mu$, and that

$$n = g_s \frac{p_F^3}{6\pi^2 \hbar^3}, \tag{2.42}$$

where p_F is the Fermi momentum defined by $\epsilon_{p_F} = \epsilon_F$.

Problem 2.6. Starting with (2.39), show that (2.42) follows by first assuming that μ/kT is arbitrarily large but finite, and by then examining the contributions to the integral for $\epsilon_p > \mu$ and $\epsilon_p < \mu$. Then take the limit $T \rightarrow 0$.

The probability $f(p)$ given by (2.38) also reduces to $\theta(p_F - p) = \theta(\epsilon_F - \epsilon_p)$ at $T = 0$.

When the fermions are nonrelativistic, $\mathbf{p} = m\mathbf{v}$ and $\epsilon_p = p^2/2m$. The energy density (2.40) is easily shown to be

$$u = \frac{3}{5} n \epsilon_F, \tag{2.43}$$

where $\epsilon_F = p_F^2/2m$ is the Fermi energy. The pressure follows from (2.41), and is

$$P = \frac{2}{3} u. \tag{2.44}$$

Problem 2.7. Verify expressions (2.43) and (2.44). Show that a nonrelativistic ideal gas at $T = 0$ is a polytrope with $\gamma = 5/3$.

For relativistic fermions $\epsilon_p = c\sqrt{p^2 + m^2c^2}$ and $v_p = v/\sqrt{1 - v^2/c^2}$. Then $\epsilon_F = c\sqrt{p_F^2 + m^2c^2}$, and u and P may again be found in analytic form. For simplicity, assume that $p \gg mc$ (extreme relativistic limit). Then $\epsilon_p \approx pc$, $v \approx c$, and it is easily shown that

$$u = \frac{3}{4} n \epsilon_F, \tag{2.45}$$

$$P = \frac{1}{3} u. \tag{2.46}$$

Problem 2.8. Verify (2.45) and (2.46). Show that an extreme relativistic ideal gas at $T = 0$ is a polytrope with $\gamma = 4/3$.

For neutrinos, $\epsilon_p = pc$ and $v_p = c$, and (2.45)–(2.46) hold exactly at $T = 0$.

The expressions (2.39)–(2.41) may be expanded at low but nonzero temperatures ($kT \ll \mu$) to obtain the leading order temperature corrections to u and P. The results are

$$u(T) = u(0) + \frac{\pi^2}{2} n \frac{(kT)^2}{\epsilon_F}, \tag{2.47}$$

$$P(T) = P(0) + \frac{\pi}{4} n \frac{(kT)^2}{\epsilon_F}. \tag{2.48}$$

The second term on the right of (2.47) represents the thermal energy density in a nearly degenerate system. The chemical potential is

$$\mu(T) = \epsilon_F - \frac{\pi^2}{12} \frac{(kT)^2}{\epsilon_F} + I_0. \tag{2.49}$$

The terms $u(0)$ and $P(0)$ are given by (2.45)–(2.46) in the extreme relativistic limit ($p \gg mc$) or by (2.43)–(2.44) in the nonrelativistic limit ($p \ll mc$).

The heat capacity c_V is, to lowest order in kT/μ,

$$c_V = \frac{3\pi^2}{2} g_s \frac{k^2 T}{mc^2} \frac{\sqrt{1 + x^2}}{x^2} n, \tag{2.50}$$

where $x = p_F/mc$, and n is given by (2.42). This may be used to construct a simple model for the cooling of white dwarfs and neutron stars.

Although a degenerate system contains a large amount of energy stored in the motion of particles in the Fermi sea (that is, in particles having energies $\epsilon_p < \epsilon_F$), only the excess energy (due to particles excited out of the Fermi sea) is measured by c_V. Only this energy is available, for example, to be radiated away at $T \rightarrow 0$.

2.6. Matter at High Temperatures

Consider an enclosure filled with electromagnetic radiation at temperature T. The pressure and energy density are given by (2.24)–(2.25) as long as the temperature does not become too great. However, when T exceeds about 2 mc^2/k, particle-antiparticle pairs, each of rest mass m, will be produced. To see how this can happen, consider the energy density associated with the radiation field. It can be written as $u_R \sim (kT)(kT/\hbar c)^3$, since $kT \sim E \sim \hbar c/\lambda$, with λ the wavelength of the photon of energy kT. The energy density of a particle-antiparticle pair, each of rest mass m and Compton wavelength $\hbar/mc$, is $\sim 2\,(mc^2) \times (mc/\hbar)^3$. When these two energy densities are comparable,

$$T \sim 2mc^2/k, \tag{2.51}$$

then quantum-mechanically the energy may be equally well associated with a photon or a particle-antiparticle pair. In general, when T approaches 2 mc^2/k, pairs may be created; when T exceeds 2 mc^2/k, pairs will dominate the system.

A typical pair-production process is

$$\gamma + \gamma \rightleftharpoons e^+ + e^-. \tag{2.52}$$

This may be handled by associating separate Fermi-Dirac distribution functions $n^-(p)$ and $n^+(p)$ with the electrons and positrons, respectively:

$$n^{\mp}(p) = \frac{1}{e^{(\epsilon_p - \mu_\pm)/kT} + 1}. \tag{2.53}$$

Thermal pair production (2.52) is an equilibrium process in which the sum of the chemical potentials of the reactants equals the sum of the chemical potentials of the products. For the reaction (2.52), it follows that the electron, positron, and photon chemical potentials satisfy

$$2\mu_\gamma = \mu_+ + \mu_-. \tag{2.54}$$

The photon chemical potential μ_γ, however, vanishes, so that

$$\mu \equiv \mu_+ = -\mu_-. \tag{2.55}$$

Substituting these into (2.53) yields the final forms for the electron and positron distributions in thermal equilibrium:

$$\begin{aligned} n^-(p) &= \frac{1}{e^{(\epsilon_p-\mu)/kT}+1}, \\ n^+(p) &= \frac{1}{e^{(\epsilon_p+\mu)/kT}+1}. \end{aligned} \tag{2.56}$$

In expressions (2.56), the fermion rest-mass energy is included in both ϵ_p and μ. It is sometimes convenient to measure energies from zero; to do so, we subtract the rest energy mc^2 from both ϵ_p and μ. Denoting quantities measured from zero energy by tildes, we have

$$\epsilon_p - \mu = \epsilon_p - mc^2 - (\mu - mc^2) = \tilde{\epsilon}_p - \tilde{\mu},$$

$$\epsilon_p + \mu = \tilde{\epsilon}_p + \tilde{\mu} - 2mc^2.$$

It follows trivially from the definition of the electron and positron number densities, n^- and n^+, that $n^+ = n^-$ when $\mu = 0$.

Problem 2.9. Find the zero temperature limits of $n^-(p)$ and $n^+(p)$ given by (2.56). Is the result for $n^+(p)$ reasonable? Explain it physically.

Problem 2.10. At high temperatures ($T \gtrsim m_ec^2/k$), the electron and positron chemical potentials are given by (2.15), with $m = m_e$, $I_0 = m_ec^2$, and $g = 2$. Show that the electron and positron number densities n_- and n_+ are related by

$$n_+n_- = 4\left(\frac{m_ekT}{2\pi\hbar^2}\right)^3 \exp\,(-2m_ec^2/kT). \tag{2.57}$$

Another pair process that is important in stellar evolution is $\gamma + \gamma \rightleftharpoons \nu_e + \bar{\nu}_e$ and $\gamma + \gamma \rightleftharpoons \nu_\mu + \bar{\nu}_\mu$ where ν_e and ν_μ denote electron and muon neutrinos, and their antiparticles are denoted by a bar. The results (2.53)–(2.56) for electrons may be applied to neutrinos in thermodynamic equilibrium. These and other processes that produce neutrinos at high temperatures constitute an efficient cooling mechanism for the cores of evolved stars, and for white dwarfs and neutron stars. They also play an important role in core-collapse models of supernovae.

2.7. Real Fluids

Under most astrophysical conditions, ideal-gas equations of state are usually adequate for describing the density and temperature of matter, because when densities are low, interparticle separations exceed the range of interparticle interactions. During the final evolutionary stages for most stars, however, interactions may lead to significant modifications in the equations of state. The interactions that have been found to be most important are the electromagnetic couplings of electrons and ions in cool matter at densities up to 10^{12} g/cm^3; and the strong interactions (nuclear forces) between baryons at densities above 10^{14} g/cm^3. A combination of electromagnetic and weak interactions is believed to be important in cooling processes in hot, dense matter.

The models developed in the introduction assumed that stellar matter could be described by ideal-gas equations of state. Such a description turns out to be an exceedingly good approximation, as a simple argument will demonstrate. The basic assumption is that whenever two sources of energy contribute to the internal energy of a fluid, one can be neglected if it is sufficiently smaller than the other. For example, take an electron gas (with background distribution of positive ions of charge Ze), and consider the energy of the electrons; the electron number density n_e and the Coulomb energy E_c of an electron due to each ion are, respectively,

$$\begin{aligned} n_e &= (Z/A)(\rho/m_H), \\ E_c &\sim Ze^2\,\rho^{1/3}/(Am_H)^{1/3}. \end{aligned} \tag{2.58}$$

The nonrelativistic and relativistic chemical potentials of the electrons, excluding rest-mass energy, are

$$\mu_{\mathrm{NR}} \simeq \frac{9}{2}\frac{\hbar^2}{m_e}\left(\frac{Z}{A}\right)^{2/3}\left(\frac{\rho}{m_N}\right)^{2/3}, \tag{2.59}$$

$$\mu_{\mathrm{ER}} \simeq 3\hbar c\left(\frac{Z}{A}\right)^{1/3}\left(\frac{\rho}{m_N}\right)^{1/3}. \tag{2.60}$$

We may now map out regions of the temperature-

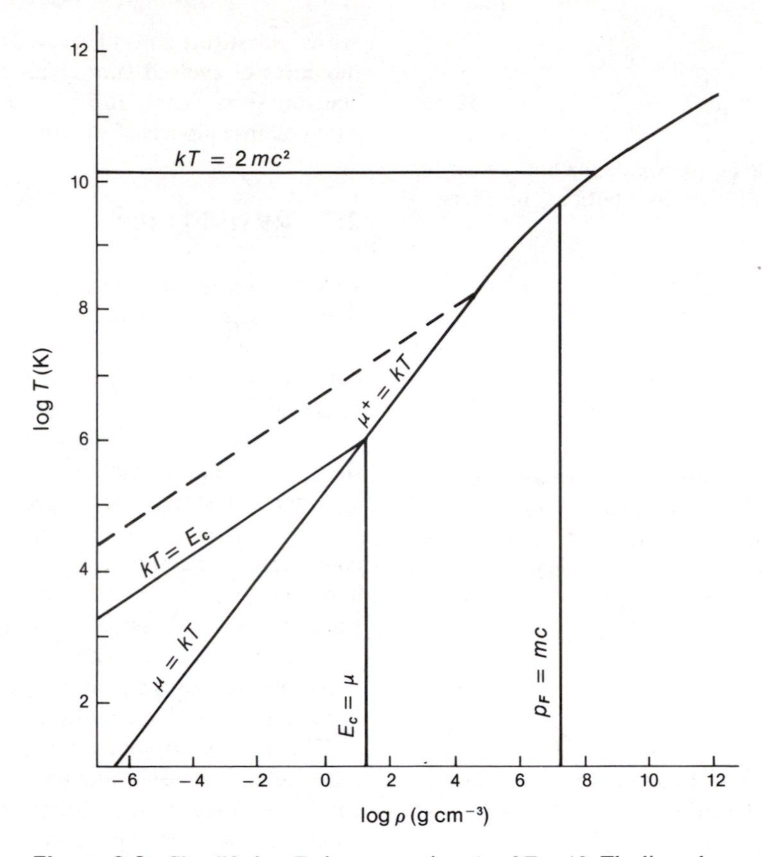

Figure 2.3. Simplified ρ, T-plane, assuming $A = 2Z = 12$. The lines show points where two contributions to the energy of a particle are equal. The chemical potential is μ; E_c is the Coulomb energy (2.58). The line bounding nondegenerate and degenerate nonrelativistic matter has been smoothly interpolated to the boundary for relativistic matter. The dashed line shows points in the solar interior.

density plane (see Figure 2.3). When $\mu \gg kT$, thermal effects are relatively small, but for $\mu \ll kT$ an ideal-gas approximation is reasonable. Similarly, the line obtained by setting $kT = E_c$ separates regions where thermal or Coulomb effects dominate; the line $\mu = E_c$ separates regions where matter is degenerate and Coulomb effects are important ($E_c \gg \mu$) from the region where Coulomb effects are negligible relative to degeneracy effects ($E_c \ll \mu$); the line $kT = 2\ m_e c^2$ bounds the pair-production region; and finally the line $p_F = m_e c$ separate the regions of nonrelativistic degenerate electrons from relativistic degenerate electrons.

Problem 2.11. The assumption that a star is in hydrostatic equilibrium and that matter obeys the ideal-gas equation of state leads to the relation between stellar central temperature and density

$$T^3/\rho \sim M^2. \tag{2.61}$$

Noting that $T_\odot \sim 2 \times 10^7$ K and $\rho_\odot \sim 40$ g/cm^3 in the Sun's core, find the regions of the ρ, T-plane occupied by stars of 0.5 $M_\odot$, 1 $M_\odot$, and 2 $M_\odot$. Note that when matter becomes degenerate, the density becomes independent of temperature.

The preceding equalities, combined with one or more of (2.58)–(2.61), may be used to find T as a function of ρ in each case (Figure 2.3). For example, the degeneracy boundary for nonrelativistic electrons is obtained from (2.59) and

$$kT = \mu_{\mathrm{NR}} = \frac{9}{2}\frac{\hbar^2}{m_e}\left(\frac{Z}{A}\right)^{2/3}\left(\frac{\rho}{m_N}\right)^{2/3},$$

which gives the relation

$$\log T = \frac{2}{3}\log \rho + 5.28 \qquad (\mu_{\mathrm{NR}} = kT)$$

if $A = 2Z$ is assumed. Similar expressions can be constructed for the other boundary lines. If we assume $A = 2Z = 12$ as typical, the following are easily shown:

$$\log T = \frac{1}{3}\log \rho + 7.48 \qquad (\mu_{\mathrm{ER}} = kT);$$

$$\log T = \frac{1}{3}\log \rho + 5.59 \qquad (kT = E_c);$$

$$\log T = 10.1 \qquad (kT = 2m_ec^2);$$

$$\log \rho = 7.07 \qquad (p_F = mc);$$

$$\log \rho = 1.083 \qquad (E_c = \mu_{\mathrm{NR}}).$$

Problem 2.12. Obtain the preceding relations for arbitrary A and Z. Comment on the sensitivity of your results to the nature of the ions involved.

Problem 2.13. Using the results of Problems 2.11 and 2.12, and assuming that more massive stars contain heavier ions in their cores, comment on how stellar mass affects the end point of stellar evolution.

Problem 2.14. Carry out a similar analysis for a system of protons. Assume that the interaction energy between protons is given by E_n, where

$$E_n = 7.04 \times 10^{-17}\,\frac{e^{-\mu_\pi r}}{r}\ \text{ergs},$$

with $\mu_\pi = m_\pi c/\hbar = 7.5 \times 10^{12}\ cm^{-1}$. Describe the principal characteristics of the matter in each region of the ρ, T-plane. The pion mass is m_π.

In cold matter at densities below about 10^{12} g/cm^3, the lowest energy state results if the ions form a crystal lattice that is penetrated by a degenerate electron gas. At higher densities it is conceivable that nucleons might also form a lattice because of the strong repulsive core of the strong interactions. In any case, when the density exceeds about 10^{14} g/cm^3, the equation of state of a degenerate system of nucleons and higher-mass baryons will deviate strongly from that for an ideal gas. For $\rho \lesssim 10^{15}$ g/cm^3, the interactions are mostly attractive; so the equation of state is softer than that for an ideal gas. Above about 10^{15} g/cm^3 the interactions are repulsive, and the matter is stiffer than an ideal gas. In these regions the electromagnetic forces between nucleons and baryons are negligible relative to the strong interactions. A typical nucleon equation of state at $T = 0$ is shown schematically in Figure 2.4.

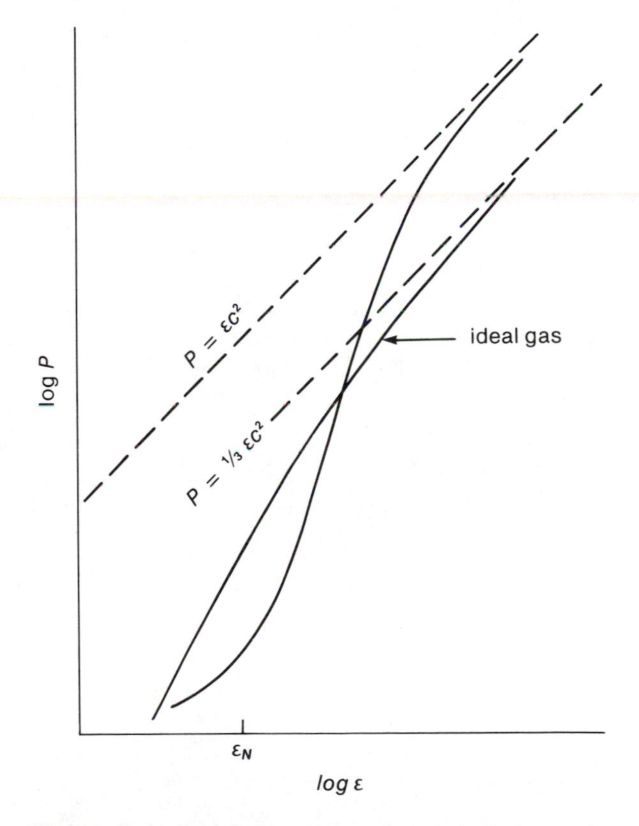

Figure 2.4. Schematic equation of state (pressure versus energy density) for matter at $T = 0$ near nuclear density, and for an ideal gas. For $\epsilon \lesssim \epsilon_N$ the attractive nuclear forces dominate, and $P < P_{\mathrm{ideal}}$. At high density the repulsive force dominates, and $P > P_{\mathrm{ideal}}$. The compressibility of the matter is proportional to the slope of $P(\epsilon)$. For any ϵ, $P(\epsilon)$ lies below the line $P = \epsilon c^2$ (causality limit), because for $P > \epsilon c^2$, sound would travel faster than light.

Chapter 3

ASPECTS OF OBSERVATIONAL ASTRONOMY

Nearly all the data we collect about stars come from a detailed analysis of the energy that they emit. Because these data play an important role in constructing and evaluating stellar models, we shall review the ways in which they are obtained from observational astronomy. In particular, we review the connection between observational data and the physical quantities and units that appear in this book.*

3.1. Systems of Brightness Measurement

In general we assume that stellar distances, positions in the sky, and proper motions are known. For the nearest stars (those within about 20 parsecs of the sun), the method of trigonometric parallaxes gives reliable distances. A star's parallax p (measured in seconds of arc) is half of the angle subtended at the star by the diameter of the Earth's orbit (diameter = two astronomical units; see Figure 3.1). The distance of stars whose parallax $p \gtrsim 0''.05$ (about 800 stars in the solar neighborhood) can be calculated accurately. The unit of distance for stars is the *parsec*, pc; a star whose parallax is $1''.0$ lies at a distance of one parsec.

To specify stellar positions, one needs a coordinate system. Galactic coordinates will be discussed in Section V; a detailed discussion of standard systems may be found in any elementary astronomy text.

Because the light we receive from the stars is our primary source of astrophysical information, we will review the terminology of brightness measurements. The absolute luminosity L of a star (which is its total rate of energy production) is not directly observable; what is measured is the apparent luminosity, l, which is defined as the rate at which light energy is received by a detector of unit area on the Earth, and has units of ergs cm^{-2} sec^{-1} in the cgs system. If we assume that light propagates with no loss of energy except that resulting from the spreading of the spherical wavefront centered on the star, then the apparent luminosity and the absolute luminosity are related by the inverse-square law,

$$l = L/4\pi r^2, \tag{3.1}$$

*The elementary terminology of astronomy can be found in any introductory astronomy text.

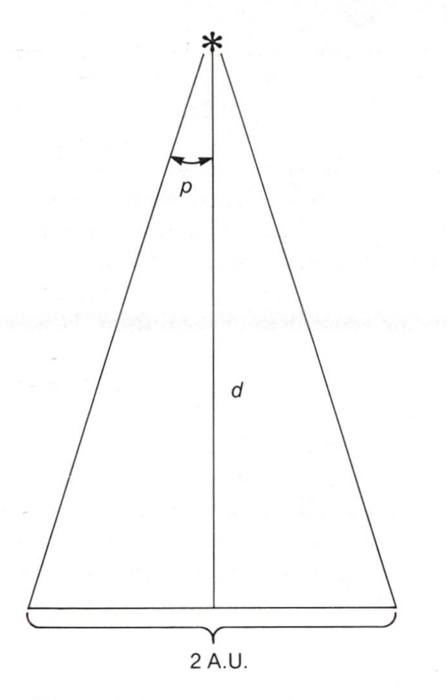

Figure 3.1. Parallax p of a star at a distance d from the Earth.

where r is the distance between the star and the Earth. If we can measure l and r, then (3.1) gives the absolute luminosity of the star. Conversely, given L, we can estimate the value of r. This is the fundamental procedure by which values are found for the distances of stars more distant than about 20 pc.

Measurements of apparent luminosity often use the magnitude system,* whose most important features are: the measure of brightness is logarithmic; the scale factor is negative, so that smaller magnitudes mean brighter stars; and, finally, the scale factor is such that a difference of five magnitudes corresponds to an intensity ratio of 100. Thus a star of magnitude 1.0 is 100 times brighter than a star of magnitude 6.0. These choices imply the following relation between apparent magnitude m and apparent luminosity l, and between absolute magnitude M and absolute luminosity L:

$$\begin{aligned} m &= -2.5 \log_{10} l + K_1, \\ M &= -2.5 \log_{10} L + K_2, \end{aligned} \qquad (3.2)$$

*In practice, the magnitude system is used primarily for optical measurements.

where K_1 and K_2 are constants established by convention. Combining (3.2) with (3.1) gives

$$m = M + 5 \log_{10} r + K_3, \qquad (3.3)$$

$$K_3 = (K_1 - K_2) + 2.5 \log_{10} (4\pi). \qquad (3.4)$$

The convention usually adopted for K_3 is that distances are to be measured in parsecs, and the absolute magnitude M is to be the apparent magnitude the star would have if placed 10 parsecs from the Earth. This yields

$$K_3 = -5 \qquad (3.5)$$

and

$$m - M = 5 \log_{10} r - 5. \qquad (3.6a)$$

Since a parsec is by definition the distance at which a star has a parallax of 1 second of arc, the simple relation

$$d = 1/p,$$

where d is in parsecs and p in arc seconds, may be used to rewrite equation (3.6a) as

$$m - M = -5 \log_{10} p - 5. \qquad (3.6b)$$

A further convention is needed to establish the value of either K_1 or K_2. In practice, K_1 is defined by specifying the apparent magnitudes of several standard stars, which are then used as calibration standards in measurements. Magnitudes can be measured to an accuracy of about one hundredth of a magnitude ($0^m.01$), corresponding to an uncertainty in the star's energy output of about 1 percent.

Actually, we never can observe the total apparent magnitude, as we have so far defined it, because of the limited range of spectral frequencies transmitted by the Earth's atmosphere and detected by measuring devices. Although measurements from above the Earth's atmosphere have partially overcome these limitations, most astronomical observations are ground-based, and specialized instruments have to be used to investigate various regions of the electromagnetic spectrum. Therefore we usually think of the apparent luminosity of a star, and its apparent magnitude, as being defined for only a limited range of frequencies.

Suppose the energy flux reaching the Earth from a star at frequency ν is l_ν erg cm^{-2} sec^{-1} Hz^{-1}. We measure at the telescope the average of this quantity over some range of frequencies that depends on both the transmission of the atmosphere and the spectral sensitivity of the instrument, which includes the effect of any filters inserted before the star's brightness was measured. We represent these effects by defining a sensitivity function S_ν, which measures the combined response of the instrument, filter, telescope, and atmosphere. The observed apparent magnitude corresponding to this particular sensitivity function is then

$$m_S = -2.5 \log_{10}\left\{\int_0^\infty S_\nu l_\nu d_\nu\right\} + K_S \qquad (3.7)$$

rather than (3.2), which holds if $S_\nu = S$, independent of frequency.

The use of colored filters in spectrophotometry yields information about the distribution of energy in the star's spectrum. Several different spectrophotometric systems are in use today. We will mention only one of them here, the *UBV* system. The letters stand for Ultraviolet, Blue, and Visual, which correspond roughly to the ultraviolet, blue, and yellow regions of the spectrum, respectively. The actual range of sensitivities of these three spectral regions is roughly

Ultraviolet	(U)	3000Å–4000Å,
Blue	(B)	3500Å–5500Å,
Visual	(V)	4800Å–6500Å.

In this system, U, B, and V denote the apparent magnitudes observed with each of the filters, and the corresponding absolute magnitudes are written as M_U, M_B, and M_V, respectively.

Also of importance in spectrophotometry are color indices. In the *UBV* system, they are defined as $U - B$ and $B - V$. Using (3.7), the indices may be written in terms of the ratio of luminosities in the two spectral regions. For example,

$$U - B = 2.5 \log_{10}\left\{\frac{\int S_{B,\nu} l_\nu d\nu}{\int S_{U,\nu} l_\nu d\nu}\right\} + C, \qquad (3.8)$$

where $S_{B,\nu}$ ($S_{U,\nu}$) is the sensitivity function in the blue (ultraviolet) region, and C is a constant chosen to make $U - B = 0.0$ for stars of spectral type A0. In particular, the hotter a star's surface, the more blue light is emitted relative to yellow, and hence the smaller $B - V$ is (recall that the scale factor in the magnitude system is negative). For a star like the sun, $B - V = +0.65$; for a somewhat hotter star such as Vega it is 0.0, and for a cooler star, such as Arcturus, $B - V$ is 1.23. Color indices are particularly useful because they can be measured for faint stars whose spectra can not be measured photographically. Note that, as long as the star's spectrum is not modified by interstellar absorption, these color indices are independent of the star's distance and represent intrinsic information about the star's spectrum. There also is a good correlation between the surface temperature of a star and its $B - V$ color index. In recent years the *UBV* system has also been successfully extended into the red and infrared regions of the spectrum.

The total apparent or absolute magnitude of the star, taken over all frequencies, is usually called the *bolometric magnitude.* The difference between the bolometric magnitude of a star and its magnitude in, for example, the V band is called the *bolometric correction,* B.C.:

$$m_{\text{bol}} = V + \text{B.C.} \qquad (3.9)$$

Because there is more energy in the whole spectrum than in a limited part of it, the bolometric correction is always negative. Table 3.1 shows the relations between some of these quantities for various types of stars.

Problem 3.1. Prove that the color index of a star is independent of its distance from the Earth. If a star emits twice as much energy in the B band as in the V band, what would be its color index $B - V$ if the constants $K_B = K_V$? (Actually, the latter condition is not satisfied by the *UBV* system in use today.)

Problem 3.2. How accurate do you think the bolometric corrections listed in Table 3.1 are? How do you think they are obtained?

Problem 3.3. Show that for a star which radiates as a black body,

$$l_\nu = \frac{2\pi h}{c^2}\left(\frac{R^2}{r^2}\right)\frac{\nu^3}{\exp(h\nu/kT) - 1},$$

where R is the stellar radius and r is its distance from the Earth. Define a sensitivity function for a pseudo-monochromatic *UBV* system by $S_\nu = \delta(\nu - \nu_i)$, with $i = U$, B, and V.

Find $U - B$ and $B - V$ for this case (this defines

Table 3.1
Parameters for stars of various spectral types. Columns give: spectral type; absolute visual magnitude, M_V; color index, $B - V$; bolometric correction, B.C.; bolometric magnitude, M_{bol}; effective surface temperature, T_{eff}; mass, radius, and luminosity in solar units; and average density in g/cm³.

Sp	M_V	B-V	B.C.	M_{bol}	T_{eff}	$\log \frac{M}{M_\odot}$	$\log \frac{R}{R_\odot}$	$\log \frac{L}{L_\odot}$	$\log \bar{\rho}$
Main sequence (V):									
O5	−5.8	−0.35	−4.0	−10	40,000	1.6	1.25	5.7	−2.0
B0	−4.1	−0.31	−2.8	− 6.8	28,000	1.25	0.87	4.3	−1.2
B5	−1.1	−0.16	−1.5	− 2.6	15,500	0.81	0.58	2.9	−0.78
A0	0.7	0.0	−0.4	0.1	9,900	0.51	0.40	1.9	−0.55
A5	2.0	0.13	−0.12	1.7	8,500	0.32	0.24	1.3	−0.26
F0	2.6	0.27	−0.06	2.6	7,400	0.23	0.13	0.8	−0.01
F5	3.4	0.42	0.0	3.4	6,580	0.11	0.08	0.4	0.03
G0	4.4	0.58	−0.03	4.3	6,030	0.04	0.02	0.1	0.13
G5	5.1	0.70	−0.07	5.0	5,520	−0.03	−0.03	−0.1	0.20
K0	5.9	0.89	−0.19	5.8	4,900	−0.11	−0.07	−0.4	0.25
K5	7.3	1.18	−0.60	6.7	4,130	−0.16	−0.13	−0.8	0.38
M0	9.0	1.45	−1.19	7.8	3,480	−0.33	−0.20	−1.2	0.4
M5	11.8	1.63	−2.3	9.6	2,800	−0.67	−0.5	−2.1	1.0
M8	16	1.8			2,400	−1.0	−0.9	−3.1	1.8
Giants (III):									
G0	1.1	0.65	−0.03	1.1	5,600	0.4	0.8	1.5	−1.8
G5	0.7	0.85	−0.2	0.5	5,000	0.5	1.0	1.7	−2.4
K0	0.5	1.07	−0.5	0.2	4,500	0.6	1.2	1.9	−2.9
K5	−0.2	1.41	−0.9	−1.0	3,800	0.7	1.4	2.3	−3.4
M0	−0.4	1.60	−1.6	−1.8	3,200	0.8		2.6	−4.0
M5	−0.8	1.85	−2.8	−3				3.0	
Supergiants (I):									
B0	−6.4	−0.25	−3	−9	30,000	1.7	1.3	5.4	−2.1
A0	−6.2	0.0	−0.5	−7	12,000	1.2	1.6	4.3	−3.5
F0	−6	0.25	−0.1	−6	7,000	1.1	1.8	3.9	−4.2
G0	−6	0.70	−0.1	−5.2	5,700	1.0	2.0	3.8	−4.9
G5	−6	1.06	−0.3	−5.2	4,850	1.1	2.1	3.8	−5.2
K0	−5	1.39	−0.7	−5.4	4,100	1.1	2.3	3.9	−5.7
K5	−5	1.70	−1.2	−6	3,500	1.2	2.6	4.2	−6.4
M0	−5	1.94	−1.9	−7		1.2	2.7	4.5	−6.7
M5		2.14	−3.2						

SOURCE: Adapted from C. W. Allen, *Astrophysical Quantities*, 3rd ed. (London: The Athlone Press, 1973).

the black-body color indices), and plot the result for 3,000 K $\leq T \leq$ 20,000 K. Calibrate the black-body color system to the solar values T_e = 5,800 K, $U - B = +0.13$, and $B - V = +0.65$, and assume that λ_U = 3700 Å, λ_B = 4450 Å, and λ_V = 5500 Å. What is the bolometric correction if we assume the Sun to be a black body?

3.2. Interstellar Absorption and Reddening

Light absorption by interstellar material complicates the description in Section 3.1. Interstellar material exists mainly in two forms: gas and dust. Interstellar gas generally absorbs radiation at discrete, well-defined frequencies, and therefore shows up as absorption lines in stellar spectra. We will defer discussion of interstellar gas to Chapter 18. Interstellar dust, however, scatters light over a much broader range of spectral wavelengths. It scatters blue light more than red light, just as dust in the Earth's atmosphere does; so absorption by interstellar dust produces an overall reddening of the star's spectrum, modifying its color index $B - V$ or $U - B$ from what it would normally be in the absence of dust. This effect is called interstellar reddening.

We will describe interstellar reddening by a simple

model in which the absorption has constant spectral characteristics, independent of the location of the dust in the Galaxy. Although this assumption is almost certainly wrong in detail, it provides a reasonably good overall picture of reddening. We define the interstellar absorption coefficient a (generally dependent on wavelength), and assume that the total loss of energy when starlight travels a distance r is proportional to

$$\exp\left[-\int_0^r a(r')\,\mathrm{d}r'\right]. \tag{3.10}$$

Problem 3.4. Justify formula (3.10). A much more detailed discussion of absorption and scattering of light is given in Chapter 25.

The quantity

$$\tau = \int_0^r a(r')\,dr' \tag{3.11}$$

is called the optical thickness of the interstellar material in the direction to the star. As we will see in Chapter 5, the photon mean free path is $1/a$; therefore the optical depth is a measure of the number of photon mean free paths along the line of sight. In the presence of absorption (3.1) must be replaced by

$$l = (L/4\pi r^2)e^{-\tau}. \tag{3.12}$$

If necessary, frequency subscripts are to be understood on l, L, and τ. Applying the definition of apparent and absolute magnitudes contained in (3.2) and (3.6a), we find

$$m - M = 5\log_{10} r - 5 + 2.5\,\tau\log_{10} e \tag{3.13}$$

in place of (3.6a). Note that the only change is the addition of the constant term

$$\Delta m = 2.5\,\tau\log_{10} e = 1.0857\,\tau. \tag{3.14}$$

Thus interstellar absorption increases the apparent magnitude of the star by an amount proportional to the total optical depth of the material between the observer and the star. The amount of interstellar absorption can be represented in this way for each of the spectral pass-bands of the *UBV* photometric system.

Problem 3.5. Assume a frequency-dependent interstellar absorption, a_ν, and a photometric system with sensitivity function S, and obtain an explicit expression for the effect of interstellar absorption.

The difference between the effects of absorption in the B pass-band and in the V pass-band increases the color index of a star, and is called the *interstellar reddening* for the $B - V$ color index:

$$\begin{aligned}\Delta(B - V) &= (B - V)_{\text{reddened}} - (B - V)_{\text{true}}\\ &= 1.0857\,(\tau_B - \tau_V).\end{aligned} \tag{3.15}$$

A similar expression can be obtained for the $U - B$ color index. The ratio of $\Delta(U - B)$ to $\Delta(B - V)$ is nearly independent of the distance of the star. Similarly, the ratio of the total absorption in, say, the V magnitude to the reddening, $\Delta(B - V)$, should be a property only of the absorption cross section of the interstellar dust. Typical values for these quantities are

$$\frac{\Delta(U - B)}{\Delta(B - V)} \simeq 0.72, \qquad R \equiv \frac{\Delta V}{\Delta(B - V)} \simeq 3. \tag{3.16}$$

These ratios may be used, for example, to estimate the amount of absorption ΔV of starlight, when some estimate of the reddening $\Delta(P - V)$ is available.

Problem 3.6. Why are these ratios independent of the properties of the star, and of its distance? How do you think the amount of interstellar reddening might be estimated, given a detailed spectroscopic observation of the star's light?

Interstellar dust is confined primarily to the Galactic plane. Although the distribution is often patchy, dust typically increases stellar magnitudes by about a magnitude per kiloparsec. Since the uncertainty in magnitude observations is of order $\pm 0^m.01$, reddening is usually important only for stars that are more than about 10 pc distant from us. Since only the brightest stars can be seen at large distances, the stars in which we most commonly see interstellar absorption effects are intrinsically bright stars in the plane of our Galaxy.

Problem 3.7. Assume that our Galaxy has a uniform optical thickness τ perpendicular to the Galactic disk. Approximately how does the amount of interstellar reddening of a distant extragalactic object vary with its Galactic latitude b in our Galaxy?

3.3. Color-Magnitude and Two-Color Diagrams

In any photometric system, the absolute magnitude and the color indices of a star are independent of the star's distance (except for effects of interstellar absorption) and are intrinsic properties of the star. We construct the color-magnitude diagram for a group of stars by plotting the position of each star, using its magnitude and color as the coordinates. We construct a color-color (or two-color) diagram by plotting, for example, $U - B$ against $B - V$ for a group of stars. These diagrams are versions of the Hertzsprung-Russell diagram, in which, originally, absolute magnitude was plotted against spectral type (which is partly a measure of the star's surface temperature).

The earliest HR diagrams were based on stars in the vicinity of the Sun. These stars are generally bright, easily observable, and close enough that their distances could be calculated by direct methods, especially by that of trigonometric parallaxes. Because their distances were known, their absolute magnitudes could be deduced directly from (3.6a). The HR diagram for stars of known distance in the solar neighborhood is shown in Figure 3.2. It reveals a group of stars forming a band that stretches from the upper left (the region of hot and bright stars) to the lower right (the region of cool and faint stars). This band, known as the *main sequence,* represents stars whose primary source of energy is the conversion of hydrogen in the stellar core into helium. The distribution of points is denser in this band than elsewhere in the diagram because nuclear fuel (hydrogen) is abundant in the stellar core. Until this fuel is exhausted in the core of these stars, the main-sequence evolution is quasistatic.

A second grouping of stars occurs in the upper right of the diagram, in the region of bright but cool stars, known as *red giants.* In these stars the original core fuel has been exhausted, and a complex (inhomogeneous) structure has developed. These stars contain relatively little fuel compared with the amount available during the main-sequence stage. As a result, stars spend far less time in the giant region of the diagram than they do on the main sequence.

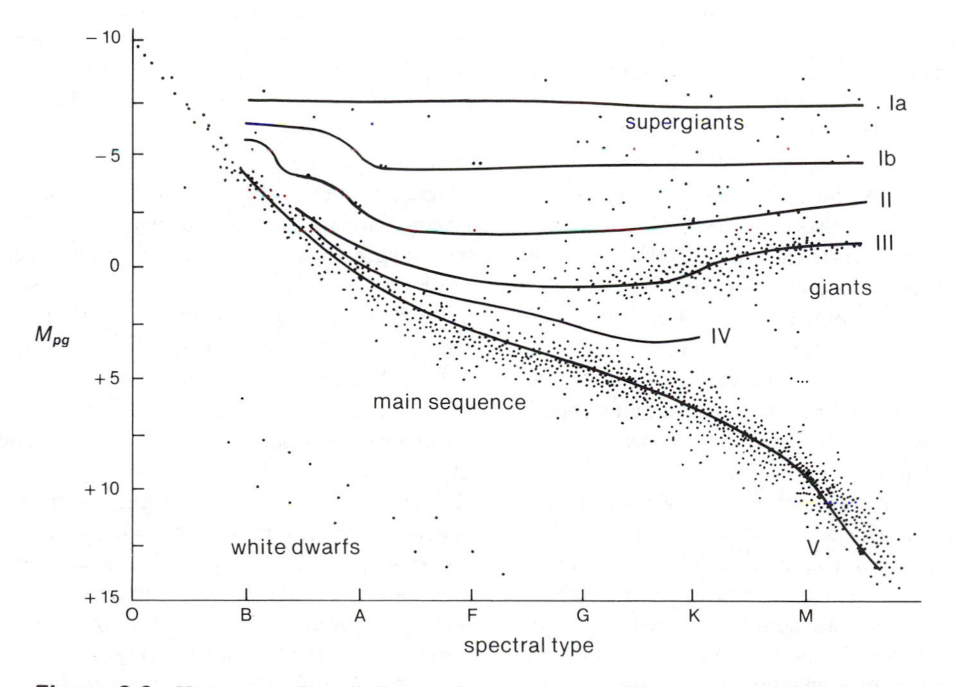

Figure 3.2. Hertzsprung-Russell diagram for stars in solar vicinity, showing absolute photographic magnitude versus spectral type. Solid lines are luminosity classes (approximate).

Problem 3.8. Explain why stars in the low-temperature/high-luminosity region of the HR diagram are called giants.

In addition, scattered around the HR diagram are a few straggling stars that are not part of these groups. These include *supergiants, subgiants, subdwarfs,* and *white dwarfs.*

Problem 3.9. Assuming that the four preceding star names describe the sizes of these objects correctly, sketch the regions of the HR diagram in which they might be expected to appear.

Problem 3.10. Why are some parts of the HR diagram more heavily populated with stars than others?

Many observational clues to understanding the structure and evolution of stars have come from the study of color-magnitude diagrams of star clusters. There are several different types of star clusters, but their important common feature is that they contain many stars physically associated with each other (and in many cases gravitationally bound to each other). It is presumed that stars in such groups were formed together at about the same time in the past, and out of the same material. We may therefore assume that each star in such a cluster had the same initial chemical composition and is of about the same age as its fellows. Therefore differences in the physical properties of the various stars in the cluster result *only* from differences in their mass (we ignore binary and multiple stars for the moment, and the effects of stellar rotation). A star cluster, therefore, represents a chemically homogeneous sample of stars, all of the same age. The differences observed between two star clusters are therefore believed to result from differences in initial chemical composition and cluster age.

Within each star cluster, the mass of a star essentially determines where it will appear in a color-magnitude diagram for that cluster. The overall shape of the HR diagram, however, depends on the age of that cluster and the initial chemical composition of the stars in it. Differences in the HR diagrams of clusters therefore represent differences in age and initial chemical composition.

A further point of practical interest is that the linear dimension of a star cluster is insignificant compared to its distance from the Earth; so we can regard all stars in the cluster as being at essentially the same distance from the Earth. Thus to study the shape of the HR diagram for that cluster, we need only plot the apparent magnitude V of each star against its color index, $B - V$; we need not use the absolute magnitude, M_V, because the pattern of the differences between the stars in the cluster will be the same for either kind of magnitude. Conversely, if the HR diagram of the cluster is of a form we recognize, we may be able to deduce the distance of the cluster by comparing its diagram with a similar HR diagram for a cluster of known distance. This is an important secondary method for discovering distances in astrophysics.

The HR diagram has become an important tool for presenting the theoretical results of calculations of stellar evolution. Therefore, in describing the evolution of a particular star, we usually describe its path in the HR diagram in terms of where it first appears, in what direction it moves, and how long it spends in different regions of the diagram.

An additional tool for discussing the characteristics of a group of stars, such as a star cluster, is the two-color diagram. In the *UBV* photometric system, for example, this is typically a plot of $(U - B)$ against $(B - V)$ for the stars in the cluster. Because star colors are mainly indicators of the surface temperature of the star, the normal two-color diagram for a star cluster is a smooth, one-parameter curve on which virtually all the stars lie. With one exception, the shape of this curve is essentially determined by the (very) roughly Planckian energy distribution corresponding to their (color) temperatures. The exception arises because the wavelength pass-band of the U magnitude straddles the Balmer discontinuity. The Balmer discontinuity is a jump in the energy distribution of the star's spectrum, corresponding to bound-free absorption by neutral hydrogen at the limit of the Balmer spectral sequence, at 3653 Å. In stars that have strong neutral hydrogen absorption (A stars, for example), the amount of energy in the U band is depressed by the Balmer discontinuity. As a result, the value of $(U - B)$ for these stars is larger than it would have been on the basis of the color temperature of the star, and the two-color $(U - B)$, $(B - V)$ diagram goes through a local maximum in $(U - B)$. Figure 3.3 illustrates the two-color relation for normal stars.

Some important astrophysical effects cause a star to deviate from this standard two-color line in the

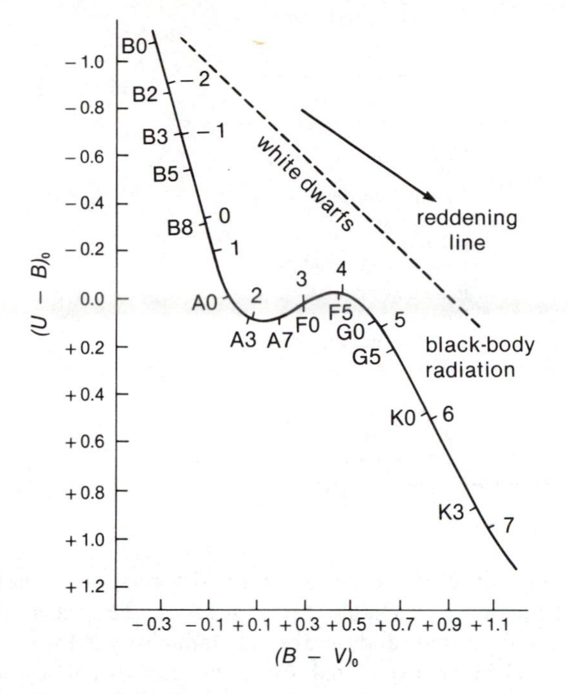

Figure 3.3. Color-color diagram for main-sequence stars. Also shown is the black-body curve (dashed) along which white dwarfs lie, and a typical interstellar reddening line for O-type stars.

diagram. One of these is interstellar reddening, as mentioned earlier. Because the ratio $\Delta\ (U - B)/\Delta\ (B - V)$ for interstellar reddening is practically a constant, the effect of reddening is to move a star away from its normal position in the two-color diagram, along a reddening line (Figure 3.3). Because this line is well separated from the standard $(U - B)$, $(B - V)$ curve, and is of known slope, we can often infer directly the amount of interstellar reddening for a star. Virtually anything else that distorts the spectrum of a star will displace it from the standard two-color curve, and may give some observable indication of its special character, for example, an anomalous chemical composition, a circumstellar gas shell giving rise to spectral emission lines, or an unseen companion of different spectral type.

3.4. Stellar Populations and Stellar Evolution

Astronomers generally think that our Galaxy evolved more or less as follows. The material from which our Galaxy was made was probably mostly hydrogen, with a small amount of helium, and minute amounts of other elements. In its early stages, when the first stars formed, the Galaxy had not yet collapsed to the flattened disk-like shape we recognize today. Hence the first stars formed out of nearly pure hydrogen, and had a more or less spherical distribution about the center of the Galaxy. We believe these stars formed in globular clusters, which have retained their spherical distribution about the center of the Galaxy. Thus the oldest objects we see today are spherically distributed about the Galactic center, and are very poor in heavy elements; that is, they are nearly pure hydrogen. These old, metal-poor stars are called *Population II* stars.

By contrast, the youngest objects we see in our Galaxy must have been recently formed out of material within the Galactic disk, since that is where most of the potential protostellar material lies. Furthermore, the process of stellar evolution causes a slow enrichment of the interstellar material from which stars are formed; hence the youngest stars we see must be relatively rich in heavy elements. These are called *Population I* stars.

A typical star cluster made up of this latter type of stars is called a *galactic cluster,* an example being the Pleiades cluster; a cluster typifying the former, older type of object is called a *globular cluster.* Table 3.2 summarizes their properties.

Figure 3.4 shows the color-magnitude diagrams of typical Population I and Population II samples. Some of the important points about these diagrams are as follows. First, the main sequence exists to a much bluer color (and presumably higher mass) in the Population I system than in the Population II system, indicating that the Population I system is younger. There are substantial numbers of giant stars in this sample, indicating that at least some stars have had time to evolve off the main sequence. In the Population II color-magnitude diagram, on the other hand, there are no upper main-sequence stars at all, because the cluster is older than the main-sequence lifetime of the more massive stars, which have subsequently evolved toward the giant region. Therefore the brightest stars in this group are red, whereas in the Population I group the brightest stars are blue. In the globular cluster color-magnitude diagram, the giant region is well populated, with two distinct branches of stars, the giant branch and the horizontal branch. Note the several stars marked with crosses in the horizontal branch; these are RR Lyrae stars, which are commonly found extensively in globular clusters, and which always lie in this part of the HR diagram.

Table 3.2
Principal characteristics of stellar populations.

Characteristic	Population I	Population II
Heavy-element content	2–3 percent	<1 percent
Dominant spectral types	O,B,A	K,M
	blue supergiants	Globular clusters
		Red giants
Variables	T Tauri	RR Lyrae
	δ Cepheids	Long-period variables
		Planetary nebulae
		Novae
Location and distribution	Galactic disk, primarily in spiralarms (patchy)	Galactic nucleus and halo (smooth)
Kinematics		
height above disk	$\lesssim$200 pc	>400 pc
average velocity	<10 km/sec	$\gtrsim$20 km/sec
Age	<1.5 times 10^9 yr	>1.5 times 10^9 yr

Several types of stars are considered special because of their variability, unusual spectrum, or other characteristics. Many of these objects are important for the theory of stellar evolution, either because they represent the response of the star to unusual conditions (such as marginal stability) or to some disturbance by a nearby star (close binary systems), or because they represent a comparatively rapid phase of stellar evolution (formation of planetary nebulae).

The reader should be familiar with the characteristics of at least the following unusual types of stars,* including their position in the HR diagram, and what makes them identifiably special: the classical variable stars (Cepheids, RR Lyrae stars, W Virginis stars, long-period variable), β Canis Majoris stars, δ Scuti variables, T Tauri stars, P Cygni stars, white dwarfs, planetary nebulae, metallic line stars, magnetic stars, novae, supernovae, and the various types of binary-star systems.

An important characteristic of the intrinsic variable stars (such as Cepheids, RR Lyrae stars, W Virginis stars, and long-period variables) is that they normally obey a period-luminosity relation. This follows because the period of free oscillation of a star under gravity is determined essentially by its mean density (Chapter 1). The mean density is an intrinsic physical property of a star, and for normal stars there is a good correlation between the intrinsic physical parameters and the extrinsic properties, such as luminosity. The practical significance of the period-luminosity relation is that the periods of variable stars can usually be measured very accurately, and the absolute luminosity of the star can then be calculated from the period-luminosity relation. Some of the pulsating variable stars, as well as other special types of stars, have been identified in star clusters, a fact that has been very useful for establishing the size of our Galaxy, and almost as useful for establishing distances within the local group of galaxies. In particular, the observation of Cepheid variables in the Magellanic clouds provided a stepping stone to the distance scale of the entire universe.

Problem 3.11. Why should the identification of some types of special (e.g., variable) stars in a star cluster help in establishing the stellar distance scale over moderate distances, such as within the Galaxy or within the local group of galaxies?

Three basic concepts emerge from the picture of stellar formation and evolution that we will cover in later chapters. The first is that normal stars spend most of their lifetime on the main sequence, their energy supplied by core hydrogen burning. The second is that more massive stars are brighter and hotter, and use up their energy supply faster than less massive stars; so that they evolve more rapidly. The third is that the time-scale for evolution in the stages that occur after the star has moved off the main sequence is relatively short.

*See, for example, Abell's *Exploration of the Universe*.

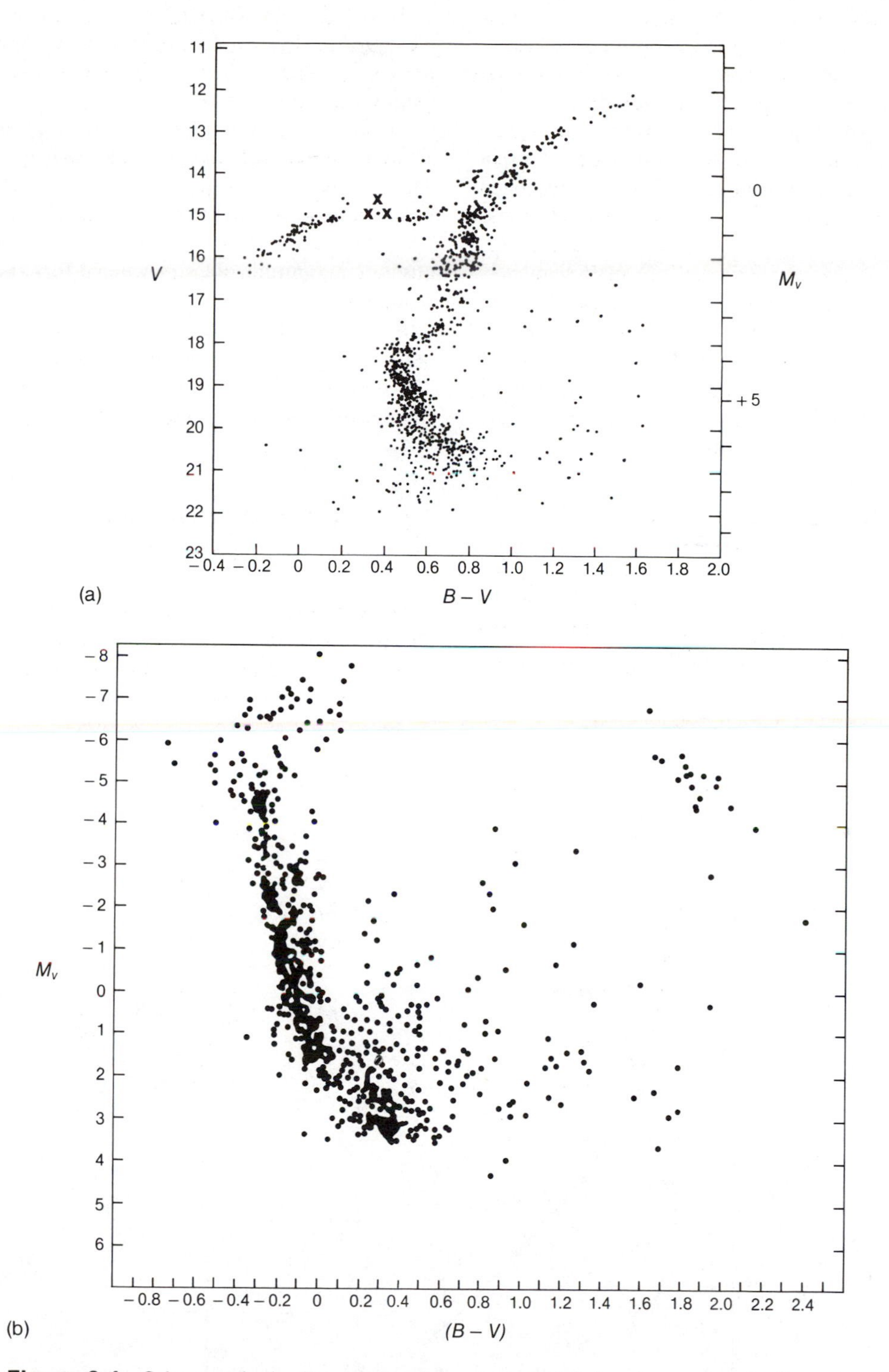

Figure 3.4. Color-magnitude diagram for: (a) a typical globular cluster, M5, in Serpens (Population II stars); and (b) a typical galactic cluster, h and χ Persei (Population I stars). $(B - V)_0$ is the observed color index. Positions of RR Lyrae variables are denoted by x.

When we observe a star cluster, we see it more or less frozen at some moment in its evolutionary history. That is, stars more massive than some transitional mass M_t will have already evolved off the main sequence, and will long ago have died. In contrast, less massive stars, with $M < M_t$, will still be on the main sequence, having a substantial supply of hydrogen remaining for core nuclear burning. Stars with mass close to M_t, probably plus or minus a few tenths $M_\odot$, will be in the process of evolving off the main sequence, and will occupy the giant and other regions of the star cluster's HR diagram. The point on the main sequence which is marked by stars of mass M_t is often called the turn-off point for the cluster. If this point can be established for a star cluster, it provides an accurate index to the age of the star cluster.

Problem 3.12. Why is it difficult to discover the turn-off point for many star clusters? (Think of observational limitations.) What observational parameter(s) would be most convenient for expressing the location of this turn-off point, and how might such parameters be calibrated in terms of age in years?

If we know the distance to a galactic cluster and can estimate the amount of interstellar reddening that its light experiences, then the member stars' absolute magnitude can be calculated. The results for a group of clusters can be used to form a composite color-absolute magnitude diagram, e.g., that in Figure 3.5. The relative age of a specific cluster is readily obtained from the position of the main-sequence turn-off point. The theory of stellar evolution predicts the age and absolute magnitude of a star when it turns off the main sequence. If its absolute magnitude can be converted into an absolute visual magnitude (as can often be done), then the cluster's absolute age can be estimated. The ordinate on the right-hand side of the figure gives theoretically predicted ages for the clusters in years.

3.5. Spectrum Analysis and Spectroscopy

A photometric color system, such as the *UBV* system, permits one form of spectral analysis of starlight, in that it yields information about the distribution of energy in the star's spectrum. More detailed spectrum

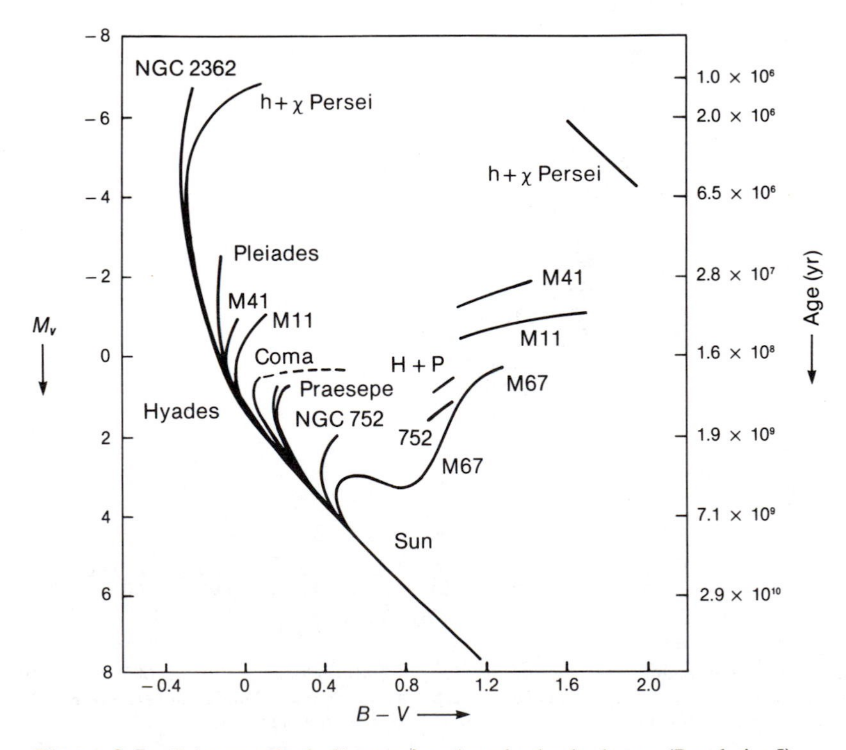

Figure 3.5. Color-magnitude diagram for selected galactic clusters (Population I).

analysis can be carried out, usually by means of a spectrograph or spectrometer. When a spectral resolution of about 1 Å is reached, the most prominent features in the spectrum of most stars are absorption lines caused by elements in the atmospheres of the stars. In some stars, emission lines are present also. These spectral lines were first discovered in the spectrum of the Sun, and are often called Fraunhofer lines. The strengths of these lines depend on the levels of excitation and ionization in the stellar atmosphere, and on the abundances of the various elements present there.

The theoretical prediction and analysis of line strengths is a major area of astrophysical theory. Nevertheless, a purely qualitative analysis of spectral line strengths can reveal much information. In particular, the visual analysis of photographically recorded stellar spectra led to the system of classification in wide use today.

We will not discuss the complexities of stellar spectral classification here, but will simply note that the strengths of lines are primarily dependent on the degree of ionization and excitation of the various atoms and ions in the stellar atmosphere. A classification of stellar spectra based on line strengths is therefore primarily a classification based on the degree of ionization of the elements present in the star's atmosphere. The degree of ionization in turn depends on physical parameters (primarily temperature and, to a lesser extent, density) in the star's atmosphere. In particular, as we will see in Chapter 6, for a given surface temperature, a giant star has a lower atmospheric density than a dwarf star, and hence a somewhat higher degree of ionization. The spectral classification sequence most common in modern astronomy is based on the intensity ratios of selected absorption lines in the star's spectrum, and depends mainly on degree of ionization. Figure 3.6 shows schematically the variation of line strengths with spectral type for some of the more important stellar lines. The more important features used to define the basic spectral classes, and the approximate temperature of each, are summarized in Box 3.1. Further classification within each group is achieved by dividing it into ten subgroups giving, for example, A0, A1, . . . , A9. Each subdivision is characterized by the appearance, dominance, or disappearance of one or more lines.

The sequence of spectral types is often modified by appending a luminosity class I–V. The luminosity classes are: I (supergiants); II (bright stars); III (giants); IV (subgiants); and V (main sequence). The elementary concepts of spectral types and luminosity classes can be found in any text on descriptive astronomy, which should be consulted for further detail.

Stellar spectral classification is normally carried out visually, on photographically recorded spectrograms at relatively low spectral resolution. For the brightest stars, however, one can obtain resolutions large enough to allow study of the fainter lines of less abundant elements, and detailed study of the shape of the absorption-line profiles of the stronger lines. To use the information contained in such high-resolution studies, one must quantify the information in some way, either by directly measuring the light intensity in the spectrum photoelectronically, using a photomultiplier arranged to scan over the spectrum and produce a direct measurement of the apparent brightness of the star as a function of frequency, or else by photographi-

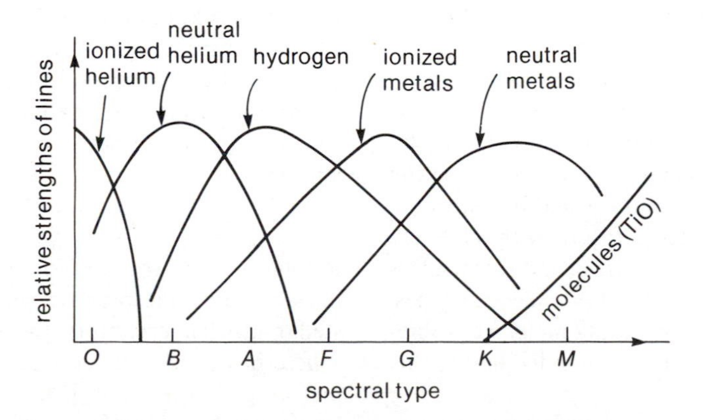

Figure 3.6. Relative line strengths for selected spectral lines used to establish spectral type (schematic).

Box 3.1
Stellar Spectral Types

O	Hot stars; strong uv continuum; HeII absorption; lines of highly ionized atoms; H lines appear weakly; $T_{eff} \gtrsim 25{,}000$ K.
B	HeI reaches maximum strength at B2; HeII vanishes beyond B0; H lines develop in late types; $12{,}000 \lesssim T_{eff} \lesssim 30{,}000$ K.
A	H lines maximum at A0; CaII increases; weak neutral metal lines appear; $7{,}500 \lesssim T_{eff} \lesssim 11{,}000$ K.
F	H lines weaken, but conspicuous; CaII becomes stronger; metal lines increase in strength (both ions and neutrals); $6{,}000 \lesssim T_{eff} \lesssim 7{,}500$ K.
G	CaII (H and K lines) become strong; solar types; Fe and metals strong; H lines weak; CH bands strengthen; $5{,}000 \lesssim T_{eff} \lesssim 6{,}000$ K.
K	Metallic lines dominate; continuum becomes weak in the blue; molecular bands (CN, CH) develop; $3{,}500 \lesssim T_{eff} \lesssim 5{,}000$ K.
M	TiO bands dominate; strong neutral metal lines; $T_{eff} \lesssim 3{,}500$ K.

cally recording the spectrum, and then using known sources of light to calibrate the spectral sensitivity of the photographic emulsions. A special scanning photometer then measures the density of the image as a function of frequency, and uses the calibration to relate photographic density to light intensity. Whichever method is used, we obtain a value for the apparent stellar luminosity as a function of wavelength, l_λ.

In most cases, this spectral measurement is further modified by rectification, which assigns the continuum level a value of unity over the whole spectrum. That is, a continuum intensity, c_λ, is defined as a function of wavelength, and the rectified spectrum is defined as

$$r_\lambda = l_\lambda / c_\lambda. \tag{3.17}$$

In practice, one establishes this continuum intensity by (in effect) looking at the spectrum and drawing in the level of the continuum one thinks would be present if there were no absorption lines. For stars with relatively few absorption lines, such as the O, B, and A stars, this is fairly straightforward, but for later spectral types, when there are many lines, including many weak lines that blend together and are not seen individually, this becomes a difficult task.

Given this rectification of the stellar spectrum, we may think of studying individual line profiles, that is, the run of r_λ with wavelength across the absorption line. Quite often, the line depth, $1 - r_\lambda$, is used instead of r_λ. If the line is quite weak, then often the most useful thing we can learn is the equivalent width of the line, W_λ, which is obtained by integrating the line depth over all wavelengths:

$$W_\lambda = \int (1 - r_\lambda)\, d\lambda. \tag{3.18}$$

For stronger lines, the whole line profile may be studied to learn about the mechanism of its formation, and hence about how the important physical properties of the star's atmosphere vary with depth.

In thinking about what we may learn from the study of spectral line intensities, we will find the following basic concept useful: weak lines, and the wings of strong lines, are formed deep in the star's atmosphere; strong lines, especially line centers, are formed high in the star's atmosphere. We will return to this idea in later chapters, but it is worth stating here. The reason for it is simply that weak lines mean weak absorption and hence low opacity in the atmosphere; so we can see more deeply into the atmosphere (that is, radiation can escape from deeper in the atmosphere). On the other hand, strong lines mean high absorption and high atmospheric opacity. Hence we do not see very far into the atmosphere; we see only the shallower layers. Finally, in most stars the upper layers of the atmosphere are cooler than the lower layers; so the intensity radiated is lower there, and the centers of absorption lines appear dark in the spectrum.

Problem 3.13. Photographs of the Sun are sometimes taken through an extremely narrow band filter, which transmits light only at a wavelength corresponding to the center of the H_α absorption line of hydrogen. What parts of the Sun's atmosphere are revealed by such photographs, and why are they of interest?

It follows from the preceding discussion that strong lines probe a wide range of physical conditions in the stellar atmosphere, from the very top of the atmosphere in the core of the line, to deeper layers in the line wings. As a result, it is often difficult to account for the whole line profile.

The end product we want from a thorough analysis of the high-resolution spectrum of a star is a physical model for the outer layers of the star, including the chemical composition of the material, the surface gravity of the star, and the run of temperature with depth in the atmosphere. Normally, we achieve these results by comparing model atmosphere calculations with the observed spectrum of the star. These models refer to the surface layers of the star, which we believe are normally not affected by the processes of thermonuclear evolution that go on in the star's interior. What we deduce about the surface chemical composition probably gives us a good guide to the original chemical composition of the material from which the star formed. This concept is important in understanding observational aspects of stellar evolution.

Stellar spectroscopy also yields a measurement of one of the basic dynamical parameters in astronomy, the radial velocity of the star. This is the component of the star's spatial motion along the observer's line of sight. This radial velocity, v_r, is calculated by measuring the apparent wavelengths λ of the lines in the spectrum of the star, and comparing them with the known wavelengths λ_0 of the lines that would be observed in the lab, assuming that the origin of the spectral lines is correctly identified. The apparent wavelength is shifted by an amount $\Delta\lambda$, which is given by

$$\frac{\Delta\lambda}{\lambda} = \frac{v_r}{c}, \qquad (3.19)$$

when $v_r \ll c$ (nonrelativistic Doppler shift). High-resolution measurements give radial velocities to within an uncertainty of about one km sec^{-1}.

In addition to basic physical properties of the ordinary stellar atmosphere, detailed stellar spectroscopy can reveal other facts about the star, such as stellar rotation, which affects the shape of all absorption-line profiles; rotational velocity can be calculated from Doppler effects that arise because one side of a rotating star moves away from the observer, while the other side moves toward him or her. Spectral analysis can also detect magnetic fields in the surface of the star, since these modify the energy levels of the atoms giving rise to the absorption, and hence modify the line profile.

Stellar spectroscopy can also reveal the binary or composite nature of some stars not resolvable as visual binaries. This may be seen as a mixed type of spectrum, containing spectral lines characteristic of both high and low temperatures (spectrum binary), or as a periodic Doppler shift of the spectral lines produced by the orbital motion of one star about the other. Sometimes the lines of only one star are seen (single-lined spectroscopic binary); sometimes both sets of lines are seen (double-lined spectroscopic binary).

The spectroscopic study of ordinary galaxies can reveal several useful types of information. A basic parameter is the redshift, which is the Doppler shift due to the expansion of the universe, plus that caused by whatever special motion the galaxy itself may have. In the study of galaxies and other extragalactic objects, it is common to express the shift in wavelength of the spectral lines in terms of the redshift parameter z, defined as

$$1 + z = \lambda_0/\lambda_e, \qquad (3.20)$$

where λ_0 is the observed wavelength of the line, and λ_e is the emitted wavelength.

There are two reasons for using (3.20) rather than (3.19). One is that for the most distant objects, we can no longer interpret a redshift simply as a velocity, since the value obtained depends on the particular cosmological model used (Chapter 26). A second, minor point is that there remains some doubt whether the observed redshift of all galaxies (especially quasars) is due entirely to the expansion of the universe or whether perhaps some other effect is operating. For example, it is known that strong gravitational fields can cause redshifts; this effect has been seen in white dwarfs and possibly in other objects. For these reasons, it is better to use a simple measure of the amount of redshift, z, which is not prejudiced in favor of any underlying theoretical model. Nevertheless, it is sometimes useful to express the very large redshifts seen in quasars in terms of a velocity of expansion expressed as a fraction of the speed of light. To do this, the special relativistic formula

$$1 + z = (1 + v/c)\,[1 - (v/c)^2]^{-1/2} \qquad (3.21)$$

is used (Chapter 27).

In addition to basic redshift information, spectroscopic studies of galaxies may reveal several other things. For example, the spectrum of a normal galaxy represents the sum total of the light emitted by all the stars in it, modified by any internal interstellar reddening. Perhaps we could synthesize the spectrum of the galaxy by mixing together in various proportions the spectra of known types of stars in our Galaxy, and thus learn something of the stellar composition of other galaxies.

Problem 3.14. What problems would you expect to encounter in performing such a study, especially for a galaxy of rather different type from our own?

A second type of information that may be revealed by composite galaxy spectra is information about the luminosity of the galaxy as a whole. This leads to a luminosity classification for galaxies, from I (supergiants) to V (dwarfs), not unlike the luminosity class system for stars, which is based on spectral-line intensity ratios. Luminosity classification schemes for galaxies are useful in establishing the extragalactic distance scale.

Spectroscopy of galaxies sometimes reveals peculiarities, especially when the galaxy is an *active galaxy*. These may have strong emission lines in their spectra, arising from emission processes in the interstellar medium of the galaxy, especially near the nucleus. Often these galaxies are peculiar in other ways as well, and may be strong radio emitters. Some types may also be related to quasars.

3.6. Binary Systems

More than half the stars in the sky are members of binary systems. We do not yet know why binary stellar systems are so common. Here we will discuss binary star systems only as a source of astronomical information.

A stellar binary system consists of two stars gravitationally bound in orbits about a common center of mass. Observations of the dynamics of the orbits yield information about the mass of the components, and are the primary source of data on stellar masses. In addition, when the orientation of the orbit of one object about the other is especially favorable, we can calculate directly other intrinsic properties of the components of the system, such as their radii. For most binary star systems, Newton's law of gravitation produces an adequate description of the binary's dynamics. In at least one system, however, the two stars are so close that the general relativistic theory of gravitation must be used to describe the orbits. Observations of this system have furnished strong evidence for the existence of gravitational radiation as predicted by Einstein's general theory of relativity (see Chapter 17).

Finally, components of binary systems presumably are the same age (assuming they were formed at the same time); so we can use information about, e.g., their relative luminosities to test theories of stellar evolution, much as we can with star clusters, except that we sometimes must consider the effects of one star on the structure and evolution of the other. When this happens, we refer to the system as a *close binary;* we will defer discussion of such systems until Chapter 17. The dynamics of close binary systems have played an important role in arguments that Cygnus X-1 is a black hole, and in identifying a class of neutron stars with X-ray pulsars.

Binary stellar systems are classified according to how we happen to be able to observe them, not according to their intrinsic characteristics. The simplest is the *visual binary,* which can be seen to be double on good-quality photographs, or sometimes by eye. More particularly, the orbital motion of one star about the other is observable on photographs taken during, say, several years. Because of observational effects, it is usually only those stars with relatively large true separations that can be seen as visual binaries, and these systems tend to have rather long orbital periods, because of Kepler's third law.

A second type of binary star is an *eclipsing binary,* in which the orbital plane of the system is nearly edge-on to the line of sight, so that periodically one star passes directly in front of the other, partially or completely eclipsing it; hence the total light we receive from the system is reduced during the eclipse. Eventually the second star emerges from behind the first, then moves in front of the first, causing a second eclipse and a second change in the brightness of the system. Figure 3.7 shows the light curve of a typical eclipsing variable (WW Aurigae). The deeper eclipse in the light curve is known as the primary eclipse, and the less deep one as the secondary eclipse.

The third, and most common, type of binary star is the *spectroscopic binary*. Here, typically, the stars are too close together to be resolved into a visual binary, and the orbit is not edge-on; so no eclipses are seen. Nevertheless, spectroscopic study reveals that the system is composite rather than a single star.

A fourth, somewhat related, type is the *spectrum*

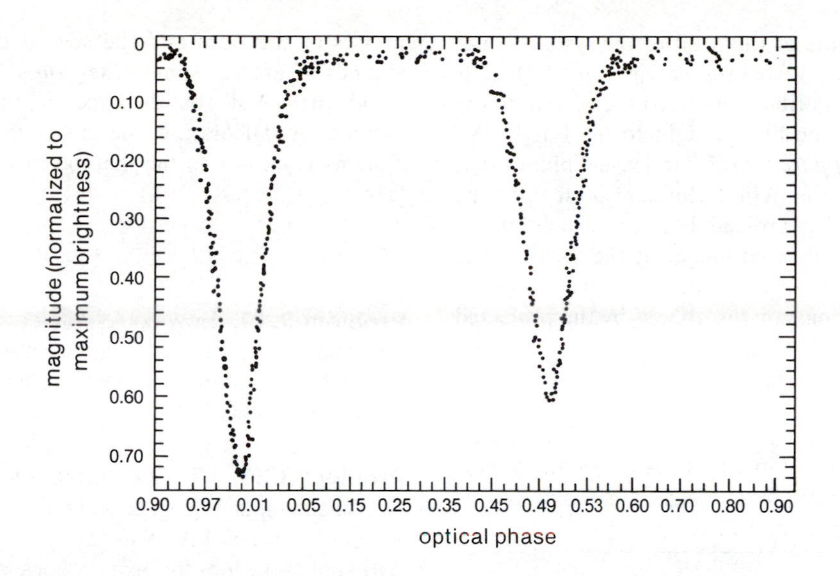

Figure 3.7. Light curve (magnitude versus orbital phase) for the eclipsing binary WW Aurigae. The orbital phase is 0.0 at primary eclipse. Note the change in orbital scale around primary and secondary eclipse.

binary. The distinction between the spectroscopic and the spectrum binary is as follows. In a spectroscopic binary, spectral lines from at least one of the components exhibit a periodic Doppler shift produced by the orbital motion of one star around the other. In a spectrum binary, on the other hand, no Doppler shift of the spectral lines is observed, but lines of both components of the system are present, indicating two different types of spectrum. Clearly, for a star to be a spectrum binary, the components must be of more or less the same brightness, in order for them to contribute comparable amounts to the star's composite spectrum. Since no orbital radial motion is seen, the orbit must be nearly face-on to the Earth, or the stars must be so far apart that no orbital motion is detectable, but then the stars would probably be observable as a visual binary. In a spectroscopic binary, it does not matter if one star is so much fainter than the other that its spectrum is not seen, provided that the orbital motion is significant. In a spectroscopic binary, if lines from only one spectrum are seen, it is called a *single-lined spectroscopic binary;* if lines from both spectra are seen, it is called a *double-lined spectroscopic binary*.

Problem 3.15. Imagine two identical stars of mass $M_\odot$ in orbit about each other at a relative distance of 100 astronomical units. Imagine placing this system at varying distances from the Earth, with a variety of orientations of the orbital plane of the system relative to the line of sight to the Earth. Under what circumstances will this system be seen as a visual binary, an eclipsing binary (sketch the light curve), a spectroscopic binary, or a spectrum binary?

Binary systems obey Kepler's laws of motion, as do the planets, but with some important qualifications. First, the masses of the two components of the system will often be comparable; so each star will follow an elliptical orbit about the center of mass of the whole system. This *absolute orbit* is not usually observed, however, except in certain astrographic studies that may reveal fluctuations in the direction of proper motion of the star, indicating the presence of an unseen companion. Instead we are normally more concerned with the *relative orbit* of one star as seen from the other. This relative orbit is also an ellipse, with the primary star at one focus of the ellipse, and is often called the *true orbit*.

Problem 3.16. Prove that the relative orbit in a binary system is an ellipse.

Second, the plane of the orbit of the system is not normally seen face on from the Earth; instead, there is some angle of inclination, i, between the normal to the orbital plane and the line of sight to the Earth. We therefore see an *apparent orbit* for a visual binary that is a projection of the actual elliptical orbit onto the plane of the sky. This projected orbit is also elliptical, but the primary star is no longer at the focus of the projected ellipse. However, Kepler's second law (areal velocity equals a constant) is obeyed by the projected ellipse also.

Problem 3.17. (a) Verify that the projected orbit is an ellipse. (b) Prove that Kepler's second law is satisfied in the apparent orbit.

Given the characteristics of the apparent orbit, relative orbit, and absolute orbit, if we know the apparent orbit and stellar distance of a visual binary, we can sometimes calculate all the absolute characteristics of the binary system, including each of the two stellar masses.

The basic physical relation that gives the masses of stars in binary systems is Kepler's third law,

$$M_1 + M_2 = a^3/P^2, \qquad (3.22)$$

where a is the semimajor axis of the true elliptical orbit measured in astronomical units, and P is the orbital period measured in years. The actual observed orbit for a visual binary star, however, gives the semimajor axis in angular measure on the sky, usually in seconds of arc. To convert this observation into a physical separation of the stars in length units, we must know the distance to the system. Fortunately, many visual binary systems are close enough to the Earth that their distances can be calculated directly by trigonometric parallaxes.

Problem 3.18. Why should visual binaries tend to be close enough to the Earth to have measurable parallaxes?

As is common in observational astronomy, we can turn this relation around and say that if we knew the masses of the stars and the size of the true orbit in seconds of arc (or some other angular measure), we could then find the distance of the system. This method of calculating the distance of binary star systems is known as the method of dynamical parallaxes.

Problem 3.19. How could you learn the masses of the stars in a visual binary system, in order to calculate the distance of the system by the method of dynamical parallaxes?

Problem 3.20. If the true orbit of the binary system has a semimajor axis s seconds of arc and a period of P years, and if the distance of the system is d parsecs, show that the exact expression for the total mass of the system is

$$M_1 + M_2 = d^3 s^3 / P^2.$$

For spectroscopic binaries, we cannot usually obtain as much information as we can for visual binaries. Our main information about a spectroscopic binary system is the radial-velocity curve for each component, $v(t)$. Since the system also has some unknown total radial velocity relative to the Earth, we have to remove this; we therefore can analyze only the change in the radial velocity. In a single-lined spectroscopic binary, we may know the radial velocity curve for only one component; so we can calculate some of the elements of the absolute orbit of only this one component about the center of mass of the system. Furthermore, the component of the star's orbital velocity in the direction of the line of sight to the Earth is proportional to sin i, where i is the angle of inclination of the orbit. Therefore, although we may be able to calculate other elements of the orbit accurately from the radial-velocity curve, our knowledge of the semimajor axis of the system is limited to $a_1 \sin i$, and the period, P, of the orbit. This is not enough data to allow us to deduce the mass of the whole system, unless we make assumptions about sin i and about the mass ratio of the two components. In a double-lined spectroscopic binary, however, we can calculate $a_1 \sin i$ and $a_2 \sin i$ separately. Since sin i is the same for both, this immediately gives us the ratio of the semimajor axes of the absolute orbits of the components, and hence the ratio of the masses of the components, since the center of gravity of the system is defined by

$$M_1 a_1 = M_2 a_2. \qquad (3.23)$$

We can also calculate the quantity

$$a' = (a_1 + a_2) \sin i = a \sin i, \qquad (3.24)$$

where a is the total semimajor axis of the system. Applying Kepler's laws to this situation yields

$$(M_1 + M_2) \sin^3 i = d^3 a'^3 / P^2; \qquad (3.25)$$

so we can find the mass of the system, except for the unknown inclination factor $\sin^3 i$. This is generally as far as we can go (unless further information on i is available, as it is if we have an eclipsing binary). However, the value of $(M_1 + M_2)\sin^3 i$ may still be useful, since it yields a lower limit to the mass of the whole system (obtained by putting $\sin^3 i = 1.0$).

Problem 3.21. Show that for a single-lined spectroscopic binary, the quantity

$$M_2^3 \sin^3 i / (M_1 + M_2) \qquad (3.26)$$

can be calculated. This is called the mass function of the system.

Finally, we mention briefly eclipsing binaries. Clearly these are systems with orbits nearly edge-on to the Earth, for which $i = 90°$, or $\sin i = 1.0$. If the system is a spectroscopic binary also, we may know a great deal about its orbit. In particular, we suppose that we know the dynamics of the orbit, that is, the rate at which the star moves along its orbit. Consider the partial eclipse of a large star by a smaller one. The amount of light that is cut off from the large star depends essentially upon the surface brightness of the large star, and the projected surface area of the smaller star. As the smaller star moves in front of the larger one, more and more light is cut off; the rate at which the light intensity changes depends on both the rate at which the projected disk of the small star crosses the limb of the larger one and the size of the smaller star. Since the rate of transverse motion of the small star is assumed known from the orbital elements of the system, we may be able to calculate the size of the smaller star directly. The analysis of the light curves of eclipsing binaries is quite complicated, depending on what assumptions, if any, are made about limb darkening in the star being eclipsed. Furthermore, since eclipsing binary systems tend to be very close pairs, many of them fall into the category of close binaries mentioned earlier, where one star exerts a distorting tidal influence upon the other, which further complicates the geometry (see Chapter 17).

Some insight can be obtained by analyzing the oversimplified model of a uniformly bright disk of radius R being partially eclipsed by a disk of radius r. The unknown parameter in this model, for any given R and r, is the angle of inclination of the orbit, which determines where on the disk of the larger star ingress and egress of the smaller star takes place (see Figure 3.8). But since even this unrealistic model is messy to analyze, we will not pursue it further.

Problem 3.22. Analyze the geometry of the eclipsing variable system in Figure 3.5. Sketch the light curve as a function of lateral position of the smaller star. How would you transform this light curve into an observed light curve for a given system (for example, what parameters of the system would you need to know?)?

Problem 3.23. Which of the techniques, discussed in this section in reference to stars, would you consider trying to apply to a system of two galaxies in orbit around each other?

3.7. Pulsating Stars

The prototype pulsating star is δ Cephei—whence Cepheids for this general class of variable stars is derived. Astrophysical questions about pulsating stars include the cause of the pulsation, the driving mechanism, and the damping mechanisms. In this chapter we will mainly be concerned with the importance of pulsating stars for observational astronomy. These stars provide valuable information because their period of pulsation is directly related to their intrinsic physical parameters, such as density, and the period of pulsation is relatively easy to measure with quite high accuracy.

As we saw in Chapter 1, for a simple radially pulsating star the period of pulsation and the mean

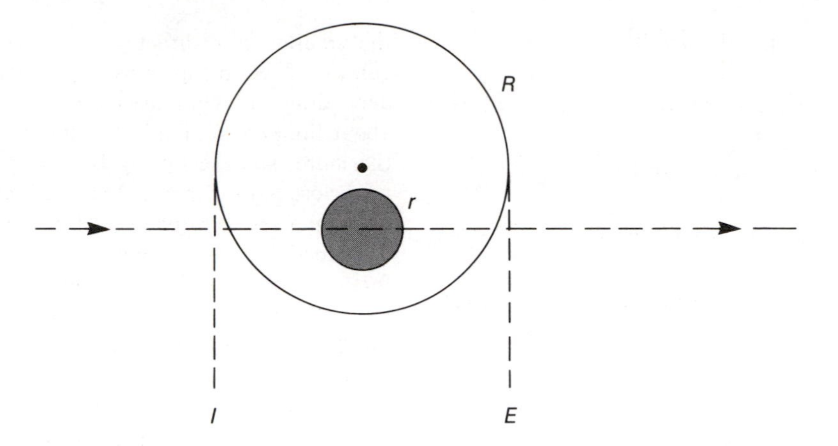

Figure 3.8. Geometry of eclipsing binary system. Motion of disk A (radius r) relative to disk B (radius R) is along dashed line. Ingress (I) and egress (E) represent the point of first and last contact, respectively, of disk A on B.

density are related by

$$P\sqrt{\rho} \approx 1/\sqrt{G}, \qquad (3.27)$$

where G is the constant of gravitation. Historically, what was first observed about Cepheid variable stars was a relation between period and luminosity, which implies a good correlation between luminosity and density for this class of stars. If luminosity and mean density are related by

$$L = K\rho^{-\alpha}, \qquad (3.28)$$

where α is a constant, then the relation $P\sqrt{\rho}$ = constant implies a period-luminosity relation

$$P \sim L^{1/2\alpha} \qquad (3.29)$$

or, expressed in magnitudes,

$$\log_{10} P = -\frac{1}{5\alpha} M + \text{constant}. \qquad (3.30)$$

The observed period-luminosity relation for Cepheid variable stars is approximately

$$\log_{10} P \simeq -0.394\, M + \text{constant}, \qquad (3.31)$$

so the index α of the luminosity-density relation for these stars is approximately 0.5.

The astrophysical aspects of pulsational variability are discussed in Chapter 11; what is important here is that, because of the period-luminosity relation, Cepheid variable stars can be used as distance indicators, since, if their period can be measured, their absolute luminosity can be calculated, and hence their absolute distance can be inferred. Furthermore, Cepheid variables are supergiants and have very high intrinsic luminosities; so they can be seen at very great distances; in fact, in nearby galaxies. Thus Cepheid variables bridge the gap between techniques for calculating distances that can be used only within our Galaxy, for relatively nearby stars, and techniques used for very large-scale distances within the universe (the extragalactic distance scale), which typically rely on the brightness of a whole galaxy or of the brightest star in it (or of other bright features, such as HII regions) as a distance indicator. To this day, the period-luminosity relation for Cepheids remains a crucial link in establishing this distance scale.*

There are some other important types of pulsational variables. Type II Cepheids, also called W Viriginis stars, are believed to be essentially the same as Cepheids, but belong to Population II, so they are poor in heavy elements. As a result, the period-luminosity relation for these stars, although similar to that for ordinary Cepheids, is displaced somewhat toward lower luminosities.

Cepheid variables are supergiants, and hence have low densities and long periods. The typical period for a Cepheid is about 10 days, and the spectral type is between F and G. A second very important type of pulsating variable includes the RR Lyrae stars, named

*In using the Cepheid period-luminosity relation, it is, of course, crucial to be sure that the variables being observed are actually Cepheids, since other pulsating variables obey period-luminosity relations that differ from (3.31). In the past, confusion of variable types has caused errors in the bridge to the extragalactic distance scale.

after the prototype of the class. These stars have spectra similar to those of the classical Cepheid variables, but are considerably less luminous. They are therefore smaller and denser objects, and have shorter periods. The typical pulsation period for an RR Lyrae star is about 0.5 to 1 day. RR Lyrae stars obey their own period-luminosity relation, which is approximately

$$\log_{10} P \simeq -0.85\, M + \text{constant}. \qquad (3.32)$$

Their periods are therefore more sensitive to luminosity than those of the Cepheid variables.

Problem 3.24. Use the differences in their periods to compare the density of an RR Lyrae star with that of a typical Cepheid variable.

Problem 3.25. What does this result imply about the luminosity-density relation for RR Lyrae stars?

RR Lyrae stars are found in great abundance in many globular clusters; so clearly they are Population II objects, and they lie in what is known as the horizontal branch of the globular cluster color-magnitude diagram. Study of the position of both the Cepheid variables and the RR Lyrae stars shows that they lie within a rather well-defined band in the HR diagram, in a spectral range $F-G$. This zone of the HR diagram has become known as the instability strip. In Chapter 11 we will discuss how this instability is associated with the ionization of hydrogen and helium in critical zones within the star's interior.

Lying in a different spectral range in the HR diagram are the long-period variables. These are red supergiants, the largest stars known, and hence they have very long periods, from 100 days to several years; so it takes quite a long time to gather full information about these objects. The prototype long-period variable star is Mira, which varies in visual brightness by more than five magnitudes during the course of about 330 days. Related to long-period variables are semiregular variables, with slightly shorter periods (around 100 days) and, as their name implies, with some irregularity of their light curves. In general, there is no absolute distinction between the long-period variables and the semiregular variables; all have light curves that do not reproduce exactly from one cycle to the next, and their periods are not exactly constant. In fact, it has been claimed that all stars in this general region of the HR diagram (red giants and supergiants) show some degree of variability.

The δ Scuti stars, a part of a general class of dwarf Cepheids, lie within the same instability strip in the HR diagram as the Cepheids and RR Lyrae stars, but are dwarf stars, and hence have much shorter periods than the Cepheids and RR Lyraes. They appear to pulsate in two or more modes simultaneously, probably including nonradial modes of oscillation. The different modes associated with each period interfere with each other and give rise to quite complicated light curves. The observational problem with these stars is to identify the various periods present and the mode of oscillation responsible for each period. Since these periods and period ratios can be used to check on physical models for the structure of the stars, they provide in principle a means of checking theory against observation.

3.8. Rotating Stars

Stellar rotation is an almost universal phenomenon. The Sun has a rotation period of about 25 days, an equatorial rotational velocity of 2 km sec^{-1}, and an angular momentum of about 10^{48} g cm^2 sec^{-1}. It is thus a relatively slow rotator, since many stars are known to have equatorial rotational velocities of one or two hundred km per sec. In general, there is a correlation between the rotational velocities of stars and their spectral types (except for pulsars, which are believed to be rapidly rotating neutron stars). In particular, the highest rotational velocities are found only among the earliest spectral types.

We usually describe stellar rotational velocities by specifying the equatorial rotational velocity in km/sec. This velocity is calculated from the Doppler broadening of spectral lines that is produced by the difference between the forward and backward motions of opposite limbs of the star. As with spectroscopic binary stars, we can calculate only the projection of the equatorial velocity in the line of sight to the Earth, that is, the quantity $v \sin i$, where i is the inclination of the rotation axis of the star to the line of sight to the Earth. For a star seen pole-on, the angle of inclination is zero, and we will see no rotational broadening of the lines even if the star is a rapid rotator.

Problem 3.26. Assume that a star is rotating with constant angular velocity, and has a constant surface brightness. What is the shape of an otherwise perfectly sharp spectral line seen from this star, if it has an inclination of 90 degrees (is rotating normal to the line of sight)?

Late-type stars, such as the Sun, have low rotational velocities and low angular momentum. However, there is more angular momentum in the planetary system than there is in the rotation of the Sun—about 10^{50} g cm^2 sec^{-1}. If the solar system remained isolated from the rest of the Galaxy during its formation and evolution, then its total angular momentum should have remained constant. This fact suggests several intriguing possibilities; one is that the Sun may have somehow lost most of its angular momentum to the planets during the formation of the solar system; another is that other slow rotators of spectral type similar to the Sun may also have planetary systems. Unfortunately, the way the solar system formed is still not well understood, and we have not yet detected (with certainty) other planetary systems.

There is a limit to the speed with which a star of given mass can rotate, or else it will become unstable, because the centrifugal force of rotation would exceed the gravitational force at the equator. This happens at a critical rotational velocity, v_c, given by

$$\frac{v_c^2}{R} \approx \frac{MG}{R^2} \tag{3.33}$$

or

$$v_c \approx (MG/R)^{1/2}. \tag{3.34}$$

For the Sun, this break-up rotational velocity is about 400 km/sec, about 200 times its actual rotational velocity. However, some stars with rotational velocities of several hundred km/sec may be approaching this limit.

Problem 3.27. As a star evolves and changes radius, its angular momentum should stay constant. Will contraction or expansion of a star cause it to become unstable if it is initially close to rotational instability?

3.9. Astronomical Statistics

The HR diagram discussed in Sections 3.3 and 3.4 represents one approach to analyzing the distribution of stars in a given region of the Galaxy. Another approach consists of describing the distribution of stars in terms of a particular set of characteristics. Such information can be useful for such topics as galactic structure, star formation, and stellar evolution. An analysis of the number of stars of a given spectral type near the Sun or within a specific galactic cluster can, in principle, be carried out observationally. Information about such distributions can be used, for example, to estimate the rate at which stars become supernovae, or to compare the rate of planetary nebula formation with the frequency of occurrence of white dwarfs.

A particular statistical distribution function is the distribution of stellar masses, or *mass function*, $\Phi(M)$, defined such that

$$dN(M) = \Phi(M)\, dM \tag{3.35}$$

is the number of stars per unit volume with mass between M and $M + dM$. We also define the *luminosity function*, which is the number of stars per unit volume with luminosity in the interval L to $L + dL$. These distribution functions are similar to the velocity distribution function of particles in thermal equilibrium (the Maxwell distribution). Furthermore, these distributions may change with time. Thus the present mass function of stars in a globular cluster, for example, differs from the initial mass function because of stellar evolution, and possibly because the least-massive stars are gradually lost by the cluster as it ages.

The empirical mass-luminosity relation can be used to relate the mass function to the luminosity function. Assuming that there is a relation of the form

$$L = K\, M^{\alpha}, \tag{3.36}$$

so that

$$dL = K\,\alpha\, M^{\alpha-1}\, dM, \tag{3.37}$$

we may write

$$\Phi_M(M)\, dM = \Phi_L(L)\, dL, \tag{3.38}$$

where $\Phi_M(M)$ is the mass distribution function, and $\Phi_L(L)$ is the luminosity distribution function. Substituting (3.37) into (3.38) allows us to express $\Phi_M(M)$ in terms of $\Phi_L(L)$, or vice versa:

$$\Phi_M(M) = \alpha K M^{\alpha-1} \Phi_L(KM^{\alpha}) \tag{3.39}$$

or

$$\Phi_L(L) = \frac{1}{K^{1/\alpha}\alpha} L^{1/\alpha-1} \Phi_M\left[\left(\frac{L}{K}\right)^{1/\alpha}\right]. \tag{3.40}$$

Problem 3.28. If the index of the mass-luminosity relation is $\alpha = 4$, and if stars have a uniform probability of being formed between masses M_1 and M_2, what is the shape of the initial luminosity function between the two mass limits?

Given an initial mass function, $\Phi_M(M)$, how does it change with time because of stellar birth and stellar evolution? We first express the number of stars in the mass interval dM per unit volume explicitly in terms of time as follows:

$$dN(t) = \Phi_M(M, t)\, dM. \tag{3.41}$$

In time, the number of stars in the mass range M to $M + dM$ will change because of several factors. First, stellar births will occur in the range M to $M + dM$ at some birth rate

$$B(t, M)\, dt\, dM$$

in the time interval dt and mass interval dM. The function $B(t, M)$ is called the *birth-rate function.* Second, because of evolutionary effects, some stars may lose mass and as a consequence leave the mass range M to $M + dM$. Often this effect is assumed to be small, and we will ignore it; in any case, it amounts to a correction of the birth-rate function to give a net effective birth rate. It follows that the mass function is related to the birth-rate function by

$$d\Phi_M/dt = B(t, M), \tag{3.42}$$

with

$$\Phi_M(t = 0) = \Phi_0(M). \tag{3.43}$$

Integrating, and using the initial value (3.43), gives

$$\Phi_M(M, t) = \Phi_0(M) + \int_0^t B(t, M)\, dt, \tag{3.44}$$

where t is the present time. Observations could, in principle, find a value for $\Phi_M(M, t)$, and theory supplies $\Phi_0(M)$ and $B(t, m)$. Thus (3.44) represents a possible observational check on the theory of star formation and stellar evolution.

Observational selection effects can be important in discussions of stellar statistics. A typical problem we face, for example, is that for any given class of star we are trying to study, the chance of discovering that star is higher if the star has a relatively high apparent brightness. Stars of intrinsically low luminosity must be relatively close to us to be observed. Observational selection severely limits our ability to do accurate statistical studies. For example, consider the problem of trying to establish the luminosity function for some class of star in the solar neighborhood. Suppose the true luminosity distribution is $\Phi(L)$, where L is the absolute luminosity. We further suppose that the stars in question are uniformly distributed in space around the Sun, so that the probability of finding a star in a given volume is just proportional to that volume. Then the number of stars that lie in the luminosity range L to $L + dL$ within a distance r to $r + dr$ is given by

$$N(r, L) = 4\pi r^2\, n\, \Phi(L)\, dL\, dr, \tag{3.45}$$

where n is the space density of these stars. We now ask for the distribution of apparent luminosity for these stars. In particular, we want to transform the distribution $N(r, L)$, which is a joint distribution in r and L, into a distribution in l and L, where l is the apparent luminosity. The result is easily obtained by noting that

$$l = L/4\pi r^2,$$

so that

$$dr = -(r/2l)\, dl. \tag{3.46}$$

We now assume further that a particular star's probability of being in a survey depends only on its apparent luminosity; that is, we introduce a selection function $s(l)$, which is unity for the brightest stars ($l \to \infty$), and goes to zero for the fainter ones ($l \to 0$). In practice, there is a fairly sharp limiting apparent magnitude (or luminosity) in most surveys. We may then say that the joint distribution of l and L is given by

$$N(l, L) = s(l)\, n \frac{L}{l} \frac{1}{2l} \left(\frac{L}{4\pi l}\right)^{1/2} \Phi_L(L). \quad (3.47)$$

The interesting thing about this result is that it is separable; that is, $N(l, L)$ may be written as the product of two factors, one of which depends only on l, the other only on L. If we were to combine all the stars in a survey to estimate the luminosity function, we would find that the luminosity function was proportional to

$$N(L) \simeq L^{3/2} \Phi_L(L). \quad (3.48)$$

$N(L)$ is obtained by integrating (3.47) over the apparent luminosity. This shows that we obtain a luminosity that is biased toward too high values for the higher-luminosity stars. This is a simplified result, based on some rather artificial assumptions; however, introducing real assumptions makes the problem worse. For example, if the probability of finding a star depends on other things than the apparent luminosity, or if the stars are not uniformly distributed in distance, then the simple relation (3.48) between the true luminosity function and the apparent one no longer holds.

For this and similar reasons, obtaining accurate stellar statistics about the solar neighborhood is difficult. Most studies of mass functions and birth-rate functions have been done on star clusters, where the observational selection problem, although still present, is somewhat easier to handle.

Star counts based on observational data (corrected for selection effects) are commonly expressed as distributions in absolute visual magnitude M_V. For example, $\Phi(M_V)$, which denotes the observed distribution of main-sequence stars with absolute visual magnitude between M_V and $M_V + dM_V$ in the solar vicinity, is shown in Figure 3.9. The integral of $\Phi(M_V)$ over M_V gives the number of main-sequence stars per cubic parsec today. Globular clusters also contain stellar distributions similar to the present luminosity function $\Phi(M_V)$ near the Sun.

Open or galactic clusters contain young, newly formed stars, most of which have not yet evolved away from the main sequence. Their luminosity function closely resembles the initial luminosity function $\psi(M_V)$, which is also shown in Figure 3.9. The two distribution functions $\Phi(M_V)$ and $\psi(M_V)$ are essen-

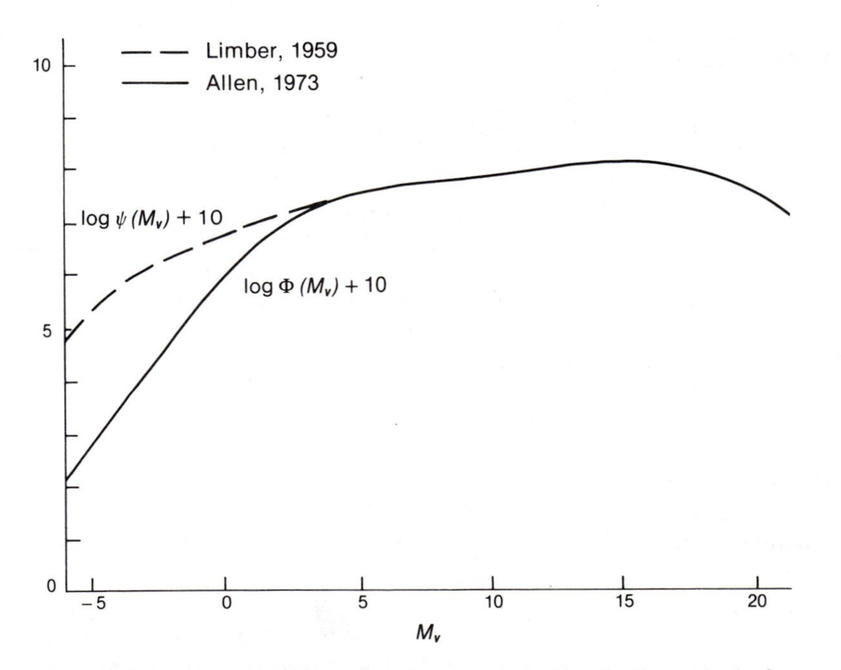

Figure 3.9. Stellar distribution functions versus absolute visual magnitude: for stars near the Sun and in globular clusters (Population II; solid curve, after Allen, 1973); and for stars in galactic clusters (Population I; dashed curve, after Limber, 1959). The ordinate is in units of stars per cubic parsec.

tially identical for $M_V \gtrsim 4$, but the observed luminosity function for stars in globular clusters and in the solar vicinity undergoes an abrupt change in slope for $M_V < 4$. The main-sequence lifetimes of stars less massive than about 1.2 $M_\odot$ (which are fainter than $M_V \sim 4$) exceed the age of the Galaxy; so these stellar types can not have evolved off the main sequence. Therefore, it is reasonable to expect $\Phi(M_V) \sim \psi(M_V)$ for $M_V > 4$.

The difference between $\psi(M_V)$ and $\Phi(M_V)$ for $M_V < 4$ can be attributed to the effects of star formation, stellar evolution, and galactic evolution (essentially the rate of change of the interstellar gas mass in the Galaxy). A simple model illustrates one of these effects (stellar evolution). Assume that the rate of star formation has been constant near the Sun since the formation of the Galaxy (age t_0 years); then only a fraction (τ_{MS}/t_0) of the stars with $M_V < M_{V,L}$ will remain on the main sequence today. The rest, having exhausted the hydrogen in their cores, have evolved off the main sequence. Therefore,

$$\psi(M_V) \simeq \Phi(M_V)\frac{\tau_{MS}}{t_0};\ M_V < M_{V,L}. \quad (3.49)$$

Assuming a main-sequence lifetime for a star of mass M and luminosity L,

$$\tau_{MS} \simeq 1.1 \times 10^{10}(M/M_\odot)/(L_\odot/L) \quad (3.50)$$

and a mass-luminosity relation, (3.49) can be written as a function of M_V. The results agree roughly with $\psi(M_V)$ as shown in Figure 3.9. The point at which the present luminosity function begins to differ from

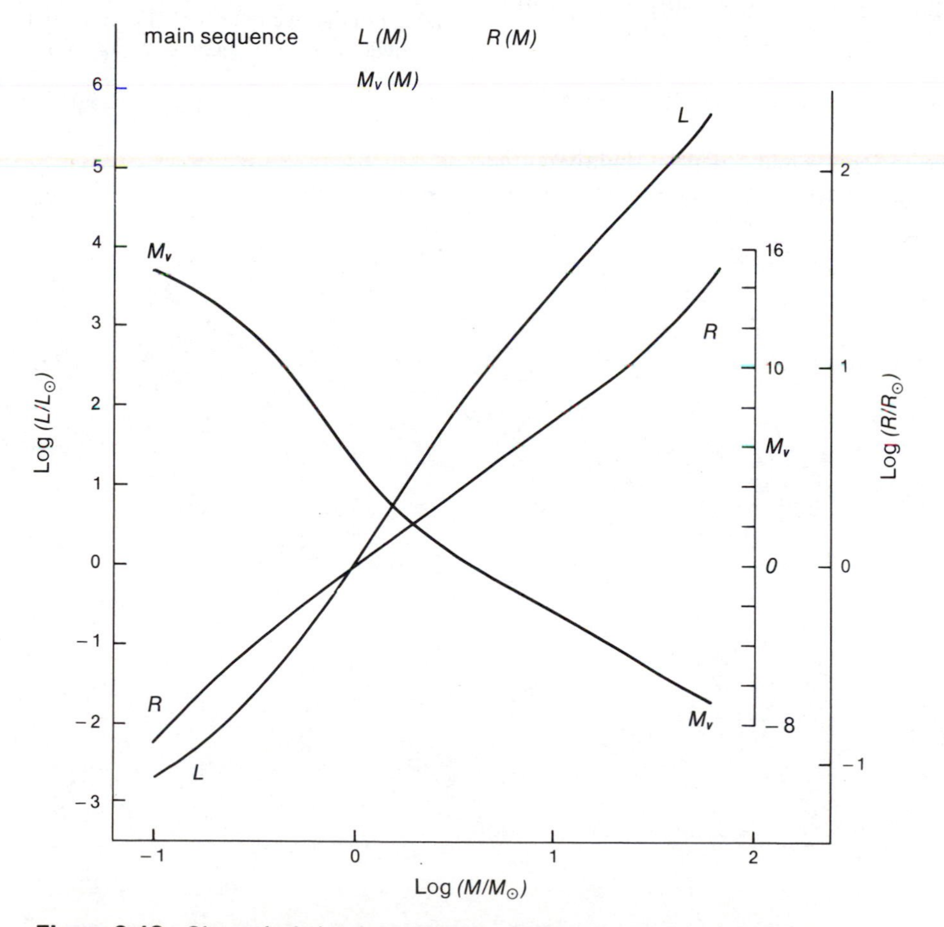

Figure 3.10. Observed relations between mass and luminosity, mass and radius, and mass and absolute visual magnitude for main-sequence stars (see Table 3.3).

$\psi(M_V)$ is roughly $M_{V,L} = 4$, corresponding to a mass $M_L \simeq 1.2\ M_\odot$ and τ_{MS} of about 6×10^9 years.

Problem 3.29. A simple model for the initial luminosity function is given by

$$\psi(M_V) = 0.03 \left(\frac{M_\odot}{M}\right)^{1.35} \frac{d\log_{10} M}{dM_V} \qquad (3.51)$$

in units of stars/pc^3. What is the density of stars having luminosities in the range $-6 \lesssim M_V \lesssim 4$? Assume a mass-luminosity relation

$$L = L_\odot (M/M_\odot)^{3.5}.$$

For stars on the main sequence, we find a smooth relationship between luminosity and mass, and between radius and mass (Figure 3.10). A simple fit to the data may be expressed in the form

$$\log (L/L_\odot) = \beta + \alpha \log (M/M_\odot), \qquad (3.52)$$

where the coefficients α and β depend slightly on the mass range (see Table 3.3.). The observed radius-mass relationship is reproduced roughly by

$$\log (R/R_\odot) = \log (M/M_\odot) + 0.10, \quad M \lesssim 0.4\ M_\odot,$$

$$\log (R/R_\odot) = 0.73 \log (M/M_\odot), \quad M > 0.4\ M_\odot. \qquad (3.53)$$

The last form underestimates the radii for $M > 30\ M_\odot$. Finally, we note that the main-sequence lifetimes as given by (3.50) become, using (3.52),

$$\tau_{MS} \simeq \tau_0 (M_\odot/M)^{\alpha - 1}, \qquad (3.54)$$

with τ_0 as given in Table 3.3.

Table 3.3
Coefficients for mass-luminosity relation (3.52) and main-sequence lifetimes (3.54).

Range	α	β	τ_0 (10^{10} yrs)
$M \lesssim 0.5\ M_\odot$	2.85	−0.15	1.55
$0.5\ M_\odot < M < 2.5\ M_\odot$	3.60	0.073	0.93
$2.5 M_\odot \lesssim M$	2.91	0.479	0.365

Part 2

STELLAR STRUCTURE

Chapter 4

STATIC STELLAR STRUCTURE

4.1. Introduction to Stellar Structure

During most of its existence a star is in a state of delicate balance both locally and globally. Small departures from this state of balance may lead to dynamic states that are stable (as in variable stars, for example), or to dynamically unstable states, such as are believed to be responsible for mass loss from red giants, novae, and supernovae. Although a star may have nearly constant macroscopic characteristics (mass, radius, and effective temperature), it is always undergoing irreversible changes. These changes are ultimately responsible for the star's evolution. Sometimes we can separate structural questions from evolutionary ones, but more often than not the two are closely related. For example, the previous state of a star determines its chemical composition, which then affects its structure during the subsequent stage.

Despite the often inseparable ties between structure and evolution, we will postpone discussion of evolution to Part 3, and concentrate here on those aspects of stellar structure that do not involve time. The systems discussed here are therefore assumed to be in hydrostatic and thermal equilibrium, and the net heat transfer from any one element of matter to another vanishes.

The general behavior of a fluid element in a star is governed in part by the hydrodynamic equation of motion, which, for a nonmagnetic spherically symmetric system, is

$$\rho \frac{d^2r}{dt^2} = -\frac{m(r)G}{r^2}\rho - \frac{\partial P}{\partial r}. \tag{4.1}$$

When d^2r/dt^2 vanishes, the system is in hydrostatic equilibrium, and

$$\frac{dP}{dr} = \frac{\partial P}{\partial r} = -\frac{m(r)G}{r^2}\rho. \tag{4.2}$$

The conditions under which (4.2) may be used in place of (4.1) follow from dimensional considerations. Suppose the term $\partial P/\partial r$ is small; then the timescale associated with (4.1) is the free-fall time $t_{ff} \sim (MG/R^3)^{1/2}$. When the pressure gradient dominates the gravitational force, then (4.1) implies that $(r/t)^2 \sim P/\rho$. This is dimensionally equivalent to the requirement that the fluid element move with the local speed of sound:

$$v \simeq c_s \equiv (\partial P/\partial \rho)_s. \tag{4.3}$$

When both conditions $v \ll c_s$ and $t_e \gg t_{ff}$ are satisfied, where t_e is the time-scale for evolutionary changes, then the system is sufficiently close to hydrostatic equilibrium that (4.2) may be used in place of (4.1). A practical approach to checking hydrostatic equilibrium is to: assume that $v \ll c_s$ and $t_e \gg t_{ff}$; calculate the stellar structure, and from it estimate the true magnitudes of v and t_e; finally, check to see that $v \ll c_s$ and $t_e \gg t_{ff}$ are indeed satisfied.

The equation of hydrostatic equilibrium states simply that pressure forces balance gravitational forces throughout a star. The extremely slow evolution that occurs when $t_e \gg t_{ff}$ can be accounted for within the static framework described by (4.2) if, after each equilibrium model has been constructed, we calculate the implied rate of fuel consumption and energy loss. These estimates make possible an approximate specification of the initial condition for the next static model. In this way, the slow evolutionary phases may be obtained without recourse to hydrodynamics. This approach is often called *quasistatic,* and is particularly suited to stars on or near the main sequence, or the cooling stages of white dwarfs and neutron stars.

Several properties of static configurations will be sensitive to the assumed chemical composition. Broadly speaking, the hydrogen (X), helium (Y), and heavy metals (Z) mass fractions are the most important parameters of the composition. Observations show that stars may be classified into a series of populations primarily in terms of the amount of heavy metals they contain. In the simplest scheme, stars are assigned to either Population I or Population II, according to which pattern of abundances in Table 4.1 they match most closely.

Stellar Structure Equations

The first half of this chapter will develop the equations of stellar structure. These are expressible as five coupled differential equations, local in nature, which describe the run of pressure, temperature, mass, and luminosity in the star. They all contain, as an independent variable, the distance measured from the center of the star. Two of the five describe alternative means of energy transport. Physically these differential equations describe local mechanical and thermal equilibrium, and conservation of mass and energy. Since the differential equations are of first order, and only four are used at any one time, four boundary conditions must be imposed. Physically, these conditions require that there be no mass or energy-source singularities at the center of the star, and that the surface temperature and pressure be consistent with the requirements imposed by a model stellar atmosphere. Finally, initial conditions are required; one fixes the total mass, and the others specify the chemical composition.

The system of equations, boundary conditions, and initial conditions outlined above does not completely specify the problem mathematically, since the four equations contain up to seven unknowns. Three of these are local: the density, opacity, and energy generation rate. Once these have been given as functions of the pressure and temperature, the formulation is complete. The relation defining the density in terms of pressure and temperature is the equation of state.

Table 4.1
Mass fractions of elements in Population I and Population II stars.

Population	Hydrogen (X)	Helium (Y)	Heavy metals (Z)
I	0.6	0.38	0.02
II	0.9	0.099	0.001

4.2. The Equation of Hydrostatic Equilibrium

Consider a spherical star, and imagine a shell of radius r, thickness dr, and density $\rho(r)$; this is sketched in Figure 4.1. The downward gravitational force on this shell is

$$\frac{m(r)G}{r^2} 4\pi r^2 \rho(r)\, dr.$$

The pressure force supporting the shell is

$$4\pi R^2\, dP,$$

where dP is the pressure difference across dr. Applying the hydrostatic equilibrium condition, and equating these two quantities,

$$\frac{dP}{dr} = -\frac{m(r)G}{r^2} \rho, \tag{4.4}$$

where the negative sign takes care of the fact that P

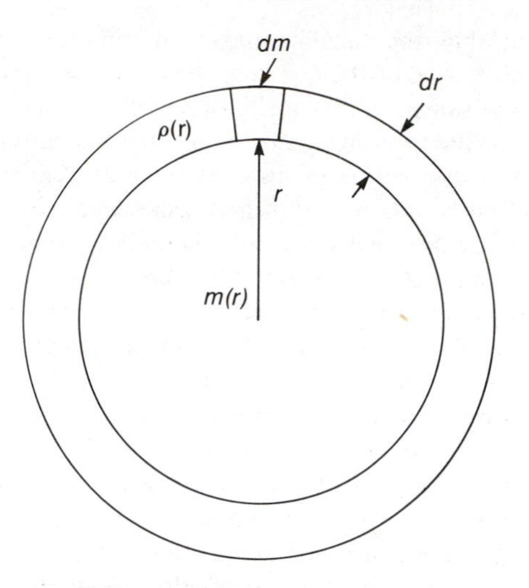

Figure 4.1. Mass shell in a spherically symmetric star. In hydrostatic equilibrium, the pressure inside r balances the gravitational force on dm due to the mass $m(r)$.

decreases as r increases, and where $m(r)$ is the mass contained within a sphere of radius r,

$$m(r) = \int_0^r 4\pi(r')^2 \rho(r')\,dr'. \tag{4.5}$$

Result (4.4) is the equation of hydrostatic equilibrium. Since it is a first-order differential equation, a boundary condition must also be specified. An obvious choice for this boundary condition is the central pressure, P_c.

To express equation (4.4) in another way, we can begin from the mass-continuity equation given by the differential form of (4.5),

$$\frac{dm}{dr} = 4\pi r^2 \rho. \tag{4.6}$$

Dividing (4.4) by (4.6) gives

$$\frac{dP}{dm} = -\frac{mG}{4\pi r^4(m)}. \tag{4.7}$$

Here m is treated as the independent variable, and r is a function of m. The boundary condition imposed on (4.6) is $M = m(R)$, where M is the total mass and R is the radius of the configuration. The advantage of form (4.7) is that the mass of a gravitating system is usually given and fixed, whereas the radius is an unknown quantity.

We clearly could have used this equation to establish our earlier approximate result that $P \approx M^2G/R^2$. We can establish an exact result from (4.7), which sets a lower limit to the central pressure, P_c, of a gravitating system. Consider the quantity

$$P + \frac{m(r)^2G}{8\pi r^4},$$

and differentiate it with respect to r:

$$\frac{d}{dr}\left[P + \frac{Gm^2}{8\pi r^4}\right] = \frac{dP}{dr} + \frac{mG}{4\pi r^4}\frac{dm}{dr} - \frac{m^2G}{2\pi r^5}$$
$$= -\frac{m^2G}{2\pi r^5},$$

since the first two terms cancel, according to the equation of hydrostatic equilibrium. This shows that the derivative of $P + m^2G/8\pi r^4$ is negative; i.e., it decreases outward in the star. Now, at the center of the star, $m^2/r^4 \to 0$, and $P = P_c$, whereas at the surface $P = 0$ (we formally adopt $P = 0$ to define the stellar surface). Hence,

$$P_c > \frac{M^2G}{8\pi R^4}. \tag{4.8}$$

This is an absolute lower limit to the central pressure of any star (or planet) in hydrostatic equilibrium. Similar inequalities can be established for a wide range of physical properties of the star: for example, central temperature, and mean temperature and pressure. The addition of other plausible criteria allows one to improve the limits somewhat. For example, if one imposes the condition that the mean density does not increase outward, then one can prove that

$$P_c > \frac{3}{8\pi}\frac{M^2G}{R^4}.$$

We may now establish the important theorem known as the virial theorem, which was mentioned in Chapter 1. Write equation (4.7) in the form

$$4\pi r^3\frac{dP}{dm} = -\frac{mG}{r}, \tag{4.9}$$

then rewrite it as

$$\frac{d}{dm}(4\pi r^3P) - 4\pi r^2 \cdot 3P\frac{dr}{dm} = -\frac{mG}{r}.$$

Now integrate this over the whole star, from $m = 0$ to $M = m(R)$. This gives

$$4\pi r^3 P \Big|_0^M - \int_0^M \frac{3P}{\rho}\,dm = -\int_0^M \frac{mG\,dm}{r}, \qquad (4.10)$$

where P, ρ, and r are considered functions of the independent variable m. The first term is zero at both limits since $r(0) = 0$ and $P(M) = 0$.

If we have a classical (nonrelativistic) gas, $3P$ is twice the thermal energy density, and so $3P/\rho$ is twice the thermal energy per unit mass. Hence the integral on the left is the total thermal (i.e., particle kinetic) energy of the star, U. The integral on the right is just the gravitational binding energy, Ω, of the star. Hence we have $-2U = \Omega$, or

$$2U + \Omega = 0. \qquad (4.11)$$

Since the total energy of the star is $E = U + \Omega$, this can also be written

$$E + U = 0. \qquad (4.12)$$

If the star were composed of relativistic particles, so that the pressure were one-third the energy density, we would have instead

$$E = U + \Omega = 0. \qquad (4.13)$$

It follows from (4.13) that a star composed of relativistic particles can be in equilibrium only if its total energy is zero. We will return to this interesting point later. The result $2U + \Omega = 0$ is known as the *virial theorem* for stars. It has a very important consequence in the early stages of the formation of a star, and in various stages of evolution. For example, suppose a star undergoes gravitational contraction because of pressure imbalance. This gravitational contraction releases a certain amount of energy, $-\Delta\Omega$. In order for hydrostatic equilibrium to be maintained, the virial theorem must be satisfied, and so the thermal energy must change by

$$\Delta U = -\frac{1}{2}\Delta\Omega. \qquad (4.14)$$

This leaves an energy excess of $-\Delta\Omega/2$, which must be lost from the star, normally in the form of radiation. Thus the effect of gravitational contraction is threefold:

(1) the star gets hotter;
(2) some energy is radiated into space; and
(3) the total energy of the star decreases (it becomes more negative, and hence more tightly bound).

Thus in the early stages of star formation, we imagine that the star has formed (somehow) a gravitationally bound system that undergoes gravitational contraction, increasing its internal temperature as it does so. This increase in temperature eventually kindles the thermonuclear conversion of hydrogen to helium. In later stages of stellar evolution, when the synthesis of H to He causes the pressure to drop (because of the increase in the molecular weight), the resultant imbalance in the pressure-gravity equilibrium will cause the stellar core to contract and the temperature to rise. These remarks are, of course, true only so long as we can consider the star to be in quasistatic equilibrium, even though undergoing contraction. In rapid phases of stellar evolution, the star is not in such equilibrium, and a full hydrodynamic treatment has to be given to the problem.*

In the outer layers of a star, the equation of hydrostatic equilibrium takes a different form, reflecting the fact that the atmosphere is generally much thinner than the radius of the star. Thus the gravitational acceleration, g, is essentially constant throughout the atmosphere. In addition, since the atmospheric thickness is much less than the radius of curvature, we can approximate the atmosphere as a plane-parallel structure (Figure 4.2). Then, measuring distance from some arbitrary level $h = 0$, we can write the equation of hydrostatic equilibrium in the atmosphere as

$$\frac{dP}{dh} = -\rho g, \qquad (4.15)$$

where $g = MG/R^2$.

Thus the surface gravity g becomes a basic physical parameter describing the structure of the atmosphere of the star. In the idealized case of an isothermal atmosphere (T = constant), whose equation of state is

*A modified version of the virial theorem that includes dynamic collapse or expansion will be discussed in Chapter 22.

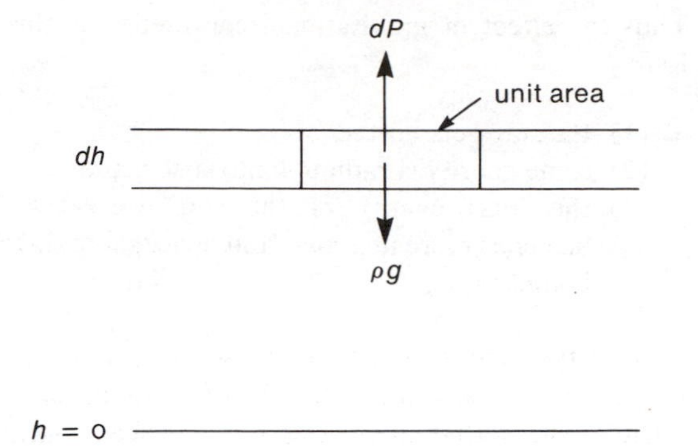

Figure 4.2. Coordinates in planar approximation to a stellar atmosphere; an arbitrary reference level is denoted by $h = 0$. The stellar surface gravity is g.

$P = \rho kT/\mu m_H$, (4.15) can be integrated immediately to give

$$P = P_0\, e^{-\mu m_H g h/kT}, \qquad \rho = \rho_0\, e^{-\mu m_H g h/kT}, \tag{4.16}$$

where P_0 is the pressure at $h = 0$, ρ_0 is the density at $h = 0$. The factor

$$H \equiv kT/\mu m_H g \tag{4.17}$$

is called the scale height of the atmosphere. It is the distance over which the pressure (or density) changes by a factor of e. Even in nonisothermal situations, it is natural to use a scale height patterned after this result, defined as

$$H \equiv -\,(dP/dh)^{-1}P = -(dh/d\ln P). \tag{4.18}$$

Problem 4.1. Estimate typical scale heights for the Earth's atmosphere ($T \sim 300$ K) and the solar atmosphere ($T \sim 6{,}000$ K).

The hydrostatic equilibrium equation can be written in a vector form that can be applied to rotating stars or stars in binary systems, i.e., systems without spherical symmetry. If the gravitational potential is denoted by ϕ, then the gravitational field is given by

$$\mathbf{g} = -\text{grad}\ \phi. \tag{4.19}$$

The pressure gradient is grad P, and the requirement that these two be in balance (hydrostatic equilibrium) becomes

$$\text{grad}\ P = -\rho\ \text{grad}\ \phi. \tag{4.20}$$

This is the general vector form of the equation of hydrostatic equilibrium. It has the following consequence: since grad P is parallel to grad ϕ, P must be constant on equipotential surfaces. Thus we may write $P = P(\phi)$. Then

$$\text{grad}\ P = (dP/d\phi)\ \text{grad}\ \phi. \tag{4.21}$$

Comparing this with (4.20) shows that the density ρ is

$$\rho = -dP/d\phi.$$

Because P is a function of ϕ, it follows that $\rho = \rho(\phi)$ is also a function only of ϕ. If the thermodynamic state of the system is determined completely by ρ, P, and T, then it follows that $T = T(\phi)$ as well. Consequently the physical state of the material is constant on an equipotential surface (Von Zeipel's theorem).

The practical application of this result to a rotating star depends on the assumption of uniform rotation (i.e., constant angular velocity throughout the star). We can then express the (pseudo) forces arising from rotation in terms of a potential.

Problem 4.2. What is the form of the potential that describes uniform rotation? Why does it not work for nonuniform rotation?

Anticipating a later result that the flow of radiation is proportional to the temperature gradient, it follows that the flow of radiation is proportional to the (gravitational plus rotational) potential gradient, and thus, for example, that a rotating star should appear brighter at the poles than at the equator.

Von Zeipel's theorem can be used to determine models of the primary component in close binary x-ray sources. If the rotational period of the primary is equal to the binary period (synchronous rotation), then the emergent local radiation flux will be proportional to the local gravitational acceleration ($T_{\text{eff}}^4 \sim |g(\mathbf{r})|$). This result may be used in conjunction with observed optical light curves to model the tidal and rotational deformations of the primary star (see Chapter 17).

4.3. Simplified Stellar Models

The equation of hydrostatic equilibrium cannot, as it stands, be integrated to give a model for the run of pressure, density, and temperature in a star; integration requires some further assumptions or some further input physics in the form of the equation of energy generation and energy transfer. However, we can obtain a fairly good idea of what happens in stars from some rather simple models. We will discuss two here. In the first it is assumed, arbitrarily, that the density varies linearly from the center of the star to the surface. That is, we assume

$$\rho = \rho_c(1 - r/R). \tag{4.22}$$

This model will be developed throughout the following chapters on stellar structure and evolution. Although the model is too simple to represent real stars, it does yield qualitatively useful results. In the second model, we will assume that the pressure and density are related throughout the star by an equation of polytropic form:

$$P = K\rho^{\gamma}. \tag{4.23}$$

We will further assume that K and γ are independent of radius, although they may depend on the system entropy. We recall that the polytropic γ need not equal c_P/c_V.

Linear Stellar Model

If we substitute (4.22) for the density in the equation of hydrostatic equilibrium, we get

$$\frac{dP}{dr} = -\frac{m(r)G}{r^2}\rho_c(1 - r/R). \tag{4.24}$$

To integrate this, we must express $m(r)$ as a function of r. We have

$$\frac{dm}{dr} = 4\pi r^2\rho_c(1 - r/R) = 4\pi r^2\rho_c - \frac{4\pi r^3}{R}\rho_c. \tag{4.25}$$

Integrating from $r = 0$ outward, we find

$$m(r) = \frac{4\pi}{3}r^3\rho_c - \frac{\pi r^4}{R}\rho_c. \tag{4.26}$$

The total mass of the star is

$$M = M(R) = \frac{4\pi}{3}R^3\rho_c - \pi R^3\rho_c$$

$$= \frac{\pi R^3}{3}\rho_c. \tag{4.27}$$

Using (4.27) to eliminate ρ_c in (4.26) yields the alternative form

$$m(r) = M(4r^3/R^3 - 3r^4/R^4).$$

Thus if we want to match a specific star (a given M, R) with a linear stellar model, we must take $P_c = P(\rho_c)$, where

$$\rho_c = \frac{3M}{\pi R^3}. \tag{4.28}$$

Inserting $m(r)$ into the equation of hydrostatic equilibrium, we get

$$\frac{dP}{dr} = -\frac{\pi G}{r^2}\rho_c^2\left(\frac{4}{3}r^3 - \frac{r^4}{R}\right)(1 - r/R)$$

$$= -\pi G\rho_c^2\left(\frac{4}{3}r - \frac{7}{3}\frac{r^2}{R} + \frac{r^3}{R^2}\right). \tag{4.29}$$

Hence

$$P = P_c - \pi G\rho_c^2\left(\frac{2r^2}{3} - \frac{7r^3}{9R} + \frac{r^4}{4R^2}\right), \tag{4.30}$$

where P_c is the central pressure. At the surface of the star, $r = R$, and we have

$$P(R) = P_c - \frac{5}{36}\pi G\rho_c^2R^2 = 0.$$

Thus the central pressure is

$$P_c = \frac{5}{36}\pi G\rho_c^2R^2 \tag{4.31}$$

or

$$P_c = \frac{5}{36}\pi G \cdot \frac{9M^2}{\pi^2 R^6} \cdot R^2$$

$$= \frac{5}{4\pi}\frac{M^2 G}{R^4}. \tag{4.32}$$

The pressure at a point r can be written, finally,

$$P(r) = \frac{5\pi}{36} G\rho_c^{\,2} R^2 \times \left(1 - \frac{24}{5}\frac{r^2}{R^2} + \frac{28}{5}\frac{r^3}{R^3} - \frac{9}{5}\frac{r^4}{R^4}\right). \tag{4.33}$$

If we ignore radiation pressure for the moment, and assume that the matter obeys the ideal gas equation of state, we have immediately the temperature variation

$$T = \frac{\mu m_H P}{k\rho}$$

$$= \frac{5\pi}{36}\frac{G\mu m_H}{k}\rho_c R^2 \times \left(1 + \frac{r}{R} - \frac{19}{5}\frac{r^2}{R^2} + \frac{9}{5}\frac{r^3}{R^3}\right). \tag{4.34}$$

Problem 4.3. Verify the formula for T. Plot graphs showing the variation of P, T, and M with r.

Problem 4.4. Find the gravitational binding energy of the linear stellar model in terms of G, M, and R.

Polytropes

To obtain the equation for a polytrope, it is first convenient to rewrite the equation of hydrostatic equilibrium and the mass-continuity equation in a different form, which is obtained by eliminating m between them. We have

$$m = -\frac{r^2}{\rho G}\frac{dP}{dr}. \tag{4.35}$$

Then the mass-continuity equation (4.6) becomes

$$\frac{1}{r^2}\frac{d}{dr}\left(\frac{r^2}{\rho}\frac{dP}{dr}\right) = -4\pi G\rho. \tag{4.36}$$

Substituting the relation $P = K\rho^\gamma$, we get

$$\frac{1}{r^2}\frac{d}{dr}\left(\frac{r^2 K}{\rho}\gamma\rho^{\gamma-1}\frac{d\rho}{dr}\right) = -4\pi G\rho, \tag{4.37}$$

which is a second-order differential equation for the density. This can be solved, subject to the boundary conditions $\rho = \rho_c$ at $r = 0$ and $\rho = 0$ at $r = R$. The total mass of the star can then be calculated by integration to express M in terms of ρ_c, or vice versa.

In order to simplify the mathematics here, we introduce the variable θ, related to the density by the equation

$$\rho = \lambda\theta^n, \tag{4.38}$$

where λ is a constant (arbitrary, for now) scaling factor. Then $\gamma - 1 = 1/n$, and (4.37) becomes

$$\frac{1}{r^2}\frac{d}{dr}\left(\frac{r^2 K(n+1)}{n}(\lambda\theta^n)^{(1-n)/n}\frac{d}{dr}\lambda\theta^n\right) = 4\pi\lambda\theta^n G,$$

$$\left[\frac{n+1}{4\pi G}K\lambda^{1/n-1}\right]\frac{1}{r^2}\frac{d}{dr}\left(r^2\frac{d\theta}{dr}\right) = -\theta^n. \tag{4.39}$$

Now let

$$\alpha = \left[\frac{n+1}{4\pi G}K\lambda^{1/n-1}\right]^{1/2}. \tag{4.40}$$

This has the dimensions of a length if λ is chosen to have the dimensions of a density, in which case θ is dimensionless.

Problem 4.5. Prove the preceding statement.

We also introduce the dimensionless variable, ξ, by the equation

$$r = \alpha\xi. \tag{4.41}$$

Using (4.40) and (4.41), we can rewrite (4.39) as

$$\frac{1}{\xi^2}\frac{d}{d\xi}\left(\xi^2\frac{d\theta}{d\xi}\right) = -\theta^n. \tag{4.42}$$

This equation is known as the Lane-Emden equation of index n.

The advantage of expressing (4.36) in this form is that once a solution $\theta(\xi)$ has been found for a given value of n, there remain two parameters, K and λ, which are still arbitrary. Thus a family of solutions has actually been found.

Some other forms are possible by transformation of variables. For example, putting

$$\chi = \xi\theta(\xi) \tag{4.43}$$

and substituting into (4.42) leads to

$$\frac{d^2\chi}{d\xi^2} = -\frac{\chi^n}{\xi^{n-1}}. \tag{4.44}$$

Alternatively, putting $x = 1/\xi$ leads to

$$x^4\frac{d^2\theta}{dx^2} = -\theta^n. \tag{4.45}$$

We may express the boundary conditions in terms of the variables θ and ξ. At the center of the star, θ has the value θ_c. Since $\rho_c = \lambda\theta_c{}^n$, it is convenient to choose $\lambda = \rho_c$; hence we have the boundary condition

$$\theta_c = 1, \qquad \xi = 0. \tag{4.46}$$

Since dP/dr is proportional to $d\theta/d\xi$, and since, from the equation of hydrostatic equilibrium, we expect dP/dr to be zero at $r = 0$, another boundary condition is

$$\frac{d\theta}{dr} = 0, \qquad \xi = 0. \tag{4.47}$$

We may integrate the Lane-Emden equation with the starting boundary conditions (4.46) and (4.47). For an arbitrary value of n, it is generally necessary to do this numerically. However, analytic solutions are known for the cases $n = 0, 1, 5$, and these are worth quoting here. If we denote the solution to (4.42) for n by θ_n, then

$$\begin{aligned} n = 0, &\qquad \theta_0 = 1 - \xi^2/6; \\ n = 1, &\qquad \theta_1 = \frac{\sin\xi}{\xi}; \\ n = 5, &\qquad \theta_5 = (1 + \xi^2/3)^{-1/2}. \end{aligned} \tag{4.48}$$

Problem 4.6. Verify that these solutions satisfy the appropriate Lane-Emden equations.

The solutions for $n = 0$ and $n = 1$ go to zero at some point, which may be taken as the boundary of the star, since it gives us the place where the density has fallen to zero. Thus the physical part of the solution stops there, and the value of ξ at the first zero (usually called ξ_1) tells us the radius of the star. Using (4.48) and solving $\theta_n(\xi_1) = 0$, we find

$$\begin{aligned} \xi_1 = \sqrt{6}, &\qquad n = 0; \\ \xi_1 = \pi, &\qquad n = 1. \end{aligned} \tag{4.49}$$

Note that $n = 0$ is a star of constant density, as follows from (4.38). For $n = 5$, $\theta(\xi)$ does not vanish for any finite value of ξ, and the star extends to infinity. For arbitrary n the Lane-Emden equation must be integrated numerically, usually out to the first zero. The gross properties of the star (M, R, P_c, T_c, etc.) may then be obtained by further integration over $\theta(\xi)$. Some of the results can be found analytically; others must be found numerically. We list here a collection of useful results that are valid for polytropes of all indices, unless stated otherwise·

Radius,

$$R = \alpha\xi_1 = \left[\frac{(n+1)K}{4\pi G}\right]^{1/2} \rho_c{}^{(1-n)/2n}; \tag{4.50}$$

Mass,

$$\begin{aligned} M &= 4\pi\alpha^3\rho_c\left[-\xi^2\frac{d\theta}{d\xi}\right]_{\xi=\xi_1} \\ &= 4\pi\left[\frac{(n+1)K}{4\pi G}\right]^{3/2}\rho_c{}^{(3-n)/2n}\left[-\xi^2\frac{d\theta}{d\xi}\right]_{\xi=\xi_1}; \end{aligned} \tag{4.51}$$

Central pressure,

$$P_c = \frac{M^2G}{R^4}\left[4\pi(n+1)\left(\frac{d\theta}{d\xi}\right)^2_{\xi=\xi_1}\right]^{-1}; \tag{4.52}$$

Binding energy,

$$\Omega = -\frac{3}{5-n}\frac{M^2G}{R}; \tag{4.53}$$

Mean density,

$$\bar{\rho} = \rho_c \left[-\frac{3}{\xi}\frac{d\theta}{d\xi} \right]_{\xi=\xi_1}. \tag{4.54}$$

The quantities ξ_1 and $(d\theta/d\xi)_{\xi=\xi_1}$ have been numerically evaluated for special values of n. Various combinations of these, which appear in (4.50) to (4.54), are given in Table 4.2. For example, the third column is proportional to the mass through (4.51), and the last column describes the central pressure.

In columns 6 and 7, the quantities

$$N_n = \frac{(4\pi)^{1/n}}{n+1} \left[-\xi_1^{n+1/n-1} \left(\frac{d\theta_n}{d\xi} \right)_{\xi=\xi_1} \right]^{1-n}$$

and

$$W_n = \frac{1}{4\pi\,(n+1) \left[\left(\frac{d\theta_n}{d\xi} \right)_{\xi=\xi_1} \right]^2}.$$

Numerical tables of $\theta(\xi)$ are also available for selected values of n. Table 4.3 gives $\theta(\xi)$ for $n = 3$, as well as $-\xi^2(d\theta/d\xi)$, which determines the mass distribution $m(r)$. These, in combination with the other forms, may be used to reproduce the radial dependence of the variables ρ, P, $\bar{\rho}$, and Ω.

Some interesting special cases might be mentioned. The case $n = 1$ gives a star whose radius is independent of the central density; and the case $n = 3$ gives a mass that is independent of the central density. The case $n = 5$ gives an infinite binding energy. For $n > 5$, the binding energy is positive, so that the gaseous configuration is not gravitationally bound. Thus a stable star with $n > 5$ cannot exist. The condition $n \geqslant 5$ corresponds to $\gamma < 6/5$. Given the solution $\theta_n(\xi)$, the run of central density and pressure are given by

$$\begin{aligned} \rho &= \rho_c\, \theta^n(\xi), \\ P &= P_c\, \theta^{n+1}(\xi). \end{aligned} \tag{4.55}$$

For a polytropic gas, which is also an ideal gas, the run in temperature may also be found:

$$T = T_c \theta(\xi). \tag{4.56}$$

The case $n = 3$ (the *Eddington standard model*) results if the star is supported by both radiation pressure

$$P_R = \frac{1}{3} a T^4 = (1 - \beta) P \tag{4.57}$$

Table 4.2
The constants of the Lane-Emden functions*

n	ξ_n	$-\xi_n^2 \left(\frac{d\theta_n}{d\xi}\right)_{\xi=\xi_1}$	$\rho_c/\bar{\rho}$	$-\xi_1^{n+1/n-1}\left(\frac{d\theta_n}{d\xi}\right)_{\xi=\xi_1}$	N_n	W_n	$-\frac{1}{(n+1)\xi_1 \left(\frac{d\theta_n}{d\xi}\right)_{\xi=\xi_1}}$
0	2.4494	4.8988	1.0000	0.33333	———	0.119366	0.5
0.5	2.7528	3.7871	1.8361	0.02156	2.270	0.26227	0.53847
1.0	3.14159	3.14159	3.28987	———	0.63662	0.392699	0.5
1.5	3.65375	2.71406	5.99071	132.3843	0.42422	0.770140	0.53849
2.0	4.35287	2.41105	11.40254	10.4950	0.36475	1.63818	0.60180
2.5	5.35528	2.18720	23.40646	3.82662	0.35150	3.90906	0.69956
3.0	6.89685	2.01824	54.1825	2.01824	0.36394	11.05066	0.85432
3.25	8.01894	1.94980	88.153	1.54716	0.37898	20.365	0.96769
3.5	9.53581	1.89056	152.884	1.20426	0.40104	40.9098	1.12087
4.0	14.97155	1.79723	622.408	0.729202	0.47720	247.558	1.66606
4.5	31.83646	1.73780	6189.47	0.394356	0.65798	4922.125	3.33100
4.9	169.47	1.7355	934800	0.14239	1.340	3.693×10^6	16.550
5.0	∞	1.73205	∞	0	∞	∞	∞

SOURCE: S. Chandrasekhar, *An Introduction to the Study of Stellar Structure* (Chicago: University of Chicago Press, 1939). Reprinted by permission of the University of Chicago Press.

*The values for $n = 0.5$ and 4.9 are computed from Emden's integrations of θ_n; for $n = 3.25$ an unpublished integration by Chandrasekhar has been used; $n = 5$ corresponds to the Schuster-Emden integral. For the other values of n the *British Association Tables*, Vol. II, has been used.

Table 4.3
Lane-Emden function θ_3 and its derivative for values of ξ between 0 and $\xi_1 = 6.9011$.

ξ	θ_3	θ_3^4	θ_3^5	$-d\theta_3/d\xi$	$-\left(\frac{\xi}{3}\right)\left(\frac{d\xi}{d\theta_3}\right)$	$-\xi^2\left(\frac{d\theta_3}{d\xi}\right)$
0.00	1.00000	1.00000	1.00000	.00000	1.00000	.0000
0.25	.98975	.96960	.95966	.08204	1.0158	.0051
0.50	.95987	.88436	.84886	.15495	1.0756	.0387
0.75	.91355	.76242	.69650	.21270	1.1754	.1196
1.00	.85505	.62513	.53451	.25219	1.3218	.2522
1.25	.78897	.49111	.38747	.27370	1.5224	.4276
1.50	.71948	.37244	.26797	.27993	1.7862	.6298
1.75	.64996	.27458	.17847	.27460	2.1243	.8410
2.00	.58282	.19796	.11538	.26149	2.5495	1.0450
2.50	.46109	.09803	.04520	.22396	3.7210	1.3994
3.00	.35921	.04635	.01665	.18393	5.4370	1.6553
3.50	.27629	.02109	.005828	.14859	7.8697	1.8203
4.00	.20942	.009185	.001923	.11998	11.113	1.9197
4.50	.15529	.003746	.000582	.09748	15.387	1.9740
5.00	.11110	.001371	.000152	.08003	20.826	2.0007
6.00	.04411	.000086	.000004	.05599	35.720	2.0156
6.80	.00471	.000001	.000000	.04360	51.987	2.0161
6.9011	.00000	.000000	.000000	.04231	54.360	2.0150

SOURCE: Adapted from A. S. Eddington, *The Internal Constitution of the Stars* (New York: Dover Publications, 1959).

and ideal gas pressure

$$P_g = \frac{\rho k T}{\mu m_H} = \beta P, \tag{4.58}$$

where the total pressure is P, and β is defined by (2.28) and is assumed to be constant. Eliminating T between (4.57) and (4.58) gives

$$\begin{aligned} \beta^4 P^4 \left(\frac{\mu m_H}{\rho k}\right)^4 &= \frac{3(1-\beta)}{a} P, \\ P &= \left(\frac{k}{\mu m_H}\right)^{4/3} \left(\frac{3(1-\beta)}{a\beta^4}\right)^{1/3} \rho^{4/3}. \end{aligned} \tag{4.59}$$

Thus, $\gamma = 4/3$ and $n = 3$.

Problem 4.7. Estimate P_c, ρ_c, and T_c for the Sun, using the Eddington standard model and the linear model. For the former, what is the value of β?

The Isothermal Gas Sphere

The equation of state for an isothermal gas sphere clearly satisfies the relation $P \sim \rho$, which is equivalent to a polytrope with $\gamma = 1$, or $n = \infty$. That is, an isothermal gas sphere is infinite in extent with positive total energy, since clearly $n > 5$. Since the results above for polytropes do not generalize when $n \rightarrow \infty$, we must go back to the equation of hydrostatic equilibrium (4.36). If we use the ideal gas equation of state to eliminate the pressure, (4.36) becomes

$$\frac{1}{r^2}\frac{d}{dr}\left(\frac{r^2}{\rho}\frac{kT}{\mu m_H}\frac{d\rho}{dr}\right) = -4\pi G\rho. \tag{4.60}$$

We now make a change of variable:

$$\rho = \rho_c e^{-\psi}, \qquad r = \left[\frac{kT}{4\pi\mu G m_H \rho_c}\right]^{1/2} \xi = \alpha\xi. \tag{4.61}$$

In terms of the new variables ψ and ξ, we have

$$\frac{1}{\xi^2}\frac{d}{d\xi}\left(\xi^2\frac{d\psi}{d\xi}\right) = e^{-\psi}, \tag{4.62}$$

with the boundary conditions

$$\psi = 0, \qquad \frac{d\psi}{d\xi} = 0, \qquad \xi = 0. \tag{4.63}$$

This equation cannot be integrated explicitly, but numerical integration reveals that the solution extends to infinity (i.e., the density never becomes zero for any value of ξ). The mass of the configuration is also infinite. Clearly, then, a finite star cannot be an isothermal gas sphere.

Isothermal gas spheres have been used to describe the distribution of stars in globular clusters, and of galaxies in clusters of galaxies (Chapter 28). Suppose a cluster of stars (or galaxies) is spherically symmetric, and suppose that the motion of a member of the cluster is determined entirely by the average or mean gravitational field due to all the other members. If the system is in statistical equilibrium, its entropy will be a maximum. The velocities of cluster members within a volume that is small compared with the cluster size will be given by

$$\tfrac{1}{2}mv^2 = E + \phi(r), \tag{4.64}$$

where E is the total energy of the individual member, and $\phi(r)$ is its gravitational potential energy. It can be shown that the distribution of these velocities is Maxwellian (2.16). The distribution of particle velocities in an ideal gas at (constant) temperature T is also Maxwellian; so the equation of state for cluster members should be of the form $P \sim \rho \sim mn$, where m is the mass of a typical member, and n is their number density. Figure 4.3 shows the density n/n_0 obtained by numerical intergration of (4.62). Observations of clusters measure the projected (surface) brightness distribution, which agrees closely with the projected number density (also shown in Figure 4.3). At large distances from the cluster's center, the observed density drops rapidly. This dropoff has been attributed to the influence of nearby objects (see Chapter 28).

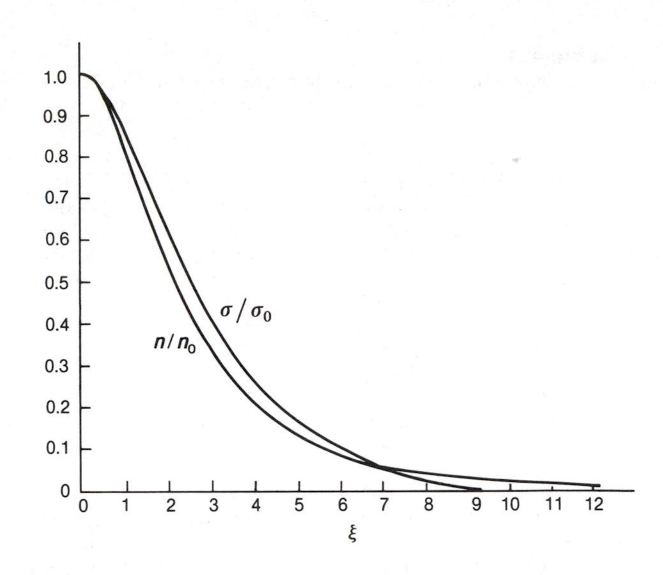

Figure 4.3. Ratio of particle (or stellar) number density n to its central value n_0 in an isothermal sphere, and surface distribution $\sigma(\xi)/\sigma(0)$. The latter may be compared directly with observed surface brightness of globular clusters or clusters of galaxies (see Part 5).

Problem 4.8. The projected stellar number density $\sigma(x)$ is defined as the integral of the volume number density n along a line of sight s through the cluster. Show that

$$\sigma(x) = \int_{-\infty}^{\infty} n(s)\, ds$$

$$= -2 \int_x^{\infty} \sqrt{r^2 - x^2}\, \frac{dn}{dr}\, dr. \tag{4.65}$$

Here $s^2 = r^2 - x^2$, and x is measured in the plane of the sky and passes through the cluster center.

Chapter 5

RADIATION AND ENERGY TRANSPORT

Electromagnetic radiation plays a central role in most stellar systems, and in many nonstellar systems. Some emit most of their energy in a limited spectral region; in others the emission may arise from several distinct processes, each of which peaks in a specific spectral region. Consequently, objects are often classified by their characteristic spectrum, e.g., as infrared, radio, or x-ray sources. Some objects may exhibit peaked emission in more than one region. Figure 5.1 shows the most frequently encountered spectral regions, along with some of the more common processes responsible for the emission. This chapter describes radiation in matter, with emphasis on the visible part of the spectrum, but most of the results are directly applicable to radiation in other spectral regions.

5.1. RADIATIVE TRANSPORT

The energy density and photon number density associated with black-body radiation at temperature T are given by aT^4 and $20T^3$ cm^{-3} (with T in K); see equations (2.36)–(2.37). The average energy per photon is thus of the order of $aT/20$ ergs/photon. For the interior of the Sun this is in the keV range, which is characteristic of x rays. The light escaping from typical stellar surfaces is in or near the visible spectrum, corresponding to energies about 10^5 times smaller than the average energy per photon in the stellar core. The source of this degradation of photon energy is the coupling between radiation and matter. Photons diffuse through most stellar matter, a process in which a given photon travels, on average, a mean free path λ before it is absorbed and reemitted in a random direction. The description of this process is similar to the statistical problem of a random walk, where n random choices for the next step (in this case, reemission in an arbitrary direction) result in a net displacement $n^{1/2}\lambda$ from the starting point. Accordingly, a photon requires, on average $n \sim (R/\lambda)^2$ reemissions to diffuse from the star's center to its radius, R. In the solar core, $\lambda \sim 1$ cm, and a typical photon escaping from the solar surface has undergone about 10^{20} encounters with matter since it was first emitted in the core. The atomic processes—scattering, absorption, and emission—that are responsible for the encounters are outlined later in this chapter. First, however, we define the radiation field macroscopically, and derive the equation of radiative transport that describes how it changes in position and time. Although the following discussion emphasizes electro-

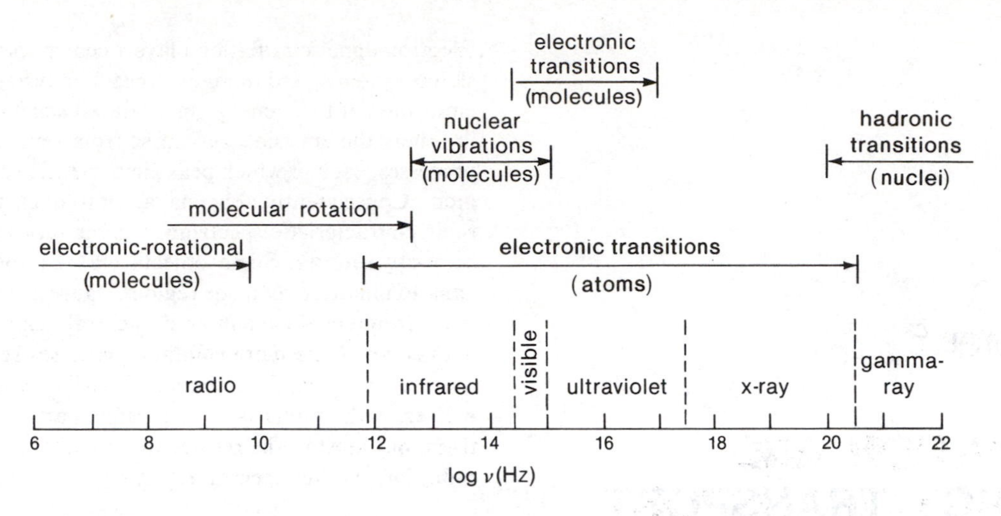

Figure 5.1. Conventional frequency intervals of the electromagnetic spectrum. Horizontal bars indicate intervals associated with important quantum transitions.

magnetic radiation, some of the results may be applied to neutrinos, which are emitted during the late stages of stellar evolution.

5.2. Description of the Radiation Field

At any point inside or outside a star, photons of various energies may be found traveling in various directions. These photons constitute the radiation field at the particular point and time. We now consider how to describe such a radiation field. Imagine a direction, specified by a unit vector $\mathbf{\Omega}$, and an area dA whose normal $\hat{\mathbf{n}}$ makes an angle θ with $\mathbf{\Omega}$ (see Figure 5.2). The projection of dA normal to $\mathbf{\Omega}$ is $dA_1 = \hat{\mathbf{n}} \cdot \mathbf{\Omega}\, dA$. Imagine further that at a distance r from dA there is another area dA_2 that subtends a solid angle $d\Omega = dA_2/r^2$ at dA. In a radiation field, some energy will flow from dA to dA_2, and if we restrict our attention to photons whose frequencies are in a range ν to $\nu + d\nu$, we may say that the energy flow from dA to dA_2 in time dt is

$$dE_\nu = I_\nu(\Omega)\, d\Omega\, dt\, d\nu\, \hat{\mathbf{n}} \cdot \mathbf{\Omega}\, dA. \tag{5.1}$$

This equation defines $I_\nu(\mathbf{r}, \mathbf{\Omega}, t)$, the monochromatic intensity of the radiation field at the point $\mathbf{r}$ and time t. The cgs units of intensity $I_\nu(\Omega)$ are erg cm^{-2} sec^{-1} ster^{-1} Hz^{-1}. It is also possible to define the intensity in terms of the wavelength, in which case the units of $I_\lambda(\Omega)$ are erg cm^{-2} sec^{-1} ster^{-1} cm^{-1}. Naturally, the numeric value of the intensity differs according to whether we use wavelength or frequency units.

Problem 5.1. What is the connection between $I_\nu(\Omega)$ and $I_\lambda(\Omega)$?

The radiation energy density associated with the beam of intensity I_ν can be found by considering an infinitesimal volume ΔV through which the beam passes. Since radiation travels at the speed of light c (we ignore dispersive effects of the medium), the beam travels a distance $dl = c\, dt$ in time dt; therefore we may rewrite (5.1) as

$$dE_\nu = I_\nu(\Omega)\, d\Omega\, d\nu\, \hat{\mathbf{n}} \cdot \mathbf{\Omega} \frac{dl}{c}\, dA$$

$$= \frac{I_\nu(\Omega)}{c}\, d\Omega\, d\nu\, dV,$$

where the volume element $dV = \hat{\mathbf{n}} \cdot \mathbf{\Omega}\, dA\, dl$. The total energy within ΔV is obtained by integrating this expression over the volume and solid angle. If ΔV is small, I_ν will be essentially constant, and may be taken outside the integral. Finally, dividing the result by ΔV

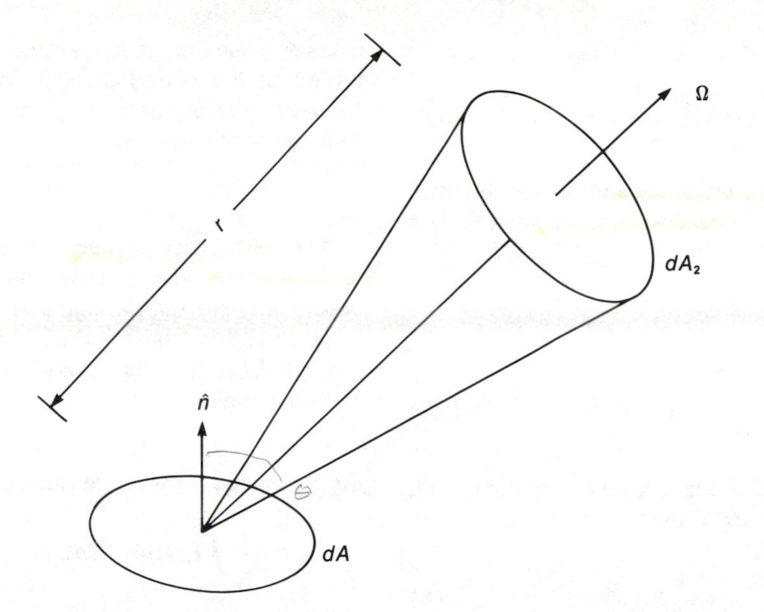

Figure 5.2. Directional vectors describing an arbitrary radiation field originating from the surface element dA, traveling in direction Ω and passing through a surface element dA_2. The distance between dA and dA_2 is r, and $\hat{n}$ is the normal to dA.

yields the energy density,

$$u_\nu \, d\nu = \frac{1}{\Delta V} \int_{\Delta V} \int_{\Omega} dE_\nu \simeq \frac{1}{c} \int I_\nu \, d\Omega \, d\nu.$$

Thus u_ν gives the density of radiation per Hz in a beam of intensity I_ν (Ω):

$$u_\nu = \frac{1}{c} \int I_\nu(\Omega) \, d\Omega. \tag{5.2}$$

Since radiation carries momentum, we may associate a pressure with the beam I_ν. The pressure is the rate at which momentum is transported across a unit area. Imagine an area dA and consider a direction making an angle θ with the normal to dA. The momentum of a photon of energy E is E/c. Thus a beam of intensity I_ν carries a momentum flux I_ν/c in the direction of the beam. The component of the momentum flux perpendicular to the area dA is $I_\nu \cos\theta/c$. The effective area of dA normal to the beam is $dA \cos\theta$. Thus the momentum flux across dA in a direction perpendicular to dA in the beam of intensity I_ν is

$$dP_\nu = \frac{I_\nu}{c} \cos^2\theta \tag{5.3}$$

per unit time, and per unit solid angle, and per unit frequency interval. The radiation pressure perpendicular to dA associated with photons of energy $h\nu$ is therefore obtained by integrating (5.3) over all solid angles:

$$P_\nu = \int \frac{I_\nu}{c} \cos^2\theta \, d\Omega$$

$$= \frac{2\pi}{c} \int_{-1}^{1} I_\nu \mu^2 \, d\mu = \frac{4\pi}{c} K_\nu, \tag{5.4}$$

where $\mu = \cos\theta$, and K_ν is defined in (5.7). The units of P_ν are dynes cm^{-2} Hz^{-1}. (Notice that some care is required with the concept of pressure in a situation where the radiation field is anisotropic, since the pressure is then no longer a scalar quantity, but has to be defined specifically as the pressure perpendicular to a given surface dA.)

Another quantity of interest is the mean intensity, usually denoted by J_ν, which is the average of I_ν over all solid angles:

$$J_\nu = \frac{1}{4\pi} \int I_\nu(\Omega) \, d\Omega. \tag{5.5}$$

In the special case where the radiation field is isotropic, $I_\nu(\Omega)$ is independent of Ω and hence $J_\nu = I_\nu$. In some cases, particularly in stellar interiors as opposed to stellar atmospheres, we are most interested in the total intensity,

$$I(\Omega) = \int_0^\infty I_\nu(\Omega)\, d\nu = \int_0^\infty I_\lambda(\Omega)\, d\lambda. \tag{5.6}$$

J_ν appears as the zero order moment of the intensity with respect to $\cos\theta$. A second order moment of I_ν is defined by

$$\begin{aligned} K_\nu &= \frac{1}{4\pi}\int I_\nu \cos^2\theta\, d\Omega \\ &= \frac{1}{2}\int_{-1}^{1} I_\nu(\mu)\, \mu^2\, d\mu, \end{aligned} \tag{5.7}$$

where $\mu = \cos\theta$. For an isotropic radiation field, $I_\nu(\mu) = I_\nu$, and the integral in (5.7) is 2/3; so

$$K_\nu = \frac{1}{3} I_\nu. \tag{5.8}$$

The quantity K_ν is related to the radiation pressure by (5.4):

$$P_\nu = \frac{4\pi}{c} K_\nu = \frac{4\pi}{3c} I_\nu. \tag{5.9}$$

However, for an isotropic radiation field, (5.2) becomes

$$u_\nu = \frac{4\pi\, I_\nu}{c}, \tag{5.10}$$

which, when compared with (5.9), gives the relation between pressure and energy density,

$$P_\nu = \frac{1}{3} u_\nu. \tag{5.11}$$

This is the expected relation for relativistic particles. Note that we have not assumed thermal equilibrium, and the radiation distribution I_ν is completely arbitrary. The total pressure and energy density follow immediately upon integration of (5.11) over frequency:

$$P = \frac{1}{3} u. \tag{5.12}$$

Problem 5.2. A neutrino is a fermion with intrinsic angular momentum ½ (like the electron) and rest mass (presumably) zero like the photon. It therefore travels at the speed of light. What is the relation between pressure and energy density for an arbitrary distribution of neutrinos?

H in lecture notes

The first order moment of I_ν is related to the radiation flux. The radiation flux associated with a beam $I_\nu(\Omega)$ passing through a unit area can be found from (5.1). By construction, the total energy flow through dA is the integral of (5.1) over all frequencies and solid angles:

$$\begin{aligned} dE &\equiv \text{(energy transported through } dA) \\ &= dt\, dA \int I_\nu(\Omega)\hat{\mathbf{n}} \cdot \boldsymbol{\Omega}\, d\Omega\, d\nu. \end{aligned} \tag{5.13}$$

The radiation flux $\mathcal{F}_\nu$ through dA follows by dividing dE by dA and dt; it is

$$\mathcal{F}_\nu = \int I_\nu(\Omega)\hat{\mathbf{n}} \cdot \boldsymbol{\Omega}\, d\Omega, \tag{5.14}$$

and has units ergs $\text{cm}^{-2}\,\text{sec}^{-1}\,\text{Hz}^{-1}$.

If we introduce spherical polar coordinates with origin at dA and θ measured from $\hat{\mathbf{n}}$, $I(\Omega) = I(\theta, \phi)$, and $d\Omega = \sin\theta\, d\theta\, d\phi$. The integral for the flux becomes

$$\mathcal{F}_\nu = \int_0^\pi d\theta \int_0^{2\pi} d\phi\, I_\nu(\theta, \phi) \sin\theta \cos\theta. \tag{5.15}$$

Note that the dependence of I_ν on $\mathbf{r}$ and t is no longer shown explicitly. In the special case of an isotropic radiation field, the net flux is zero, since there is no net energy flow. Nevertheless, it is useful to split the flux into two parts, $\mathcal{F}_\nu^+$ in the outward direction, and $\mathcal{F}_\nu^-$ in the inward direction:

$$\mathcal{F}_\nu^+ = \int_0^{\pi/2} d\theta \int_0^{2\pi} d\phi\, I_\nu(\theta, \phi) \sin\theta \cos\theta, \tag{5.16}$$

$$\mathcal{F}_\nu^- = -\int_{\pi/2}^{\pi} d\theta \int_0^{2\pi} d\phi\, I_\nu(\theta, \phi) \sin\theta \cos\theta, \tag{5.17}$$

$$\mathcal{F}_\nu = \mathcal{F}_\nu^+ - \mathcal{F}_\nu^-. \tag{5.18}$$

Notice that $\mathcal{F}_\nu^- \geq 0$ by construction. For an isotropic radiation field, I is independent of θ and ϕ; so

$$\mathcal{F}_\nu^+ = \mathcal{F}_\nu^- = I_\nu \int_0^{\pi/2} \sin\theta \cos\theta \, d\theta \int_0^{2\pi} d\phi$$
$$= \pi I_\nu. \qquad (5.19)$$

The net flux $\mathcal{F}_\nu \equiv 0$, as expected. The flux is defined here in terms of some particular element of area, dA, and hence in terms of some particular direction $\hat{\mathbf{n}}$, the normal to that area. However, the net flux of radiation is in fact a vector quantity, in some direction $\hat{\mathbf{n}}_F$ (where $\hat{\mathbf{n}}_F$ is a unit vector in the direction of $\mathbf{F}$). The flux across area dA, with normal unit vector $\hat{\mathbf{n}}_{dA}$, is then $\pi \mathbf{F} \cdot \hat{\mathbf{n}}_{dA}$.

In stellar atmospheres, one may usually assume that the radiation field is axially asymmetric about the stellar radius, so that I is independent of ϕ, and $I_\nu(\Omega) = I_\nu(\theta)$. In this case

$$J_\nu = \frac{1}{4\pi} \int_0^{\pi} I_\nu(\theta) \sin\theta \cos\theta \, d\theta \int_0^{2\pi} d\phi$$
$$= \frac{1}{2} \int_{-1}^{1} I_\nu(\mu) \, d\mu. \qquad (5.20)$$

Similarly,

$$\mathcal{F}_\nu = \pi F_\nu = 2\pi \int_{-1}^{1} I_\nu(\mu)\mu \, d\mu,$$
$$\mathcal{F}_\nu^+ = \pi F_\nu^+ = 2\pi \int_0^{1} I_\nu(\mu)\mu \, d\mu, \qquad (5.21)$$
$$\mathcal{F}_\nu^- = \pi F_\nu^- = 2\pi \int_{-1}^{0} I_\nu(\mu)\mu \, d\mu.$$

The quantities I, J, F, and K are sufficient for an approximate analysis of the radiation field, and are the quantities traditionally used to describe it inside a star or stellar atmosphere. However, we have neglected one potentially important property of electromagnetic radiation: its polarization. In principle, the quantities we have introduced can be modified to take polarization into account, as may be necessary in some problems. Also, a kinetic-theory description of the radiation field can be given in terms of the density of photons in phase space, by defining a distribution function $n(\mathbf{p}, \mathbf{x})$ such that $n(\mathbf{p}, \mathbf{x})\, d\mathbf{p}\, d\mathbf{x}$ is the number of photons with $\mathbf{x}$ in $(\mathbf{x}, \mathbf{x} + d\mathbf{x})$ and momentum $\mathbf{p}$ in $(\mathbf{p}, \mathbf{p} + d\mathbf{p})$. We can express the traditional radiation quantities I, J, F, K in terms of this distribution function.

Problem 5.3. Express I_ν in terms of the distribution function $n(\mathbf{p}, \mathbf{x})$ in phase space. (Hint: transform the Cartesian momentum components p_x, p_y, p_z into spherical polar momentum components.)

Problem 5.4. Prove that the intensity in a beam of radiation is conserved; it does not drop off with distance in the absence of absorption.

Problem 5.5. Is it possible to define the intensity of the radiation from a point source?

Problem 5.6. When the radiation field is very nearly isotropic, we can expand $I(\theta)$ in a Taylor series about $\theta = 0$ [where $\cos\theta = 1$], and write

$$I(\theta) = I_0 + I_1 \cos\theta + \cdots, \qquad (5.22)$$

where I_0 and I_1 are independent of θ. Relate I_0 and I_1 to the flux and energy density of the field. Under what conditions on the energy density and the flux is (5.19) likely to be valid?

5.3. Opacity and Emissivity

Photons passing through matter may be scattered or absorbed by atoms, ions, and molecules. They may also be emitted as a result of charged particle motion, or from excited atomic and molecular states. These processes, taken collectively, result in modifications of the radiation field I_ν passing through matter. When this happens the matter and radiation are coupled. Later we will discuss absorption, scattering, and emission microscopically. Here, however, we consider the macroscopic aspects of these processes. In this Section we will show how they affect I_ν.

Consider the effect of scattering, absorption, and emission on a beam of photons described by $I_\nu(\Omega)$ passing through matter. Absorption removes photons from the beam and heats the gas. Scattering of a photon initially moving along Ω to a new direction Ω' removes energy from $I_\nu(\Omega)$ but adds it to some other beam $I_{\nu'}(\Omega')$. Conversely, photons from all other beams $I_{\nu'}(\Omega')$ may scatter into $I_\nu(\Omega)$. Emission results in the addition of photons to $I_\nu(\Omega)$. The details of absorption and scattering on the macroscopic level are contained in the material opacity. The details of emission are contained in the material emissivity. We consider the opacity first.

Imagine a beam of intensity I_ν passing through a

very thin slab of material whose thickness in the direction parallel to the beam is ds. Because the interaction of radiation with matter is essentially statistical, there is a certain probability per unit path length that a photon will be scattered or absorbed. This probability is called the (volume) opacity of the material or the total extinction coefficient. The volume opacity k_ν has units of area per unit volume (or cm^{-1}), and is inversely proportional to the photon's mean free path. We may also define a specific opacity, κ_ν, such that $\kappa_\nu \rho$ is equal to the volume opacity. The specific opacity has units of $\mathrm{cm}^2\ \mathrm{g}^{-1}$. Finally, we note that the opacity per particle, or cross section σ_ν, is defined by $\sigma_\nu = \kappa_\nu \rho / n$, where n is the number density of absorbing particles per unit volume. The cross section is most useful in developing the microscopic properties of absorbing matter. The volume opacity is most useful in discussing stellar atmospheres.

In addition to the total opacity k_ν we could define the scattering opacity k_ν^s and the absorption opacity k_ν^a, each of which could be developed independently. However, since $k_\nu = k_\nu^a + k_\nu^s$, we need not do so at this point in the discussion. We note that

$$l_\nu = k_\nu^{-1}$$

is a measure of the mean free path, that is, of the average distance a photon travels between encounters with matter. Similarly, $(k_\nu^s)^{-1}$ and $(k_\nu^a)^{-1}$ give the scattering and absorption mean free paths denoted by l_ν^s and l_ν^a, respectively.

An alternative way of describing the situation is to say that each atom has associated with it an interaction cross section, σ_ν, such that the total interaction cross section presented to the beam by a unit area of material containing n atoms per unit volume of thickness ds is

$$n\sigma_\nu\, ds. \tag{5.23}$$

The fraction of the photons passing through this unit area that will be absorbed or scattered is the same. Thus the probability of absorption or scattering in ds is $n\sigma_\nu\, ds$, and the opacity (the probability per unit length) is

$$k_\nu = \kappa_\nu \rho = n\sigma_\nu. \tag{5.24}$$

Thus σ_ν can also be called the opacity per atom (or per particle). Although k_ν, σ_ν, and κ_ν are functions of frequency, they are not distributions in frequency (like I_ν or F_ν), and so do not contain frequency units in their definitions. In particular, $k_\nu = k_\lambda$, where $\lambda = c/\nu$.

The total energy extracted from a beam of intensity I_ν by absorption and scattering after it travels a distance ds in a gas is given by the product of I_ν (the flux of photons of energy $h\nu$ per Hz) times the probability that a photon will be scattered or absorbed:

$$dI_\nu(\Omega) = -\, I_\nu(\Omega) \mathrm{k}_\nu\, ds \equiv -I_\nu(\Omega)\, d\tau_\nu. \tag{5.25}$$

The last step defines the dimensionless quantity

$$d\tau_\nu = k_\nu\, ds = \frac{ds}{l_\nu}. \tag{5.26}$$

This is called the *optical thickness* of the material to radiation of frequency ν, and is the ratio of the distance traveled in the medium to the photon's mean free path. When

$$\tau_\nu = \int_{s_0}^{s} k_\nu\, ds = \int_{x_0}^{x} \frac{ds}{l_\nu} \approx 1, \tag{5.27}$$

the photons in the beam will have traveled far enough to be scattered or absorbed. The relative transparency of a gas is in fact determined by τ_ν. When $\tau_\nu \gg 1$, the gas is optically thick, since radiation there will undergo absorption or scattering many times before traveling a distance $\Delta s = x_1 - x_0$. When $\tau_\nu \ll 1$, however, the photon's mean free path $l_\nu \gg \Delta s$ and there will be virtually no scattering or absorption within the distance Δs.

When absorption and scattering are the only processes occurring, one finds the change in I_ν over a distance s by formally integrating (5.25) from 0 to s:

$$\begin{aligned} I_\nu(s) &= I_\nu(0) \exp\left(-\int_0^s k_\nu(s')\, ds'\right) \\ &= I_\nu(0) \exp\left(-\tau_\nu(s)\right), \end{aligned} \tag{5.28}$$

$$\tau_\nu(s) = \int_0^s k_\nu(s')\, ds'. \tag{5.29}$$

When $\tau_\nu(s) \approx 1$, the intensity is reduced by a factor of e^{-1}. In order to relate $\tau_\nu(s)$ to the physical distance traveled by the beam, one must specify k_ν. In general, it will depend on ρ and T. The evaluation of (5.28)–(5.29) is complicated when T and ρ depend on position in the medium.

Problem 5.7. Electron scattering is the primary source of opacity in the solar core, with $\sigma_\nu \sim .6 \times 10^{-24}$ cm^2. Assume a density $\rho = 100$ g/cm^3 in the core, and estimate the distance traveled by a photon between scatterings.

Matter, in addition to absorbing radiation, may itself be the source of radiation. The atoms may convert thermal energy into radiation, or they may have been excited by the previous absorption of a photon. Macroscopically, we may imagine that a unit volume of material emits an amount of radiation $j_\nu'(\theta, \phi)$ per unit time, per unit frequency interval, and per unit solid angle, in direction specified by (θ, ϕ). This quantity is called the volume emissivity and has units erg cm^{-3} sec^{-1} ster^{-1} Hz^{-1}. If a beam of intensity I_ν passes through material of thickness ds, then the amount of radiation added to the beam because of the emission is given by

$$dI_\nu(\Omega) = j_\nu'(\Omega)\, ds. \tag{5.30}$$

We note here that emission may be attributed to one of two causes. The first, spontaneous emission, is the result of a quantum system's natural tendency to reach its lowest energy state, and is independent of the nature of the local radiation field I_ν. The second, induced emission, is related to the number of photons, and hence to I_ν, in the vicinity of an excited atom or molecule capable of emitting another photon of energy $h\nu$. Both aspects of emission must be included in the definition of j_ν'. We will return to this later.

In a macroscopic description of absorption, the energy of the absorbed photon loses its identity and is given to the general thermal-energy store of the stellar material. This is often termed "true" absorption. Similarly, "true" emission results when the energy of an emitted photon comes from the material's thermal-energy store. In photon scattering, the photon energy is returned more or less directly to the radiation field as photon energy (Compton scattering is an example). In the simplest case, we assume that the scattering is isotropic, so that the direction of the scattered photon bears no relation to the direction of the incident photon. We also assume that the scattering is coherent (or elastic), so that the frequency of the scattered photon is the same as that of the incident one. This is the form in which we will use scattering in the rest of this chapter. The increase in I_ν caused by scattering is then given by

$$dI_\nu = \int_{\Omega'} k_\nu^s\, I_\nu(\Omega') \frac{d\Omega'}{4\pi}\, ds,$$

where k_ν^s is the volume scattering coefficient, in cm^{-1}. Note that since k_ν^s is assumed independent of Ω' (isotropic), this can also be written

$$dI_\nu = k_\nu^s\, J_\nu\, ds$$

where J_ν is the mean intensity. This is the usual form of the scattering equation in stellar structure calculations.

It is possible to account formally for nonisotropic scattering by allowing k_ν^s to be a function of the directions of the incident and scattered photons, in which case we have

$$dI_\nu(\Omega) = \int_{\Omega'} k_\nu^s(\Omega,\Omega') I_\nu(\Omega') \frac{d\Omega'}{4\pi}\, ds.$$

It is also possible to account formally for incoherent or inelastic scattering, in which the scattered photon has a frequency (and hence energy) different from this incident photon; here we also have

$$dI_\nu(\Omega) = \int_{\nu'} \int_{\Omega'} k^s(\nu,\Omega,\nu',\Omega') I_{\nu'}(\Omega') \frac{d\Omega'}{4\pi}\, d\nu'\, ds. \tag{5.31}$$

Notice, however, that this is not strictly a case of pure scattering, for the frequency difference between the two photons is equivalent to an energy difference, which must be supplied by (or taken up by) the scattering atom in other forms, presumably thermal.

5.4 Equation of Radiative Transfer

The results discussed in the preceding sections describe the changes in I_ν due to the coupling between radiation and matter. The change in I_ν when it passes through a slab of matter of thickness ds is

$$\frac{dI_\nu}{ds} = \left(\frac{dI_\nu}{ds}\right)_{\text{absorption}} + \left(\frac{dI_\nu}{ds}\right)_{\text{scattering}} + \left(\frac{dI_\nu}{ds}\right)_{\text{emission}}. \tag{5.32}$$

The change caused by absorption and scattering out of

the beam is given by (5.25); the change caused by scattering into the beam is given by (5.31); and the change caused by emission by (5.30). Thus

$$\frac{dI_\nu(\Omega')}{ds} = -(k_\nu^s + k_\nu^a)\, I_\nu(\Omega) + \int k^s(\nu, \Omega, \nu', \Omega')\, I_{\nu'}(\Omega') \frac{d\Omega}{4\pi}\, d\nu' + j_\nu'(\Omega), \tag{5.33}$$

which is the equation of radiative transfer. It can be written in a more compact form that can be formally integrated:

$$\frac{dI_\nu(\Omega)}{ds} = -k_\nu I_\nu(\Omega) + j_\nu(\Omega), \tag{5.34}$$

where k_ν is the total opacity. Note that j_ν contains the last two terms in (5.33), the first of which describes scattering into the beam.

At the beginning of Section 5.2 we noted that I_ν depends on $\mathbf{r}$ and t in general. Since dI_ν/ds gives the change in I_ν along the beam's path, we have (summation over i is implied)

$$\frac{dI_\nu}{ds} = \left(\frac{\partial I_\nu}{\partial t} + \frac{\partial I_\nu}{\partial x^i}\frac{dx^i}{dt}\right)\frac{dt}{ds} = \frac{1}{c}\frac{\partial I_\nu}{\partial t} + \boldsymbol{\Omega} \cdot \nabla I_\nu, \tag{5.35}$$

where the path parameter s has been rescaled so that $ds/dt = c$, which implies $dx^i/dt = c\Omega^i$. The equation of radiative transfer is therefore

$$\frac{1}{c}\frac{\partial I_\nu}{\partial t} + \boldsymbol{\Omega} \cdot \nabla I_\nu = -k_\nu I_\nu + j_\nu. \tag{5.36}$$

In steady state, $\partial I_\nu/\partial t = 0$, and an alternative form may be obtained if we replace ds by the optical depth $d\tau_\nu = k_\nu\, ds$ defined by the total opacity; dividing (5.34) by k_ν, we have

$$\frac{dI_\nu}{d\tau_\nu} = -I_\nu + \frac{j_\nu}{k_\nu} = -I_\nu + S_\nu, \tag{5.37}$$

where

$$S_\nu = \frac{j_\nu}{k_\nu}$$

is called the source function. This may be formally integrated if we multiply through by e^{τ_ν}. Measuring τ_ν from $\tau_\nu = 0$, we find

$$e^{\tau_\nu}\left(\frac{dI_\nu}{d\tau_\nu} + I_\nu\right) = \frac{d}{d\tau_\nu} e^{\tau_\nu} I_\nu = e^{\tau_\nu}\frac{j_\nu}{k_\nu};$$

integrating over $d\tau_\nu$ yields upon rearrangement

$$I_\nu(\tau_\nu) = I_\nu(0)e^{-\tau_\nu} + \int_0^{\tau_\nu} \frac{j_\nu}{k_\nu} e^{(\tau_\nu' - \tau_\nu)}\, d\tau_\nu'. \tag{5.38}$$

The arguments leading to (5.37) and (5.38) involve no assumptions about the thermodynamic state of the radiation field or the gas. In fact, both results are applicable to systems that are not in thermal equilibrium. Restriction to local thermodynamic equilibrium (LTE) will be used later to relate k_ν and j_ν, but has not yet been incorporated into the transfer equation.

Equation (5.38) has a clear physical interpretation: the intensity at the point τ_ν is due to (a) the original intensity, attenuated by a factor $e^{-\tau_\nu}$ because of absorption and scattering; and (b) the integral of the source function j_ν/k_ν over all optical depths $d\tau_\nu'$, attenuated by a factor $e^{-(\tau_\nu - \tau_\nu')}$, where $\tau_\nu - \tau_\nu'$ is the optical depth between points of emission and observation.

When the radiation field is constant in time, the term $\partial I_\nu/\partial t$ vanishes. If the field is also axially symmetric, the intensity (5.36), measured relative to a surface element normal to the symmetry axis $\hat{\mathbf{n}}$ ($\hat{\mathbf{n}} \cdot \boldsymbol{\Omega} = \cos\theta$), (5.36) reduces to

$$\cos\theta \frac{dI_\nu}{dx} = -k_\nu I_\nu + j_\nu, \tag{5.39}$$

with x measured along $\hat{\mathbf{n}}$ (Figure 5.3). Let us substitute the source function $S_\nu = j_\nu/k_\nu$, use $\mu = \cos\theta$, and rewrite (5.39) in terms of the optical depth τ_ν; then

$$\mu \frac{dI_\nu}{d\tau_\nu} = I_\nu - S_\nu. \tag{5.40}$$

Now multiply by the element of solid angle $d\Omega = \sin\theta\, d\theta\, d\phi$, and integrate, assuming axial symmetry. We have

$$\begin{aligned} \frac{d}{d\tau_\nu}\int_{-1}^{1} \mu I_\nu\, d\mu &= \int_{-1}^{1} I_\nu\, d\mu - \int_{-1}^{1} S_\nu\, d\mu, \\ -\frac{1}{2}\frac{dF_\nu}{d\tau_\nu} &= 2J_\nu - \int_{-1}^{1} S_\nu\, d\mu. \end{aligned} \tag{5.41}$$

Suppose the source function itself is spherically sym-

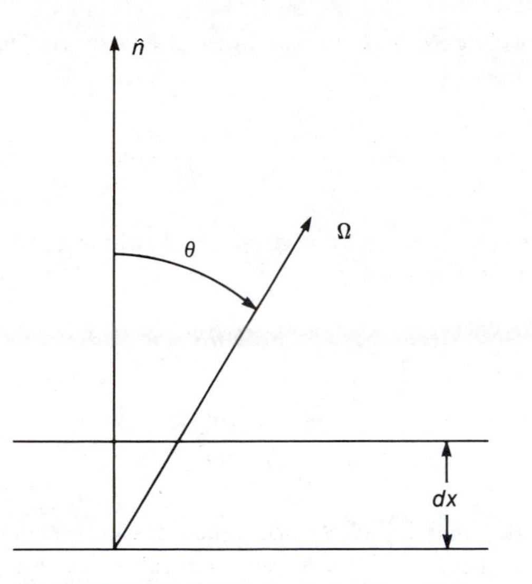

Figure 5.3. Cylindrically symmetric radiation field.

metric; i.e., the emissivity has no preferred direction. Then S_ν is independent of μ, and we get

$$-\frac{1}{4}\frac{dF_\nu}{d\tau_\nu} = J_\nu - S_\nu. \tag{5.42}$$

We will find that this equation has some useful applications in simple types of stellar atmospheres.

A further equation of interest can be obtained by multiplying the equation of transfer (5.40) by μ before integrating over a solid angle. This gives us

$$\frac{d}{d\tau_\nu}\int_{-1}^{+1}\mu^2 I_\nu\, d\mu = \int_{-1}^{+1}\mu I_\nu\, d\mu - \int_{-1}^{+1}\mu S_\nu\, d\mu,$$

$$-2\frac{dK_\nu}{d\tau_\nu} = \frac{F_\nu}{2} - \int_{-1}^{+1} S_\nu \mu\, d\mu.$$

If the source function is isotropic, the last integral vanishes and

$$-\frac{dK_\nu}{d\tau_\nu} = \frac{1}{4}F_\nu. \tag{5.43}$$

Because of the connection of K_ν with the radiation pressure, we can also write this as

$$\pi F_\nu = 4\pi\frac{dK_\nu}{d\tau_\nu} = -c\frac{dP_\nu}{d\tau_\nu}. \tag{5.44}$$

Thus, perhaps not surprisingly, the flux of radiation is proportional to the gradient of the radiation pressure. In deep stellar interiors, we are especially concerned with the total energy flux, which is πF_ν integrated over all frequencies. Formally, then, this is

$$\int_0^\infty \pi F_\nu\, d\nu = \int_0^\infty \frac{1}{k_\nu}\frac{dP_\nu}{ds}\, d\nu.$$

It would be very convenient if one could somehow write this equation in terms of the total radiation pressure, $P_R = \int_0^\infty P_\nu\, d\nu$. The following purely formal device enables us to do this. We define a mean opacity, $\bar{k}$ (averaged over frequency), by the equation

$$\frac{1}{\bar{k}}\int_0^\infty \frac{dP_\nu}{ds} = \int \frac{1}{k_\nu}\frac{dP_\nu}{ds},$$

so that we have

$$\pi F = -\frac{c}{\bar{k}}\frac{dP_R}{ds}. \tag{5.45}$$

This form of the equation of transfer is the most useful one for the study of stellar interiors, as we will see in the next section.

The preceding discussion has emphasized the response of the radiation field to the presence of matter. However, if the intensity is great enough, the energy absorbed or emitted by the gas may change the state of the gas itself. Since the gas properties determine k_ν and j_ν, dynamic equations describing the behavior of the gas must be solved together with the equation of transfer. Sometimes one must also include additional terms in the equation of transfer, as shown in Problem 5.8.

Problem 5.8. When the intensity is large, energy may be added to or lost from the radiation field as it does work on the gas. Show that the change in I_ν in this case is

$$dI_\nu(\Omega) = k_\nu I_\nu(\Omega)\frac{\mathbf{v}\cdot\Omega}{c}\,ds, \tag{5.46}$$

where $\mathbf{v}$ is the velocity of the gas through which the beam passes. Hint: show that the force exerted on the gas is $\mathbf{\Omega}\,(I_\nu/c)$.

5.5. Black-Body Radiation

A distribution of photons contained within an enclosure whose walls are maintained at constant temperature T will eventually reach a state of thermodynamic equilibrium at the same temperature. This equilibrium distribution is referred to as black-body radiation.

In a thermodynamic enclosure at temperature T the intensity I_ν of this radiation is found to be independent of direction, and is given by the Planck function

$$I_\nu \equiv B_\nu(T) = \frac{2h\nu^3}{c^2} \frac{1}{e^{h\nu/kT} - 1} \tag{5.47}$$

or, in wavelength units,

$$I_\lambda \equiv B_\lambda(T) = \frac{2hc^2}{\lambda^5} \frac{1}{e^{hc/\lambda kT} - 1}. \tag{5.48}$$

The other radiation quantities in thermodynamic equilibrium are

$$\begin{aligned} J_\nu &= B_\nu(T), & K_\nu &= \frac{1}{3} B_\nu(T), \\ F_\nu^+ &= F_\nu^- = B_\nu(T), & P_\nu &= \frac{4\pi}{3c} B_\nu(T), \\ F_\nu &= 0, & u_\nu &= \frac{4\pi}{c} B_\nu(T). \end{aligned} \tag{5.49}$$

By integrating over all frequencies, one finds

$$\begin{aligned} B(T) &= \int_0^\infty B_\nu(T)\, d\nu = \frac{2k^4\pi^4}{15h^3c^2} T^4 \\ &= \frac{\sigma}{\pi} T^4 = \frac{uc}{4\pi}, \end{aligned} \tag{5.50}$$

where σ is the Stefan-Boltzmann constant, and u is the energy density.

Now, in the deep interior of a star, the temperatures are very high, and thus the radiation density is high, but the net outward flux is small compared to the general level of radiation. Thus we might feel justified in assuming that, in the deep interior, the radiation field is close to isotropic and close to the Planck black-body radiation spectrum in form, corresponding to the local temperature of the gas. If we make this assumption, for the moment, then (5.49)–(5.50) show that $P_\nu = 4\pi B_\nu(T)/3c$. Substituting this into (5.44), replacing $d\tau_\nu$ by $k_\nu\, ds$, and integrating over frequency, we find

$$\pi F = -\frac{4\pi}{3} \int_0^\infty \frac{1}{k_\nu} \frac{dB_\nu}{dT} \frac{dT}{ds}\, d\nu. \tag{5.51}$$

We now define an average opacity $\bar{k}$, which is independent of frequency, as

$$\frac{1}{\bar{k}} \equiv \frac{\displaystyle\int_0^\infty \frac{1}{k_\nu} \frac{dB_\nu}{dT}\, d\nu}{\displaystyle\int_0^\infty \frac{dB_\nu}{dT}\, d\nu}. \tag{5.52}$$

Notice that ρ and dT/ds, which are independent of frequency, have been removed from the integral. Noting that $dB_\nu/dT = 4\sigma T^3/\pi$, we may combine (5.51) and (5.52) to obtain

$$\pi F = -\frac{4ac}{3\bar{k}} T^3 \frac{dT}{ds}. \tag{5.53}$$

In a stellar interior, we replace s by the radius r and define a quantity $L(r)$ as the total radiation energy crossing a sphere of radius r in unit time. Clearly $L(R)$ is the total luminosity, L, of the star. By definition, $L(r) = 4\pi r^2 \pi F$; so

$$L(r) = -\frac{16\pi ac}{3\bar{k}} r^2 T^3 \frac{dT}{dr}. \tag{5.54}$$

This is the equation of radiative transfer appropriate to deep stellar interiors, where the radiation field can be taken as nearly isotropic and close to Planckian. The mean opacity defined by equation (5.52) is called the *Rosseland mean opacity*. We will discuss its form later.

In stellar interiors, we may set $aT^4 = u$, the radiation energy density, and then rewrite (5.53) in the form

$$\mathcal{F} = -\frac{c}{3\bar{k}} \frac{du}{dr}.$$

The quantity $D = c/3\bar{k}$ is called the diffusion coefficient, which relates the radiation flux to the gradient of the energy density. This is known as the diffusion approximation.

When radiation is absorbed, an element of matter

of density ρ will be accelerated. This follows from (5.54) and the definition of the radiation pressure (2.25). Rewriting (5.54), we find

$$\frac{\bar{k}L}{4\pi r^2 c} = -\frac{d}{dr}\frac{1}{3}aT^4 = -\frac{dP_R}{dr}. \qquad (5.55)$$

An analysis like that leading to (4.4) shows that the force exerted per unit volume by a radiation pressure gradient across a spherical shell of matter is

$$f_R = -dP_R/dr. \qquad (5.56)$$

Denote the radiation acceleration by a_R; then it follows from (5.55) and (5.56) that

$$a_R = \frac{\bar{\kappa}L(r)}{4\pi r^2 c}. \qquad (5.57)$$

where we have introduced the mean specific opacity $\bar{\kappa} = \bar{k}/\rho$.

The radiation acceleration is usually small; however, if a region of a star becomes extremely opaque (large κ), or if the luminosity is large, then a_R may become important. In fact, the existence of radiation acceleration places a limit on how luminous a star in hydrostatic equilibrium can be. When the radiation field is intense, an element of matter in a stellar envelope will be in equilibrium as long as $a_R \lesssim m(r)G/r^2$. Combining (5.55) to (5.57) and rewriting gives as a limit on the luminosity for an equilibrium envelope:

$$L(r) = \frac{4\pi cGM_\odot}{\bar{\kappa}}\left(\frac{m(r)}{M_\odot}\right). \qquad (5.58)$$

Since $L(r) \simeq L$ and $M \simeq m(r)$ in a stellar envelope, we may rewrite this as

$$L_{\text{Ed}} = \frac{5.03 \times 10^{37}}{\bar{\kappa}}\left(\frac{M}{M_\odot}\right) \text{ergs/sec}. \qquad (5.59)$$

When electron scattering is the primary source of opacity,

$$\bar{\kappa} \simeq 0.2(1 + X),$$

and

$$L_{\text{Ed}} = 1.5 \times 10^{38}\,(M/M_\odot) \text{ ergs/sec}.$$

Therefore, if the luminosity in a stellar envelope exceeds L_{Ed}, the outer layers will not be in hydrostatic equilibrium, but must be accelerating outward. Equation (5.59) is the Eddington limit, and will be encountered when we discuss mass loss from stars.

Equation (5.53) for the flux is equivalent to the heat-conduction equation $\mathbf{F} = -\kappa_c \nabla T$, where the heat conductivity $\kappa_c = (4acT^3/3\pi\bar{k})$. The average energy of a photon in the equilibrium distribution $B_\nu(T)$ is kT. The factor dB_ν/dT multiplying k_ν^{-1} in (5.52) peaks at an energy of about 4 kT. Consequently, the Rosseland mean opacity used in the flux equation implies that heat is conducted primarily by the high-energy photons ($h\nu \gtrsim 4kT$) in the equilibrium distribution $B_\nu(T)$.

What we have done in deriving (5.54) is typical of what must often be done in analyzing stellar atmospheres and stellar interiors. We have, on the one hand, assumed that the radiation field is locally describable by a temperature T, but we have also admitted that there is a net flux of radiation in the star, because this temperature is macroscopically nonuniform. Thus we require that a temperature be definable throughout a region that is large enough that the thermodynamic enclosure approximation can be used, but small enough that the properties of the region are uniform. This is one version of the general hypothesis called local thermodynamic equilibrium or LTE. It is not easy to define precisely what one means by LTE in a general form. One can instead pinpoint the assumptions made in each application, and this we will do. The term LTE, and its opposite, non-LTE, are not always used consistently by different authors. Here our LTE assumption is that the radiation field is locally Planckian, with temperature T.

Finally note that some of these equations can be written in vector form (if the radiation pressure can be considered isotropic, and thus representable by a scalar quantity). Then

$$\begin{aligned} \pi\,\mathbf{F}_\nu &= \mathcal{F}_\nu = -\frac{c}{\bar{k}_\nu}\nabla P_\nu, \\ \pi\,\mathbf{F} &= -\frac{c}{\bar{k}}\nabla P = -\frac{16ac}{3\bar{k}}T^3\nabla T. \end{aligned} \qquad (5.60)$$

5.6. Radiative Equilibrium

Suppose, through thermonuclear or other processes, an amount of energy is being generated per unit mass of the star at distance r from the center. In traversing a shell of thickness dr, then, the additional amount of energy added to the radiation field is

$$dL = 4\pi r^2 \rho \epsilon \, dr. \tag{5.61}$$

The energy passing through a spherical surface of radius r is $L(r)$. The increase in energy passing through a shell of thickness dr and radius $r + dr$ must be caused by energy produced locally, according to (5.61). Therefore

$$L(r + dr) - L(r) \simeq \frac{dL}{dr} dr = 4\pi r^2 \rho \epsilon \, dr$$

or

$$\frac{dL}{dr} = 4\pi r^2 \rho \epsilon. \tag{5.62}$$

In its application to stellar interiors, this is usually called the energy-generation equation. The energy-generation rate, ϵ, depends on the physical conditions of the material at the given radius. One may write, formally,

$$\epsilon = \epsilon(\rho, T, C) \text{erg g}^{-1} \text{sec}^{-1}, \tag{5.63}$$

where C stands for the chemical composition. In real stars, energy may also be transported by means other than radiation, especially by convection. In these circumstances, the total energy crossing radius r, $L(r)$, is made up of a radiative part, $L_{\rm rad}(r)$, a convective part $L_{\rm conv}(r)$, and perhaps some others. That is

$$L(r) = L_{\rm rad}(r) + L_{\rm conv}(r) + \ldots \tag{5.64}$$

Naturally only the radiative component of (5.64) is given by the radiative transfer equation, whereas the total luminosity appears in the energy-generation equation. We will discuss convective energy transport later, but will ignore it here.

When a star, or a portion of it, undergoes quasi-static evolution, heat transfer to or from a unit mass of the gas may affect the luminosity. This process may be incorporated by adding a term to (5.62) to account for heat transfer $T\,ds$, where s is the total entropy per gram of the gas. The corresponding rate of heat transfer to the gas is $T\,ds/dt$; so the change in luminosity is the negative of this quantity, and (5.62) becomes

$$\frac{dL}{dr} = 4\pi r^2 \rho \left(\epsilon - T\frac{ds}{dt} \right), \tag{5.65}$$

where ϵ now includes only energy generated by nuclear processes. The last term can be ignored for static configurations, but is critical when considering quasi-static evolution. We have already related the change in entropy to the pressure and temperature or density of an ideal gas (2.13).

Consider a region of a star in which no energy generation is taking place. Then, if energy is being transported solely by means of radiation, the net energy flow into the region must be zero, or else the star would not be in equilibrium. Formally this means that $L(r + dr) = L(r)$, or that

$$\frac{dL}{dr} = 0; \tag{5.66}$$

when (5.66) holds, $L(r) = \text{const.}$

In the atmosphere of a star, in the plane-parallel approximation, we have simply $F = \text{constant.}$* Since the flux leaving the surface of a star is, by definition, $\sigma T_{\rm eff}^4$, we have simply, in stellar atmospheres,

$$\pi F = \sigma T_{\rm eff}^4. \tag{5.67}$$

This total flux has been generated by processes in the interior of the star. Once we are outside the energy-generating region, the radiation simply filters through the outer part of the star until it finally emerges, admittedly redistributed in wavelength, but with a total flux that has already been determined. Therefore this constant flux, or equivalently the effective temperature, is an important physical parameter of stellar atmospheres.

5.7. Simple Stellar Atmospheres

Energy transport in the atmospheres of most stars is radiative. The principal exceptions to this rule are probably white dwarfs and low-temperature stars, where the role of convective transport may be signifi-

*This is specifically the total flux, integrated over all wavelengths. Naturally the energy may be redistributed in wavelength as the radiation makes its way through the star.

cant. The atmospheres problem is simplified by the absence of energy sources, and because the underlying structure of the star (M, R, and L) is known from interior calculations. It is complicated by the fact that the atmosphere may not be in LTE, and because I_ν must clearly be non-Planckian near the surface. Furthermore, the radiation passing through the stellar atmosphere contains a wealth of detailed information about local conditions (pressure, composition, presence of turbulence or magnetic fields) imprinted in the number, strength, and shape of atomic and molecular spectral lines. Finally, for cool stars that have deep convective zones immediately below the atmosphere, the interior problem can only be solved in conjunction with model atmospheres. In this section we will consider simple model atmospheres where transport is radiative.

Consider a stellar atmosphere assumed to be in radiative equilibrium, with a constant flux $F = \sigma T_{\text{eff}}^4/\pi$. Usually the scale height in the atmosphere is much less than R, so we may replace the actual atmosphere by a plane layer of infinite extent. Measuring the optical depth from the top of the atmosphere inward (Figure 5.4), the flux equation (5.45) becomes

$$c\frac{dP_R}{d\tau} = \sigma T_{\text{eff}}^4. \tag{5.68}$$

Since the physical distance is measured in the outward direction,

$$d\tau = -\bar{k}\,dh,$$

where $\bar{k}$ is the mean opacity (5.52). The quantity τ is the mean optical depth. Since the flux is constant, (5.68) can be integrated immediately to give

$$P_R = \frac{\sigma}{c} T_{\text{eff}}^4 (\tau + q), \tag{5.69}$$

where q is a constant determined by the boundary condition $P_R = P_R$ (surface) at $\tau = 0$:

$$q = \frac{c}{\sigma T_{\text{eff}}^4} P_R\,(\text{surface}). \tag{5.70}$$

Equation (5.68) contains the explicit assumption that the source function is isotropic. We now make the additional assumption of strong LTE, that is, that the radiation field is locally Planckian. Although this is

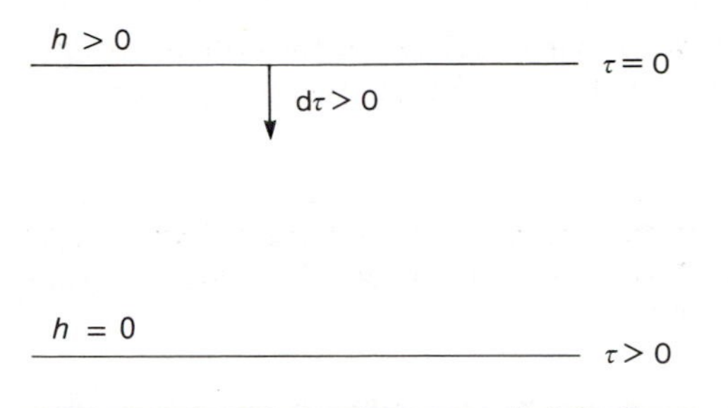

Figure 5.4. Relation between optical depth τ and physical depth in a planar layer of absorbing matter. Optical depth increases into the layer. The radiation moves upward in direction of increasing h.

probably a very good approximation in stellar interiors, it is probably a very bad approximation in stellar atmospheres; so the results will be only a first guide to what to expect.

Now, at the surface of a star, even though it is at temperature T, radiation is flowing in one direction only, outward; there is usually no incident radiation. If we assume that the outward radiation has black-body character with temperature T_{eff}, the surface radiation pressure should be half that given by the Planck formula at temperature T_{eff}. That is,

$$P_R(\text{surface}) = \frac{2\sigma}{3c} T_{\text{eff}}^4. \tag{5.71}$$

Problem 5.9. Prove (5.71) under the assumption of LTE, using the moments of the transport equation discussed in section 5.2.

Therefore we have

$$q = 2/3 \tag{5.72}$$

and

$$T^4 = \frac{3}{4} T_{\text{eff}}^4(\tau + 2/3), \tag{5.73}$$

where the mean opacity to be used now is, of course, the Rosseland mean opacity. This is our first, simple model stellar atmosphere; it gives the variation of temperature with Rosseland mean optical depth. Of course, we don't know what the Rosseland mean is yet, so we cannot yet relate it to the distribution of temperature with physical depth, but that will come later.

Problem 5.10. Make a table showing the variation of T with τ for the atmosphere of the Sun. Take $T_{\text{eff}} = 6{,}000$ K.

Note that in this simple model, the effective temperature is the temperature actually existing at optical depth $\tau = 2/3$:

$$T(2/3) = T_{\text{eff}}. \tag{5.74}$$

Also note that the surface temperature of the star at $\tau = 0$ is

$$T(0) = T_{\text{eff}}/2^{1/4} = 0.841\, T_{\text{eff}}. \tag{5.75}$$

In spite of the many crude approximations made, this simple model does quite well in accounting for the general features of stellar atmospheres.

Problem 5.11. List as many of the explicit or implicit assumptions or approximations made in getting to formula (5.75) as you can. Which do you think produce the most serious errors?

The only information about stellar interiors used so far has been the total luminosity L, or flux. In order to complete a model stellar atmosphere, we must use the hydrostatic equilibrium condition to find the pressure and density distribution. This introduces the star's surface gravity g, which depends on the star's mass M and radius R. For most atmospheres, the variation in h is small compared with R, and the mass of the atmosphere $M_{\text{atm}} \ll M$. Therefore we may set $M(r) = M$ and $r = R$ in the equation of hydrostatic equilibrium, obtaining

$$\frac{dP}{dh} = -\frac{MG}{R^2}\rho = -g\rho. \tag{5.76}$$

We divide both sides by $\bar{k}$ to get

$$\frac{dP}{d\tau} = \frac{g\rho}{\bar{k}}. \tag{5.77}$$

Unfortunately, we cannot integrate this equation until we know how the Rosseland mean opacity varies with the physical properties of the atmosphere.* As a crude approximation, we suppose that the Rosseland mean opacity does not vary in the atmosphere. This amounts to assuming that the (mean) photon absorption cross section per particle is independent of T and ρ; then (5.77) can be integrated formally to obtain

$$P = P_0 \exp\left[\frac{g\mu m_H}{\bar{k}} \int_0^\tau \frac{d\tau'}{kT(\tau')}\right]. \tag{5.78}$$

The pressure would have an exponential dependence on optical depth if T were constant. The density (assuming gas pressure predominates) is then

$$\rho = P_0 \frac{\mu m_H}{k} \frac{1}{T(\tau)} \exp\left[\frac{g\mu m_H}{\bar{k}} \int_0^\tau \frac{d\tau'}{kT(\tau')}\right]. \tag{5.79}$$

This is not a very good approximation to real stellar atmospheres, but it does show conceptually how a model atmosphere is constructed.

Problem 5.12. Using the simple $T(\tau)$ model, tabulate P, ρ, T as a function of τ for the solar atmosphere. Take $\bar{\kappa} = \bar{k}/\rho = 1\ \text{cm}^2\ \text{g}^{-1}$.

Another approach to a simple model atmosphere is the grey atmosphere, in which opacity is assumed to be independent of frequency. (Note that this is not the same thing as introducing a mean opacity). If opacity is independent of frequency, then the transport equation

$$-\frac{1}{4}\frac{dF_\nu}{d\tau} = J_\nu - S_\nu \tag{5.80}$$

(we have dropped the ν subscript on τ) can be integrated over frequency to get

$$-\frac{1}{4}\frac{dF}{d\tau} = J - S. \tag{5.81}$$

*Modern treatments use the column mass $\rho\, dh = dm$ as independent variable, so that $P = mg + c$, and $\tau(m) = -\int \kappa\, dm$. This is more convenient for numerical work where m is the Lagrangian mass variable.

Now, in radiative equilibrium, $dF/d\tau = 0$; hence for a grey atmosphere in radiative equilibrium,

$$J = S, \tag{5.82}$$

and the (integrated) source function is equal to the (integrated) mean intensity. The equation of transfer (integrated over frequency) now becomes

$$\mu \frac{dI}{d\tau} = I - J = I - \frac{1}{2}\int_{-1}^{1} I\, d\mu. \tag{5.83}$$

This integro-differential equation for the intensity I can in principle be solved. The solution yields $I(\tau,\mu)$, but does not say anything specifically about the physical parameters, such as temperature. In order to make that connection, we must adopt some LTE hypothesis. Also, since the equation of transfer has a formal solution (5.38) that expresses $I(\mu,\tau)$ as an integral over the source function, the solution to the grey-atmosphere problem can be given simply as the function $J(\tau)$. The solution to (5.83) does not have a simple analytic form. However, an analytic approximation may be constructed to give exact values at $\tau = 0$ and $\tau = \infty$:

$$\begin{aligned} J(\tau) &= \frac{3}{4} F\,(\tau + 0.7104 - 0.1331\, e^{-3.449\tau}) \\ &= \frac{3}{4} F\,[\tau + q(\tau)]. \end{aligned} \tag{5.84}$$

The value of $q(\tau)$ thus varies from 0.5773 at $\tau = 0$ to 0.7105 at $\tau = \infty$, which is not that different from the value $q = 2/3$ in (5.72). Indeed, we can obtain $q = 2/3$ by various approximations (usually called Eddington's first approximation) to the solution of (5.83), which has come to be known as the Milne-Schwarzschild equation. If we were to add the strong LTE assumption $J = \sigma T^4/\pi$, we would obtain from (5.84) a slightly improved (but not much!) model for the $T(\tau)$ variation in a stellar atmosphere.

We noted at the beginning of this section that we can derive information about compositions, for example, from model atmospheres. To do so, we need to solve the transfer equation for individual energy groups; that is, we need I_ν rather than just I. To reach it, we could start with (5.80), with τ replaced by the optical depth τ_ν for each energy group. In order to proceed further, however, we need to know the dependence of κ_ν and j_ν on frequency. Then I_ν yields the continuous radiation spectrum of the star, which may be used in turn to investigate the strength and form of emission and absorption lines. The latter are often sensitive to the composition and ionization state of the gas. By investigating the effects of various elements at the temperatures and pressures corresponding to the continuum spectrum, we can often identify elements in a stellar atmosphere whose concentrations are many orders of magnitude less than the dominant constituents H and He. We will return to these points after we discuss the microscopic nature of emission and absorption processes.

5.8. True Absorption and Scattering

We have said that radiation interacts with matter statistically; there is some probability that a photon will be absorbed by an atom (or by a molecule, or by some other effective "particle"). We have also described this interaction probability by means of a macroscopic cross section. What happens microscopically when an atom absorbs a photon of energy $h\nu$? The answer is simple for an isolated atom: an electron in the atom (almost always an outer valence electron) is raised to a higher-energy state. The energy difference ΔE between the two states is given by the Planck relation, $\Delta E = h\nu$. If ΔE is less than the ionization energy of the electron in the atom, the atom is now in an excited state. What happens to this energy of excitation?

There are several possibilities (see Figure 5.5). The simplest is that the electron may simply return (usually on the order of 10^{-6} sec) to its initial energy state (case **a**). If it does so, it re-emits a photon of the same frequency as it absorbed; so there is no net transfer of energy from the radition field to the matter, just a temporary exchange. We call this process *scattering,* since it does not change the nature of the photon, but merely redirects it. On the other hand, if collisions between atoms are very frequent (**c**), or if there are many other energy levels between the excited level and the lower level (**d**), the energy of excitation may be transferred to kinetic energy of the atoms (**c**), or may be degraded into lower-energy photons (**d**). Also, a further photon might excite the atom from the upper energy level to a still higher one (**b**), from which it cascades down to the first level. Likewise, a collision between two atoms may transfer some kinetic energy from the collision to excitation energy of the atoms, which may then convert it to photons by radiative

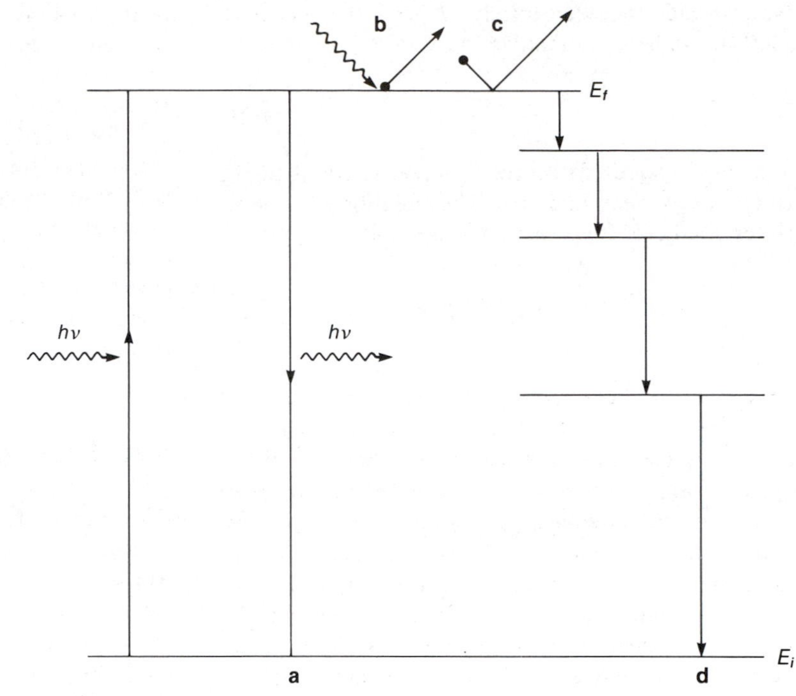

Figure 5.5. Excitation and decay processes between quantum states. At extreme left, an electron initially in level E_i absorbs a photon ($h\nu$) and is excited to level E_f. Subsequent decays (**a, d**) and excitation (**b, c**) processes are described in the text.

transitions. In an extreme version of this conversion, the photons emitted from a given atom or group of atoms may have no particular relation to the ones absorbed (other than the general ones of statistical equilibrium in thermodynamics). This process is described as true absorption and true emission, and in this form can be handled by the laws of thermodynamics. We will discuss later the conditions under which we may expect to find cases of pure scattering and cases of pure absorption and emission. What actually happens to excited atoms in a radiation field is somewhere between the two extremes.

We shall suppose for the moment that the real process of absorption and emission in a star is a combination of these processes. That is, we shall write

$$k_\nu = k_\nu{}^a + k_\nu{}^s \text{ cm}^{-1}, \tag{5.85}$$

where $k_\nu{}^a$ is the true opacity, and $k_\nu{}^s$ is the scattering opacity (i.e., cross section per unit volume or per unit mass). Likewise, we write the emission coefficient j'_ν as

$$j'_\nu = j_\nu{}^e + j_\nu{}^s \text{ ergs cm}^{-3}\text{ sec}^{-1}\text{ Hz}^{-1} \tag{5.86}$$

where $j_\nu{}^e$ is the true emission and $j_\nu{}^s$ is the scattering emission (i.e., the rescattered radiation). Now we make the simplifying assumption that the scattered radiation is re-emitted isotropically. [In practice, this is generally not true; instead, the re-emission of scattered radiation follows a so-called phase function, in which it is distributed over an angle ϕ between the incident photon along $\mathbf{\Omega}'$ and the re-emitted photon along $\mathbf{\Omega}$, according to a distribution function

$$f(\phi) = f(\mathbf{\Omega} \cdot \mathbf{\Omega}'). \tag{5.87}$$

To include this complication here would mean that the source function, for significant scattering, would not be isotropic, as we have so far assumed it to be. This is a complication we will not discuss.] The energy absorbed from a beam of intensity I_ν in a direction (θ, ϕ) by the scattering opacity $k_\nu{}^s$ is

$$k_\nu{}^s I_\nu (\theta, \phi)\, d\Omega \text{ ergs sec}^{-1}\text{ cm}^{-3}\text{ Hz}^{-1}. \tag{5.88}$$

The total energy absorbed per unit volume is therefore obtained by integrating this over a solid angle, and is clearly

$$k_\nu{}^s J_\nu \text{ ergs cm}^{-3}\text{ Hz}^{-1}\text{ sec}^{-1}, \tag{5.89}$$

where J_ν is the mean intensity at frequency ν. This energy is re-emitted isotropically, according to our assumption. Thus

$$j_\nu^s = k_\nu^s J_\nu. \tag{5.90}$$

Therefore, for pure scattering (no true emission and absorption), we have

$$S_\nu = \frac{j_\nu^s}{k_\nu^s} = J_\nu; \tag{5.91}$$

so the source function is equal to the mean intensity at each frequency. Obviously there is no redistribution of the radiation in frequency in this model. Then, as we remarked earlier, the mathematics of the solution to the grey-atmosphere problem can be carried over to this one, and we have, at each frequency,

$$J_\nu(\tau_\nu) = \frac{3}{4} F_\nu [\tau_\nu + q(\tau_\nu)], \tag{5.92}$$

where $q(\tau_\nu)$ is the same function as in equation (5.84).

In the other extreme (pure absorption), the source function is given by

$$S_\nu = j_\nu^e / k_\nu^a. \tag{5.93}$$

Notice that this quantity is a property only of the material of the star and not of the radiation field; so we can use a thermodynamic argument to obtain this ratio, assuming that the matter is in thermodynamic equilibrium. We imagine a unit volume of the matter to be surrounded by a thermodynamic enclosure at the temperature T of the matter. Since the mean intensity of the radiation field in this enclosure is $B_\nu(T)$, the amount of energy absorbed by the unit volume is

$$k_\nu^a B_\nu(T) \text{ ergs sec}^{-1} \text{ cm}^{-3} \text{ Hz}^{-1}. \tag{5.94}$$

The amount of energy emitted by the unit volume is

$$j_\nu^e \text{ ergs sec}^{-1} \text{ cm}^{-3} \text{ Hz}^{-1}. \tag{5.95}$$

In (thermodynamic) equilibrium these must be equal, because there must be no net gain or loss of energy to the unit volume. Hence we must have

$$\begin{aligned} j_\nu^e &= k_\nu^a B_\nu(T), \\ j_\nu^e / k_\nu^a &= B_\nu(T), \\ I_\nu &= B_\nu(T). \end{aligned} \tag{5.96}$$

This is one form of Kirchhoff's law. Since the left-hand side is independent of the nature of the radiation field, we can write

$$\frac{j_\nu^e}{k_\nu^a} = B_\nu(T), \qquad I_\nu \neq B_\nu(T), \tag{5.97}$$

or, for true absorption,

$$S_\nu = B_\nu(T),$$

as long as the matter can be considered to be in thermodynamic equilibrium. The assumption that the emission and absorption processes for true emission can be described in terms of a matter temperature in this way is called *the hypothesis of (weak) local thermodynamic equilibrium.* This is what is most commonly meant by LTE. However, some authors also extend LTE to the assumption that no scattering is taking place, so that the source function is equal to the Planck function. This is clearly a stronger assumption, but not as strong as what we have called strong LTE, where the radiation field is taken as Planckian. We will call this hypothesis of no scattering *moderate LTE* (although this is not a standard term).

In the general case, with both absorption and scattering we have, in weak LTE,

$$\begin{aligned} S_\nu &= \frac{j_\nu^e + j_\nu^s}{k_\nu^a + k_\nu^s} = \frac{k_\nu^a B_\nu(T) + k_\nu^s J_\nu}{k_\nu^a + k_\nu^s} \\ &= \frac{k_\nu^a}{k_\nu^a + k_\nu^s} B_\nu(T) + \frac{k_\nu^s}{k_\nu^a + k_\nu^s} J_\nu \\ &= \lambda_\nu B_\nu(T) + (1 - \lambda_\nu) J_\nu, \end{aligned} \tag{5.98}$$

where λ_ν is simply the ratio of the true absorption coefficient to the total absorption (i.e., true plus scattering). Since λ_ν is in principle determined from the physics of the photon-atom interactions (we will come to this later), this equation is the general equation for the source function under conditions of weak LTE.

Kirchhoff's law may be used in the equation of transfer to estimate the rate at which an arbitrary radiation distribution I_ν approaches the black-body distribution $B_\nu(T)$, that is, the strength of the coupling between the radiation field and matter. For simplicity, consider a uniform density medium of infinite extent at temperature T, which is suddenly permeated by a uniform radiation field I_ν. We may ignore gradients and, assuming that the gas is in weak LTE, so that

(5.98) applies, the equation of transfer for an isotropic field is

$$\frac{1}{c}\frac{\partial I_\nu}{\partial t} + \mathbf{\Omega} \cdot \nabla I_\nu = \frac{1}{c}\frac{\partial I_\nu}{\partial t}$$
$$= -k_\nu I_\nu + j_\nu$$
$$= -\lambda_\nu k_\nu [I_\nu - B_\nu(T)]. \quad (5.99)$$

This may be integrated to give the time-development of I_ν,

$$I_\nu(t) = \exp(-\lambda_\nu k_\nu tc)\, I_\nu(0) + [1 - \exp(-\lambda_\nu k_\nu tc)]\, B_\nu(T). \quad (5.100)$$

For times $t \gg (\lambda_\nu k_\nu c)^{-1}$, the initial spectrum will have become essentially Planckian.

Problem 5.13. Scattering in the solar core is due primarily to electron scattering, for which $\bar{\kappa} = 0.2\,(1 + X)\ \mathrm{cm^2/g}$, where X is the hydrogen abundance. Assuming core densities $\rho \simeq 10^2\ \mathrm{g/cm^3}$, how long should it take for an arbitrary radiation field to become Planckian? Assume $\lambda_\nu \simeq 1$.

5.9. Radiation in the Solar Atmosphere

The Sun is the only star for which we see a visible disk, and therefore it is the only star for which we can investigate the distribution of the intensity of the emergent radiation in terms of both frequency and angle θ. When we look at a portion of the solar disk, we are looking at radiation emerging at an angle θ to the outward normal (Figure 5.6). The center-to-limb variation of intensity, as we observe it, is then the same as the variation of I_ν with θ at a single point on the Sun, assuming that the properties of the solar atmosphere are uniform (ignoring sunspots, solar rotation, etc.). Furthermore, we can write down a formula for the emergent radiation at a point determined by θ, using the formal solution of the equation of transfer (5.38). We may immediately write

$$I_\nu\,(\text{surface}) = I_\nu(0)\, e^{-\tau_{\max}} + \int_0^{\tau_{\max}} S(\tau_\nu')\, e^{-\tau_\nu'}\, d\tau_\nu',$$

where τ_ν' is the optical depth measured inward along the line of sight. Since the origin of the radiation is deep in the interior of the star, at an extremely large optical depth we may for all practical purposes set $\tau_{\max} = \infty$; so the emergent radiation is given by

$$I_\nu(0) = \int_0^\infty S(\tau_\nu')\, e^{-\tau_\nu'}\, d\tau_\nu', \quad (5.101)$$

or, transforming to the radial optical depth, with $d\tau_\nu' = d\tau_\nu/\cos\theta$,

$$I_\nu\,(0, \mu) = \int_0^\infty S\,(\tau_\nu)\, e^{-\tau_\nu/\cos\theta} \sec\theta\, d\tau_\nu. \quad (5.102)$$

We see that the emergent intensity is given by integrating the source function S over τ_ν and multiplying by an exponential. That is,

$$I_\nu\,(\mu) = \int_0^\infty S(\tau_\nu)\, w\,(\tau_\nu/\mu)\, d\tau_\nu/\mu, \quad (5.103)$$

where

$$w(x) = e^{-x}. \quad (5.104)$$

The exponential weighting function decreases as we go

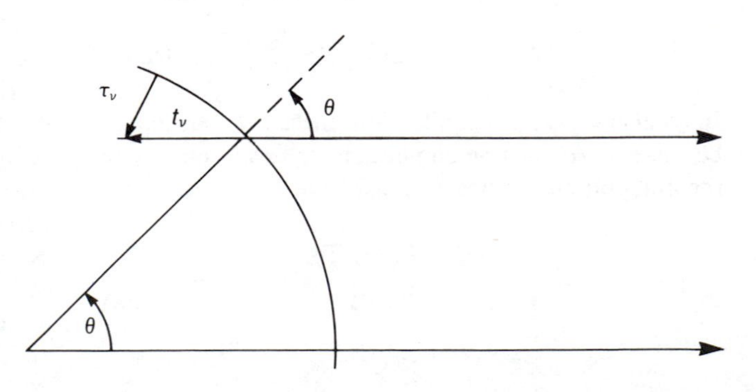

Figure 5.6. Limb darkening: variation in optical depth because of the curvature of the solar disk. The optical depth τ_ν is measured normal to the solar surface, where t_ν is measured along the line of sight.

further into the star. However, the source function will generally increase with temperature, and therefore will increase as we go into the star. For example, imagine that the source function is a linear function of optical depth:

$$S = a + b\tau_\nu. \tag{5.105}$$

Then we have

$$\begin{aligned} I_\nu(\mu) &= \int_0^\infty (a + b\tau_\nu)\, w(\tau_\nu/\mu)\, d\tau_\nu/\mu \\ &= \int_0^\infty (a + bt_\nu\mu)\, w(t_\nu)\, dt_\nu \end{aligned} \tag{5.106}$$

with $t_\nu = \tau_\nu/\mu$. Performing the integration, we get

$$\begin{aligned} I_\nu(\mu) &= a\int_0^\infty e^{-t}dt + b\mu\int_0^\infty te^{-t}\, dt \\ &= a + b\mu. \end{aligned} \tag{5.107}$$

Therefore, as μ goes from 1 at the center of the solar disk to 0 at the limb, the intensity drops from $a + b$ to a. The ratio $I_\nu(\mu)/I_\nu(1)$ is usually called the *limb-darkening law*, and says that the limb of the Sun should have a lower emergent intensity than the center.

Problem 5.14. The apparent luminosity of an astronomical source of radiation is defined as the energy falling on a unit area of a detector in a unit time (and a unit frequency interval, if necessary). If the surface brightness of an astronomical source is defined as the ratio of its apparent luminosity, ℓ_ν, to its apparent area A, measured in square degrees, steradians, etc., prove that the surface brightness of a small piece of the Sun is equal to the emergent intensity at that point (apart from numerical factors to take care of units):

$$S_\nu = \frac{\ell_\nu}{A} = I_\nu \text{ erg sec}^{-1}\text{ cm}^{-2}\text{ ster}^{-1}\text{ Hz}^{-1}.$$

Thus, for a linear source function, the limb-darkening law is

$$\begin{aligned} \frac{I_\nu(\theta)}{I_\nu(0)} &= \frac{a + b\cos\theta}{a + b} \\ &= 1 - \beta + \beta\cos\theta, \end{aligned} \tag{5.108}$$

where $\beta \equiv b/(a + b)$ is the limb-darkening coefficient. For example, suppose we observe the Sun in integrated radiation, of all wavelengths, and assume that the opacity of the solar atmosphere is grey and that moderate LTE is valid, so that $S = B(T) = (\sigma/\pi)T^4$. Then

$$S = \frac{\sigma T^4}{\pi} = \frac{3}{4}\frac{\sigma}{\pi} T_{\text{eff}}^4 (\tau + 2/3). \tag{5.109}$$

Since this is of the form $a + b\tau$, the limb-darkening coefficient is then

$$\beta = \frac{b}{b + a} = (1 + 2/3)^{-1} = 0.6, \tag{5.110}$$

so that for a grey model atmosphere, in total radiation,

$$\frac{I(\cos\theta)}{I(0)} = 0.4 + 0.6\cos\theta. \tag{5.111}$$

Suppose now that a value for the function $I_\nu(\mu)$ has been found by observation of solar limb darkening. Write it now as a function of $s = \sec\theta = 1/\mu$. Then we can write

$$f(s) = \frac{I_\nu(s)}{s} = I_\nu(\mu)\mu = \int_0^\infty S_\nu(t)\, e^{-ts}\, dt. \tag{5.112}$$

Thus the limb-darkening function, expressed in appropriate form, is the Laplace transform of the source function. In particular, since a Laplace transform can be inverted by standard mathematical techniques, we can use observations of solar limb darkening as a function of frequency as a direct probe of the behavior of the source function in the solar atmosphere and thus, with some additional LTE-type assumptions, as a probe of the temperature structure of the solar atmosphere.

The presence of limb darkening in the Sun is familiar from photographs of the Sun. For stars, limb darkening is rarely, if ever, observable. (The few possible exceptions are eclipsing binary stars, where the form of the light curve may depend on the limb-darkening characteristics of the star). In fact, the emergent intensity from a star is not directly observable; instead we see the integrated radiation over the whole stellar surface. Obviously we see some integrated version of the emergent intensity. Assume for

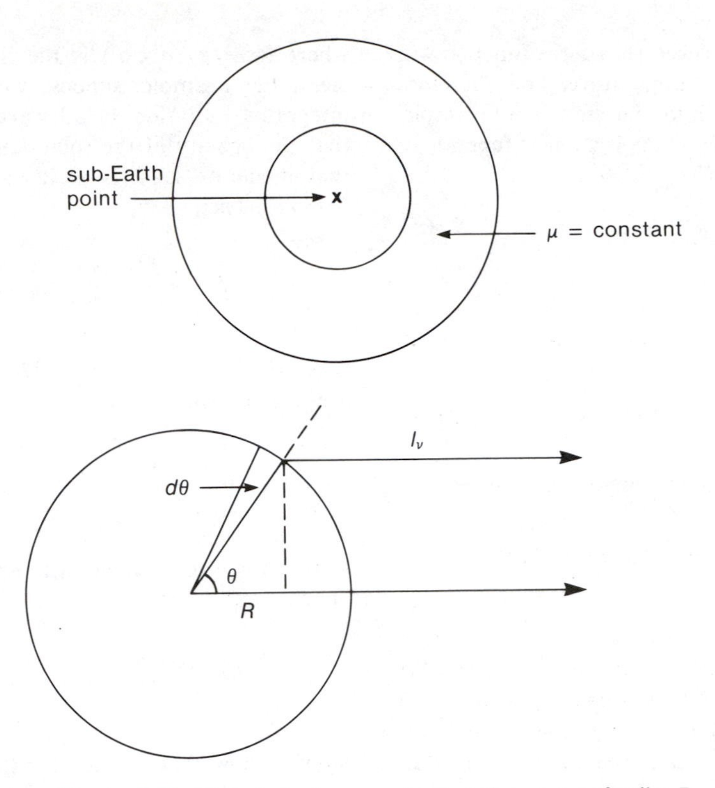

Figure 5.7. Limb-darkening geometry for a distant star of radius R.

the moment that the star is perfectly spherical. Then the loci of constant μ on the star are circles around the "sub-Earth" point (Figure 5.7): the area of the star having θ in the range $\theta, \theta + d\theta$, is $dA = 2\pi R^2 \sin\theta\, d\theta$. The projection of this area in the direction toward Earth is

$$\cos\theta\, dA = 2\pi R^2 \sin\theta \cos\theta\, d\theta.$$

The intensity emitted in that direction is $I_\nu(\theta)$. Hence the total radiation obtained from the star is given by integrating over θ:

$$\begin{aligned} l_\nu &= \int_0^{\pi/2} 2\pi R^2 I_\nu(\theta) \sin\theta \cos\theta\, d\theta \\ &= 2\pi R^2 \int_0^1 I_\nu(\mu)\mu\, d\mu \\ &= \pi R^2 F_\nu^+, \end{aligned} \tag{5.113}$$

where F_ν^+ is the outward flux of radiation at the surface of the star. Thus observations of stellar spectra give the flux at the stellar surface, whereas observations of solar surface brightness give the emergent intensity.

Problem 5.15. Reconcile the result (5.113) for the amount of radiation emitted in the direction of Earth with the usual formula for the apparent luminosity of a star,

$$l_\nu = \frac{L_\nu}{4\pi d^2},$$

where d is the distance of the star, and L_ν is the absolute luminosity at frequency ν.

If we take equation (5.102) or (5.103), multiply by μ, and integrate, we find

$$\begin{aligned} F_\nu^+ &= 2\int_0^1 I_\nu(\mu)\mu\, d\mu \\ &= 2\int_0^\infty S(\tau_\nu) \int_0^1 \mu e^{-\tau_\nu/\mu} \frac{d\mu}{\mu}\, d\tau_\nu \\ &= 2\int_0^\infty S(\tau_\nu)\, w_F(\tau_\nu)\, d\tau_\nu, \end{aligned} \tag{5.114}$$

where now the weighting function for the flux is

$$w_F(\tau) \equiv \int_0^1 e^{-\tau/\mu}\, d\mu. \tag{5.115}$$

It is usual to make the transformation $s = 1/\mu$ in this expression. This gives

$$w_F(\tau) = \int_1^\infty \frac{e^{-s\tau}}{s^2} ds \equiv E_2(\tau), \qquad (5.116)$$

where $E_n(\tau)$ is the exponential integral of order n defined by

$$E_n(\tau) = \int_1^\infty \frac{e^{-x\tau}}{x^n} dx. \qquad (5.117)$$

For a linear source function, we have

$$\begin{aligned} F &= 2a \int_0^\infty E_2(\tau)\, d\tau + 2b \int_0^\infty \tau E_2(\tau)\, d\tau \\ &= a + \frac{2}{3} b. \qquad (5.118) \end{aligned}$$

This result could also be obtained directly by integrating the emergent intensity for the linear source function: $I = a + b\mu$.

Problem 5.16. Verify that (5.118) gives the correct expression for the flux from a grey atmosphere.

Later we will use the fact that the flux at some depth τ in the star, not at the surface, is given by the sum of two contributions, one from optical depths greater than τ, and one from optical depths less than τ. In fact,

$$\begin{aligned} F(\tau) &= 2 \int_\tau^\infty S(\tau')\, E_2\,(\tau' - \tau)\, d\tau' \\ &\quad - 2 \int_0^\tau S(\tau')\, E_2\,(\tau - \tau')\, d\tau'. \qquad (5.119) \end{aligned}$$

(For convenience we have dropped the frequency subscripts on F, τ', τ, and S in these equations.) It can also be shown that the mean intensity, J, at any depth is given by

$$\begin{aligned} J(\tau) &= \frac{1}{2} \int_0^\infty S(\tau') E_1(|\tau' - \tau|)\, d\tau' \\ &= \frac{1}{2} \int_\tau^\infty S(\tau') E_1(\tau' - \tau)\, d\tau' \\ &\quad + \frac{1}{2} \int_0^\tau S(\tau') E_1(\tau - \tau')\, d\tau' \qquad (5.120) \end{aligned}$$

and also that

$$\begin{aligned} K(\tau) &= \frac{1}{2} \int_\tau^\infty S(\tau') E_3(\tau' - \tau)\, d\tau' \\ &\quad + \int_0^\tau S(\tau')\, E_3\,(\tau - \tau')\, d\tau'. \qquad (5.121) \end{aligned}$$

We have seen that the radiation from a stellar surface is given by an integral of the source function over the second exponential integral as a weighting function. Although obviously the observed radiation from a star does not all come from a particular optical depth, it is nevertheless useful sometimes to think of $\tau \simeq 2/3$ as a depth of formation of the radiation. In particular, when we translate this into geometrical (i.e., physical) depth in the atmosphere, we can get some insight into the properties of stellar spectra. For example, if we observe at a frequency or wavelength where the opacity is low, then we see physically further into the star, because to reach an optical depth of $2/3$, we must penetrate deeper into the atmosphere. In particular, we shall be seeing to regions of relatively high temperature. Now, if we observe at a frequency where the opacity is high, we do not see so far into the atmosphere, because the optical depth $2/3$ is reached rather rapidly. Therefore the temperature in that region of the star is relatively low. This, in effect, explains why we see absorption lines in stellar spectra. Where the opacity is high we see higher levels in the stellar atmosphere, and consequently lower temperatures; so we receive less radiation from the star than in the neighboring continuum (i.e., the part of the spectrum away from the absorption line), where we see to higher temperatures because the opacity is low. In addition, we can see that a strong absorption line will be "formed" higher in the atmosphere (and consequently at lower temperatures and pressures) than a weak absorption line (Figure 5.8). Furthermore, within one line, the core of the line, i.e., its central, deep part, is formed higher in the atmosphere than the wings of the line. These remarks obviously do not depend on the specific value $\tau = 2/3$, but are more in the nature of a scaling change on the optical depth as we go from low opacity to high opacity. This kind of argument is also sometimes given as a simple explanation of solar limb darkening: we see to optical depth $2/3$, but at the limb this value corresponds to a higher point in the atmosphere, and hence to a lower temperature, than it does at the center of the disk.

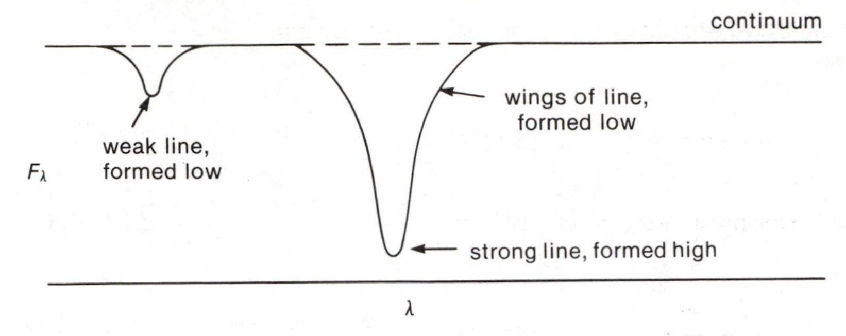

Figure 5.8. Absorption line profiles arising in stellar atmosphere (schematic). Weak lines (left) are formed deeper in the atmosphere than are strong lines (right). The wings are formed deep in the atmosphere; line centers form in upper layers.

Problem 5.17. A homogeneous nebula of radius R emits radiation with a constant emissivity j. If the volume opacity k is a constant, find the intensity $I(\theta)$ at the nebula's surface, and the flux corresponding to this intensity. Show that the limits $F \rightarrow j/k$ obtained when $\tau \rightarrow 0$ (nebula thin to its own radiation) and $F \rightarrow (4/3)(j\tau/k)$ obtained when $\tau \rightarrow \infty$ (nebula thick to its own radiation) are physically reasonable.

Problem 5.18. In an optically thin object (such as a gaseous nebula) a photon emitted within the body has a high probability of escaping without further interaction with the matter. Show that the mean opacity $\overline{k}_p$ characteristic of such a system is given by

$$\overline{k}_p = \frac{\int k_\nu u_\nu \, d\nu}{\int u_\nu \, d\nu}.$$

This is the Planckian mean opacity. The corresponding Planck mean free path is $\ell_p = 1/\overline{k}_p$. How does it compare to the Rosseland mean free path for optically thick matter?

Problem 5.19. Consider a spherical shell of matter of infinitesimal thickness dr in a star. Use the diffusion approximation discussed following (5.54) to show that the net rate of change of the shell's thermal energy density u is given by the diffusion equation

$$\frac{\partial u}{\partial t} = \frac{1}{r^2}\frac{\partial}{\partial r} r^2 \frac{c}{3\overline{k}}\frac{\partial u}{\partial r} \equiv \nabla \cdot D\nabla u.$$

Use dimensional arguments to show that the timescale for photons to diffuse a distance ℓ is of order $t_D \sim \ell^2/D \sim \ell^2 \overline{k}/c$.

5.10 Summary of Results on Radiative Stellar Structure

Table 5.1 summarizes our results so far; for stellar interiors the main simplification we can make is to assume strong LTE, so that we need consider only the Rosseland mean opacity. The Rosseland mean optical depth is used to parametrize stellar atmospheres, but this is neither necessary nor current practice. Instead, what we have written as $\overline{k}$ can be taken as the opacity at any standard frequency or wavelength, usually 5000 Å $= 0.5\mu$. This optical depth is then used as the basic independent variable in terms of which the structure of the atmosphere is described. As examples, two stellar atmosphere models are set forth in Tables 5.2 to 5.4.

In order to calculate such an atmosphere, we must know the details of the so-called "constitutive" equations for opacity and the equation of state. In addition, the appropriate boundary conditions must be satisfied. In stellar atmospheres, one of the most important of these is the flux constancy in radiative equilibrium. Because the opacity in stellar atmospheres is markedly non-grey (i.e., it varies considerably with frequency), the process of satisfying this particular condition is usually one of trial and error. One takes a possible model, calculates the opacity and optical depth at each frequency as a function of the optical depth $\overline{\tau}$, calculates the flux at each frequency, integrates over frequency, and then checks to see if it is constant. If it is not, then one must develop techniques of successive iteration involving adjustments of the model until the condition is satisfied. Stellar interior calculations do not have this particular problem; instead, they have the problem that some of the boundary conditions are

Table 5.1
Summary of stellar structure equations (first column) that apply in the stellar interior (second column) and in stellar atmospheres (third column). C denotes the chemical composition.

Equation	Stellar interiors	Stellar atmospheres
Hydrostatic equilibrium	$\frac{dP}{dr} = -\frac{m(r)G}{r^2}\rho(r)$	$\frac{dP}{dh} = -g\rho, \quad \frac{dP}{d\bar{\tau}} = \frac{g}{\bar{\kappa}}$
Mass conservation	$\frac{dm(r)}{dr} = 4\pi r^2 \rho(r)$	———
Radiative transport	$\frac{dT}{dr} = \frac{-3\bar{\kappa}\rho\, L(r)}{16\pi\, acr^2 T^3}$	$F_\nu(\tau_\nu) = 2\int_{t_\nu}^{\infty} S_\nu(t_\nu)\, E_2(t_\nu - \tau_\nu)\, dt_\nu$ $- 2\int_0^{\tau_\nu} S_\nu(t_\nu)\, E_2(\tau_\nu - t_\nu)\, dt_\nu,$ $S_\nu(\tau_\nu) = \lambda_\nu B_\nu(t(\tau_\nu)) + (1 - \lambda_\nu) J_\nu(\tau_\nu),$ $\lambda_\nu(P, \rho, T, C) = a_\nu/(a_\nu + S_\nu),$ $d\tau_\nu = (\kappa_\nu/\bar{\kappa})\, d\bar{\tau}$
Radiative equilibrium	$\frac{dL}{dr} = 4\pi r^2 \epsilon(r)\rho(r)$	$\frac{dL}{dr} = 0, \frac{dF}{d\bar{\tau}} = 0,$ $F(\bar{\tau}) = \int_0^{\infty} F_\nu(\bar{\tau})\, d\nu = \frac{\sigma}{\pi} T_{eff}^4$
Opacity	$\bar{\kappa} = \bar{\kappa}(\rho, T, P, C)$	$\bar{\kappa} = \bar{\kappa}(\rho, T, P, C), \frac{\kappa_\nu}{\bar{\kappa}} = \frac{\kappa_\nu}{\bar{\kappa}}(\rho, T, P, C)$
Energy generation rates	$\epsilon = \epsilon(\rho, T, C)$	
Equation of state	$P = P(\rho, T, C)$	$P = P(\rho, T, C)$
Parameters	M = mass C = composition	g = surface gravity (cm s^{-2}) T_{eff} = effective temperature C = composition
Interior boundary conditions	$r = 0$; $m(r) = 0$ $L(r) = 0$	$F = \frac{\sigma}{\pi} T_{eff}^4$
Surface boundary conditions	$r = R$: $T = 0$ $P = 0$	$\bar{\tau} \rightarrow 0$ as $P, T \rightarrow 0$
Model calculations give	$P, \rho, T, (C)$ as functions of $r, L, T_{eff}, (t)$	$P, \rho, T, (C)$ as functions of $\bar{\tau}, L_\nu$ or $F_\nu, (t)$

center conditions, but others are surface conditions. Normally, one must begin two numerical integrations, one starting at the center and one at the surface, and hope they meet in the middle. In general they will not meet, and an iterative procedure must again be used to obtain a consistent solution.

One approach is to choose central values P and T, and integrate the stellar structure equations outward, to an arbitrary radius r_0. Surface values L and R are also chosen, and the equations integrated inward to r_0. The procedure is then repeated two more times, using the initial choices $P_c + \Delta P$, T_c, $L + \Delta L$, R and P_c, $T_c + \Delta T_c$, L, $R + \Delta R$. Analysis of the mismatch at r_0 usually allows new initial values to be obtained within a few iterations. Table 5.4 sets forth a complete model for a 1 $M_\odot$ star like the sun.

In Table 5.3, g_{eff} is an effective surface gravity, introduced in hot stars to allow for the effect of radiation pressure in the following way. The equation of hydrostatic equilibrium for stellar atmospheres is

$$\frac{g}{\bar{\kappa}} = \frac{dP}{d\tau} = \frac{dP_g}{d\tau} + \frac{dP_R}{d\tau}.$$

Table 5.2
Model atmosphere for main-sequence star of spectral type G0, assuming $T_{\text{eff}} = 6000$ K and $g = 10^4$ cm sec^{-2}. Columns give: (1) optical depth for λ = 5000 Å; (2) temperature; (3) gas pressure (dynes/cm^2); (4) electron number density (cm^{-3}); (5) specific absorption coefficient at λ = 5000 Å; (6) density (g/cm^3); (7) ratio of proton to total hydrogen concentration; (8) geometric depth for an arbitrary zero at τ = 1; (9) fraction of energy transported by convection. Numbers in parentheses are powers of 10 by which entries are multiplied.

τ_{5000}	T(K)	P_g	n_e	κ_{5000}	ρ	$H^+/\Sigma H$	x(km)	F_C/F
.0010	4,900	2.186 (+03)	3.005 (+11)	7.781 (−03)	6.898 (−09)	2.869 (−05)	−1,166	.0000
.0016	4,914	2.862 (+03)	3.757 (+11)	9.505 (−03)	9.007 (−09)	2.525 (−05)	−1,081	.0000
.0025	4,931	3.738 (+03)	4.712 (+11)	1.165 (−02)	1.172 (−08)	2.250 (−05)	−996	.0000
.0040	4,950	4.869 (+03)	5.926 (+11)	1.432 (−02)	1.521 (−08)	2.042 (−05)	−911	.0000
.0063	4,974	6.324 (+03)	7.483 (+11)	1.765 (−02)	1.966 (−08)	1.902 (−05)	−828	.0000
.0100	5,004	8.194 (+03)	9.485 (+11)	2.180 (−02)	2.533 (−08)	1.824 (−05)	−744	.0000
.0158	5,044	1.059 (+04)	1.214 (+12)	2.703 (−02)	3.247 (−08)	1.852 (−05)	−661	.0000
.0251	5,096	1.364 (+04)	1.571 (+12)	3.371 (−02)	4.140 (−08)	2.008 (−05)	−578	.0000
.0398	5,155	1.752 (+04)	2.042 (+12)	4.213 (−02)	5.256 (−08)	2.232 (−05)	−495	.0000
.0631	5,231	2.241 (+04)	2.711 (+12)	5.329 (−02)	6.624 (−08)	2.691 (−05)	−412	.0000
.1000	5,339	2.845 (+04)	3.753 (+12)	6.926 (−02)	8.242 (−08)	3.689 (−05)	−331	.0000
.1585	5,473	3.572 (+04)	5.412 (+12)	9.281 (−02)	1.009 (−07)	5.461 (−05)	−251	.0000
.2512	5,652	4.407 (+04)	8.454 (+12)	1.323 (−01)	1.206 (−07)	9.124 (−05)	−176	.0000
.3981	5,889	5.297 (+04)	1.471 (+13)	2.054 (−01)	1.391 (−07)	1.716 (−04)	−107	.0000
.6310	6,197	6.164 (+04)	2.883 (+13)	3.514 (−01)	1.538 (−07)	3.580 (−04)	−48	.0000
1.000	6,600	6.926 (+04)	6.417 (+13)	6.669 (−01)	1.622 (−07)	8.372 (−04)	0	.0000
1.585	7,119	7.536 (+04)	1.585 (+14)	1.389 (+00)	1.634 (−07)	2.168 (−03)	37	.0005
2.512	7,830	7.968 (+04)	4.485 (+14)	3.356 (+00)	1.564 (−07)	6.586 (−03)	64	.0856
3.981	8,455	8.263 (+04)	9.671 (+14)	6.857 (+00)	1.491 (−07)	1.505 (−02)	84	.4954
6.310	8,861	8.529 (+04)	1.509 (+15)	1.078 (+01)	1.456 (−07)	2.412 (−02)	102	.7649
10.00	9,179	8.812 (+04)	2.085 (+15)	1.530 (+01)	1.440 (−07)	3.376 (−02)	121	.8724
15.85	9,456	9.137 (+04)	2.723 (+15)	2.070 (+01)	1.436 (−07)	4.427 (−02)	144	.9437
25.12	9,710	9.524 (+04)	3.441 (+15)	2.722 (+01)	1.443 (−07)	5.572 (−02)	171	.9294
39.81	9,958	9.994 (+04)	4.286 (+15)	3.539 (+01)	1.461 (−07)	6.862 (−02)	203	.9560
63.10	10,205	1.057 (+05)	5.297 (+15)	4.579 (+01)	1.488 (−07)	8.328 (−02)	242	.9724

SOURCE: E. Novotny, *Introduction to Stellar Atmospheres and Interiors* (London: Oxford University Press, 1973).

Now, from the equation of transfer (5.45), $dP_R/d\tau$ is given by

$$\frac{dP_R}{d\tau} = \frac{\pi F}{c} = \frac{\sigma T_{\text{eff}}^4}{c}$$

for an isotropic source function. Hence we have

$$\frac{dP_g}{d\tau} = \frac{g\rho}{\bar{k}} - \frac{\sigma T_{\text{eff}}^4}{c} \equiv \frac{g_{\text{eff}}\rho}{\bar{k}},$$

where g_{eff}, the effective surface gravity, is defined by

$$g_{\text{eff}} \equiv g - \bar{k}\frac{\sigma T_{\text{eff}}^4}{\rho c} = \frac{MG}{R^2} - \frac{\bar{k}\sigma T_{\text{eff}}^4}{\rho c}.$$

Stable stellar atmospheres presumably cannot exist if $g_{\text{eff}} \leq 0$. This represents a limit to the existence of stable stars in the Hertzsprung-Russell diagram. Some stars, the so-called P Cygni stars, are thought to be close to this limit. These stars are literally blowing themselves apart with their own radiation pressure.

5.11. Nonradiative Energy Transport

Radiation is a common means of energy transport in stars, but there are also others. We will consider these other means of energy transport roughly in the order of their importance or incidence in "normal" stars.

Convection

We may write the radiative transfer equation for stellar interiors in the form

$$\frac{dT}{dr} = -\frac{3\bar{k}}{16\pi ac}\frac{L(r)}{r^2T^3}. \tag{5.122}$$

Although this may not be the way one normally thinks

Table 5.3
Model atmosphere for main-sequence star of spectral type B0.5, assuming $T_{eff} = 25{,}000$ K and $g = 10^4$ cm sec^{-2}. Columns are labeled as in Table 5.2, and g_{eff} is in cm sec^{-2}; see text for a discussion of the latter.

τ_{5000}	T(K)	P_g	n_e	κ_{5000}	ρ	log g_{eff}	x(km)
.0010	15,155	2.440 (+01)	5.833 (+12)	4.683 (−01)	1.245 (−11)	3.9824	11,714
.0016	15,319	3.561 (+01)	8.399 (+12)	5.302 (−01)	1.793 (−11)	3.9805	10,936
.0025	15,572	5.111 (+01)	1.186 (+13)	6.086 (−01)	2.533 (−11)	3.9787	10,171
.0040	15,859	7.230 (+01)	1.647 (+13)	7.078 (−01)	3.517 (−11)	3.9770	9,423
.0063	16,166	1.009 (+02)	2.256 (+13)	8.319 (−01)	4.812 (−11)	3.9753	8,686
.0100	16,478	1.391 (+02)	3.051 (+13)	9.866 (−01)	6.511 (−11)	3.9735	7,960
.0158	16,791	1.897 (+02)	4.083 (+13)	1.179 (+00)	8.713 (−11)	3.9713	7,242
.0251	17,109	2.562 (+02)	5.413 (+13)	1.416 (+00)	1.155 (−10)	3.9684	6,528
.0398	17,449	3.434 (+02)	7.111 (+13)	1.703 (+00)	1.518 (−10)	3.9648	5,815
.0631	17,838	4.572 (+02)	9.266 (+13)	2.043 (+00)	1.977 (−10)	3.9604	5,097
.1000	18,309	6.066 (+02)	1.198 (+14)	2.432 (+00)	2.554 (−10)	3.9555	4,363
.1585	18,893	8.039 (+02)	1.539 (+14)	2.865 (+00)	3.280 (−10)	3.9507	3,601
.2512	19,618	1.068 (+03)	1.969 (+14)	3.341 (+00)	4.198 (−10)	3.9461	2,797
.3981	20,505	1.426 (+03)	2.517 (+14)	3.856 (+00)	5.357 (−10)	3.9422	1,937
.6310	21,568	1.915 (+03)	3.215 (+14)	4.415 (+00)	6.836 (−10)	3.9390	1,009
1.000	22,810	2.591 (+03)	4.112 (+14)	5.029 (+00)	8.747 (−10)	3.9366	0
1.585	24,227	3.526 (+03)	5.269 (+14)	5.710 (+00)	1.121 (−09)	3.9347	−1,098
2.512	25,838	4.829 (+03)	6.764 (+14)	6.474 (+00)	1.439 (−09)	3.9328	−2,293
3.981	27,663	6.644 (+03)	8.695 (+14)	7.326 (+00)	1.849 (−09)	3.9306	−3,595
6.310	29,745	9.174 (+03)	1.117 (+15)	8.270 (+00)	2.375 (−09)	3.9276	−5,017
10.00	32,160	1.270 (+04)	1.432 (+15)	9.316 (+00)	3.036 (−09)	3.9231	−6,576
15.85	35,016	1.758 (+04)	1.828 (+15)	1.058 (+01)	3.845 (−09)	3.9158	−8,293
25.12	38,449	2.415 (+04)	2.312 (+15)	1.231 (+01)	4.757 (−09)	3.9041	−10,183
39.81	42,547	3.284 (+04)	2.884 (+15)	1.425 (+01)	5.754 (−09)	3.8938	−12,289
63.10	47,319	4.495 (+04)	3.577 (+15)	1.560 (+01)	7.023 (−09)	3.8937	−14,732
100.0	52,814	6.292 (+04)	4.498 (+15)	1.671 (+01)	8.780 (−09)	3.8999	−17,632
158.5	59,030	8.990 (+04)	5.757 (+15)	1.805 (+01)	1.121 (−08)	3.9075	−20,963

SOURCE: E. Novotny, *Introduction to Stellar Atmospheres and Interiors* (London: Oxford University Press, 1973).

of the radiation field, equation (5.122) is a correct way of viewing what happens in a star, for a certain amount of energy is generated in the interior by nuclear reactions. Somehow this energy must escape the star. In the outer parts of the star, which (we assume for the moment) are in radiative equilibrium, a certain temperature gradient must be set up, according to this equation, to maintain the requisite energy flow, L. From (5.122) it is clear that if the opacity of the stellar material becomes rather high (because of the physical origin of the opacity and its dependence on temperature; see Chapter 6), then a rather steep temperature gradient must be set up to maintain the energy flow. A very steep temperature gradient is unstable in a star, just as it is in the Earth's atmosphere.

To see why this is so, consider a small piece of material of the star at density ρ_1, pressure P_1. Imagine that this element of material rises a little in the star, and that it does so adiabatically, with no change in its internal energy (no heat lost or gained). If its initial density and pressure were P_1, ρ_1, and its density and pressure after rising the short distance are P_2, ρ_2, then for adiabatic motion

$$P_1 \rho_1^{-\gamma} = P_2 \rho_2^{-\gamma}, \qquad (5.123)$$

where γ is the ratio of the specific heats at constant pressure and at constant volume. We now assume that the element of material maintains the same pressure as its surroundings, otherwise it would expand or contract until pressure equilibrium was restored. In effect, we assume that the time-scale for removing pressure imbalance is very short compared to the time-scale for the establishment of thermal equilibrium. We know that time-scales for adjusting periodic imbalances over the whole star are $\simeq$30 min for the Sun. We will show later that typical convective time-scales in the solar envelope are $\simeq$1 month. Now, if the small element of material finds itself at a lower density than its surroundings, it will continue to rise because of buoyancy forces in the gravitational field of the star. In a gravitational field g, the buoyancy force per unit

Table 5.4

Model of 1 $M_\odot$ star on the main sequence when the central hydrogen abundance is 0.5. Columns give: (1) radius; (2) mass fraction; (3) luminosity; (4) hydrogen mass fraction; (5) pressure (dynes cm^{-2}); (6) temperature (K); (7) density ($g\ cm^{-3}$); (8) specific absorption coefficient ($cm^2\ g^{-1}$); (9) $3\ \rho(r)/\bar{\rho}$; (10) $-d \ln P/d \ln r$; (11) polytropic index.

r/R	$\frac{M_r}{M}$	$\frac{L_r}{L}$	X	$\log P$	$\log T$	$\log \rho$	κ	U	V	$n+1$
0.00	0.000	0.000	0.494	17.351	7.165	+2.128	1.07	3.000	0.000	3.264
0.02	0.001	0.006	0.498	17.335	7.162	+2.113	1.07	2.976	0.043	3.281
0.04	0.006	0.042	0.520	17.307	7.154	+2.084	1.10	2.903	0.166	3.284
0.06	0.018	0.124	0.545	17.265	7.141	+2.046	1.16	2.802	0.350	3.312
0.08	0.040	0.244	0.571	17.205	7.123	+1.995	1.24	2.680	0.597	3.350
0.10	0.073	0.396	0.611	17.135	7.102	+1.932	1.34	2.534	0.873	3.401
0.12	0.113	0.538	0.643	17.058	7.080	+1.867	1.45	2.398	1.159	3.448
0.14	0.162	0.668	0.670	16.970	7.054	+1.796	1.60	2.257	1.480	3.513
0.16	0.217	0.774	0.694	16.874	7.027	+1.721	1.76	2.115	1.748	3.592
0.18	0.276	0.854	0.714	16.774	7.000	+1.642	1.93	1.973	2.160	3.681
0.20	0.337	0.909	0.723	16.667	6.971	+1.561	2.12	1.838	2.524	3.776
0.22	0.399	0.945	0.728	16.554	6.942	+1.476	2.32	1.702	2.896	3.875
0.24	0.460	0.968	0.733	16.438	6.912	+1.389	2.55	1.565	3.265	3.972
0.26	0.519	0.981	0.737	16.319	6.882	+1.298	2.78	1.431	3.630	4.064
0.28	0.574	0.989	0.741	16.196	6.852	+1.205	3.02	1.303	3.988	4.149
0.30	0.626	0.994	0.744	16.072	6.823	+1.109	3.23	1.180	4.332	4.226
0.32	0.672	0.997	0.744	15.944	6.793	+1.011	3.47	1.066	4.676	4.293
0.34	0.716	0.998	0.744	15.816	6.763	+0.913	3.73	0.959	5.009	4.353
0.36	0.753	0.999	0.744	15.690	6.734	+0.816	3.99	0.862	5.327	4.403
0.38	0.788	1.000	0.744	15.562	6.705	+0.717	4.25	0.771	5.640	4.449
0.40	0.818	1.000	0.744	15.432	6.676	+0.616	4.51	0.687	5.944	4.488
0.42	0.844	1.000	0.744	15.302	6.648	+0.514	4.74	0.613	6.243	4.520
0.44	0.867	1.000	0.744	15.174	6.619	+0.415	5.04	0.543	6.534	4.549
0.46	0.887	1.000	0.744	15.045	6.591	+0.314	5.31	0.482	6.825	4.573
0.48	0.904	1.000	0.744	14.917	6.563	+0.214	5.60	0.426	7.111	4.595
0.50	0.919	1.000	0.744	14.788	6.535	+0.113	5.89	0.375	7.401	4.613
0.52	0.932	1.000	0.744	14.660	6.507	+0.013	6.21	0.330	7.692	4.628
0.54	0.943	1.000	0.744	14.531	6.480	−0.089	6.47	0.290	7.989	4.641
0.56	0.953	1.000	0.744	14.403	6.452	−0.189	6.83	0.254	8.291	4.653
0.58	0.961	1.000	0.744	14.274	6.424	−0.290	7.19	0.222	8.602	4.662
0.60	0.967	1.000	0.744	14.144	6.397	−0.393	7.48	0.193	8.926	4.669
0.62	0.973	1.000	0.744	14.015	6.369	−0.494	7.87	0.167	9.261	4.675
0.64	0.979	1.000	0.744	13.885	6.341	−0.596	8.27	0.145	9.612	4.679
0.66	0.982	1.000	0.744	13.755	6.313	−0.698	8.69	0.125	9.982	4.679
0.68	0.985	1.000	0.744	13.622	6.285	−0.803	9.08	0.107	10.33	4.676
0.70	0.988	1.000	0.744	13.489	6.256	−0.907	9.59	0.092	10.79	4.667
0.72	0.989	1.000	0.744	13.355	6.228	−1.013	10.01	0.078	11.24	4.648
0.74	0.992	1.000	0.744	13.218	6.198	−1.120	10.60	0.066	11.73	4.611
0.76	0.994	1.000	0.744	13.079	6.168	−1.229	11.17	0.056	12.27	4.547
0.78	0.995	1.000	0.744	12.937	6.136	−1.339	11.96	0.047	12.88	4.427
0.80	0.996	1.000	0.744	12.792	6.103	−1.451	12.86	0.039	13.57	4.199
0.82	0.997	1.000	0.744	12.642	6.056	−1.563	14.40	0.032	14.47	3.790
0.84	0.998	1.000	0.744	12.484	6.017	−1.673	17.53	0.027	15.77	2.905
0.86	0.998	1.000	0.744	12.312	5.947	−1.775	conv.	0.023	17.84	2.500
0.88	0.999	1.000	0.744	12.119	5.870	−1.891	conv.	0.018	20.83	2.500
0.90	0.999	1.000	0.744	11.898	5.782	−2.024	conv.	0.015	24.97	2.500
0.92	1.000	1.000	0.744	11.631	5.675	−2.184	conv.	0.011	31.25	2.500
0.94	1.000	1.000	0.744	11.297	5.541	−2.384	conv.	0.006	41.60	2.500
0.96	1.000	1.000	0.744	10.832	5.355	−2.663	conv.	0.003	62.33	2.500
0.98	1.000	1.000	0.744	10.063	5.046	−3.123	conv.	0.000	125.0	2.500
1.00	1.000	1.000	0.744	———	———	———	conv.	0.000	∞	2.500

SOURCE: M. Schwarzschild, *The Structure and Evolution of the Stars* (Princeton: Princeton University Press, 1958).

volume on a density fluctuation $\Delta\rho$ is

$$f = -g\Delta\rho$$

per unit volume. Thus if the actual density gradient in the star is less than that which the element would experience in rising adiabatically, the star will be unstable against convection. Since we have assumed the pressure of the element and the surroundings to be the same, we could also state this instability criterion by saying that the actual temperature gradient should exceed the adiabatic temperature gradient that would be experienced by an element rising adiabatically; that is, we have instability if

$$\left|\frac{dT}{dr}\right|_{\text{rad}} > \left|\frac{dT}{dr}\right|_{\text{ad}} \tag{5.124}$$

or

$$\left|\frac{d\ln T}{dr}\right|_{\text{rad}} > \left|\frac{d\ln T}{dr}\right|_{\text{ad}} \tag{5.125}$$

or, dividing through by $d \ln P/dr$,

$$\left|\frac{d\ln T}{d\ln P}\right|_{\text{rad}} > \left|\frac{d\ln T}{d\ln P}\right|_{\text{ad}} \tag{5.126}$$

or finally

$$\left|\frac{d\ln P}{d\ln T}\right|_{\text{rad}} < \left|\frac{d\ln P}{d\ln T}\right|_{\text{ad}}. \tag{5.127}$$

From the adiabatic relation,

$$PT^{\gamma/(1-\gamma)} = \text{constant}, \tag{5.128}$$

we see that the gradient $d \ln P/d \ln T$ is given by

$$\left|\frac{d\ln P}{d\ln T}\right| = \frac{\gamma}{\gamma - 1} = n + 1. \tag{5.129}$$

Thus the instability criterion can be written

$$\left|\frac{d\ln P}{d\ln T}\right| < \frac{\gamma}{\gamma - 1}. \tag{5.130}$$

Hence the ratio of specific heats, γ, is an important quantity in determining the stability of a star against convection. Clearly convective instability is likely to occur when (a) the opacity is high, (b) the luminosity is high in the deep stellar interior, where r is small (i.e., the central regions of bright stars), or (c) the value of $n = 1/\gamma - 1$ is high, i.e., γ is close to unity.* For a perfect gas,

$$c_P - c_V = \frac{k}{m_H} \tag{5.131}$$

and

$$\frac{\gamma}{\gamma - 1} = \frac{c_P}{c_P - c_V} = \frac{c_P}{k} m_H. \tag{5.132}$$

Hence $d \ln P/d \ln T$ is high (leading to instability) when c_P is high. For a classical monatomic gas, $\gamma = 5/3$ and $\gamma/(\gamma - 1) = 2.5$. It is sometimes easier to think about the physical adiabatic temperature gradient, rather than the dimensionless logarithmic gradient $d \ln P/d \ln T$. The adiabatic gradient is the one followed if the P, ρ, T relation is adiabatic. Writing $P = K\,T^{n+1}$, where K is a constant, we have

$$\rho = \frac{K\mu m_H}{k} T^n,$$

and so the equation of hydrostatic equilibrium requires that

$$K(n + 1)\,T^n \frac{dT}{dr} = -\frac{g\mu m_H}{k} KT^n. \tag{5.133}$$

(This result is correct anywhere in the star, not just where g is constant.) The adiabatic gradient is therefore

$$\begin{aligned}\left(\frac{dT}{dr}\right)_{\text{ad}} &= -\frac{\mu m_H}{k}\frac{g}{n+1} = -\frac{\mu m_H g}{k}\frac{\gamma - 1}{\gamma} \\ &= -\frac{\mu m_H g}{k}\frac{c_P - c_V}{c_P} \\ &= -\frac{g}{c_P}.\end{aligned} \tag{5.134}$$

Equation (5.134) is particularly instructive in showing how a high specific heat leads to a low adiabatic

*This can occur in regions where ionization takes place, or in protostars in regions of molecular dissociation.

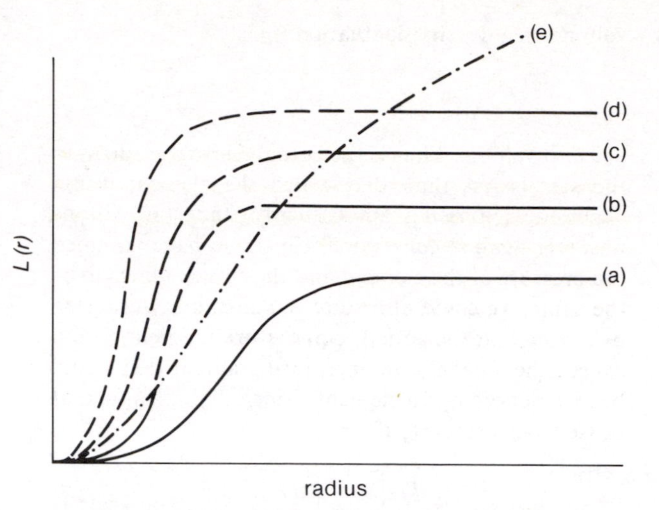

Figure 5.9. Luminosity versus radius (**a** – **d**) and the maximum luminosity versus radius that can be sustained by radiative transport (curve **e**). Dashed segments correspond to regions that are convectively unstable.

gradient, and hence to convective instability. Another way to interpret the stability criterion for radiative transport is to examine the maximum luminosity $L_{\max}$ at a point r that can be attributed to radiative transport. Combining (5.122) and (2.19) with the equality $(dT/dr)_{\rm rad} = (dT/dr)_{\rm ad}$ and using the equation of hydrostatic equilibrium yields the relation

$$L_{\max}(r) = 1.23 \times 10^{-8} \mu \frac{m(r)}{\bar{\kappa}} \left(\frac{T^3}{\rho}\right)$$

$$= \frac{m_H}{k} \frac{16\pi acG}{3} \mu \frac{m(r)}{\bar{\kappa}} \left(\frac{T^3}{\rho}\right). \quad (5.135)$$

Radiative transport is stable as long as $L(r) < L_{\max}(r)$. Several useful conclusions follow from (5.135). For simplicity, consider those cases in which T^3/ρ is strictly constant throughout the star. In the envelope $m(r) \simeq M$ and $L_{\max}$ is proportional to $\bar{\kappa}^{-1}$. When the specific opacity $\bar{\kappa}$ is large (as in ionization zones), $L_{\max}$ becomes small and convection may take over.

Figure 5.9 shows several schematic luminosity profiles $L(r)$ and a possible form of $L_{\max}(r)$ as (curve **e**). Curve (**a**) represents a completely radiative star; (**b**) a star with a radiative core and outer envelope with an intermediate convective zone; (**c**) a star with a convective core and radiative envelope.

The instability criterion requires that the real temperature gradient exceed the adiabatic temperature gradient. The difference is called the superadiabatic gradient,

$$\frac{d\Delta T}{dr} \equiv \frac{dT}{dr} - \left(\frac{dT}{dr}\right)_{\rm ad}.$$

We now consider the relative amounts of energy carried by convection and by radiation when a region becomes convectively unstable. Strictly speaking, the approximation we have made to adiabatic convection demands that there be very little additional transfer of energy by radiation, otherwise the convection would, by definition, not be adiabatic. We will try to show that, at least in the deep interior of a star, convection is an extremely efficient means of energy transport, so efficient, in fact, that when convection does take place, we can ignore the superadiabatic gradient and take the adiabatic gradient as the actual temperature gradient existing in the star. Unfortunately, this is not quite true for the outer layers of the star, and an adequate theory of convection in stellar atmospheres is still lacking.

We will limit our discussion to a relatively simple (but useful) model of convective energy transport known as mixing-length theory. In this model, convective elements are considered to have characteristic sizes $\Lambda(r)$ at each point r within a convective zone (Figure 5.10). The flow of one element (a rising one, for example) past nearby elements (presumably descending) produces unstable growth of any irregularities in the element's shape, which leads to turbulence and eventual mixing of the rising element with its surroundings. Thus the hotter elements rise, traveling a characteristic distance of order Λ before mixing into the surrounding material, whereas the cooler ones (shown shaded) sink about the same distance before mixing in a similar way into their surroundings. Clearly in a convective zone the density and temperature will actually depend on angle (for fixed r). How-

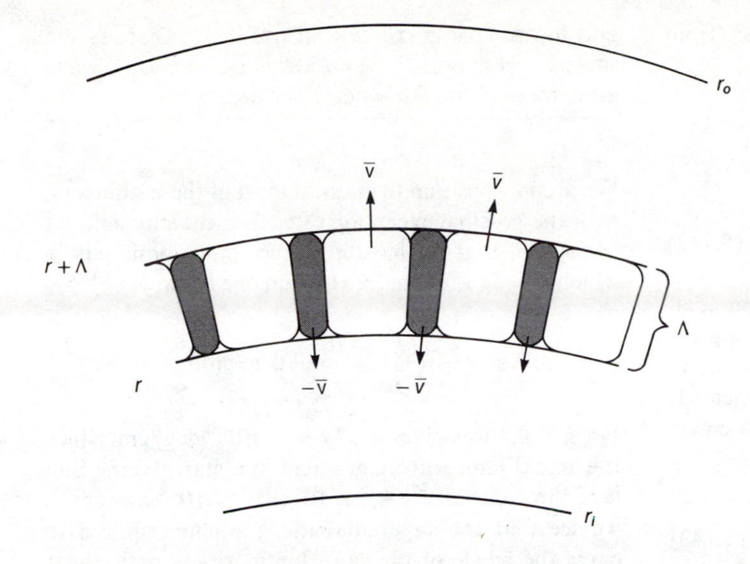

Figure 5.10. Shell of convectively unstable elements at r in a convective zone of inner radius r_i and outer radius r_0. The velocity of the elements (averaged over Λ) is $\bar{v}$. The shaded elements are sinking; the open elements are rising.

ever, we may define average values of these quantities (averaged over the shell of thickness Λ at r). Thus an element at the average density ρ and temperature T of a shell of thickness Λ in a convective zone will be stable. In fact, the stability analysis developed above may be thought of as comparing the temperature and density within an element to these average values.

If a turbulent element of material is rising, it is less dense than its surroundings and therefore at a higher temperature (since $P \sim \rho T$ and it maintains pressure equilibrium). It therefore transports thermal energy upward. If it has a temperature excess ΔT relative to its surroundings, it has a density deficiency

$$\Delta\rho \simeq -\frac{\rho}{T}\Delta T \tag{5.136}$$

and is thus subject to a buoyancy force per unit volume:

$$f = g\,\Delta\rho = \frac{g\rho}{T}\Delta T. \tag{5.137}$$

The equation of motion of this rising fluid element is

$$\frac{d^2r}{dt^2} = g\frac{\Delta T}{T}. \tag{5.138}$$

If we put in some numbers typical for the solar interior [$g \approx \frac{1}{2} g_\odot$ (surface) $\simeq 10^4$ cm/sec^2 and $T \approx 10^7$ K] and for calculation purposes suppose $\Delta T \simeq 1$ K, then the time it takes a convective element to travel a fraction α of the solar radius (we set $\alpha R_\odot \sim \Lambda$) is roughly

$$t \sim \left(\frac{\alpha R_\odot T}{g\,\Delta T}\right)^{1/2} \simeq \alpha^{1/2} \times 10^7 \text{ sec}$$
$$\simeq \frac{\alpha^{1/2}}{3} \text{yrs.} \tag{5.139}$$

The value of α is not well-known, since we do not know the size of the convection cells; that is, we do not know how far the element moves before it mixes with its surroundings. Assuming $\alpha \simeq 1/10$,* then the time-scale for the convection, for a one-degree temperature excess, is about a month. This is, as we will see, a rather rapid rate of energy transport.

The convective energy flux is given by the product of the thermal energy per gram carried by the fluid element moving at constant pressure, $c_P\,\Delta T$, times the mass flux $\frac{1}{2}\,\rho\,\bar{v}$ (where the factor $\frac{1}{2}$ appears because only half the matter moves outward at any one time),

$$\mathcal{F}_{\text{conv}} = \frac{1}{2}\rho\bar{v}\,c_P\,\Delta T, \text{ erg cm}^{-2}\text{ sec}^{-1}, \tag{5.140}$$

where the averaging is carried out over the vertical components of the velocities of the turbulent elements.

*This is characteristic of stellar cores. In the outer parts of a star, a value proportional to the pressure scale height $(d \ln P/dr)^{-1}$ is usually used.

A crude estimate of a typical velocity is (from (5.138))

$$\bar{v} = (g\alpha R_\odot \Delta T/T)^{1/2}. \qquad (5.141)$$

The convective energy flux is therefore

$$\mathcal{F}_{\text{conv}} = \tfrac{1}{2}\,(g\alpha R_\odot \Delta T/T)^{1/2}\,\rho\, c_P \Delta T. \qquad (5.142)$$

This equation can be written in terms of the superadiabatic temperature gradient, since only if the temperature gradient really is superadiabatic will there actually be a temperature excess in the rising element. If the superadiabatic gradient is $d(\Delta T)/dr$, then we may write

$$\Delta T \simeq \alpha R_\odot \frac{d\,\Delta T}{dr} \qquad (5.143)$$

and

$$\mathcal{F}_{\text{conv}} = \frac{1}{2}\rho\, c_P \left(\frac{g}{T}\right)^{1/2} (\alpha R_\odot)^2 \left(\frac{d\,\Delta T}{dr}\right)^{3/2}. \qquad (5.144)$$

Now we ask ourselves what superadiabatic gradient would be needed if the convective flux were to supply all the energy transport of the star. That is, we set the left side equal to $L/4\pi r^2$, and we get

$$L(r) = \frac{4\pi r^2}{2} c_P\, \rho \left(\frac{g}{T}\right)^{1/2} (\alpha R_\odot)^2 \left(\frac{d\,\Delta T}{dr}\right)^{3/2}. \qquad (5.145)$$

Problem 5.20. The efficiency of convective energy transport Γ_c may be defined as the ratio of a convective element's excess heat content (relative to its surroundings just before it mixes with its surroundings) to the energy it radiates during its lifetime. Show that (apart from numerical factors)

$$\Gamma_c \simeq \frac{c_P}{6ac}\frac{\bar{k}\,\bar{v}\,\alpha\, R}{T^3}.$$

Evaluate this for the solar core, assuming $\bar{k} \simeq 10^4$ cm^{-1}.

Problem 5.21. When convection occurs in stars, part of the energy flux, $\mathcal{F}_c$, is carried by convection, and part, $\mathcal{F}_R$, is carried by radiative transport. Express the fractional flux carried by convection, $\mathcal{F}_c/(\mathcal{F}_c + \mathcal{F}_R)$, in terms of the convective efficiency, Γ_c, and the gradients dT/dr and $(dT/dr)_{\text{ad}}$. Discuss the limits $\Gamma_c \to 0$ and $\Gamma_c \to \infty$. Does convection always carry most of the flux when Γ_c is large?

We are in a position to guess at most of these numbers, with the possible exception of c_P. For the moment, let us assume that we have a perfect monatomic gas in which $\gamma = 5/3$. Then, for the Sun one finds

$$\left(\frac{d\,\Delta T}{dr}\right)_\odot \simeq \frac{10^{-18}}{\alpha}\ \text{deg/cm}.$$

For $\alpha \simeq 0.1$ this gives $d\,\Delta T/dr \simeq 10^{-10}$ deg/cm. Since the actual temperature gradient in a star like the Sun is of the order of $T_0/R_\odot \simeq 10^7/10^{11} = 10^{-4}$ deg cm^{-1}, we see that the superadiabatic gradient required to carry the whole of the sun's luminosity is only about one millionth of the total gradient. Thus the real temperature gradient is extremely close to the adiabatic one, even when all the energy is being carried by convection. In stellar interiors, then, it is a very good approximation to say that, if convection takes place, the temperature gradient is equal to the adiabatic one. We may then replace (5.122) by the relation

$$\frac{dT}{dr} = \frac{\gamma - 1}{\gamma}\frac{T}{P}\frac{dP}{dr}, \qquad (5.146)$$

which follows immediately from the relation $P = KT^{\gamma/(\gamma-1)}$ with K a constant.

A major source of uncertainty in the preceding analysis is the quantity α, or αR, the length of the path of the turbulent element before it is destroyed or loses its identity. This characteristic distance is usually called the *mixing length,* and the rather simpleminded theory we have been using is called the *mixing-length theory*. It is obviously not an adequate theory, since, among other things, it does not predict a value for the mixing length, which has to be guessed at. In addition, if convection takes place near the surface of a star, one finds that a larger excess of the real temperature gradient over the adiabatic gradient is required, since convection near stellar surfaces usually occurs because the specific heat is high (this happens in the region of the star where, say, hydrogen is partially ionized, that is, where an input of heat energy goes almost entirely into ionization energy and very little into raising the temperature). When the specific heat is high, the adiabatic gradient is low; so convection, by itself, is relatively less efficient at transporting energy.

In addition, the mixing length is presumably not significantly greater than the local scale height, and the velocity appropriate to the turbulent convection is usually taken to be limited by the velocity of sound, $c_s = (\gamma P/\rho)^{1/2}$; otherwise supersonic motion would result, and shock waves would develop. There is some evidence that the solar corona is in fact heated by shock waves generated in the convective zone near the surface of the Sun.

In principle, a star could be wholly convective, that is, could have convective energy transfer throughout. In the region where the hydrogen is fully ionized, such a star would behave very much like a classical, monatomic, perfect gas with $\gamma = 5/3$, and would have the structure of an $n = 1.5$ polytrope. In the outermost layers of such a star, there must be radiative energy transfer. Therefore the results of the $n = 1.5$ polytrope cannot be carried quite to the surface of such a star. We will see later how to handle this surface condition. Stars that are gravitationally bound, but have not yet reached the ignition temperature for hydrogen (around 10^7 K), are probably close to being wholly convective, because they are very cool and contain appreciable zones of high heat capacity where H_2 is being dissociated, and H is being ionized. Such stars are contracting onto the main sequence.

Conduction

The thermal conductivity λ_c of a medium is defined by the equation

$$\text{heat flow} = \lambda_c \, dT/dr$$

or

$$\mathcal{F}_{\text{cond}} = -\lambda_c \nabla T, \tag{5.147}$$

and has units erg cm^{-2} sec^{-1} (deg cm^{-1})$^{-1}$. In ordinary circumstances, the thermal conductivity of a gas is negligible compared to the radiative conductivity. However, in the very highly compressed cores of some stars, particularly in late stages of evolution, the density may be high enough that significant heat conduction takes place. We note that a (specific) conductive opacity may be introduced by defining

$$\bar{\kappa}_c = \frac{4acT^3}{3\rho\lambda_c} \text{ cm}^2/\text{g}, \tag{5.148}$$

so that the conductive heat flow is

$$\mathcal{F}_{\text{cond}} = -\frac{4ac}{3\bar{\kappa}_c\rho} T^3 \frac{dT}{dr}. \tag{5.149}$$

This means that the conductive flux can be formally combined with the radiative flux if a total (specific) opacity is defined by

$$\frac{1}{\kappa_{\text{total}}} = \frac{1}{\bar{\kappa}_c} + \frac{1}{\bar{\kappa}_{\text{rad}}}. \tag{5.150}$$

This can also be written in the form

$$\lambda_{\text{total}} = \lambda_c + \lambda_{\text{rad}},$$

where the radiative conductivity λ_{rad} is

$$\lambda_{\text{rad}} = \frac{4ac}{3\rho} \frac{T^3}{\bar{\kappa}_{\text{rad}}}. \tag{5.151}$$

Circulation

We remarked in section 4.2 that in a rotating star (with uniform rotation) or in a component of a binary that is tidally affected by its companion, we do not have spherical symmetry; nevertheless, the physical state of the material must be constant on equipotential surfaces. In particular, the temperature is constant on equipotential surfaces, $T = T(\phi)$. The radiative flux is thus given by (5.60):

$$\pi F = -\frac{16ac}{3\bar{\kappa}\rho} T^3 \frac{dT}{dr} = -\frac{16ac}{3\bar{\kappa}\rho} T^3 \frac{dT}{d\phi} \nabla\phi$$
$$= \pi f(\phi) \nabla\phi, \tag{5.152}$$

where $f(\phi)$ is a function only of ϕ. Thus the radiative flux is parallel to the potential gradient. Now let us try to apply the condition of radiative equilibrium, $\nabla \cdot \mathbf{F} = 0$. We have

$$\nabla \cdot \mathbf{F} = \nabla \cdot (f \nabla\phi) = \nabla f \cdot \nabla\phi + f \nabla^2\phi$$
$$= \frac{df}{d\phi} |\nabla\phi|^2 + f(\phi) \nabla^2\phi. \tag{5.153}$$

For a star in a binary system, the potential is purely gravitational; so $\nabla^2\phi$ is given by Poisson's law

$$\nabla^2\phi = 4\pi G\rho. \tag{5.154}$$

If there is also a rotational component to the potential,

then one easily finds

$$\nabla^2\phi = -2\omega^2 = \text{constant}.$$

Thus, in both cases, $\nabla^2\phi$ is proportional to physical variables and is constant on equipotential surfaces. However, it is clear that $\nabla\phi$ is not constant on equipotential surfaces, for the potential gradient is greater near the poles than near the equator, for example. It is therefore, in general, impossible for the left-hand side of (5.153) to be zero, for a quantity constant on equipotentials cannot be equal to a quantity not constant on equipotentials. Hence a nonspherical star cannot be in radiative equilibrium.

What must happen instead is that radiative equilibrium holds on average, over the equipotential, but in various local regions the radiative flux is more or less than that required for equilibrium. Because the equipotentials are closer together along the rotation axis than toward the equator, the heat flux toward the poles is greater than it would be in the absence of rotation, and the flux toward the equator less. As a result, along the polar axis material heats up, its density decreases, and it begins to rise. Conversely, near the equator material is inadequately heated, becomes more dense than its surroundings, and begins to sink. Thus circulation currents are set up, as sketched in Figure 5.11. This process is sometimes called *meridional circulation.* It has the potentially important consequence, not only of significant energy transfer by means of the circulation currents, but also of rotational mixing.

The normal pattern of stellar evolution, as we remarked earlier and will see in more detail later, is the development of chemical inhomogeneities by means of thermonuclear reactions in the interior, especially $H \rightarrow He$. In principle, the circulation currents set up in a rotating star can remove these inhomogeneities, and hence presumably can cause a significant difference between stellar structure and evolution in rotating stars from that in normal stars. The same remarks, of course, apply to stars in binary systems where there is enough tidal distortion to make the stars nonspherical. In addition, because of the centrifugal force of rotation, one expects the effective gravitational field to be reduced in the rotating star, so that interior pressures, and hence temperatures, will be less than for the corresponding nonrotating stars. This will affect the luminosities and the surface temperatures of the stars.

Estimates for the circulation velocity in the Sun,

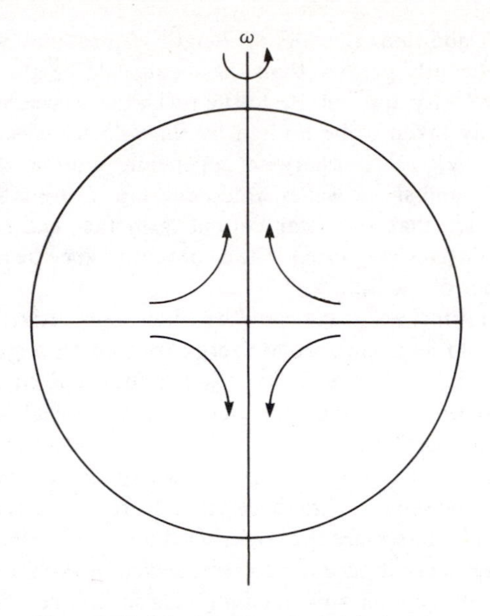

Figure 5.11. Meridional circulation in a rotating star. Material sinks along the equator and rises along the rotation axis.

which is not a particularly fast rotator (period = 27 days), have been given at about 10^{-9} cm sec^{-1}. At this speed, the time-scale for a complete circular path is of the order of $R/10^{-9} = 10^{20}$ sec $= 3 \times 10^{12}$ yrs. Since this is much longer than what we think the evolutionary time-scale of the Sun is (about 10^9 yrs), we may conclude that rotational mixing is not important in the Sun. However, many stars rotate much faster than the Sun, and we cannot rule out rotational mixing for them.

No completely satisfactory treatment of problems of rotational mixing (or of binary-induced mixing) seems to exist. The best attitude seems to be to point to the circulation and mixing problems as reasons for suspecting that the structure and evolution of binary and rotating stars may not be the same as for single stars. Unfortunately, most of our knowledge of fundamental stellar parameters, particularly masses, comes from binary stars. Furthermore, a very large percentage of the stars in the sky, perhaps as many as 50 percent, are binary. We can only hope that if we stick to binary systems in which the tidal distortion seems to be small, or to stars with low rotation velocities, then the spherically symmetric approximation should be good.

Chapter 6

ATOMIC PROPERTIES OF MATTER

In the preceding chapters we have established much of the general framework for the study of stellar interiors and atmospheres. The cornerstones of this framework are the concepts of hydrostatic equilibrium, radiative transfer and radiative equilibrium, thermonuclear energy generation (to be introduced in Chapter 7), and the complications of convective and conductive energy transfer where applicable. These concepts describe stellar matter from the macroscopic viewpoint. Within this framework there appear some constitutive equations that formally express the essential physical properties of the stellar material. These fundamental physical properties, which are microscopic in origin, are

(1) equation of state [$P = P(\rho, T)$],
(2) opacity,
(3) energy generation rates,
(4) specific heats, and
(5) thermal conductivity.

To these we might add

(6) nucleosynthesis,

by which we mean how the chemical composition of the star will change as a result of nuclear reactions. The chemical composition appears as a parameter in the other quantities. We now begin a study of these properties of stellar material. We consider first the atomic properties of stellar matter that affect energy transport.

6.1. The Hydrogen Atom

Hydrogen is the most abundant element in the universe, by a large factor, and makes up the largest part of interstellar matter and all normal stars. The physical properties of stellar material are therefore dominated by the properties of the hydrogen atom. In this section we will emphasize the electronic structure of hydrogen, the energy levels, and the associated spectrum, and will introduce the language of astrophysical spectroscopy.

The principal energy levels of the hydrogen atom can be found from semiclassical arguments applied to an electron moving in a circular orbit of radius a around a proton. Classically, the Coulomb force is balanced by the centripetal force:

$$\frac{mv^2}{a} = \frac{e^2}{a^2},$$

where m is the reduced mass of the system. The action integral $\oint p\,dq$ can have any value in a classical system, but for bound quantum systems it must equal some multiple of Planck's constant. Quantization of the action integral is equivalent to the requirement that a bound system can only change its state in units of h. For the hydrogen atom the generalized momentum p is the angular momentum, and $dq = d\phi$; thus the action integral becomes

$$\oint p\,dq = mva \int_0^{2\pi} d\phi = 2\pi mva = nh,$$

or

$$mva = n\hbar,$$

where n is a positive integer. Eliminating the velocity between the equations above gives the orbital radius

$$a_n = \frac{n^2\hbar^2}{me^2} = n^2 a_0, \qquad n = 1, 2, \ldots$$

where a_0 is the Bohr radius for hydrogen. The total energy of the bound pair of charges is one-half the potential energy, or, using the results above

$$E_n = -\frac{e^2}{2a_n} = -\frac{me^4}{2n^2\hbar^2}$$

Note that $m \approx m_e$ for most purposes. In the absence of external forces, there are $2n^2$ states with energy E_n.

In the absence of an external magnetic field, the energy levels of the single electron in the hydrogen atom are well represented by the formula

$$E_n = -Z^2\frac{m_e e^4}{2n^2\hbar^2} \equiv -\frac{Z^2\mathcal{R}}{n^2}. \tag{6.1}$$

Here Z is the atomic charge on the nucleus in units of the charge of the electron. This is unity for the hydrogen atom, but we include Z explicitly because many ions, such as He^+, have largely hydrogen-like spectra, which (6.1) describes with good accuracy. Here n is the principal quantum number, and is used to label the energy levels. The energy given by (6.1) is negative, going to zero as $n \rightarrow \infty$. E_n represents the binding energy of the electron to the ion. The most tightly bound state (the ground state) and the lowest energy level occur for $n = 1$, and have a binding energy

$$\begin{aligned} -E_1 &= \frac{1}{2}\frac{Z^2 m_e e^4}{\hbar^2} \\ &= 2.18 \times 10^{-11}\, Z^2 \text{ ergs} \\ &= 13.6\, Z^2 \text{ eV} = \mathcal{R}Z^2. \end{aligned} \tag{6.2}$$

The quantity $\mathcal{R}$ is the Rydberg constant. The state $n = 1$ is the ground state, and $13.6\,Z^2$ volts is the ionization potential (or $13.6\,Z^2$ eV is the ionization energy) from the ground state. The negative energy levels corresponding to bound states for hydrogen are shown in Figure 6.1. Positive energies correspond to a free electron. However, such an electron moves in the electromagnetic field of the proton; so the free energy states must be considered along with the bound energy states as part of the total energy structure of the hydrogen atom. The energy levels of the free electron are not quantized, and any energy is permitted. These levels are sometimes called the energy continuum. Transitions between energy levels are accompanied by the emission ($E_m > E_n$) or absorption ($E_n > E_m$) of a photon of frequency ν, and its energy $h\nu$ is given by the usual Planck relation

$$h\nu = |E_m - E_n|. \tag{6.3}$$

The spacing of energy levels in the hydrogen atom is such that transitions with the same lower level give rise to photons in the same general region of the spectrum. Thus, the spectrum of hydrogen is classified according to series of transitions, all having the same lower level n. The most important of these are shown in Box 6.1.

Box 6.1
Transition series.

Lower level series	Principal transition
$n = 1$, Lyman	$L_\alpha = 1216$ Å;
$n = 2$, Balmer	$H_\alpha = 6563$ Å;
$n = 3$, Paschen	$P_\alpha = 18750$ Å.

The series are designated L, H, and P respectively, and the α notation indicates the first of the series. That is H_α is the transition between the levels $n = 3$ and $n = 2$; H_β is the transition between the levels $n = 4$ and $n = 2$; and so on. The Balmer series is the only one lying in the visible part of the spectrum, and is the characteristic spectrum of hydrogen seen in laboratory discharge

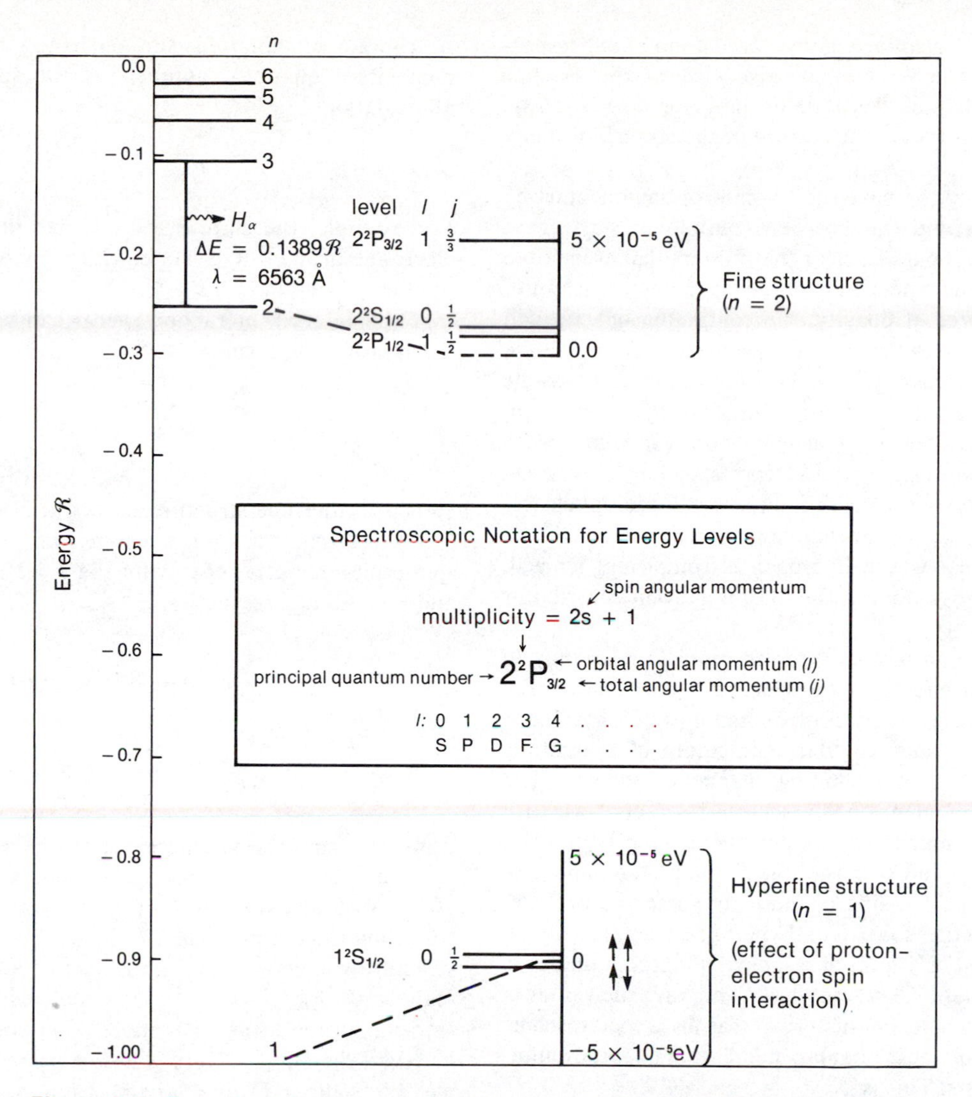

Figure 6.1. Energy levels in atomic hydrogen. The upper-right inset shows the fine structure for the $n = 2$ state. The Lamb shift between the $2\,^2S_{1/2}$ and $2\,^2P_{1/2}$ states corresponds to $\Delta E = 4.4 \times 10^{-6}$ eV. The lower inset shows the hyperfine structure of the $n = 1$ state. The energy difference between the two states is $\Delta E = 5.8 \times 10^{-6}$ eV. The photon emitted by a transition between these two states has $\nu = \Delta E/h = 1420$ MHz, corresponding to $\lambda = 21.11$ cm.

tubes and in the spectra of most stars. The first few lines in the Balmer series are

$$
\begin{aligned}
&H_\alpha, && \lambda = 6563\ \text{Å},\\
&H_\beta, && \lambda = 4861\ \text{Å},\\
&H_\gamma, && \lambda = 4340\ \text{Å},\\
&H_\delta, && \lambda = 4101\ \text{Å}.
\end{aligned}
$$

These transitions give rise to absorption or emission of photons at discrete wavelengths, and thus correspond to absorption or emission lines in the spectrum of hydrogen. A transition from $n = 2$ to the continuum, however, produces absorption or emission over a wide range of wavelengths, starting at the wavelength corresponding to the ionization energy from the $n = 2$ level. This is 3.4 eV, and the corresponding wavelength is 3647 Å. Thus absorption into the continuum of energy states produces a continuous opacity at wavelengths shorter than 3647 Å. Correspondingly, a transition from the continuum to level 2 causes emission of a photon of wavelength shorter than 3647 Å, i.e., continuous emission. These two processes are often also called *photoionization* or *bound-free absorption,* and (radiative) *recapture* or *free-bound emission,* respectively. For $n = 2$, they give rise to the Balmer continuum, and the Balmer jump or discontinuity, which is an important feature of the spectra of many stars.

Similarly, we have the Lyman continuum, starting at 912 Å, and the Paschen continuum, starting at 8204 Å. It should be clear that in stars that are not too hot, and where neutral hydrogen is therefore an appreciable source of opacity, the continuum opacity will vary quite irregularly with wavelength, as we will see when we discuss opacity. Transitions are also possible that involve only the continuum of the hydrogen atom. These are called *free-free transitions* (or *Bremsstrahlung* transitions), and the corresponding absorption *free-free opacity,* which is, as we will see, much less frequency-dependent than bound-free opacity. In stellar interiors, where hydrogen is completely ionized, free-free absorption is the most important contributor to the opacity.

The energy levels expressed by (6.1) refer only to the orbital energy of the electron in the field of the proton. However, an electron and a proton both have an intrinsic spin angular momentum of magnitude $\hbar/2$, and a corresponding magnetic moment. The magnetic moment of the electron interacts both with the orbital magnetic moment of the atom (spin-orbit interaction) and with the spin of the proton (spin-spin interaction). The first of these gives rise to fine structure, and the second to hyperfine structure, in the spectrum. For a level of principal quantum number n, the orbital quantum number, l, can have the values 0, 1, 2, . . . , $n - 1$. For historical reasons, a spectroscopic notation for these angular momentum quantum numbers persists. This is

$$\text{S, P, D, F}, \ldots,$$

$$l = 0, 1, 2, 3, \ldots.$$

The total angular momentum corresponding to a state specified by a quantum number l is $[l(l+1)]^{1/2}\hbar$. The spin of the electron has a quantum number ½; so its spin angular momentum is $[s(s+1)]^{1/2}\,\hbar = \sqrt{3}\hbar/2$. The orbital angular momentum and the spin angular momentum interact to produce a total angular momentum quantum number j, given by $j = l \pm \frac{1}{2}$. Thus we have for the first few states

S states ($l = 0$)	$j = \frac{1}{2}$,
P states ($l = 1$)	$j = \frac{1}{2}, \frac{3}{2}$,
D states ($l = 2$)	$j = \frac{3}{2}, \frac{5}{2}$.

A convenient notation summarizing the angular momentum quantum numbers is the spectroscopic abbreviation

$$n^{2j+1}l_s.$$

For example, the state $3\,^2\text{S}_{1/2}$ is one in which the electron is in an $l = 0$ (S) state, with principal quantum number $n = 3$ and spin $s = \frac{1}{2}$.

The spin-orbit interaction energy can be written in the approximate form

$$\Delta E_{n,j} = \frac{\alpha^2 \mathcal{R} Z^4}{n^3}\left(\frac{3}{4n} - \frac{1}{j + \frac{1}{2}}\right). \qquad (6.4)$$

Here α is the fine-structure constant: $\alpha = e^2/\hbar c \approx (137)^{-1}$. From this, for example, we see that the spin-orbit interaction energy for the $n = 2, l = 1, j = \frac{1}{2}$ and $\frac{3}{2}$ states are, respectively,

$$\Delta E_{j=1/2} = -\frac{5}{64}\alpha^2 \mathcal{R} = -5.66 \times 10^{-5}\ \text{eV},$$

$$\Delta E_{j=3/2} = -\frac{1}{64}\alpha^2 \mathcal{R} = -1.1 \times 10^{-5}\ \text{eV}.$$

Thus, the spin-orbit interaction splits the $n = 2$ energy level into two levels, separated in energy by 4.5×10^{-5} eV. Since the value of $E_m - E_n$ for the H_α transition (for example) is 1.2 eV, we see that the fine structure will show up in the H line at a wavelength scale $\sim 10^{-5}\lambda_{H\alpha} = 0.06$ Å. It is thus visible only at rather high resolution. For more complicated atoms, however, the fine structure is easily visible; for example, the sodium D lines are about 6 Å apart. The difference arises because the valence electron in sodium is subject to a different (nonhydrogenic) potential, caused by the presence of other electrons in the atom.

The astrophysics of interstellar matter is strongly influenced by transitions in which the electron changes its spin orientation, even though these are of very low probability. For example, the transition from $2\,^2\text{S}_{1/2}$ to $1^2\text{S}_{1/2}$ would not change the orbital angular momentum of the electron. However, the photon emitted in the transition will carry off one unit of angular momentum. Since the total angular momentum must be conserved by the transition, the only way in which this could happen would be for the electron to flip its spin. This transition is "forbidden" (i.e., takes place with extremely low probability). A hydrogen atom in a

$2\,^2S_{1/2}$ state will typically remain in that state for quite a long time before it spontaneously decays to the lower state. At the densities existing in stars, an atom in the $2\,^2S_{1/2}$ state will collide with other particles many times before it can spontaneously decay. As a result of these collisions, excitation energy may be transferred to the colliding partner as kinetic energy. The result is known as *collisional de-excitation,* and the reverse process (where an atom is promoted to an excited state) is *collisional excitation.*

In the tenuous regions of interstellar space, however, or in low-density gaseous nebulae, collisions may be rather infrequent; and the most probable process will then be a two-photon emission whose transition rate is about 10 $\sec^{-1}$ (compared to 10^8 $\sec^{-1}$ for normal "allowed" transitions). This two-photon emission is an important source of continuous radiation in gaseous nebulae at wavelengths longer than Lyman α. Atomic energy levels such as this one, which have no high-probability transition out of them are called *metastable levels,* and are very important in conditions where extremely low densities prevent collisional de-excitation.

The hyperfine structure that arises from the interaction of the electron's spin with the photon's spin produces a splitting of the lowest ($1\,^2S_{1/2}$) energy level in hydrogen by an amount $\Delta E = 5.9 \times 10^{-6}$ eV. A transition between these two energy levels is highly forbidden, because the electron spin would have to change direction. Its probability is so low that the lifetime of the hydrogen atom in this level is around 1.1×10^7 years. Nevertheless, because of the large amount of low-density hydrogen in our Galaxy and in neighboring galaxies, this transition can be easily observed by radio astronomers at a frequency of 1420 MHz, or a wavelength of 21 cm. This transition is used extensively to map the density of neutral hydrogen in regions in the Galaxy and to investigate Galactic disk structure.

A neutral hydrogen atom has a significant electromagnetic field in its vicinity, though it is not a Coulomb field. It is usually called the van der Waals field, and has approximately an r^{-3} radial dependence. This is of importance in two astrophysical areas.

First, if a stellar atmosphere is fairly cool, so that most of the hydrogen is neutral, the van der Waals field will be the most common interatomic field that, say, a sodium atom will experience. If the densities are reasonably high, so that typical interatomic distances are small, then this van der Waals field will perturb the energy levels in the sodium atom. Hence, statistically, the sodium atoms in the stellar atmosphere will not all have precisely the same energy levels, and thus the frequency or wavelength of the D lines will be slightly different for each atom. This will broaden the spectral line seen in the spectrum of the star. This is the most important type of line-broadening mechanism in cool stars (at least for most lines), and is usually called *pressure broadening.*

Second, the van der Waals field is attractive, and is strong enough that the neutral hydrogen atom can capture another electron and become a negative hydrogen ion. There is only one bound state in the H^- ion, with the small binding energy of 0.75 eV, and all optical transitions involve the continuum, and thus bound-free absorptions from this state give rise to continuous opacity at wavelengths shorter than 16,502 Å, the wavelength corresponding to 0.75 eV. Free-free transitions can also take place in the negative hydrogen ion, out to longer wavelengths. This is a very significant source of opacity in cool stars similar to the Sun. In fact, the continuous opacity in the solar atmosphere is believed to be mostly due to negative hydrogen, a fact only recognized within the last 35 years.

In the presence of magnetic fields, the energy-level structure is further split by the interaction of the magnetic moment of the orbital motion and the electron spin with the magnetic field. The quantum unit of magnetic moment is the Bohr magneton, which for the electron is

$$\mu_B \equiv \frac{e\hbar}{2m_e c} = 0.927 \times 10^{-20}\ \text{ergs/gauss} = 5.79 \times 10^{-9}\ \text{eV/gauss}, \qquad (6.5)$$

and the component of the magnetic moment in the direction of the magnetic field is described by the magnetic quantum number m. The interaction energy of a state of magnetic quantum number m with a magnetic field H is

$$E_H = \mu_B\, g\, m\, H, \qquad (6.6)$$

where g is the (dimensionless) Landé g factor, which is a function of the quantum numbers (j, s, l) of the state involved. In effect, g measures the ratio of the magnetic moment of the state to the angular momentum of the state. Magnetic fields encountered in astrophysics usually conform to the weak-field approximation, in which the magnetic interaction energy is less than the spin-orbit interaction energy. In this case, the possible values of the magnetic quantum number m are determined by the possible orientations of the total angular momentum, j, in the magnetic field. Thus m can have the values $-j, -j+1, \ldots, 0, \ldots, j-1, j$, that is, $2j+1$ values all together. The weak-field approximation gives rise to the *Zeeman effect*, where the Landé g factor is

$$g = 1 + \frac{j(j+1) + s(s+1) - l(l+1)}{2j(j+1)}.$$

For example, the Landé g factor for the 2 $P_{3/2}$ level of hydrogen is 4/3. This enables us to estimate what we mean by a weak field, for we want the magnetic interaction to be less than the spin-orbit interaction. Since the maximum magnetic interaction occurs when $m = j$, we can say that the field is weak if

$$\mu_B \frac{4}{3}\frac{3}{2} H < 4.5 \times 10^{-5} \text{ eV}$$

or $H < 3{,}900$ gauss. For other atoms, such as sodium, the fine-structure separation is greater than for hydrogen, and the weak-field limit for H is higher. This is a large magnetic field, although fields as high as 30,000 gauss have been detected in some peculiar A stars. Magnetic fields constitute yet one more "peculiarity" that must ultimately be taken into account by the theory of stellar structure. The Sun has only a very weak average magnetic field, but local fields attain values of a few thousand gauss in the vicinity of sunspots. The intense fields are, however, confined to narrow flux tubes, whose typical diameters are of order 100 km.

The full specification of an atomic energy level involves the four quantum numbers, n, l, s, m. The energy "level" described by the principal quantum number alone is thus a composite of several different levels that have almost the same energy. In the language of quantum mechanics, the level is degenerate. For example, the $n = 2$ level of hydrogen is made up of the following sublevels:

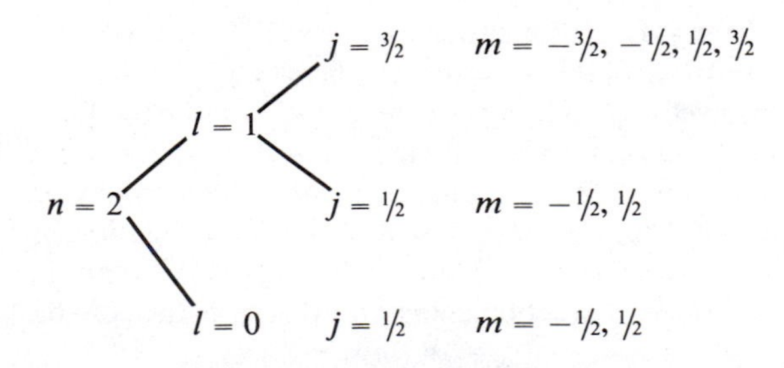

This gives a total of eight levels. One can show that generally the number of sublevels of principal quantum number n is $2n^2$. This is called the *statistical weight* of level n, denoted by g_n. Obviously the concept of statistical weight is applicable only when the energy differences between the sublevels can be neglected. In effect, the statistical weight of an energy level n is the number of distinct quantum states it contains.

When discussing the spectra of ions, it is conventional to denote the ionization state by giving the chemical symbol, followed by a roman numeral specifying the degree of ionization. The neutral state is denoted by roman numeral I. Thus OI, OII, and OIII refer to the neutral atom, the singly ionized ion, and the doubly ionized atom, respectively, of oxygen. An ion giving rise to forbidden transitions is enclosed in square brackets; for example, [NII].

6.2. Thermal Excitation and Ionization

The properties of stellar and interstellar matter are determined not only by the composition parameters, but also by their degree of ionization and state of excitation. These characteristics involve the atomic or molecular nature of matter, as discussed in the previous section, and the thermodynamic parameters (ρ, P, and T) of the gas.

Thermal ionization becomes significant when kT is comparable to the binding energy. In H (E_B = 13.6 eV), this occurs for $T \simeq 10^5$ K; for He (E_B = 24.6 eV) it occurs above 4×10^5 K. Excitation requires a few eV, corresponding to $T \simeq 10^4$ K. At lower temperatures, ionization or excitation will be incomplete, and the gas will contain a mixture of states. We now consider these effects in greater detail.

Designate the energy of an atomic level n by E_n and the number of atoms in this level by N_n. In general, E_n contains the ionization and excitation energy of the state. Using Boltzmann's law, and denoting the number of particles in the state n by g_n, we have

$$N_n = A\, e^{-E_n/kT} g_n. \tag{6.7}$$

The total number of atoms in all levels is then

$$N = \sum_{n=1}^{\infty} N_n = A \sum_{n=1}^{\infty} g_n e^{-E_n/kT} \equiv A\, Z(T), \tag{6.8}$$

where $Z(T)$ is called the partition function. Using (6.8) to eliminate the constant A in (6.7) gives

$$N_n = \frac{N}{Z(T)} g_n e^{-E_n/kT}, \tag{6.9}$$

$$Z(T) = \sum_n g_n e^{-E_n/kT}. \tag{6.10}$$

The energies E_n are positive, and are measured from the ground state $E_0 = 0$. Actually, $Z(T)$ as given by (6.10) is infinite, because $(E_n)_{max} = \chi_n$, the binding energy; and g_n increases with n (there are an infinite number of states from which the electron could be removed). This difficulty is removed in principle by observing that in a gas the presence of neighboring ions and electrons spreads the atomic energy levels over a narrow energy range ΔE. As a result, all levels whose separation from their neighboring levels is less than ΔE merge into the continuum, thereby removing from the discrete spectrum all levels with n greater than a finite value n_c. Another way of phrasing this is to say that interparticle interactions depress the continuum. Finding a value for n_c in practice may be quite complicated. Usually only the lowest states enter, and n_c is small.

The relative populations of two levels of the same atom follow directly from (6.9), as

$$\frac{N_n}{N_m} = \frac{g_n}{g_m} e^{-(E_n - E_m)/kT}. \tag{6.11}$$

The partition function cancels in this case.

Formula (6.11) has immediate application to stellar spectra. For example, consider the H_α absorption line. Since this transition is from the $n = 2$ level, we will have an appreciable number of H atoms in that level capable of absorbing H_α, if the Boltzmann factor $e^{-E_2/kT}$ is not negligibly small. Therefore we must have $kT \simeq E_2$, where E_2 is the excitation energy of the $n = 2$ level above the ground state. Since $E_2 = 10.2$ eV, and Boltzmann's constant is $k = 8.6167 \times 10^{-5}$ eV deg^{-1}, we have $T \simeq 1.2 \times 10^5$K. Since a temperature of 120,000 degrees is somewhat higher than we find in the atmosphere of the Sun, we do not expect to see very strong H_α absorption there, and indeed we do not, although it is certainly present. However, if we look at hotter stars, we do find the H_α absorption strength increasing. It reaches a maximum strength in the spectra of A stars such as Sirius, which have atmospheric temperatures in the neighborhood of 10,000 K. At higher temperatures, the strength declines again.

Problem 6.1. What fraction of the neutral hydrogen atoms are in the $n = 2$ level at a temperature of 10,000 K?

The decline in strength at higher temperatures is due to thermal ionization. To see how this happens, one can (in principle) apply Boltzmann's law to electrons in the continuum of energy levels, and obtain

$$\frac{N_i(v)}{N_n} = \frac{g(v)}{g_n} e^{-(\chi_n + 1/2 m_e v^2)/kT}$$
$$= \frac{g(v)}{g_n} e^{-\chi_n/kT} e^{-p^2/2m_e kT}. \tag{6.12}$$

Here $\chi_n = E_\infty - E_n$ is the ionization energy of the electron originally in the state n, $E_\infty + \frac{1}{2} m_e v^2$ is the energy of the continuum electron, $N_i(v)$ stands for the density of ions whose free electron has velocity v, and $g(v)$ is the statistical weight of the free electron of velocity v.

This last quantity is somewhat tricky to evaluate. To obtain it, one must go back to the foundations of the quantum mechanics of unbound systems, where one finds that the state of a free particle is specified by its three position coordinates, x, y, z, and its three momentum coordinates, p_x, p_y, p_z. According to quantum mechanics, the number of quantum states (i.e., states for which the statistical weight is unity) in an element of phase space $dx\, dy\, dz\, dp_x\, dp_y\, dp_z = dN$ is

$$\frac{dN}{h^3}, \tag{6.13}$$

which is the statistical weight of this element of phase space. In addition, the electron has two possible spin orientations; so the total statistical weight of this element of phase space for a free electron is

$$\frac{2\, dN}{h^3}.$$

The statistical weight per unit volume is then

$$\frac{2\, dp_x\, dp_y\, dp_z}{h^3}.$$

A complication now enters, in that there may already be some free electrons that have originated in other atoms, which occupy some of the available states in the element of phase space dN. The usual argument made in allowing for such electrons is that if the electron number density is n_e, then the effective volume available to a thermally ionized electron is reduced from unit volume to a volume $1/n_e$. Thus the statistical weight available to the "ionizing electron" is

$$g(v) = 2\frac{dp_x\, dp_y\, dp_z}{n_e\, h^3}$$

per unit volume. Therefore,

$$\frac{n_e N_i(v)}{N_n} = \frac{2}{h^3}\frac{e^{-\chi_n/kT}}{g_n} e^{-p^2/2m_e kT}\, dp_x\, dp_y\, dp_z. \quad (6.14)$$

We now integrate (6.14) over all values of the total momentum, using the variable change $p^2 = p_x^2 + p_y^2 + p_z^2$, and get

$$\frac{n_e N_i}{N_n} = \frac{2}{h^3}\frac{e^{-\chi_n/kT}}{g_n} 4\pi \int_0^\infty p^2 e^{-p^2/2m_e kT}\, dp$$

$$= \frac{2}{h^3}\frac{e^{-\chi_n/kT}}{g_n}(2\pi m_e kT)^{3/2}. \quad (6.15)$$

There is one more correction to take care of: the ion that remains may have its own statistical weight, corresponding to whatever state it is left in. Hence we multiply by a factor g_i, the statistical weight of the ion. We thus arrive at

$$\frac{n_e N_i}{N_n} = \frac{2g_i}{g_n}\frac{(2\pi m_e kT)^{3/2}}{h^3} e^{-\chi_n/kT}, \quad (6.16)$$

which is known as Saha's equation. In most applications, we take $n = 1$, so that we refer everything to the density of atoms in the ground state. For some purposes, however, it is more useful to have the total density of neutral atoms in all states, in which case we get essentially the same result with the partition function (measured from the ground state with $E_1 = 0$) replacing g_1. In a similar way, we can allow for the distribution of the ion over its possible energy states by writing the partition function of the ion instead of g_i. Thus we have, in general

$$\frac{n_e N_i}{N_0} = \frac{2Z_i}{Z_0}\frac{(2\pi m_e kT)^{3/2}}{h^3} e^{-\chi_0/kT}. \quad (6.17)$$

In practice, it is rarely necessary to worry about this distinction, since there will always be other, more significant things to worry about. Finally, another way of writing equation (6.16) is to multiply both sides by kT and use the electron pressure $P_e = n_e kT$ to get

$$\frac{N_i}{N_0} = \frac{2Z_i}{Z_0}\frac{(2\pi m_e)^{3/2}}{h^3}\frac{(kT)^{5/3}}{P_e} e^{-\chi_0/kT}. \quad (6.18)$$

We can learn several things of astrophysical significance from this equation. First, other things being equal, atoms with low ionization potentials will be relatively more ionized at any given temperature than will atoms of higher potential. To illustrate some of the consequences of this almost trivial remark, examine Table 6.1, which shows ionization potentials for the chemical elements. We see that, of the common elements, He and Ne are the hardest to ionize (χ in excess of 20 eV), H, C, N, O, F, P, S, Cl, Ar, are the next (χ between 10 and 20 eV), and the easiest are Li, Na, Mg, Al, K, Ca, Si, etc. (χ around 5 eV). To get some idea of this effect, compare the exponential factor alone in the Saha equation for sodium and for hydrogen, in the solar atmosphere, with T $\simeq$ 6,000 K. We have

$$N(\mathrm{Na}^+) \simeq e^{-5.16\,\mathrm{eV}/kT},$$

$$N(\mathrm{H}^+) \simeq e^{-13.6\,\mathrm{eV}/kT}.$$

The ratio of these is thus $N(\mathrm{Na}^+)/N(\mathrm{H}^+) \simeq 10^7$. If we now examine the table of abundances in the solar atmosphere, Table 6.2, we see that the exponential factor in the ionization equation just about makes up for the abundance deficiency of sodium relative to hydrogen (which is $\approx 10^{-6}$). Thus, in the solar atmosphere, the free electrons present are almost all from atoms of low ionization potential, such as sodium, Ca, Al, etc. These free electrons, from the metals, will determine the ionization of hydrogen through n_e in the Saha equation. We see therefore that certain species of atoms can be very significant in stellar atmospheres, even though their abundance is very low compared to that of hydrogen. In particular, variations in metallic abundance from one star to another may have major effects on the physical state of the material, especially its ionization equilibrium.

One case of practical interest occurs in stars whose metal abundance is low compared to the Sun's, such as Population II stars. Consider the effect on the spectrum of a star like the Sun if the abundance of the

Table 6.1

Ionization potentials (eV). The first two columns give atomic numbers and chemical elements. The remaining columns give the ionization potentials for the neutral atom (I), first ionized state (II) and so on.

Atomic number	Element	Ionization potential (electron volts)												
		I	II	III	IV	V	VI	VII	VIII	IX	X	XI	XII	XIII
1	H	13.598												
2	He	24.587	54.416											
3	Li	5.392	75.638	122.451										
4	Be	9.322	18.211	153.893	217.713									
5	B	8.298	25.154	37.930	259.368	340.217								
6	C	11.260	24.383	47.887	64.492	392.077	489.981							
7	N	14.534	29.601	47.448	77.472	97.888	552.057	667.029						
8	O	13.618	35.116	54.934	77.412	113.896	138.116	739.315	871.387					
9	F	17.422	34.970	62.707	87.138	114.240	157.161	185.182	953.886	1,103.089				
10	Ne	21.564	40.962	63.45	97.11	126.21	157.93	207.27	239.09	1,195.797	1,362.164			
11	Na	5.139	47.286	71.64	98.91	138.39	172.15	208.47	264.18	299.87	1,465.091			
12	Mg	7.646	15.035	80.143	109.24	141.26	186.50	224.94	265.90	327.95	367.53	1,648.659		
13	Al	5.986	18.828	28.447	119.99	153.71	190.47	241.43	284.59	330.21	398.57	1,761.802	1,962.613	
14	Si	8.151	16.345	33.492	45.141	166.77	205.05	246.52	303.17	351.10	401.43	442.07	2,085.983	2,304.08
15	P	10.486	19.725	30.18	51.37	65.023	220.43	263.22	309.41	371.73	424.50	476.06	523.50	2,437.67
16	S	10.360	23.33	34.83	47.30	72.68	88.049	280.93	328.23	379.10	447.09	479.57	560.41	611.85
17	Cl	12.967	23.81	39.61	53.46	67.8	97.03	114.193	348.28	400.05	455.62	504.78	564.65	651.63
18	Ar	15.759	27.629	40.74	59.81	75.02	91.007	124.319	143.456	422.44	478.68	529.26	591.97	656.69
19	K	4.341	31.625	45.72	60.91	82.66	100.0	117.56	154.86	175.814	503.44	538.95	618.24	686.09
20	Ca	6.113	11.871	50.908	67.10	84.41	108.78	127.7	147.24	188.54	211.270	564.13	629.09	714.02
21	Sc	6.54	12.80	24.76	73.47	91.66	111.1	138.0	158.7	180.02	225.32	591.25	656.39	726.03
22	Ti	6.82	13.58	27.491	43.266	99.22	119.36	140.8	168.5	193.2	215.91	249.832	685.89	755.47
23	V	6.74	14.65	29.310	46.707	65.23	128.12	150.17	173.7	205.8	230.5	265.23	291.497	787.33
24	Cr	6.766	16.50	30.96	49.1	69.3	90.56	161.1	184.7	209.3	244.4	255.04	308.25	336.26
25	Mn	7.435	15.640	33.667	51.2	72.4	95	119.27	196.46	221.8	248.3	270.8	298.0	355
26	Fe	7.870	16.18	30.651	54.8	75.0	99	125	151.06	235.04	262.1	286.0	314.4	343.6
27	Co	7.86	17.06	33.50	51.3	79.5	102	129	157	186.13	276	290.4	330.8	361.0
28	Ni	7.635	18.168	35.17	54.9	75.5	108	133	162	193	224.5	305	336	379
29	Cu	7.726	20.292	36.83	55.2	79.9	103	139	166	199	232	321.2	352	384
30	Zn	9.394	17.964	39.722	59.4	82.6	108	134	174	203	238	266	368.8	401
31	Ga	5.999	20.51	30.71	64							274	310.8	419.7
32	Ge	7.899	15.934	34.22	45.71	93.5								
33	As	9.81	18.633	28.351	50.13	62.63	127.6							
34	Se	9.752	21.19	30.820	42.944	68.3	81.70	155.4						
35	Br	11.814	21.8	36	47.3	59.7	88.6	103.0	192.8					
36	Kr	13.999	24.359	36.95	52.5	64.7	78.5	111.0	126	230.9				
37	Rb	4.177	27.28	40	52.6	71.0	84.4	99.2	136	150	277.1			
38	Sr	5.695	11.030	43.6	57	71.6	90.8	106	122.3	162	177			
39	Y	6.38	12.24	20.52	61.8	77.0	93.0	116	129	146.2	191	324.1		
40	Zr	6.84	13.13	22.99	34.34	81.5						206	374.0	
41	Nb	6.88	14.32	25.04	38.3	50.55	102.6	125						
42	Mo	7.099	16.15	27.16	46.4	61.2	68	126.8	153					
43	Te	7.28	15.26	29.54										
44	Ru	7.37	16.76	28.47										
45	Rh	7.46	18.08	31.06										
46	Pd	8.34	19.43	32.93										
47	Ag	7.576	21.49	34.83										
48	Cd	8.993	16.908	37.48										
49	In	5.786	18.869	28.03	54									
50	Sn	7.344	14.632	30.502	40.734	72.28								
51	Sb	8.641	16.53	25.3	44.2	56	108							
52	Te	9.009	18.6	27.96	37.41	58.75	70.7	137						
53	I	10.451	19.131	33										
54	Xe	12.130	21.21	32.1										
55	Cs	3.894	23.1											
56	Ba	5.212	10.004											
57	La	5.577	11.06	19.175										

SOURCE: Adapted from E. Novotny, *Introduction to Stellar Atmospheres and Interiors* (London: Oxford University Press, 1973). Original data from C. E. Moore, *Ionization Potentials and Ionization Limits from the Analysis of Optical Spectra,* NSRDS–NBS34 (Washington, D.C.: Department of Commerce, 1970).

metals were reduced. Since these contribute substantially to the electron pressure, the electron pressure will go down. An important consequence is the increased ionization of H^-, that is, a reduction in the H^- abundance. Since H^- is the major contributor to the continuous opacity in the solar spectrum, the continuous opacity is reduced as a result of reduced metal abundance. A further effect in solar-type spectra is a reduction in *metallic line blanketing,* the general effect of the combined absorption of all the

Table 6.2
Solar abundances of chemical elements, normalized such that the abundance of Si is 10^6. The abundances are typical of Population I stars.

Element	Abundance	Element	Abundance	Element	Abundance
1 H	3.2×10^{10}	29 Cu	540	58 Ce	1.18
2 He	2.1×10^{10}	30 Zn	1244	59 Pr	0.149
3 Li	49.5	31 Ga	48	60 Nd	0.78
4 Be	0.81	32 Ge	115	62 Sm	0.226
5 B	350	33 As	6.6	63 Eu	0.085
6 C	1.18×10^7	34 Se	67.2	64 Gd	0.297
7 N	3.74×10^6	35 Br	13.5	65 Tb	0.055
8 O	2.15×10^7	36 Kr	46.8	66 Dy	0.36
9 F	2450	37 Rb	5.88	67 Ho	0.079
10 Ne	3.44×10^6	38 Sr	26.9	68 Er	0.225
11 Na	6.0×10^4	39 Y	4.8	69 Tm	0.034
12 Mg	1.061×10^6	40 Zr	28	70 Yb	0.216
13 Al	8.51×10^4	41 Nb	1.4	71 Lu	0.036
14 Si	1.00×10^6	42 Mo	4.0	72 Hf	0.21
15 P	9600	44 Ru	1.9	73 Ta	0.021
16 S	5.0×10^5	45 Rh	0.4	74 W	0.16
17 Cl	5700	46 Pd	1.3	75 Re	0.053
18 Ar	1.172×10^5	47 Ag	0.45	76 Os	0.75
19 K	4200	48 Cd	1.48	77 Ir	0.717
20 Ca	7.21×10^4	49 In	0.186	78 Pt	1.4
21 Sc	35	50 Sn	3.6	79 Au	0.202
22 Ti	2775	51 Sb	0.316	80 Hg	0.4
23 V	262	52 Te	6.42	81 Tl	0.192
24 Cr	1.27×10^4	53 I	1.09	82 Pb	4
25 Mn	9300	54 Xe	5.38	83 Bi	0.143
26 Fe	8.3×10^5	55 Cs	0.387	90 Th	0.058
27 Co	2210	56 Ba	4.8	92 U	0.0262
28 Ni	4.80×10^4	57 La	0.445		

SOURCE: Adapted from A.G.W. Cameron, "A Critical Discussion of the Abundances of Nuclei," in *Explosive Nucleosynthesis,* ed. D. N. Schramm and W. D. Arnett (Austin: University of Texas Press, 1973).

metallic lines in the spectrum, which in effect run together to increase the effective continuum opacity. This effect is also reduced when the metallic abundance is reduced.

Overall, then, the metallic lines weaken and the continuum opacity goes down relative to the HI lines, which seem stronger as a result. Because of this spectral change, such stars may easily be misclassified as being of an earlier spectral type than corresponds to their true atmospheric temperature, or as subluminous stars or subdwarfs.

A similar effect occurs for two stellar atmospheres of identical effective temperatures, but differing surface gravities. This is most easily seen in the grey atmosphere approximation, where the $T(\tau)$ relation is fixed, so that the pressure at a given τ is essentially proportional to the value of g, as is the density. Hence the overall densities will be lower in a stellar atmosphere of lower surface gravity. The electron densities will also be lower, and hence the amount of ionization of a given atomic species will be greater. This is one important example of how density affects the physical state of the stellar material, and is one reason why we see spectral differences between stars of the same surface (or effective) temperatures, but differing gravities (e.g., between giant and dwarf stars of the same T_{eff}).

We may also make a remark at this stage about the sequence of spectral classification, described in all basic astronomy texts. This sequence (O, B, A, F, G, K, M) is a classification based on the strengths of the lines in the spectrum. The strength of an absorption line depends on the number of atoms in the state (lower energy level) capable of absorbing that line. In stars like the Sun, most of these lines are from neutral atoms. Now, a giant star of the same temperature as a dwarf star has a lower atmospheric density and hence a higher degree of ionization than the dwarf. Thus the density of neutral atoms is somewhat less. In order to make the density of neutral atoms that are in the state

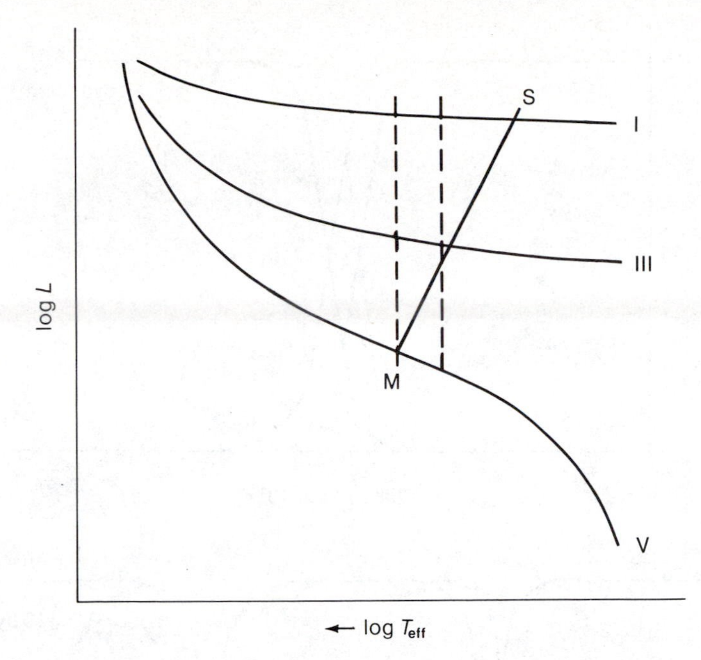

Figure 6.2. Dependence of spectral type on luminosity class for main sequence stars (V), giants (III) and supergiants (I). The line MS is line of constant spectral type.

able to absorb a given line the same as it is in a dwarf, we must lower the temperature slightly (Figure 6.2). Hence the statement in elementary astronomy texts that the spectral classification sequence is a temperature sequence is not quite correct. A giant star will generally have a somewhat lower temperature than a dwarf star of the same spectral class.

Finally, we note that the degree of ionization, x, is defined as the ratio of the ion density to the total density of the species of atom:

$$x = \frac{N_i}{N_i + N_0}.$$

Thus the Saha equation can be written

$$\frac{x}{1-x} = \frac{2Z_i}{n_e Z_0} \frac{(2\pi m_e kT)^{3/2}}{h^3} e^{-\chi_0/kT}. \qquad (6.19)$$

Figure 6.3 shows the degree of ionization of H and He calculated from this formula, with a constant electron pressure $P_e = 10$ dyne cm^{-2}. An interesting feature is immediately apparent. Except in a narrow range of temperatures, which we may describe as the ionization zone, hydrogen is either completely ionized ($x = 1$) or completely neutral ($x = 0$). The hydrogen ionization zone occurs at around 10^4 K, for this particular electron pressure. Similarly, there is a helium ionization zone at 15,000 K, and a second ionization zone at 30,000 K.

Problem 6.2. How does the temperature of the hydrogen ionization zone vary with n_e? Hint: define the temperature of the ionization zone by the maximum rate of change of x; hence set $d^2x/dt^2 = 0$ there.

The Saha formula defines a relation between the degree of ionization and the electron density and temperature, $x(n_e,T)$. We may combine the Saha and Boltzmann formulas to give the density of atoms in a given energy state of a given state of ionization of an atomic species. For example, the density of neutral hydrogen atoms in the $n = 2$ state is given by

$$N_2 = \frac{g_2}{Z_0(T)} N_0 e^{-E_2/kT}$$

$$= \frac{g_2}{Z_0(T)} e^{-E_2/kT} [1 - X(n_e, T)]\, N_H$$

$$= \frac{g_2}{Z_0(T)} e^{-E_2/kT} \left[1 + \frac{2Z_1}{n_e Z_0} \frac{(2\pi m_e kT)^{3/2}}{h^3} e^{-\chi_0/kT}\right]^{-1} N_H,$$

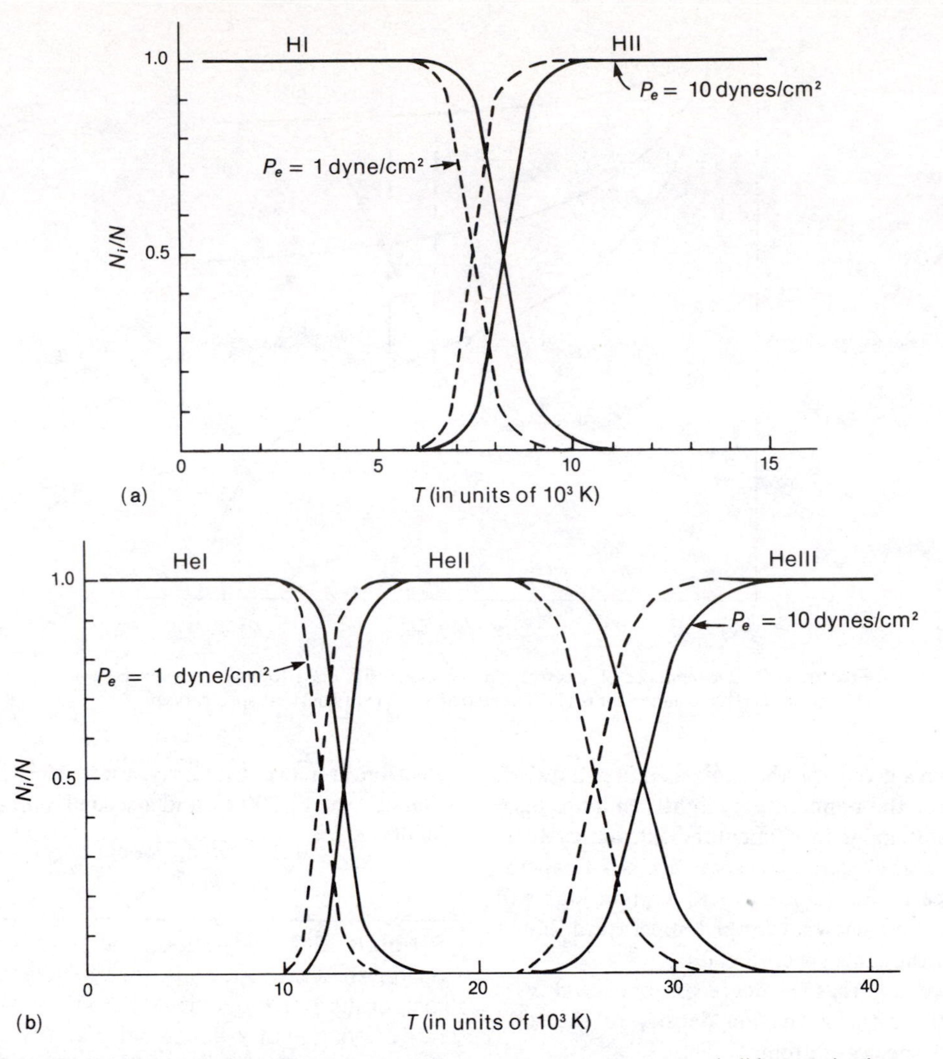

Figure 6.3. (a) Fractional ionization of hydrogen versus temperature (solid curve is electron pressure $P_e = 10$ dynes/cm^2; dashed curve is $P_e = 1$ dyne/cm^2). (b) Fractional ionization of He (for $P_e = 1$ and $P_e = 10$ dynes/cm^2).

where $N_H = N_i + N_0$. This shows that the strength of the H_α absorption declines at temperatures higher than around 10,000 K as ionization sets in.

Given the Saha ionization equation, and a list of total abundances, we can find the degrees of ionization of all atomic species. Since we then have the density of all particles as a function of n_e and T, we can calculate one of the basic parameters in the equations of stellar structure: the mean molecular weight as a function of n_e and T. Actually, since the stellar material must be electrically neutral, there is an additive relation between n_e and the densities of the ionic species that in effect determine n_e. Therefore, we have the mean molecular weight as a function of T for a given total chemical composition. The calculation is in principle straightforward, though tedious. Tables 5.2 to 5.4 give the electron density, mean molecular weight, and gas pressure as a function of T and P_e for three model atmospheres.

The Saha equation was derived in the preceding from purely statistical arguments, assuming only that the ionization and excitation processes were in equilibrium; for example, denoting an arbitrary ion by A_i, ionization may be written as a reaction

$$A_i \rightleftharpoons A_{i+1} + e^- + h\nu. \tag{6.20}$$

In particular, no basic assumptions were made about the interactions involved in the reaction. This suggests that a Saha-type equation could be obtained for other nonelectromagnetic reactions occurring in equilibrium. This is indeed the case.

Consider the reaction, assumed to occur in equilibrium at temperature T,

$$A + B \rightleftharpoons C + D. \tag{6.21}$$

A basic law of thermodynamics states that in equilibrium the sum of the chemical potentials of the reactants counted once for each particle must equal the sum of the products counted once for each particle. Denoting the chemical potentials by μ, the reaction (6.21) is described by

$$\mu_A + \mu_B = \mu_C + \mu_D. \tag{6.22}$$

Since the chemical potentials may be expressed as functions of T and the number density of particles n_A, etc., direct substitution into (6.22) yields the Saha equation.

Problem 6.3. The chemical potential for a nondegenerate ion, electron, or atom can be written as

$$\mu_i = -kT \ln\left[\frac{g_i}{n_i}\left(\frac{m_i T}{2\pi\hbar^2}\right)^{3/2}\right] + \chi_i, \tag{6.23}$$

where i designates the ion, atom, or electron, and χ_i is nonzero only for the electron, in which case it is the ionization energy. Derive the Saha equation from the equilibrium condition (6.22). The chemical potential of the photon is zero.

This method may be used to obtain, for example, the number densities of nuclei in the photodisintegration reaction $Fe^{56} + \gamma \rightleftharpoons 13\, He^4 + 4n$, or the particles in the weak interaction process $p + e^- \rightleftharpoons n + \nu_e$, given the chemical potentials for the various particles.

6.3. Detailed Balancing, Transition Probabilities, and Line Opacities

The results of Section 6.2 on thermal ionization and excitation derive from the application of classical thermodynamic results, in particular the Boltzmann factor, $\exp(-E/kT)$. We can also consider the detailed quantum processes going on which, in equilibrium, must give rise to the same distribution of atoms over their possible quantum states as predicted by the general laws of thermodynamics.

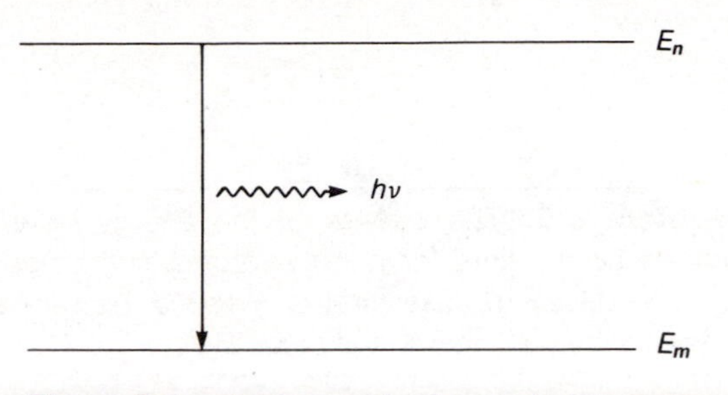

Figure 6.4. Atomic transition from excited state n to state m (radiative decay).

The quantum processes that take place in matter are essentially statistical in nature, and are characterized by a probability per unit time that a certain process will take place. For example, consider two energy levels, n and m, in an atom, and imagine that the atom is in the upper state n (Figure 6.4). The probability that the atom will spontaneously make the transition to the lower level, with the emission of a photon of frequency $\nu = (E_n - E_m)/h$ is an atomic constant, depending only on the structure of the atom. (Quantum-mechanically, it depends on the overlap integral of the wave functions corresponding to the two energy states.) We introduce the Einstein transition probability A_{nm} to describe this process. As an example, for H_α, $A_{nm} = 10^8\ \mathrm{sec}^{-1}$. If we have $N_n(0)$ atoms in level n at time $t = 0$, the number of atoms in that level at time t will be given by the exponential law

$$N_n(t) = N_n(0)\, e^{-A_{nm}t}. \tag{6.24}$$

Problem 6.4. Prove (6.24), using the definition of A_{nm}.

This is similar to the law of radioactive decay, which arises in the same way: the nuclei have a constant probability of decay per unit time. It follows from (6.24) that the mean life of the atoms in the upper energy state against transitions to the lower state n is A_{nm}^{-1}. This ignores the possibility of transitions to other levels. The total transition probability out of level n to all other states of lower energy ($m < n$) is

$$A_n = \sum_m A_{nm}. \quad (6.25)$$

Problem 6.5. The half-life of the energy level is defined as the time for half the atoms in the upper state to decay to lower states. What is the relation between the half-life and the mean life?

Now consider the reverse process, the absorption of a photon by an atom in the lower level, m. The rate at which this process takes place will depend on the density of photons, as well as on the atomic properties. We introduce Einstein's B_{mn} coefficient, which is also an atomic constant defined so that its product with the energy density of photons of frequency ν,

$$u_\nu B_{mn} = B_{mn} \frac{8\pi}{c^3} \frac{h\nu^3}{e^{h\nu/kT} - 1}, \quad (6.26)$$

is the probability that an atom in the lower level m will be excited to the upper level n by the absorption of a photon of frequency $\nu = (E_n - E_m)/h$. Note that we define B_{mn} with respect to the energy density u_ν in the radiation field.

These two processes alone can not produce the relative energy-level populations predicted by thermodynamics, since, if N_n and N_m are the populations of the upper and lower levels, respectively, the requirement that the two processes be in statistical equilibrium (detailed balancing) would imply that

$$N_n A_{nm} = N_m u_\nu B_{mn} = N_m B_{mn} \frac{8\pi}{c^3} \frac{h\nu^3}{e^{h\nu/kT} - 1}.$$

If A and B are to be atomic constants, this relation can not lead to $N_n/N_m = e^{-h\nu/kT}$ as required by thermodynamics. The missing process is *stimulated* (or *induced*) *emission,* in which a photon incident on an atom in its upper state may induce a transition to the lower state before the transition takes place spontaneously. Induced emission produces a new photon, of the same frequency and in the same quantum state as the old one. We introduce another atomic constant, B_{nm}, to describe this stimulated emission. The transition rate induced by photons of frequency ν and energy density u_ν from atoms in the initial state n to the final state m is then

$$N_n u_\nu B_{nm}. \quad (6.27)$$

Statistical equilibrium then requires that the rate of absorption (6.25) exactly balance the rate of spontaneous and induced emission (6.20) and (6.27), respectively:

$$N_n A_{nm} + N_n u_\nu B_{nm} = N_m u_\nu B_{mn}. \quad (6.28)$$

By rearrangement, we find

$$u_\nu = \frac{N_n A_{nm}}{N_m B_{mn} - N_n B_{nm}} = \frac{A_{nm}/B_{nm}}{\dfrac{N_m B_{mn}}{N_n B_{nm}} - 1}. \quad (6.29)$$

By comparing this with the radiation energy-density formula

$$u_\nu = \frac{8\pi h}{c^3} \frac{\nu^3}{e^{h\nu/kT} - 1}, \quad (6.30)$$

we see that we must have $B_{mn} = B_{nm}$,

$$\frac{N_m}{N_n} = e^{h\nu/kT}, \quad (6.31)$$

and

$$A_{nm} = \frac{8\pi h\nu^3}{c^3} B_{nm}. \quad (6.32)$$

We have performed this calculation assuming that the upper and lower levels of the atom were simple, i.e., had a statistical weight of unity. In principle, the inclusion of complex levels is straightforward; in practice, it is very tricky, because some of the transitions between pairs of sublevels may be forbidden, or of very low probability. Consider the two levels n, m, as broken up into sublevels, which we label by i and j, respectively (Figure 6.5). Let the transition probability from sublevel (n, i) to sublevel (m, j) be $A(n, i; m, j)$. What is the relation between the quantity A_{nm} we have previously defined and these quantities?

We are imagining here, of course, that we have split up the levels into all their composite levels, so that the

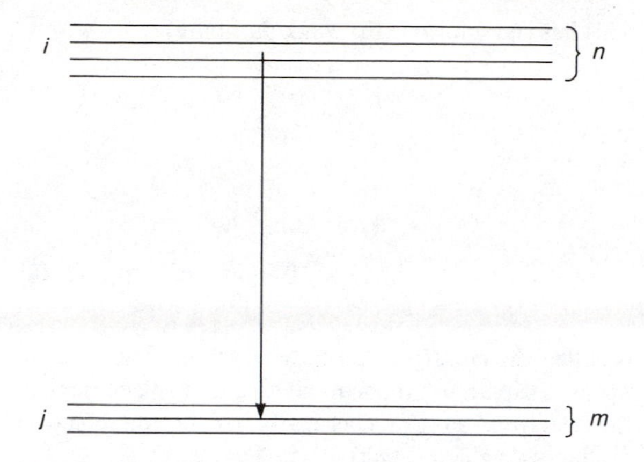

Figure 6.5. Degenerate atomic levels. The difference in energy between levels of fixed n or m is small compared to $E_{n,i} - E_{m,j}$.

statistical weight of each is unity. If the population of level $N(n, i)$ is N_{ni}, then the transition rate from all levels in n to all levels in m is

$$\sum_{i,j} N_{ni} A(n, i; m, j) = \frac{N_n}{g_n} \sum_{i,j} A(n, i; m, j). \quad (6.33)$$

This must be equal to the quantity we have previously called $N_n A_{nm}$. Hence

$$A_{nm} = \frac{1}{g_n} \sum_{i,j} A(n, i; m, j); \quad (6.34)$$

g_n is, of course, the statistical weight of level n; and if we assume the energy differences between the sublevels is negligible, each level has the same population; hence we could write $N_{n,i} = N_n / g_n$. In the special case where the transition probabilities to all the sublevels of m are the same, say, A, then we have

$$A_{nm} = g_m A / g_n, \quad (6.35)$$

which is the formula usually given in texts. Now consider the Einstein coefficient. The total transition rate from m to n is clearly given by

$$u_\nu \sum_{i,j} N_{m,j} B(m, j; n, i) = u_\nu \frac{N_m}{g_m} \sum_{i,j} B(m, j; n, i), \quad (6.36)$$

where, of course, $B(m, j; n, i)$ is related to $A(n, i; m, j)$ by (6.32). This expression must equal $N_m B_{mn}$ as we have defined it earlier. Hence

$$B_{mn} = \frac{1}{g_m} \sum_{i,j} B(m, j; n, i)$$
$$= \frac{c^3}{8\pi h\nu^3} \frac{1}{g_m} \sum_{i,j} A(n, i; m, j), \quad (6.37)$$

and hence

$$B_{mn} = \frac{c^3}{8\pi h\nu^3} \frac{g_n}{g_m} A_{nm}. \quad (6.38)$$

Note that this result does not require all the individual A_{nm} to be equal. In a similar way, one may conclude that

$$B_{nm} = \frac{g_m}{g_n} B_{mn}. \quad (6.39)$$

Problem 6.6. Prove equation (6.39).

We now consider values for the transition probabilities. Since the values of B_{mn} can be derived from A_{nm} according to the thermodynamic relations, we concentrate only on A_{nm}. Ultimately, finding these values is a quantum-mechanical problem, and must be handled by the full use of quantum-mechanical wave functions, overlap integrals, and the like. Nevertheless, there are some things we can say.

First, the probability of an event generally depends on the number of ways in which it can happen. Suppose that photons in the frequency range $(\nu, \nu + d\nu)$ are involved in the transition. In how many ways can photons have this frequency? To put it another way, what is the statistical weight of the frequency range $(\nu, \nu + d\nu)$? The statistical weight of a free particle with position in the range $(\mathbf{x}, \mathbf{x} + d\mathbf{x})$ and momentum in the range $(\mathbf{p}, \mathbf{p} + d\mathbf{p})$ is (6.13). The statistical weight per unit volume of a particle with total momentum p, in the direction specified by a solid angle $d\Omega$, is then

$$\frac{p^2\, dp\, d\Omega}{h^3}. \quad (6.40)$$

Since the momentum of a photon is $p = h\nu/c$, the

statistical weight per unit volume of photons in the frequency range ν, $\nu + d\nu$, is

$$\frac{\nu^2\, d\nu\, d\Omega}{hc^2}. \tag{6.41}$$

Thus transitions at high frequencies tend to have higher transition probabilities than transitions at low frequencies. For example, optical transition probabilities are generally greater than radio transition probabilities.

It is possible to give a semiclassical treatment of the transition probability that falls into the right order-of-magnitude range, provided that specific quantum-mechanical effects, such as forbidden transitions, do not come in. In this treatment, one imagines the excited atom to be an oscillating electric dipole in which an electron oscillates about the nucleus.

If the dipole oscillated without losing energy, the classical equation of motion of the electron would be

$$m\frac{d^2r}{dt^2} = -\,4\pi^2\nu_0^2 r, \tag{6.42}$$

where ν_0 is the frequency of oscillation. Classically, an electric dipole radiates energy at the rate

$$I = \frac{2}{3}\frac{e^2}{c^3}\left|\frac{d^2r}{dt^2}\right|^2. \tag{6.43}$$

When energy is lost because of this radiation, there must be a further term in the equation of motion of the electron that corresponds to a damping force, tending to slow down the electron. It can be shown that the damping force is given by

$$F = \frac{2}{3}\frac{e^2}{c^3}\frac{d^3r}{dt^3}. \tag{6.44}$$

When the damping is small, one can assume that the motion is nearly simple harmonic; that is,

$$r = r_0 \cos 2\pi\nu_0 t. \tag{6.45}$$

Hence the third derivative is $\dddot{r} = -4\pi\nu_0^2 r^2\dot{r}$, and the equation of motion of the electron becomes

$$m\frac{d^2r}{dt^2} = -\,4\pi^2\nu_0^2\frac{dr}{dt} - 4\pi^2\nu_0^2\,\frac{2}{3}\frac{e^2}{c^3}\frac{dr}{dt}. \tag{6.46}$$

This has the solution (for weak damping)

$$r = r_0 e^{-\gamma t/2}\cos 2\pi\nu_0 t, \tag{6.47}$$

where

$$\begin{aligned}\gamma &= (8\pi^2 e^2/3m_e c^3)\nu_0^2 \\ &= 2.47 \times 10^{-22}\,\nu_0^2\ \text{sec}^{-1}\end{aligned} \tag{6.48}$$

is called the *classical damping constant*. The field set up by a dipole is proportional to the displacement of the electron; so the classically oscillating electron produces an electric field

$$E(t) = E_0 e^{-\gamma t/2}\cos 2\pi\nu_0 t. \tag{6.49}$$

We see from this that after a time of the order of $(\gamma/2)^{-1}$, the electric field has dropped to $1/e$ of its initial value. This characterizes the time-scale for decay of the oscillation, and must be of the same order as $A_{nm}{}^{-1}$. Thus we have the "classical" value of the transition probability as

$$A_{\text{classical}} \simeq \gamma. \tag{6.50}$$

Now for H_α, $\lambda = 6563$Å and $\nu_0 = c/\lambda$, which, when substituted into (6.48) yields $\gamma \simeq 5 \times 10^7\ \text{sec}^{-1}$.

Second, because of the exponential decay of the electric field, the emitted radiation is not strictly monochromatic. In order to find its spectrum, or effective range of frequencies, one must do a Fourier analysis, which yields the spectrum of the electric field as a function of frequency. The square of this gives the intensity of the field. The result is

$$\begin{aligned}I(\nu) &= I_0\frac{\gamma}{4\pi^2(\nu-\nu_0)^2 + (\gamma/2)^2} \\ &= I_0\frac{\gamma/4\pi^2}{(\nu-\nu_0)^2 + (\gamma/4\pi)^2}.\end{aligned} \tag{6.51}$$

From this we easily see that the intensity has a natural width of the order of γ. In fact, $\gamma/2\pi$ is the width of the line at half intensity (Figure 6.6). This is our first example of line broadening, in this case, natural line broadening. Quantum-mechanically, this natural width arises from the uncertainty principle. Since the uncertainty in the time during which an atom is in a state is its lifetime in that state, which on average is

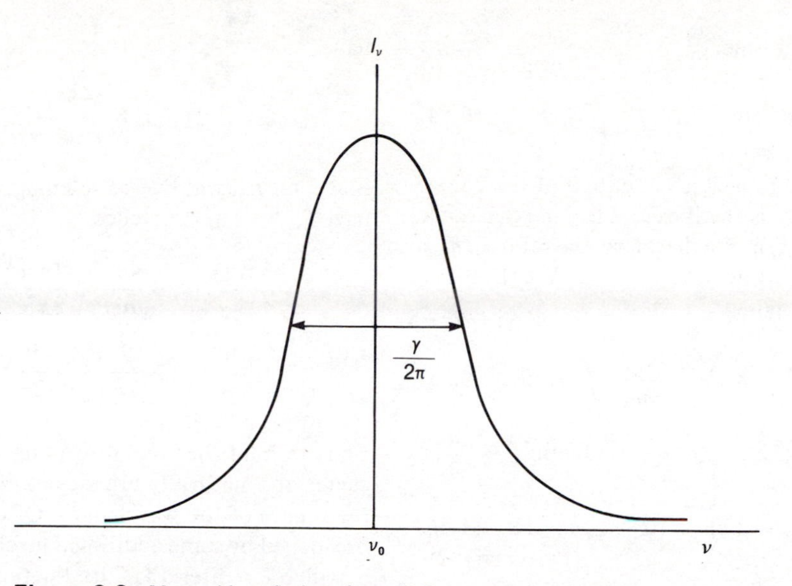

Figure 6.6. Natural broadening of atomic line centered at frequency ν_0.

equal to A^{-1}, we see that the energy uncertainty is $\Delta E = h/\Delta t \approx Ah$; so the frequency spread is $\Delta\nu \simeq A \approx \gamma$. The line shape given in Figure 6.6 is called the Lorentzian or damping profile.

Because the transition probability is usually of the same order of magnitude as the classical damping constant, it is useful to express exact values of A_{nm} in terms of the classical value. The so-called oscillator strength f_{nm} of a transition $n \rightarrow m$ is defined by the equation

$$A_{nm} = 3\frac{g_m}{g_n} f_{nm}\,\gamma = \frac{g_m}{g_n}\frac{8\pi^2 e^2 \nu^2}{m_e c^3} f_{nm} \qquad (6.52)$$

or

$$B_{mn} = \frac{\pi e^2}{m_e h\nu} f_{nm}.$$

The oscillator strengths for the Balmer series of hydrogen are ($f_\alpha = f_{32}$; $f_\beta = f_{42}$; etc.):

H_α, $f_\alpha = 0.641$,
H_β, $f_\beta = .119$,
H_γ, $f_\gamma = .044$,
H_δ, $f_\delta = .021$,
H_ϵ, $f_\epsilon = .012$.

Because hydrogen is a relatively simple atom, we can give an analytic approximation to the oscillator strengths, by what is known as *Kramers' formula:*

$$f_{nm} = \frac{2^6}{3\sqrt{3}\,\pi}\frac{1}{g_m}\frac{1}{\left(\dfrac{1}{m^2} - \dfrac{1}{n^2}\right)^3}\frac{g_I}{n^3 m^3}. \qquad (6.53)$$

Here, as before, n refers to the upper level and m to the lower level. The quantity g_I is a quantum-mechanical correction factor of the order of magnitude of unity. It is called the *Gaunt factor for bound-bound transitions.*

We can now find the absorption coefficient per atom a_ν, in the vicinity of an absorption line centered at ν_0, defined so that the probability of absorption per unit path length of a photon is Na_ν. Here N is the number density of atoms capable of absorbing the photons. Consider a beam of intensity I_ν traversing a unit cross-sectional area. In time dt, these photons have traveled a distance $c\,dt$. Hence the energy removed from the beam in the range ν to $\nu + d\nu$ is $I_\nu\, Na_\nu\, c\, dt$. Alternatively, the number of transitions is $I_\nu\, Na_\nu\, c\, dt/h\nu$. Since these photons occupy a volume $c\,dt$, the number of transitions per unit volume per unit time per unit frequency due to this beam is $I_\nu\, Na_\nu/h\nu$. Integrating over all directions (solid angles) and assuming the radiation to be in thermal equilibrium ($4\pi\, I_\nu = cu_\nu$), the number of absorptions is given by $cu_\nu\, Na_\nu\, h\nu$. We now integrate this over frequency, over the whole line profile, to obtain the total number of transitions from the lower level to the upper per-unit volume and per-unit time:

$$\text{transition rate} = c\,N \int_0^\infty \frac{u_\nu}{h\nu}\, a_\nu\, d\nu. \qquad (6.54)$$

This must be equal to $u_{\nu_0} B_{nm} N$. Because of the nature of line absorption, a_ν is small except in a narrow range near $\nu_0 = (E_n - E_m)/h$. Furthermore, the ratio $u_\nu/h\nu$ is usually slowly varying across such lines. We then find, for all practical purposes,

$$u_{\nu_0} B_{nm} N = cN \int_0^\infty \frac{u_\nu}{h\nu}\, a_\nu\, d\nu \simeq \frac{c u_{\nu_0} N}{h\nu_0} \int_0^\infty a_\nu\, d\nu. \qquad (6.55)$$

Equating this to (6.52), for $\nu = \nu_0$ we obtain

$$\int_0^\infty a_\nu d\nu = \frac{\pi e^2}{m_e c} f_{nm} = 2.65 \times 10^{-2} f_{nm}\ \mathrm{cm^2\ Hz}, \qquad (6.56)$$

which is Ladenburg's relation. Note that the oscillator strength f_{nm} is dimensionless. The integral of a_ν over all frequencies can be thought of as the total absorption cross section per atom initially in the lower level. The quantity a_ν can be rewritten as

$$a_\nu = \frac{\pi e^2}{m_e c} f_{nm} \phi_\nu, \qquad (6.57)$$

where $\phi_\nu\, d\nu$ is the probability that a photon in the frequency range ν to $\nu + d\nu$ is absorbed in the line. Note that $\int \phi_\nu\, d\nu = 1$. The quantity ϕ_ν is called the *line-broadening function.*

The variation of a_ν with ν is expected to follow the Lorentz profile. If we write

$$a_\nu = a_0 \frac{\gamma/4\pi^2}{(\nu - \nu_0)^2 + (\gamma/4\pi)^2}, \qquad (6.58)$$

then, since a_ν/a_0 has a unit integral, the absorption cross section within the line is given by

$$a_\nu = \frac{\pi e^2}{m_e c} f_{nm} \frac{\gamma/4\pi^2}{(\nu - \nu_0)^2 + (\gamma/4\pi)^2}. \qquad (6.59)$$

In principle, the factor γ is proportional to f_{nm}, if only natural line broadening is being considered. However, in practice natural line width is rarely, if ever, significant; instead, other broadening mechanisms dominate, some of which also produce a Lorentz profile, but with the value of γ determined by some other physical process. From equation (6.59), we may estimate the opacity at the line center. For $\nu = \nu_0$, we have

$$a_\nu = \frac{4\pi e^2}{m_e c} \frac{f_{nm}}{\gamma}. \qquad (6.60)$$

Since, for natural line broadening, $\gamma \simeq A \simeq 3 f_{nm} \gamma_{\mathrm{cl}}$, we have $f_{nm}/\gamma = 3/\gamma_{\mathrm{cl}}$. Hence

$$a_\nu \simeq \frac{4\pi e^2}{m_e c}\, 3\, \frac{3 m_e c^3}{8\pi^2 e^2} \frac{1}{\nu_0^2}$$

$$= \frac{9}{2\pi} \frac{c^2}{\nu_0^2} = \frac{9}{2\pi} \lambda_0^2. \qquad (6.61)$$

Thus we find that the absorption cross section at the center of a naturally broadened line is approximately the square of the wavelength. Of course, if the line is broadened by some additional mechanism, the value of γ will be greater than its "natural" value, and the line-center opacity will be correspondingly less.

Finally, we may correct the absorption coefficient to include stimulated emission. In the preceding discussions, we have tacitly assumed that the emission (and hence the source function) is isotropic. On the other hand, stimulated emission is very strongly correlated in the direction of the original photon. Since our treatment of the incident photons in radiative transfer problems did take account of their directional properties, a simple way to allow for stimulated emission correctly is to regard it as negative absorption, and to reduce the absorption coefficient by a suitable factor. In a radiation field of density u_ν, the number of absorptions per unit time is $N_m u_\nu B_{mn}$, and the number of stimulated emissions is $N_n u_\nu B_{nm}$. Subtracting the emissions from the absorptions, the net number of absorptions is

$$N_m u_\nu B_{mn} - N_n u_\nu B_{nm} = N_m u_\nu B_{mn} \left(1 - \frac{N_n B_{nm}}{N_m B_{mn}}\right)$$

$$= N_m u_\nu B_{mn} (1 - e^{-h\nu/kT}).$$

Hence we can allow for the effect of stimulated emission by multiplying the ordinary absorption coefficient (6.57) by the factor $(1 - e^{-h\nu/kT})$ to obtain

$$a_\nu = (1 - e^{-h\nu/kT}) \frac{\pi e^2}{m_e c} f_{nm} \phi_\nu. \qquad (6.62)$$

Problem 6.7. What is the order of magnitude of this correction in the solar atmosphere?

We note that the absorption coefficient k_ν that appears in the transfer equation is related to a_ν for lines by

$$k_l(\nu) = n\, a_\nu, \qquad (6.63)$$

where n is the number density of absorbing atoms.

6.4. Continuous Opacity in Stars

An important factor governing the rate at which energy escapes from a star is the opacity of the gas. The opacity is a measure of the extent to which radiation couples to matter. Physically the source of this coupling is the electromagnetic interactions between photons and electrons, ions, or sometimes neutral atoms. The principal sources of opacity include:

(1) bound-bound transitions;
(2) bound-free (photoionization and its inverse, recombination) transitions;
(3) free-free (*Bremsstrahlung*) transitions;
(4) electron-scattering (Thompson and Compton scattering); and
(5) molecular transitions.

In addition to these, electron conduction (in which the transfer of heat is via electrons rather than photons) may be described by an opacity (see Section 5.11). We will here consider the first four of the processes above, as well as electron conduction.

Besides regulating the flow of radiation from the stellar core, where energy generation typically occurs, the opacity leads to a gradual degradation of photon energy. For example, a photon released during the process 4H $\rightarrow$ He in the Sun's core has an energy of about 13 MeV, and a mean free path $l \simeq 1$ cm. After traveling a distance of order l, a typical photon is scattered, or absorbed and then re-emitted, possibly at a lower energy. Since the average energy of light emerging from the atmosphere is several eV, a typical photon will have had its energy reduced by a factor of about 10^6. Because photons diffuse outward in a random-walk manner, they will undergo as many as $(R_\odot/l)^2$ collisions with matter before reaching the surface and escaping. The time required for this to occur is roughly $R_\odot{}^2/lc$, which is about 10^3 yrs for the Sun, and the number of collisions involved can exceed 10^{21}.

The variation of the opacity with ρ and T in H and He ionization zones in stellar envelopes can induce local instabilities associated with stellar pulsations observed in Cepheid and R-R Lyrae variables, and with thermal instabilities believed to contribute to the underlying mechanism responsible for recurrent novae. These will be discussed in Chapter 11.

The absorption or emission of a photon is fundamentally a quantum process, involving the transition of an electron from one energy level to another (see Section 6.1). The transitions responsible for processes (1–3) and (5) in the preceding list are shown schematically in Figure 6.7. Bound-bound transitions occur only between the discrete atomic or molecular energy levels, whereas bound-free transitions involve only one discrete level. In both bound-free and free-free transitions, one or both levels are in the continuum; however, an atom or ion must be present so that the transitions will conserve both energy and momentum. Electron scattering differs from these, in that the initial and final states involve an electron and a photon. In fact, the process may be thought of as the absorption of a photon by an electron, followed by the re-emission of another photon a time $t < \hbar/\Delta E$ later, where ΔE is the electron-photon energy difference.

We will consider formulas that apply to hydrogen, or hydrogen-like atoms. For more complex atoms, a hydrogenic approximation can sometimes be used, often quite satisfactorily; if not, one must resort to quantum-mechanical calculations, or sometimes extrapolations from known results. Fortunately, hydrogen is the most abundant element, and in cool stars, negative hydrogen bound-free and free-free absorption is dominant. In hotter stars, bound-free and free-free absorption in atomic (ordinary) hydrogen dominates, and in stellar interiors the material is completely ionized; so free-free absorption in H is the most important effect. At very high temperatures, the absorption processes we have described become less efficient, because each must be multiplied by the stimulated emission correction $(1 - e^{-h\nu/kT})$. Clearly, as $T \rightarrow \infty$, this goes to zero. Therefore another source of continuous opacity is important at high temperatures, that of simple electron scattering. Since at high temperatures the material is fully ionized, there are plenty of electrons around to scatter light.

In principle an opacity calculation begins with a quantum-mechanical description of electron-photon interactions, yielding a cross section for absorption or scattering of a photon of energy $h\nu$, which we denote by $a_i(\nu)$. The corresponding specific opacity (g/cm^2) for the process is given by

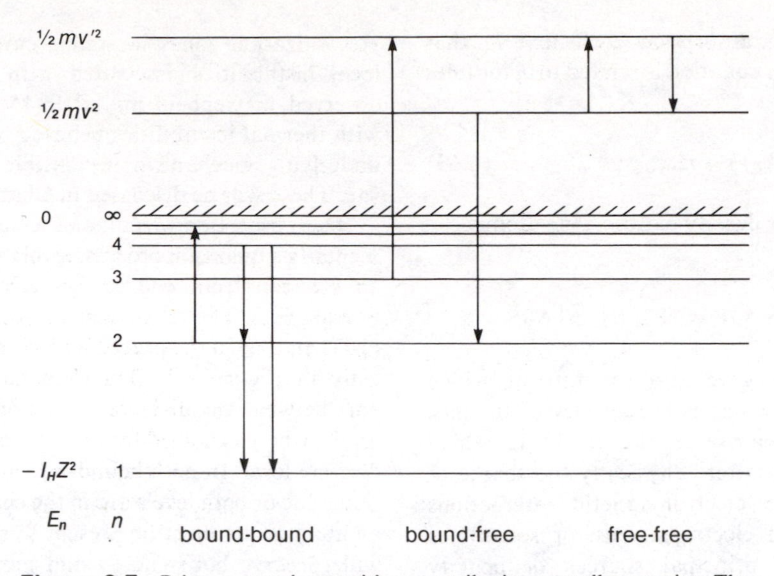

Figure 6.7. Primary atomic transitions contributing to stellar opacity. The leftmost column gives the electron's energy E_n; the next contains the principle quantum number n of the bound state. The ionization limit is shown cross-hatched.

$$\kappa_i(\nu) = \frac{n_i a_i(\nu)}{\rho} \text{ cm}^2\text{g}^{-1}, \tag{6.64}$$

where n_i is the number density of scatterers or absorbers. The index i specifies the exact process (1–5) in the list. The total opacity of the matter to radiation is given by

$$\kappa_{\text{tot}}(\nu) = \sum_i \kappa_i(\nu). \tag{6.65}$$

Frequency-dependent opacities are needed for detailed models of stellar atmospheres and especially for studies of line formation. In stellar interiors, particularly during slow evolutionary stages, the Rosseland mean opacity is usually sufficient, in which case $k_\nu = \kappa(\nu)\rho$ in (5.52) is replaced by $\rho\kappa_{\text{tot}}(\nu)$. Often one source of opacity in (6.65), say, $\kappa_j(\nu)$, exceeds all others, and we may calculate $\bar{\kappa}$ from $\kappa_j(\nu)$:

$$\frac{1}{\bar{\kappa}} = \langle \kappa_{tot}(\nu)^{-1} \rangle = \left\langle \left[\sum_i \kappa_i(\nu)\right]^{-1} \right\rangle$$

$$\simeq \langle \kappa_j(\nu)^{-1} \rangle = \frac{1}{\bar{\kappa}_j}. \tag{6.66}$$

Problem 6.8. In general $\bar{\kappa} \neq \Sigma_i \bar{\kappa}_i$. However, if all $\kappa_i(\nu)$ have the same frequency dependence, then

$$\bar{\kappa} = \sum_i \bar{\kappa}_i. \tag{6.67}$$

Assume that the frequency dependence of $\kappa_i(\nu)$ is the same for all i, and show that (6.67) is obtained.

Electron Scattering

When a photon collides with an electron, energy and momentum of the whole system must be conserved. This leads, as is well known, to the Compton effect. In photon scattering, the wavelength of the photon is increased* by the amount $2\lambda_c \sin^2\theta/2$, where $\lambda_c = h/m_e c = 0.024$Å is the Compton wavelength of the electron. Here θ is the scattering angle of the photon

*If the electron's energy exceeds the photon energy, the photon wavelength will decrease. Although this is not typical in stellar interiors or atmospheres, it may occur in interstellar matter, where relativistic electrons (produced, for example, by supernovae or pulsars) occasionally scatter off starlight. The process contributes to the background gamma-ray flux in the Galaxy.

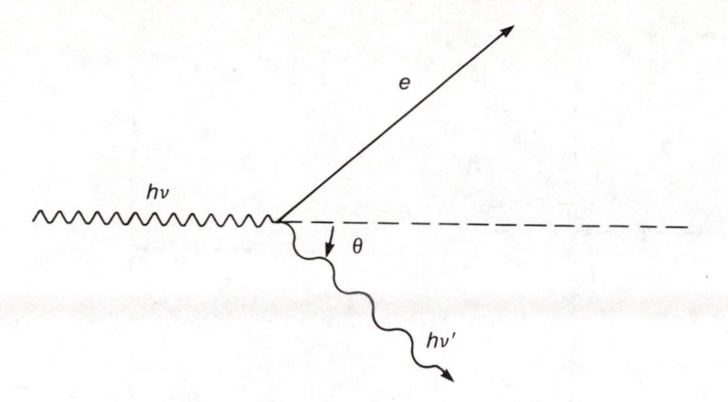

Figure 6.8. Electron scattering. The electron is initially at rest.

(Figure 6.8). The Compton shift comes about because the photon has transferred some of its energy and momentum to the electron. Since this shift in wavelength is very small, it is important only for very high-energy photons, whose wavelength is very small. The cross section for low-energy electron scattering is given by the following classical argument.

Since photons are the quanta of the electromagnetic field, we ask for the cross section σ_T for an electron to absorb energy from an electromagnetic field. For low-energy electrons ($v \ll c$), we may ignore the magnetic component of the field. The energy absorbed by an electron per second is then given classically by

$$\frac{dE}{dt} = \frac{2}{3}\frac{e^2}{c^3}a^2. \tag{6.68}$$

Here a is the magnitude of the electron's acceleration, given by

$$a = \frac{e\,\mathcal{E}}{m_e}, \tag{6.69}$$

if $\mathcal{E}$ is the magnitude of the electric field. Since the energy density in the field is $\langle \mathcal{E}^2/4\pi \rangle$, where the brackets denote a time average, the energy flux per electron will be $c\langle \mathcal{E}^2/4\pi \rangle$. Taking the time average of (6.68), the energy absorbed per electron per second must be

$$\left\langle \frac{dE}{dt} \right\rangle = \sigma_T c\, \langle \mathcal{E}^2 \rangle / 4\pi = \frac{2}{3}\frac{e^2}{c^3}\frac{e^2}{m_e^2}\langle \mathcal{E}^2 \rangle. \tag{6.70}$$

This defines σ_T, the Thomson scattering cross section for the electron,

$$\sigma_T = \frac{8\pi}{3}\left(\frac{e^2}{m_e c^2}\right)^2 = 6.65 \times 10^{-25}\ \mathrm{cm}^2, \tag{6.71}$$

which is proportional to the classical cross sectional area of the electron $(e^2/m_e c^2)^2$. At higher electron energies, or for $T \gtrsim 10^9$ K, the magnetic field becomes nonnegligible, and relativistic and quantum effects become important. These effects tend to lower the cross section, which approaches zero as $h\nu/m_e c^2$ becomes large.

When (6.71) is valid, the electron-scattering opacity κ_e is given by the number density of free electrons n_e in the gas times σ_T/ρ:

$$\kappa_e = \frac{\sigma_T n_e}{\rho} = 0.2(1 + X), \tag{6.72}$$

where X is the mass fraction of hydrogen. Notice that electron scattering as described by the Thomson cross section is isotropic and independent of frequency.

Photons also scatter off free protons (H^+) and other ions. However, since the cross section for these processes is proportional to $(1/Am_H)^2$, the ratio of ion to electron scattering is $(m_e/Am_H)^2 < 10^{-6}$. Therefore such processes make a negligible contribution to the absorption.

A slight modification of the preceding arguments yields the Rayleigh scattering cross section σ_R, which describes the scattering of a photon of low energy by a bound atomic electron. This differs from conventional bound-bound transitions. Classically, Rayleigh scattering results when a photon of energy less than atomic energy spacings ($\Delta E \sim$ eV) is absorbed by an electron

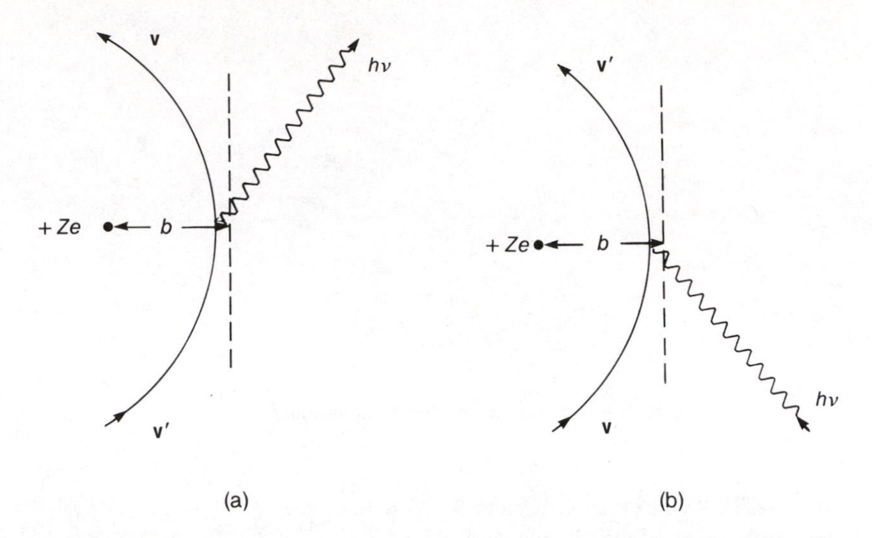

Figure 6.9. Free-free absorption and emission (*Bremsstrahlung*). (a) Emission of a photon by an electron accelerated in vicinity of ion. (b) Absorption of a photon by an electron in the field of an ion. The distance of closest approach of the electron and ion is *b*. The process is inelastic ($|\mathbf{v}| \neq |\mathbf{v}'|$).

that oscillates harmonically about its unperturbed energy level. The electron reradiates the photon, but remains in the same energy state. The cross section is easily shown to be

$$\sigma = \sigma_T \frac{1}{[1 - (\nu_0/\nu)^2]^2}, \tag{6.73}$$

where $h\nu$ is the energy of the photon, and $h\nu_0$ is a measure of the restoring force acting on the harmonically bound electron. When the photon energy $h\nu$ is small ($\nu \ll \nu_0$), we obtain the classical Rayleigh cross section

$$\sigma_R = \sigma_T(\nu/\nu_0)^4 = \sigma_T(\lambda_0/\lambda)^4. \tag{6.74}$$

Problem 6.9. The equation of motion for a harmonically bound electron driven by an electric field $\mathscr{E}$ is

$$ma + m\omega_0^2 r = e\mathscr{E}. \tag{6.75}$$

Show that the cross section for scattering of a photon of energy $h\nu = \hbar\omega$ is given by (6.73).

The requirement $\nu \ll \nu_0$ corresponds to $kT \ll \Delta E$, implying temperatures of the order of several times 10^3 K or less. Although Rayleigh scattering is clearly unimportant for stellar interiors, it is important in the outer layers of cool stars. The fourth-power dependence on wavelength is also responsible for the blue color of the daytime sky.

Free-Free Absorption

Another form of free-free scattering, *Bremsstrahlung*, occurs when an electron moving in the Coulomb field of an ion of charge Ze emits or absorbs a photon. We may construct the free-free opacity by the following semiclassical argument. Imagine an electron moving in the Coulomb field of the ion. As it accelerates in the vicinity of the ion it radiates a photon of energy $h\nu$ (see Figure 6.9, a). The reverse process involves the absorption of a photon by the electron. In either case, the presence of the ion is crucial if both energy and momentum are to be conserved. Consider the emission process. The initial and final electron velocities are $\mathbf{v}'$ and $\mathbf{v}$ respectively. Then, by conservation of energy,

$$\frac{1}{2} m_e v^2 + h\nu = \frac{1}{2} m_e v'^2; \tag{6.76}$$

the energy absorbed (or radiated) by the electron per second is given by (6.68). Therefore the energy emitted per ion per electron because of the entire scattering process is

$$E_{\rm abs} = \int_{-\infty}^{\infty} \frac{dE}{dt}\, dt = \frac{2}{3}\frac{e^2}{c^3}\int_{-\infty}^{\infty} a^2\, dt. \qquad (6.77)$$

Most of the energy is emitted during time $t \simeq b/v'$ when the electron is near the ion, and its acceleration is of order $Ze^2/m_e b^2$. Thus

$$E_{\rm abs} \simeq \frac{2}{3}\frac{e^2}{c^3}\left(\frac{Ze^2}{m_e b^2}\right)^2 \frac{b}{v'} = \frac{2}{3}\frac{Z^2 e^6}{m_e^2 c^3}\frac{1}{bv'}. \qquad (6.78)$$

In order to approximate the frequency dependence of this process, we expand dE/dt in a Fourier series

$$\frac{dE(\nu)}{dt} = \frac{1}{2\pi}\int_{-\infty}^{\infty} \frac{dE(t)}{dt} e^{-2\pi i \nu t}\, dt. \qquad (6.79)$$

The principal contribution to the integral comes from frequencies satisfying $2\pi\nu t \approx 1$ during the time $t \approx b/v'$. This implies that the frequency range into which most of the radiation is emitted comes from electrons having an impact parameter b and initial velocity v':

$$2\pi\nu \simeq \frac{v'}{b}. \qquad (6.80)$$

It then follows that the energy absorbed per electron per ion from the frequency interval $d\nu$ is

$$\begin{aligned} dq_\nu &= 2\pi b\, db\, E_{\rm abs} \\ &= \frac{8\pi^2}{3}\frac{Z^2 e^6}{m_e^2 c^3 v'^2}\, d\nu \end{aligned} \qquad (6.81)$$

using (6.78) and (6.80). The total energy emitted in *Bremsstrahlung* is therefore given by the product of the ion number density, n_i; the flux of electrons of initial velocity v', $n_e v' f(v')\, dv'$; and the energy emitted per electron per ion per unit frequency dq_ν:

$$n_i n_e v' f(v')\, dv'\, dq_\nu. \qquad (6.82)$$

The reverse process (shown in Figure 6.9) defines the *Bremsstrahlung* absorption coefficient a_ν giving the absorption per ion per electron of velocity $\mathbf{v}$ from the radiation field. In thermodynamic equilibrium, the photon energy density $u_{\nu p}$ is given by (5.49). Then the net energy absorbed is given by the product of the flux of photons of energy $h\nu$, $cu_{\nu p}\, d\nu$; $n_i n_e f(v)\, dv\, a_\nu$; and the factor $(1 - e^{-h\nu/kT})$, which takes into account stimulated emission:

$$n_i n_e f(v)\, dv\, a_\nu (1 - e^{-h\nu/kT}) c u_{\nu p}\, d\nu. \qquad (6.83)$$

For a system in thermal equilibrium, the principle of detailed balance asserts that the processes (6.82) and (6.83) must be equal, and

$$a_\nu = \frac{\pi}{3}\frac{Z^2 e^6}{hcm_e^2 v \nu^3}. \qquad (6.84)$$

This differs from the exact classical result at high photon energies by a factor of $4/\sqrt{3}$.

Problem 6.10. Equating (6.82) and (6.83), and assuming that the electron velocities are Maxwellian at temperature T, show that a_ν is given by (6.84).

The *Bremsstrahlung* opacity is given by

$$\begin{aligned} \kappa_{\rm ff}(\nu) &= \int_0^{\infty} n_i n_e f(v)\, a_\nu\, dv\, g_{\rm ff}(\nu, v) \\ &= \frac{4}{3}\left(\frac{2\pi}{3m_e kT}\right)^{1/2} \frac{Z^2 e^6}{hcm_e \nu^3} n_i n_e \bar{g}_{\rm ff}(\nu), \end{aligned} \qquad (6.85)$$

where the factor $4/\sqrt{3}$ has been included. The Gaunt factor $\bar{g}_{\rm ff}(\nu)$, which is of order unity and varies slowly with frequency, takes into account the deviations between the classical and quantum-mechanical calculations.

The Rosseland mean opacity follows when $\kappa_{\rm ff}(\nu)$ is substitiuted into (5.52). The result can be shown to be

$$\bar{\kappa}_{\rm ff} = \kappa_0 \rho T^{-3.5}, \qquad (6.86)$$

where κ_0 is a constant that is a function of chemical composition.

Problem 6.11. Verify the general form of (6.86), assuming that

$$\kappa_{\rm ff}(\nu) \sim \nu^{-3} T^{-1/2}.$$

The constant κ_0 in (6.86) may be written in the form

$$\kappa_0 = 3.68 \times 10^{22} (1 + X)(1 - Z)\bar{g}_{\rm ff}, \quad (6.87)$$

where X and Z are the mass abundances of hydrogen and of heavy metals ($A > 4$), respectively.

Problem 6.12. Present physical arguments motivating the dependence of κ_0 on X and Z shown in (6.87).

The *Bremsstrahlung* absorption coefficient decreases rapidly with ν. The $\rho T^{-7/2}$ dependence of the mean opacity was obtained originally by Kramers, and this formula is often referred to as *Kramers' opacity*. It is a very useful formula for obtaining some general order-of-magnitude results about stellar atmospheres and interiors, and we will use it quite extensively later.

Bound-Free Opacity

Bound-free transitions occur when the absorption or emission of a photon involves the continuum and a discrete atomic energy level E_m (see Figure 6.10). The absorption coefficient and opacity for this process can be calculated by using some of the preceding results. Denote the energy of the $n^{\rm th}$ discrete state by

$$E_n = -\frac{m_e Z^2 e^4}{2\hbar^2 n^2} \equiv -\frac{I_H Z^2}{n^2}. \quad (6.88)$$

Then electron capture or its reverse, photoionization, requires that the photon energy satisfy

$$\frac{1}{2} m_e v^2 - E_n = h\nu, \quad (6.89)$$

where v is the velocity of the ejected or captured electron. The primary difference between *Bremsstrahlung* and electron capture is the discrete nature of the final state. Consider the capture process semiclassically. The electron's initial energy ($m_e v^2/2$) is positive, but decreases as it accelerates in the field of the ion of charge Ze. Most of the electron's energy is radiated away prior to capture (the acceleration going into a bound state is smaller than $Ze^2/m_e b^2$ estimated in the discussion of *Bremsstrahlung*). Therefore, we may approximate the energy loss per captured electron per ion by dq_ν (6.81). The cross section for the emissions of photons into a frequency interval $d\nu$ is defined by $dq_\nu = h\nu \, d\sigma_\nu = h\nu(d\sigma_\nu/d\nu)\, d\nu$. We may introduce the discrete nature of the final state of the electron by defining the cross section σ_{cn} for capture into a quantum state E_n characterized by n in the range n to n +

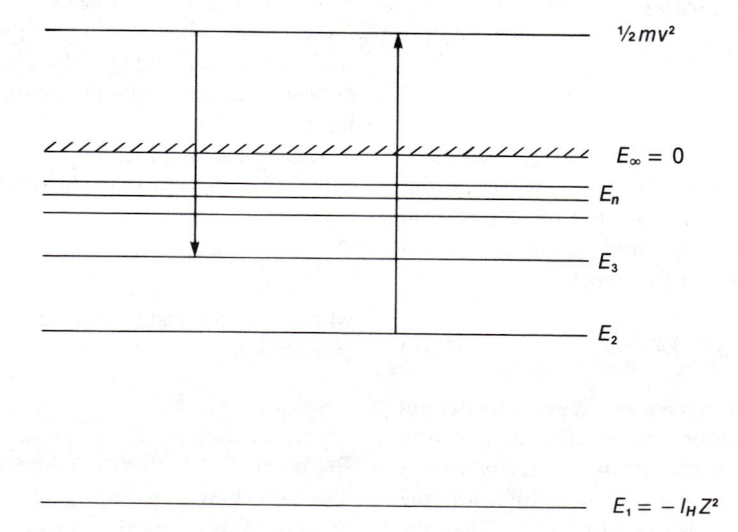

Figure 6.10. Bound-free transitions. Atomic bound states lie below cross-hatched level (ionization limit), and continuum states (unbound electrons) lie above it.

dn, so that $d\sigma_\nu = \sigma_{cn}\, dn$. We then find

$$h\nu \frac{d\sigma_\nu}{d\nu} = h\nu\sigma_{cn} \frac{dn}{d\nu} = \frac{dq_\nu}{d\nu}.$$

Solving for σ_{cn}, and using (6.88) to evaluate $dn/d\nu$, we find

$$\sigma_{cn} = \frac{2I_H Z^2}{h^2 \nu n^3} \frac{dq_\nu}{d\nu} = \frac{32}{3}\pi^4 \frac{Z^4 e^{10}}{m_e c^3 h^4 v^2 \nu n^3}. \qquad (6.90)$$

Problem 6.13. Verify the expression for (6.90).

The capture process just described may be related to its reverse, photoionization, by detailed balance. If the photoionization cross section is $\sigma_{\nu n}$, then the number of photons absorbed of energy $h\nu$ per second with emission of electrons of energy $\frac{1}{2}m_e v^2 - E_n$ from the nth atomic state is

$$\frac{c u_{\nu p}}{h\nu} d\nu\, \sigma_{\nu n} N_n (1 - e^{-h\nu/kT}). \qquad (6.91)$$

Problem 6.14. Derive (6.91).

Here N_n is the number of atoms in the state n, and spontaneous emission has been included. In the reverse process, the number of electrons with initial velocities v captured per second into the atomic state n is given by

$$N_i \sigma_{cn} n_e v f(v)\, dv. \qquad (6.92)$$

The semiclassical arguments in the preceding are applicable to hydrogen and hydrogen-like ions. For hydrogen-like atoms, $Z > 1$, the ion has charge Z, and the $Z - 1$ electrons of the "neutral" ion are not affected by bound-free processes. Applying Boltzmann's relation, the ratio of atoms in the nth excited state to those in the ground state is

$$\frac{N_n}{N_1} = \frac{g_n}{g_1} \exp[-I_H Z^2(1 - 1/n^2)/kT], \qquad (6.93)$$

where $g_1 = 2$, and $g_n = 2n^2$ for hydrogen-like atoms.

Next, the Saha equation may be used to relate the numbers of atoms N, ions N_i, and the number density of electrons n_e:

$$\frac{N_i N_e}{N} = 2\frac{g_i}{g}\left(\frac{2\pi m_e kT}{h^2}\right)^{3/2} e^{-I_H Z^2/kT}. \qquad (6.94)$$

Problem 6.15. Show that the number of atoms N is related to the number in the ground state N_1 by

$$N/N_1 = g/g_1, \qquad (6.95)$$

where

$$g = \sum_n g_n \exp\,[-I_H Z^2(1 - 1/n^2)/kT] \qquad (6.96)$$

is the electronic partition functions of the atoms.

The principle of detailed balance asserts that (6.91) and (6.92) must be equal. Equating these, using (6.93) and (6.95) to eliminate N_1, and assuming that the electrons are Maxwellian and that $u_{\nu p}$ is given by (5.49) in thermodynamic equilibrium, we can straightforwardly show that

$$\sigma_{\nu n} = \frac{g_i}{g_n}\left(\frac{m_e v c}{h\nu}\right)^2 \sigma_{cn}. \qquad (6.97)$$

Since we have assumed that the $Z - 1$ electrons in the atom do not participate in the bound-free process, we set $g_i = 1$. Combining the preceding results, and including the bound-free Gaunt factor $g_{\text{bf}}(\nu, n)$, which corrects for quantum-mechanical effects, we find

$$\sigma_{\nu n} = \frac{64\pi^4}{3\sqrt{3}} \frac{m_e Z^4 e^{10}}{h^6 c \nu^3 n^5} g_{\text{bf}}(\nu, n). \qquad (6.98)$$

The Gaunt factor is of order unity, and is a slowly varying function of frequency. The capture cross section follows immediately from (6.96). When using $\sigma_{\nu n}$ or σ_{cn}, we must remember that the photon involved must have an energy in excess of the binding energy in the nth level, E_n. Therefore the cross sections vanish for $h\nu < I_H Z^2/n^2$. Conversely, the only atomic states that are coupled to photons of frequency ν by bound-free processes have energy levels E_n for $n > n^*$, where

$$n^* = (I_H Z^2/h\nu)^{1/2} \qquad (6.99)$$

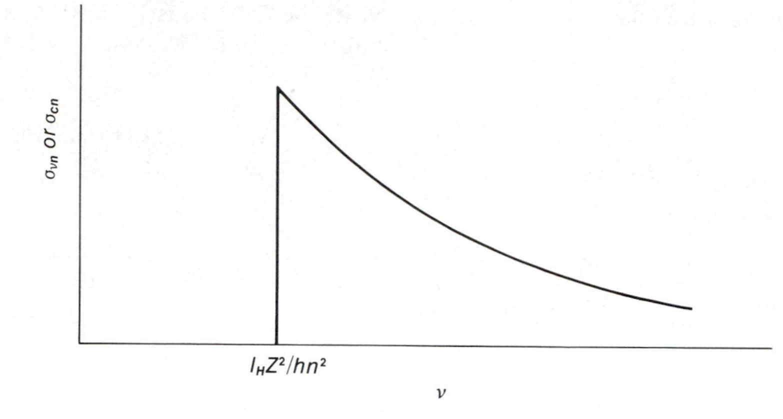

Figure 6.11. Absorption edge. The absorption cross section for bound or free states is zero for frequencies less than $I_H Z^2/hn^2$.

(see Figure 6.11). This defines the absorption edge for electrons in state n. Since $h\nu \approx 1/n^2$ at the absorption edge, the peak value of the absorption coefficient is proportional to $(n^5\nu^3)^{-1} \sim n^6/n^5 \sim n$.

A contribution to bound-free absorption of the form (6.98) will occur whenever there are enough atoms in the nth excited state. Denoting their number density by N_n, the total opacity for bound-free transitions is

$$\kappa_{\text{bf}}(\nu) = \sum_n \frac{N_n}{\rho} \sigma_{\nu n}. \tag{6.100}$$

When bound-free absorption is important, most atoms are unionized and $N \approx N_1$. Then (6.93) shows that $N_n \approx 2n^2$ for large n. Since $\sigma_{\nu n} \sim n$ at the absorption edge, the terms in the sum (6.100) increase as n^{+3}, so that the major contribution to κ_{bf} is from highly excited states. The dependence of $\kappa_{\text{bf}}(\nu)$ on frequency is shown schematically in Figure 6.12. The dashed curves give the contribution from a single excited state. The absorption edges, designated by the principle quantum number of the electron shells, are also shown. For photon energies $h\nu > h\nu_1$, the absorption is $\sim\nu^{-3}$, and no further edges occur.

Problem 6.16 (straightforward, but long). Calculate the absorption cross section for a hydrogen atom in the third principal quantum level ($n = 3$) at the absorption edge, i.e., where the atom is just capable of absorbing into the continuum. Then calculate the opacity of this atom at a wavelength equal to that of H_α. Next calculate (we've done it already) the absorption coefficient at the center of the H_α line per atom in the second level ($n = 2$). Taking a temperature of 6,000 K, estimate the relative opacity of the line and continuum at H_α. Assume that the continuum opacity is nearly the same as the Rosseland mean opacity. Use the simple model atmosphere $T^4 = T_{\text{eff}}^4$ $(\tau = 2/3)$, and estimate the strength of the H_α line as measured by the flux emitted from the center of H_α, relative to that emitted in the surrounding continuum; this is known as the central depth of the line. (This problem contains a number of erroneous assumptions. In particular: the continuum opacity is not primarily the Paschen continuum; the H_α is not broadened only by natural line broadening; the continuum opacity is not equal to the Rosseland mean; and the simple Eddington model atmosphere is not very good. Nevertheless, this is the method and procedures one must follow in more sophisticated calculations.)

Once the population of excited states N_n is known, $\kappa_{\text{bf}}(\nu)$ may be substituted into the expression for the Rosseland mean to obtain κ_{bf}. When this is done, we find that

$$\kappa_{\text{bf}} = \kappa_0 \rho T^{-3.5}, \tag{6.101}$$

where for bound-free absorption

$$\kappa_0 = 4.34 \times 10^{25} Z(1 + X) \frac{\overline{g}_{\text{bf}}}{t}. \tag{6.102}$$

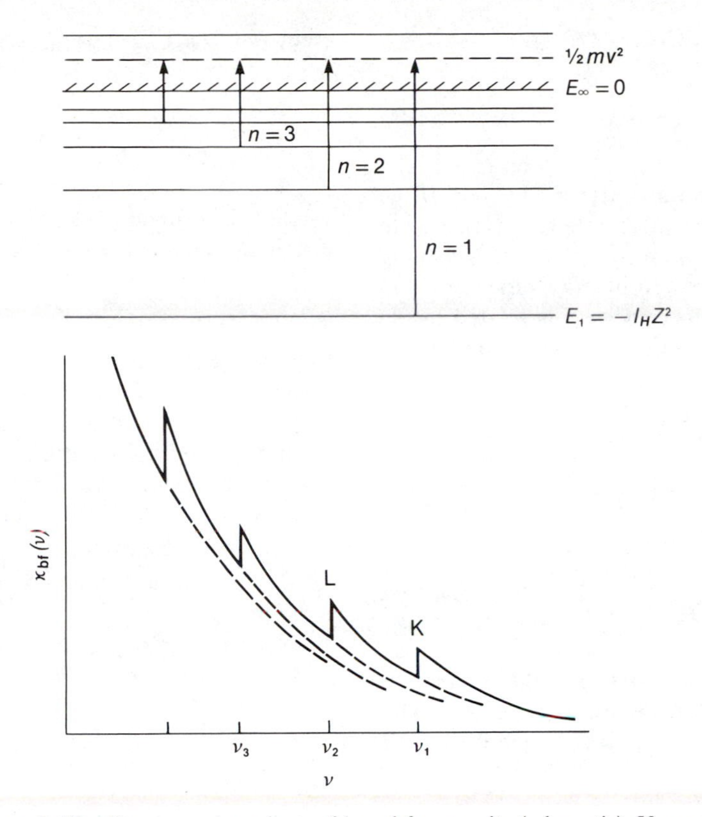

Figure 6.12. Frequency dependence of bound-free opacity (schematic). Upper figure shows energy levels. Also shown are K and L absorption edges corresponding to initial states $n = 1$ and $n = 2$.

Here $\bar{g}_{bf}$ is a frequency-averaged Gaunt factor, and the quantity t is the guillotine factor. This last factor corrects for an overestimate of the number of bound electrons per atom arising at low temperatures from approximations used in obtaining (6.101).

Problem 6.17. Show why κ_0 is proportional to the heavy-metal abundance Z for bound-free absorption.

Problem 6.18. Compare the contributions of bound-free and free-free absorption to the opacity of Population I and II stars, using (6.87) and (6.101) with $\bar{g}_{bf} = \bar{g}_{ff} = 1$, and $t = 10$.

Electron Conduction

We noted in Section 5.11 that we could incorporate heat transfer by electron conduction into the radiative transfer equation by introducing the electron-conduction opacity κ_c, which is related to the conductivity λ_c by (5.148). Consider a nondegenerate plasma containing electrons and ions. The average velocity $\bar{v}$ of the electrons is much greater than that of the ions; so the electrons are responsible for heat transfer. The Coulomb field of the ions, however, is strong enough to scatter the electrons.

First, consider the electron energy-transfer process. The energy flux is given by the product of the electron flux $n_e\bar{v}$ and the electron energy difference over the distance of an electron's mean free path, ℓ_e:

$$\begin{aligned}\mathcal{F}_{\text{cond}} &\simeq -n_e\bar{v}[\epsilon(x+\ell_e)-\epsilon(x)] \\ &\simeq -n_e\bar{v}\ell_e\frac{\partial\epsilon}{\partial x}\,.\end{aligned} \tag{6.103}$$

For nonrelativistic electrons, $\epsilon = m_e\bar{v}^2/2 \simeq 3kT/2$, so that

$$\partial\epsilon/\partial x \simeq (3k/2)\,\partial T/\partial x.$$

Substituting into $\mathcal{F}_{\text{cond}}$ and comparing with (5.147), we find the thermal conductivity,

$$\lambda_c = \frac{3}{2} n_e \bar{v} \ell_e k. \tag{6.104}$$

Second, consider electron scattering by ions. If the cross section is σ, then the mean free path is related to the ion number density n_i by $\ell_e \simeq (\sigma n_i)^{-1}$. Assuming that the scattering is due to Coulomb interactions, and that the average electron velocity $\bar{v}$ is related to the ion-electron separation r_0 during scattering by

$$\frac{1}{2} m_e \bar{v}^2 = \frac{Ze^2}{r_0},$$

we find for the cross section

$$\sigma = \pi r_0^2 \simeq \frac{4\pi}{9} \frac{Z^2 e^4}{k^2 T^2}, \tag{6.105}$$

where we have used $\bar{v} = \sqrt{3kT/m_e}$. Eliminating the mean free path from (6.104) and using (6.105) with $n_e = \rho(1 + X)/2m_H$ and $n_i = \rho/Am_H$, we find

$$\lambda_c = \frac{27\sqrt{3}}{16\pi} \frac{k^{7/2}}{m_e^{1/2} e^4} \frac{A}{Z^2} (1 + X) T^{5/2}. \tag{6.106}$$

When this is substituted into (5.148) for the opacity, we find

$$\kappa_c \sim \frac{4ac}{3} \frac{16\pi}{27\sqrt{3}} \frac{m_e^{1/2} e^4}{k^{7/2}} \frac{Z^2/A}{1 + X} \frac{T^{1/2}}{\rho}$$

$$\simeq 5 \times 10^3 \frac{Z^2/A}{1 + X} \frac{T_{(7)}^{1/2}}{(\rho/10)}, \tag{6.107}$$

where $T_{(7)} = T \times 10^{-7}$ and ρ is in g/cm^3.

This simple model is good to a factor of 2 or 3 in the nonrelativistic and nondegenerate regime. Electron conduction is also important when densities are high enough that the plasma is degenerate and the electrons relativistic. The primary change when the electrons become relativistic is an increase in $\bar{v}$ in (6.104). Degeneracy reduces scattering, since there will be few available final electron states unoccupied. This in turn increases the mean free path ℓ_e. Both effects increase λ_c and decrease κ_c. Therefore, electron conduction becomes extremely efficient at high densities. For nonrelativistic electrons in the limit of extreme degeneracy ($kT \ll \mu_e$), the opacity is given by

$$\kappa_c \simeq 5 \times 10^{-3} \frac{(Z^2/A)}{(1 + X)^2} \frac{T_{(7)}^2}{(\rho/10^5)^2} \Theta, \tag{6.108}$$

where Θ is a function of order unity that corrects for distant encounters between electrons and ions.

Problem 6.19. Show that (6.108) has the correct dependence on Z, A, T, and ρ for nonrelativistic degenerate electrons. First, show that the average thermal energy per electron $\epsilon \approx T^2 n_e^{-2/3}$. Relate the average velocity $\bar{v}$ of an electron to the Fermi energy to show that $\bar{v} \approx n_e^{1/3}$, and then show that (6.104) is

$$\sigma_{\text{deg}} \simeq \pi r_0^2 \simeq \frac{\pi Z^2 e^4}{n_e^{4/3}} \tag{6.109}$$

Combine these results to show that $\lambda_c \simeq n_e^2 T/Z^2 n_i$ and that $\kappa_c \simeq (Z^2/A) \cdot (T/\rho)^2$.

Total Opacity

Combining the contributions from all sources of continuous opacity in stellar atmospheres is somewhat long and tedious. For given T, ρ, P, and chemical composition, one must first calculate the degrees of ionization and excitation of all (actually, all relevant) atomic and ionic species. One must then insert the bound-bound, bound-free, and free-free absorption cross sections, weighted according to the number of atomic species in each level. One adds electron scattering and Rayleigh scattering, and then totals the whole thing to give the variation of opacity with wavelength or frequency.

The situation is simpler in stellar interiors, where the Rosseland mean opacity may be used in place of frequency-dependent values. Furthermore, in stellar interiors the atoms are mostly ionized; so the major sources of opacity are free-free and bound-free absorption, and electron scattering. For simple analyses the analytic approximations given by (6.86), (6.87), (6.101), and (6.102) may often be used.

The principal sources of opacity for Population I stellar interiors are summarized in Figure 6.13, which shows the locus of points in the $\rho - T$ plane where two opacities are equal. For example, above the line $\kappa_{bf} = \kappa_e$ the electron opacity dominates the bound-free opacity.

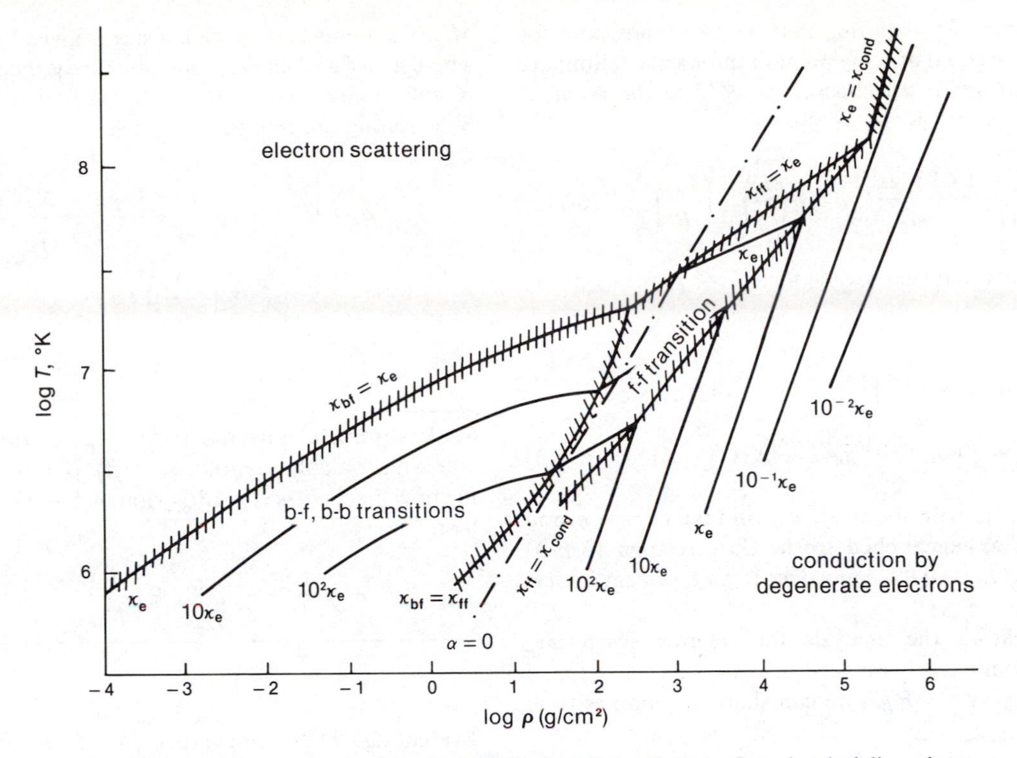

Figure 6.13. Stellar opacity in the ρ,T-plane for Population I stars. Cross-hatched lines denote boundaries at which the contributions from the two atomic process shown are equal.

6.5. SIMPLIFIED STELLAR MODELS

The pressure and temperature of the linear model were discussed in Section 4.3. We return to this model to illustrate some aspects of radiative transport and convection. Some of the results to be developed here will be useful when we discuss stellar evolution in Part III. We begin by constructing the mass-radius-luminosity relation for stars in radiative equilibrium.

The temperature $T(r)$ is given by (4.34). The temperature gradient may be found by differentiating this expression with respect to radius:

$$\frac{dT}{dr} = \frac{5\pi}{36}\frac{G\mu m_H}{k}\rho_c R\left(1 - \frac{38}{5}\frac{r}{R} + \frac{27}{5}\frac{r^2}{R^2}\right). \quad (6.110)$$

The star's luminosity $L(r)$ is related to dT/dr by (5.54) when energy transport is radiative. Assuming this to be the case, we might be tempted to obtain $L(r)$ and finally $L = L(R)$ in this way. Unfortunately, the resulting luminosity at the surface vanishes. The difficulty is traceable to the surface boundary conditions $P(R) = 0$ and $\rho(R) = 0$.

One way around this difficulty is offered by the following simple model. Define the temperature gradient as in (6.110), but take $L(r) = L =$ constant throughout the star:

$$\frac{dT}{dr} = -\frac{3\bar{\kappa}\rho}{16\pi ac}\frac{L}{r^2T^3}. \quad (6.111)$$

This presents two difficulties: (1) the gradient becomes infinite at the origin, which implies a point source for the luminosity; (2) the two expressions for dT/dr can be equal at only one point in the star, which implies that the system is not in thermal equilibrium. This last point is not surprising; since $\rho(r)$ was arbitrarily specified, the star can be adjusted to be in mechanical equilibrium. This is what gave $P(r)$ and, via the ideal gas law, $T(r)$. Once $L(r)$ is imposed, however, there is no way to readjust the system into thermal equilibrium. Nevertheless, the model does produce useful estimates of the mass-radius-luminosity relation. By requiring that the two expressions for dT/dr be equal at some point in the star, we eliminate temperature. If we choose $r = R/2$ as the point, it follows that for Kramers' opacity

$$\frac{L}{L_\odot} = \frac{1.43 \times 10^{25}}{\kappa_0} \mu^{7.5} \left(\frac{M}{M_\odot}\right)^{5.5} \left(\frac{R_\odot}{R}\right)^{0.5}. \quad (6.112)$$

An approximate expression for κ_0 for Kramers' opacity is

$$\kappa_0 = 4.34 \times 10^{25} \left(\frac{\bar{g}_{bf}}{t}\right) Z(1 + X) + 3.68 \times 10^{22} \bar{g}_{ff}(1 + X)(1 - Z). \quad (6.113)$$

Unless otherwise specified, we will take $\bar{g}_{bf}/t = 1/3$ and $\bar{g}_{ff} = 2$. When applied to the Sun, relation (6.113) yields $L/L_\odot \approx 0.92$, equivalent to an 8 percent underestimate.

Repeating the analysis for electron scattering, which dominates in massive stars $\kappa = 0.2(1 + X)$, and matching at $r = R/2$, we can show the luminosity to be

$$\frac{L}{L_\odot} = \frac{179}{1 + X} \mu^4 \left(\frac{M}{M_\odot}\right)^3. \quad (6.114)$$

The luminosity is independent of radius for electron scattering, and is less sensitive to mass.

Stars with highly temperature-sensitive nuclear-energy sources in their cores will become convective, with outer radiative envelopes. Typically, this occurs in upper main-sequence stars, wherein H burns by the CN cycle and the CNO bi-cycle. The size of the core may be estimated by noting that at the boundary between the convective core and the radiative envelope, $(dT/dr)_{ad} = (dT/dr)_{rad}$. Equating the two temperature gradients and using $aT^4/3 = P_{rad} \equiv (1 - \beta) P$ to eliminate the temperature, we find immediately that

$$q_c \equiv \frac{M_c}{M} = \frac{\Gamma_2}{\Gamma_2 - 1} \frac{\kappa L}{16\pi c G M(1 - \beta)}. \quad (6.115)$$

The core mass $M_c = M(r_c)$, and we assume that all energy is generated inside r_c, so that $L(r_c) = L$. Although not shown explicitly, κ is also evaluated at the core. We will obtain q_c for a star whose principal source of opacity is electron scattering, as is typical of upper main-sequence stars more massive than about 3 $M_\odot$. The luminosity of such a star is given by (6.114), where $\mu \rightarrow \beta\mu$ when radiation pressure is included, and X and μ are evaluated in the radiative envelope. Substituting this into (6.115) yields

$$q_c = 6.93 \times 10^{-4} \frac{\Gamma_2}{\Gamma_2 - 1} \frac{1 + X_c}{1 + X_e} \times \frac{(\beta\mu_e)^4}{1 - \beta} \left(\frac{M}{M_\odot}\right)^2. \quad (6.116)$$

Problem 6.20. Assume that (T^3/ρ) is constant in a star, and evaluate the ratio $(1 - \beta)/\beta$. Use the result to show that if T_c and ρ_c are given by the linear model, then

$$\frac{(\beta\mu)^4}{1 - \beta} = 207.9 \left(\frac{M_\odot}{M}\right)^2. \quad (6.117)$$

Evaluating μ in the core in (6.117) and substituting the result into q_c in (6.116) yields

$$q_c = 0.14 \frac{\Gamma_2}{\Gamma_2 - 1} \frac{1 + X_c}{1 + X_e} \left(\frac{\mu_e}{\mu_c}\right)^4. \quad (6.118)$$

When the gas is completely ionized, as in stellar cores, $\Gamma_2 = 5/3$, and, from (2.34), it follows that

$$\frac{\Gamma_2}{\Gamma_2 - 1} = \frac{32 - 24\beta - 3\beta^2}{8 - 6\beta}. \quad (6.119)$$

It follows from (6.117) and (6.119) that q_c depends on the star's mass only through β, and it is easily shown that (6.119) varies from 4 when $\beta = 0$ to 2.5 when $\beta = 1$. At the beginning of main-sequence evolution, when $\mu_e \approx \mu_c$, q_c lies between 0.35 and 0.56, depending on the amount of radiation pressure present.

Problem 6.21. Explain physically why stars with appreciable radiation pressure have larger convective cores than ones with less. Describe the change in mass of the convective region as the star evolves. Do not assume β to be constant during evolution.

Fully Convective Stars

Most stars contain both radiative and convective zones. However, there are gravitating configurations in which the entire system, except for the atmosphere, is convective. One of the most important categories is stars that have not yet reached the main sequence. In these stars, the temperature is still low enough that nuclear-energy generation has not begun, and all the radiated energy comes from contraction, that is, the conversion of gravitational potential energy into internal energy and radiation. Simple models of fully convective stars may be divided into three groups: (1) stars with low surface density; (2) stars with high surface density; and (3) stars with high surface density and with ionization zones below the radiative photosphere. We will consider these three cases separately.

In fully convective stars, the luminosity depends on the energy flux at the boundary of the photosphere. Figure 6.14(a) shows a convective envelope just below the radiative atmosphere. In Figure 6.14(b), the convective envelope and radiative atmosphere are separated by a hydrogen ionization zone. The base of the photosphere is at $\tau = 2/3$. We will assume that the convective interior ends at the base of the photosphere, and that the opacity in the photosphere may be approximated by the relation

$$\kappa = \kappa_0 P^a T^b, \tag{6.120}$$

where both a and b are positive. Taking the bottom of the photosphere to be at optical depth $\tau = 2/3$ and applying the results of Section 5.7, we can readily show that (see Problem 6.31)

$$P_{\text{ph}} = \frac{2}{3}(a + 1)\frac{g}{\kappa_{\text{ph}}(T_{\text{eff}}, P_{\text{ph}})}. \tag{6.121}$$

These serve as boundary conditions for the fully convective star. For numerical purposes, we use the values for κ_0, a, and b in Table 6.3. These values approximate the opacity due to H^- ions when the temperature is around 3,500 K.

The luminosity is fixed by specifying an additional boundary condition on P_{ph} and T_{eff} at the base of the photosphere, namely, that the radiative flux entering the photosphere equals the convective flux leaving the stellar interior (energy conservation):

$$F_{\text{rad}} = F_{\text{conv}}. \tag{6.122}$$

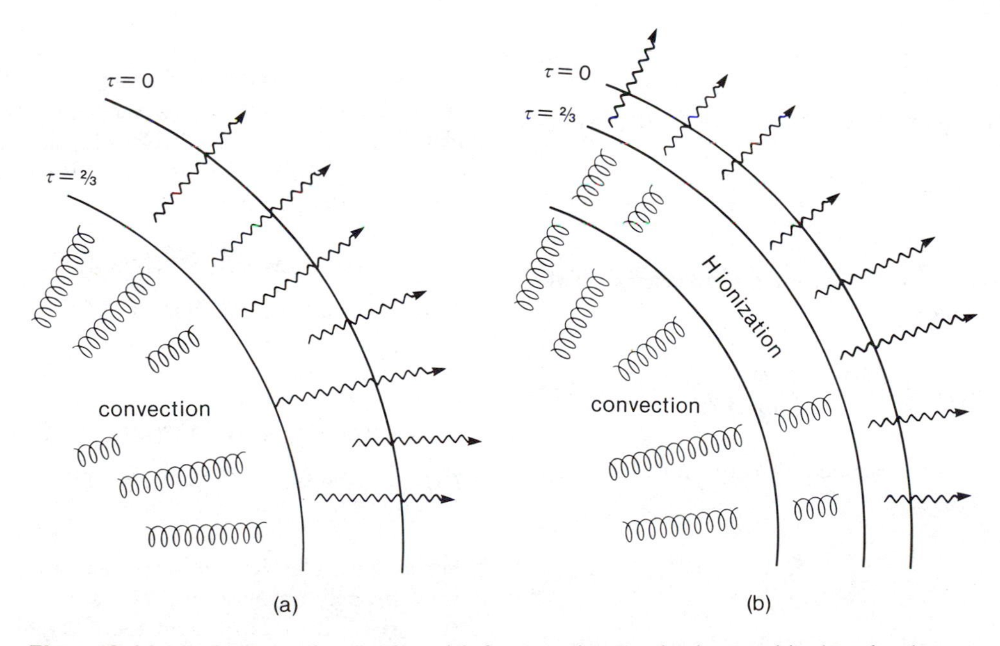

Figure 6.14. Surface layers in a simple model of a convective star, showing transition boundary between convective interior and a surface layer in which τ varies from 2/3 to zero. Essentially all observed radiation arises from this layer. (a) Transition layer directly above neutral H; (b) H ionization zone at base of transition layer.

Table 6.3
Coefficients for the opacity approximation (6.120) for Population I and Population II stars. The opacity is in $cm^2\ g^{-1}$.

Population	κ_0	a	b
I	6.9×10^{-26}	0.7	5.3
II	6.1×10^{-40}	0.6	9.4

Therefore, we construct approximate expressions for F_{rad} and F_{conv}. The former is trivial:

$$F_{\text{rad}} = \sigma\, T_{\text{eff}}^{\ 4} = \frac{ac}{4}\, T_{\text{eff}}^{\ 4}. \tag{6.123}$$

The convective flux is more difficult, and is treated later.

Stars with Low Surface Density

In stars having low surface densities, convection is inefficient near the photosphere because of the large temperature gradient that develops when ρ is small, so that there are substantial radiative losses from the convective elements. According to (5.140) the convective flux is given by

$$F_{\text{conv}} = \tfrac{1}{2}\rho c_P \bar{v}\, \Delta T. \tag{6.124}$$

Let us approximate the radial-velocity fluctuation by

$$\bar{v} \simeq \frac{1}{2} c_s = \frac{1}{2}\left(\frac{\gamma k T}{\mu m_H}\right)^{1/2}, \tag{6.125}$$

and approximate $\rho c_P \Delta T$ by the internal energy of the convective element:

$$\rho c_P \Delta T \simeq \frac{3}{2}\gamma\rho\left(\frac{k}{\mu m_H}\right)\Delta T. \tag{6.126}$$

In stars with low surface density, $\Delta T \simeq T$, and (6.124)–(6.126) give

$$F_{\text{conv}} \simeq \frac{3\gamma}{8}\left(\gamma\frac{k}{\mu m_H}\right)^{1/2} P T^{1/2}, \tag{6.127}$$

where the ideal gas law was used to eliminate ρ. Equating (6.127) and (6.123), and denoting the pressure by P_{ph}, we find that

$$P_{\text{ph}} = \frac{8}{3\gamma}\left(\frac{\mu m_H}{\gamma k}\right)^{1/2} \sigma\, T_{\text{eff}}^{\ 3.5} \tag{6.128}$$

for fully convective stars with low surface density. The two conditions (6.128) and (6.121) may be used to eliminate P_{ph} and give T_{eff} in terms of the stellar mass and radius:

$$T_{\text{eff}} \sim \left[\frac{M}{\mu^{(a+1)/2}\kappa_0 R^2}\right]^{1/[b+3.5(1+a)]} \tag{6.129}$$

Problem 6.22. Derive (6.129), including the constant of proportionality for arbitrary a, b, and γ.

The constant factor in (6.129) depends on a, b, the adiabatic index γ, and physical constants; for $n = 1.5$ and a composition typical of Population II or Population I stars, the constant is between 6 and 8×10^3, respectively, with M and R in solar units. Finally, the star's luminosity is obtained from

$$L = 4\pi\sigma R^2 T_{\text{eff}}^{\ 4}. \tag{6.130}$$

Assuming the opacity coefficients given in Table 6.3 and the chemical composition for Population I and Population II stars given in Table 4.1, it is straightforward to show that (6.129) and (6.130) reduce to the following. For Population I stars

$$\begin{aligned} T_{\text{eff}} &= 7.5 \times 10^3 (M/M_\odot)^{0.089}(R/R_\odot)^{-0.178}, \\ L/L_\odot &= 2.86 (M/M_\odot)^{0.356}(R/R_\odot)^{1.288}; \end{aligned} \tag{6.131}$$

for Population II stars,

$$\begin{aligned} T_{\text{eff}} &= 6.2 \times 10^3 (M/M_\odot)^{0.066}(R/R_\odot)^{-0.133}, \\ L/L_\odot &= 1.34 (M/M_\odot)^{0.267}(R/R_\odot)^{1.466}. \end{aligned} \tag{6.132}$$

Whereas in radiative stars the luminosities often depend strongly on mass, (6.131), and (6.132) show that in these convective stars there is only a weak dependence of luminosity on mass, and a much stronger dependence on radius.

Problem 6.23. Explain physically why the surface opacity of Population II stars should be more temperature-dependent than that of Population I stars. Assume that H^- ions are a major source of opacity. Then explain why the surface temperature of Population II stars is less sensitive to radius than that of Population I stars.

Stars with High Surface Density

A high surface density implies that convective transport is very efficient. The temperature gradient is therefore nearly adiabatic. It can then be shown that the temperature fluctuations are small compared to the temperature itself. Approximating the temperature gradient by $(dT/dr)_{\text{ad}}$, the pressure and temperature are related by

$$P = KT^{\gamma/(\gamma-1)}. \tag{6.133}$$

For fully convective stars, equation (6.133) holds throughout. Evaluating it at the center of the star and base of the photosphere yields (if $\gamma = 5/3$ throughout)

$$\frac{P_c}{T_c^{2.5}} = \frac{P_{\text{ph}}}{T_{\text{eff}}^{2.5}}, \tag{6.134}$$

since K is a constant. This gives one relation between P_{ph} and T_{eff}. The second is (6.121); when (6.121) and (6.134) are combined, along with a model relating P_c and T_c to the mass and radius, we obtain the effective temperature $T_{\text{eff}}(M, R)$.

Problem 6.24. Use the linear model for the convective interior of a star with high surface density, and solve for T_c as a function of M and R. Assume κ as given by (6.120), with the parameters given in Table 6.3. Compare results with (6.131)–(6.132). Take $\gamma = 5/3$.

The effective temperature and luminosity, which are obtained when κ is given by (6.120) are, for Population I stars,

$$\begin{aligned} T_e &= 2.2 \times 10^3 (M/M_\odot)^{0.194} (R/R_\odot)^{0.0576}, \\ L/L_\odot &= 0.02 (M/M_\odot)^{0.776} (R/R_\odot)^{2.23}, \end{aligned} \tag{6.135}$$

and for Population II stars,

$$\begin{aligned} T_{\textit{eff}} &= 2.5 \times 10^3 (M/M_\odot)^{0.172} (R/R_\odot)^{0.0298}, \\ L/L_\odot &= 0.035 (M/M_\odot)^{0.686} (R/R_\odot)^{2.119}. \end{aligned} \tag{6.136}$$

The temperature T_{eff} of convective stars with high surface density increases with radius, whereas T_{eff} for those with low surface density decreases with radius. The mass dependence of the luminosity is somewhat stronger here than in (6.131)–(6.132), though it is still much weaker than for main-sequence stars that combine radiation and convection.

Ionization Zones

Stars that have central temperatures above 2×10^5 K will have zones containing partially ionized hydrogen. On either side of this layer the preceding methods are applicable. The variables P and T at the inner and outer boundaries of the ionization zone must be related. From (6.133) it follows that for $\gamma = 5/3$

$$\frac{P_c}{T_c^{2.5}} = \frac{P_b}{T_b^{2.5}}, \tag{6.137}$$

where P_b is the pressure at the base of the ionization zone. Since the ionization zone is convective, the entropy across it is constant. For a mixture of hydrogen and helium of mass fractions X and Y, where x denotes the fraction of hydrogen ionized, the entropy per unit mass is

$$\begin{aligned} S = \frac{Xk}{m_H}\Bigg[&\frac{5}{2}(1 + x + \delta) + \frac{\chi}{kT} + \ln\left(\frac{2\pi m_H}{h^2}\right)^{3/2} \\ &+ \delta \ln\left(\frac{2\pi m_{He}}{h^2}\right)^{3/2} + x \ln\left(\frac{2\pi m_e}{h^2}\right)^{3/2} \\ &+ (1 + x + \delta)\ln(kT)^{5/2}\frac{(1 + x + \delta)}{P}\Bigg]. \end{aligned} \tag{6.138}$$

The ionization energy of hydrogen is denoted by χ, and $\delta = Y/4X$. The fraction of hydrogen ionized, x, varies from $x = 1$ at the base to $x = 0$ at the top of the zone. Continuity of S implies that $S(1) = S(0)$, which yields a relation between the boundary conditions at the center of the star and at the base of the photosphere. Following the steps of Section 6.4, we may use (6.137) and (6.138), along with (6.121) to relate T_{eff} and L to the stellar mass and radius. In this way we may obtain, for Population I stars,

$$T_{\text{eff}} = 5.7 \times 10^3 (M/M_\odot)^{0.27} (R/R_\odot)^{0.288},$$
$$L/L_\odot = 0.93 (M/M_\odot)^{1.08} (R/R_\odot)^{3.15}, \tag{6.139}$$

and for Population II stars,

$$T_{\text{eff}} = 4.6 \times 10^3 (M/M_\odot)^{0.193} (R/R_\odot)^{0.204},$$
$$L/L_\odot = 0.42 (M/M_\odot)^{0.77} (R/R_\odot)^{2.82}. \tag{6.140}$$

The existence of a H ionization zone leaves the mass dependence of T_{eff} about the same, but greatly increases the dependence on stellar radius.

6.6. Line Broadening and Line Opacity

Line absorption and emission occur whenever enough neutral or partially ionized atoms are present. Energy transport, however, is nearly always dominated by continuous absorption processes, as may be understood by considering the energy emitted from a layer of gas in thermal equilibrium at temperature T. The continuous spectrum $B_\nu(T)$ carries off an energy per unit area per second given by

$$cB(T) = c\int B_\nu(T)\, d\nu = \sigma T^4.$$

Line emission from matter at temperature T, which is transparent to continuum radiation but is opaque to lines, will produce a spectrum I_ν shown by the spikes in

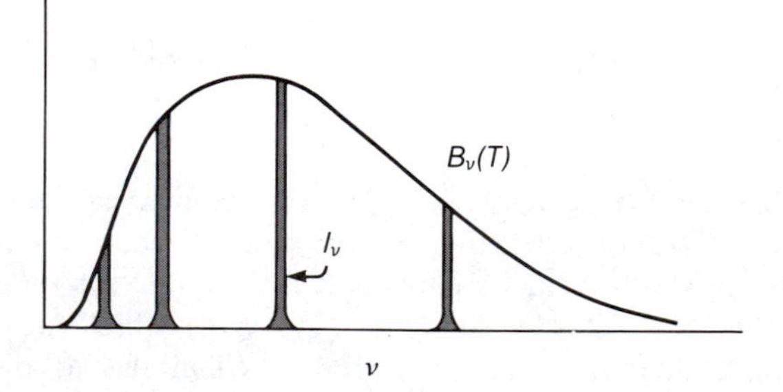

Figure 6.15. Line emission. Solid line is Planckian spectrum at temperature T. Individual lines (shaded) are emitted by matter at temperature T. Total intensity in lines is given by the shaded area.

Figure 6.15, all of which lie on the black-body curve $B_\nu(T)$.

For lines in the visible part of the spectrum, $\nu \approx 10^{14}$ sec^{-1} and their widths $\Delta\nu \approx 10^8$ sec^{-1}, so that $\Delta\nu/\nu \approx 10^{-6}$. It is therefore evident that the total energy density absorbed per second $I = \int I_\nu\, d\nu$ by lines in normal stars is small compared to $B(T)$.

Nevertheless, line emission and absorption are extremely important in stellar astrophysics, because they are sensitive to local conditions in the gas. In fact, a detailed study of line shapes can give information on ρ, P, and atomic abundances.

Finally, we note that line processes may actually dominate the energy-loss rate from nebulae, and play an important role in the astrophysics of interstellar matter.

The mechanism of line absorption in stellar atmosphere may be understood by examining the total (continuum plus line) absorption coefficient, an example of which is shown schematically in Figure 6.16. Strong absorption implies that the mean free path for photons near ν_0, $l_l(\nu_0) \approx (\rho\kappa_l(\nu_0))^{-1}$, is small compared with the mean free path of photons in the continuum, $l_c(\nu) \approx (\rho\kappa_c(\nu))^{-1}$. As a result, line absorption will occur at an optical depth τ_a in the stellar atmosphere, which is smaller than τ_c, where the continuum originates.

Absorption at τ_a is followed by emission, which will be characterized by the gas temperature $T(\tau_a)$, assuming that scattering is unimportant, $k^a(\nu) \gg k^s(\nu)$. Because $T(\tau_c) > T(\tau_a)$, the continuum energy flux $\sim B_{\nu_0}(T_c)$ will exceed the line intensity $I_{\nu_0}(\tau_a)$, and the spectrum will exhibit an absorption feature at ν_0. Before turning to a more detailed consideration of stellar line formation, we note that the continuous spectrum is not strictly Planckian, as is shown schematically in Figure 6.17. The differences may be explained in part by the fact that the escaping radiation originates from a region within which the temperature is not constant.

Problem 6.25. Using arguments similar to the preceding, explain the difference between the continuum flux of a star (solid curve) and the flux from a black body whose temperature is chosen to give the same energy density (broken line in Figure 6.17).

The preceding remarks suggest, as one approach to the study of line formation in stars, that we adopt the continuum radiation predicted by a model stellar

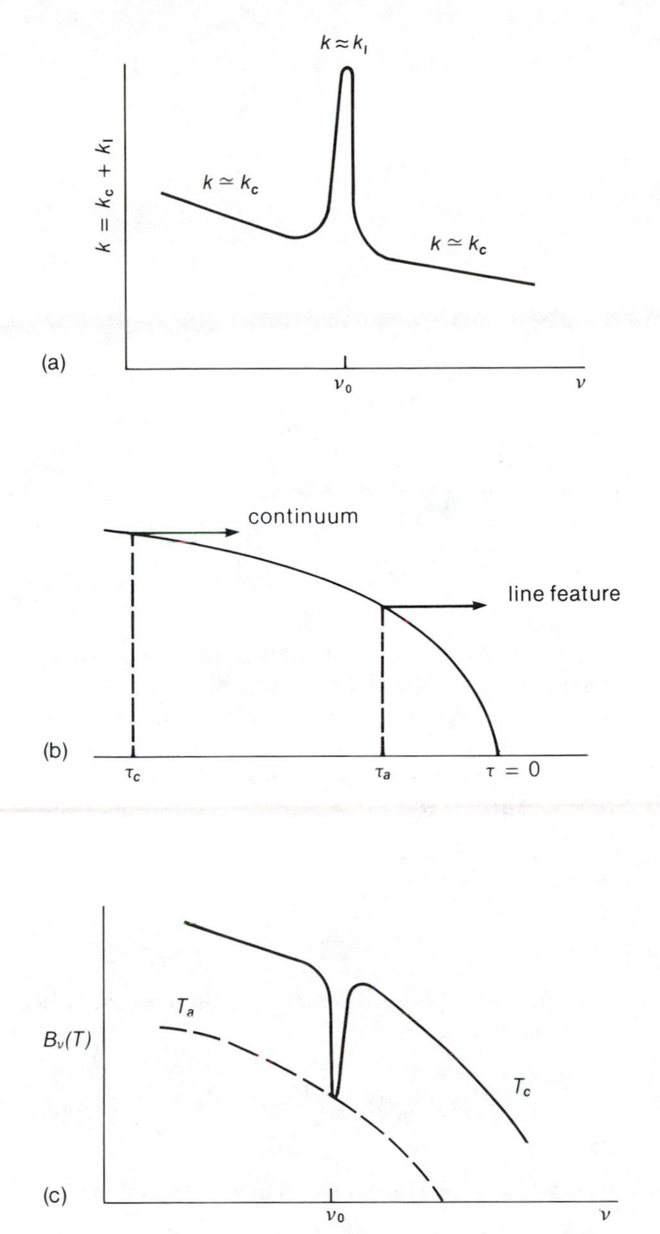

Figure 6.16. Line absorption: (a) total absorption coefficient (line plus continuum); (b) temperature profile showing location of continuum and line emission; (c) observed spectrum near line feature, with $B_\nu(T_a)$ shown as broken line.

atmosphere, and then investigate the absorption that results when radiation of given wavelengths passes through it. The composition parameters of the gas may then be adjusted to reproduce observed features in actual stellar atmospheres. Conversely, given the spectral features from observations (from which the composition of absorbing atoms is in principle obtained), we may infer characteristics of the atmosphere, such as density, temperature, and surface gravity. We therefore consider first line opacity.

In Section 6.3 we discussed natural line broadening, which arises from the uncertainty principle. This leads to the characteristic Lorentz shape for the absorption coefficient for a line:

$$a(\nu) = a_\nu = a_0 \frac{\gamma/4\pi^2}{(\nu - \nu_0)^2 + (\gamma/4\pi)^2}, \quad (6.141)$$

where

$$a_0 = (\pi e^2/m_e c)f.$$

Although natural broadening is always present, other sources of line broadening also occur in stellar spectra. One of the most important is the Doppler effect. Consider the emission and absorption of a photon by two atoms moving with relative velocity v. If the photon's frequency is ν_0 relative to the emitting atom, then in the frame of the absorbing atom it will be Doppler-shifted to the value ν given by

$$\frac{v}{c} = \frac{\nu_0 - \nu}{\nu_0}, \quad (6.142)$$

when $v/c \ll 1$.

Therefore a moving atom absorbs radiation from a slightly different range of photon frequencies than a stationary atom. If the absorption coefficient at frequency ν for a stationary atom is $a(\nu)$, then an atom moving at velocity v will absorb at frequency $\nu' = \nu + \nu(v/c)$ instead of ν. At a given frequency, then, the absorption coefficient is given by the product of the absorption coefficient of an atom moving at velocity v, $a(\nu - \nu v/c)$, times the relative fraction of atoms moving with this velocity, $f(v)$, integrated over all velocities:

$$a(\nu) = \int_0^\infty a(\nu - \nu v/c) f(v)\, dv. \quad (6.143)$$

Here the velocity v is in the direction of the photon path, and is defined so that a positive velocity causes a redshift.

The velocity distribution function, $f(v)$, is made up of at least two components, in principle. First, there is the thermal motion of the atoms, given presumably by Maxwell's law. Since only one component of the

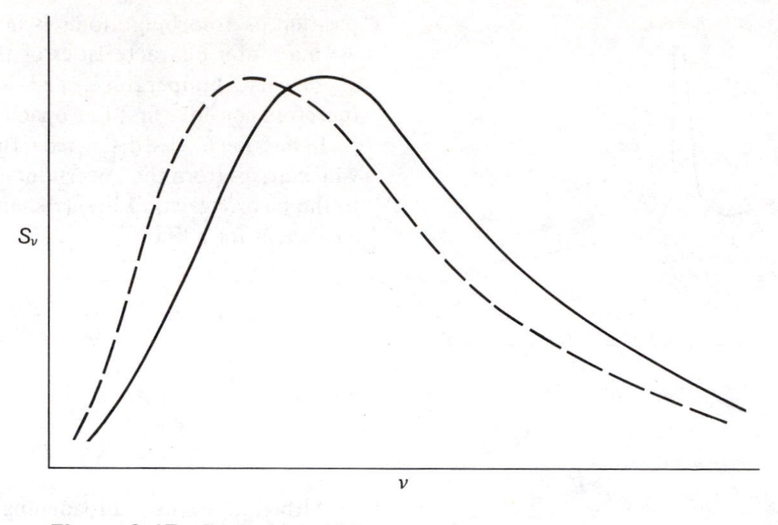

Figure 6.17. See Problem 6.25.

velocity appears in (6.143), the distribution of the velocity is simple:

$$f(v) = \frac{1}{\sqrt{\pi}}\left(\frac{m}{2kT}\right)^{1/2} e^{-mv^2/2kT}. \qquad (6.144)$$

The quantity

$$v_D = \sqrt{\frac{2kT}{m}}$$

is sometimes known as the *thermal Doppler velocity.* If this expression is used for $f(v)$ in (6.143), we have

$$a(\nu) = \frac{1}{\sqrt{\pi}}\frac{1}{v_D}\int_0^\infty a(\nu - \nu v/c)e^{-mv^2/2kT}dv$$

$$= \frac{1}{\sqrt{\pi}}\frac{1}{\Delta\nu_D}\int_0^\infty a(\nu - \nu')e^{-\nu'^2/\Delta\nu_D{}^2}d\nu'. \qquad (6.145)$$

Here we have defined the quantity

$$\Delta\nu_D = \frac{\nu_0 v_D}{c} = \frac{\nu_0}{c}\sqrt{\frac{2kT}{m}} \qquad (6.146)$$

as the Doppler width of the line in frequency units. In practice, $\Delta\nu_D$ is much larger than the natural line width. For example, for a sodium atom in the solar atmosphere, $T = 6{,}000$ K, we have $v_D = 2.1$ km/sec. The sodium D lines at 5900 Å have frequencies 5×10^{14} Hz, for which $\Delta\nu_D = 3 \times 10^9$ Hz. The natural line width is $\gamma \simeq 10^8$ Hz. Thus, the Doppler broadening dominates the natural line width. In these circumstances it is permissible to neglect the natural line width entirely, and the effective absorption coefficient per atom (6.141) can be approximated by a delta function, $a(\nu) = a_0\delta(\nu - \nu_0)$. If this is substituted into (6.145) we find

$$a(\nu) = \frac{a_0}{\sqrt{\pi}}\frac{1}{\Delta\nu_D}e^{-(\Delta\nu/\Delta\nu_D)^2}. \qquad (6.147)$$

Here $\Delta\nu = \nu - \nu_0$ is the distance from the line center. In wavelength units, we have

$$a(\lambda) = \frac{a_0(\lambda)}{\sqrt{\pi}}\frac{1}{\Delta\lambda_D}e^{-(\Delta\lambda/\Delta\lambda_D)^2}, \qquad (6.148)$$

where $\Delta\lambda_D$ is the Doppler width in wavelength units.

Problem 6.26. Imagine that the H_α line in the Sun's spectrum is broadened only by thermal Doppler effect. What is the opacity per atom at the line center?

A second contribution to the velocity field may come from turbulent motions in the stellar atmosphere. We know, for example, that the Sun has a convective zone near its surface. If the typical scale of these turbulent motions is smaller than the region of the stellar atmosphere over which the line is formed, it may be permissible to assume a Gaussian distribution for the turbulent velocities, in which case they will simply add to the effective Doppler width. That is, we may introduce a microturbulent velocity, v_T, so that the effective Doppler width is given by

$$v_D^2 = \frac{2kT}{m} + v_T^2. \tag{6.149}$$

If turbulent motions take place over a region that is large compared to a typical region of line formation (macroturbulence), then they must be handled in a different way, as we will see later. The intermediate case is quite complicated and not well understood. In practice, the Doppler velocity appears as a fitting parameter in predicting and determining line shapes.

Another important source of line broadening in stars is interatomic collisions. The word "collision" is to be interpreted somewhat loosely here. We imagine that an atom in an excited state is influenced by the passage nearby of another atom. Two extreme approximations are often used. First, the impact approximation assumes that the time-scale for the collision is short compared to the natural lifetime of the excited level (A^{-1}). This process may force the atom to de-excite sooner that it otherwise would. The net result is that the effective mean lifetime of the atom in the level is shortened, since the energy of the level is correspondingly less well defined (uncertainty principle). The line profile is a damping profile, but with the value of γ increased by the effect of collision de-excitation. This process can also be analyzed quasi-classically, in terms of an oscillator that randomly suffers discrete phase changes because of the collisions. This also predicts a damping profile with an increased value of γ.

If the number of collisionally induced transitions per unit time is Γ, the effective damping constant is $\gamma + \Gamma$.

Let σ_c be the cross section for collisional de-excitation, which depends on the force field of the colliding particle, as well as the detailed quantum-mechanical properties of the atomic energy levels. Since typical thermal velocities are of the order of the Doppler velocity v_D, the number of collisionally induced transitions per second will be of the form

$$\Gamma \sim n_c \sigma_c v_D, \tag{6.150}$$

where n_c is the number density of colliding particles (most often neutral hydrogen atoms, since these are the most common atomic species in cooler stars, where collisional broadening is most important). The ratio of the damping half-width to the Doppler half-width is therefore

$$\frac{\Gamma}{\Delta\nu_D} \sim \frac{n_c \sigma_c c}{\nu_0}.$$

If we examine the typical densities in the solar atmosphere, and suppose that σ_c is of the order of atomic dimensions, say 10^{-16} cm^2, then for a transition in the visible region of the spectrum, we find that Γ is much less than $\Delta\nu_D$.

Problem 6.27. Verify the preceding statement.

For situations in which we want to take care of both Doppler broadening and damping, the net absorption cross section as a function of frequency is given by the convolution of the two effects:

$$a(\nu) = \frac{\pi e^2}{m_e c} f \int_{-\infty}^{\infty} \frac{1}{\sqrt{\pi}} \times \frac{1}{\Delta\nu_D} e^{-(\Delta\nu - \Delta\nu')^2/\Delta\nu_D^2} \frac{\gamma/4\pi^2}{\Delta\nu'^2 - (\gamma/4\pi)^2} \, d\Delta\nu'.$$

By making the substitutions

$$v \equiv \frac{\Delta\nu}{\Delta\nu_D}, \qquad u \equiv \frac{\Delta\nu'}{\Delta\nu_D}, \qquad \alpha \equiv \frac{\gamma}{4\pi\Delta\nu_D} = \frac{\gamma}{2\Delta\omega_D},$$

we can write this in the form

$$a(\nu) = \frac{\pi e^2}{m_e c} f \frac{1}{\sqrt{\pi}} \frac{1}{\Delta\nu_D} \left[\frac{\alpha}{\pi} \int_{-\infty}^{\infty} \frac{e^{-(v-u)^2}}{u^2 - \alpha^2} du \right]. \tag{6.151}$$

The quantity in square brackets is known as the *Voigt function,* and is denoted by $H(\alpha, v)$. At the center of the line, $v = 0$, and

$$a(\nu_0) = \frac{\pi e^2}{m_e c} f \frac{1}{\sqrt{\pi}} \frac{1}{\Delta\nu_D} H(\alpha, 0). \tag{6.152}$$

In situations of importance in astrophysics, $\alpha \ll 1$; that is, the Doppler broadening is more important than damping. In this case, $H(\alpha, 0) = 1$; so the opacity at the line center is

$$a(\nu_0) = \frac{\pi e^2}{m_e c} f \frac{1}{\sqrt{\pi}} \frac{1}{\Delta\nu_D}.$$

The line opacity can also be written

$$\frac{a(\nu)}{a(\nu_0)} = H(\alpha, v), \qquad v = \Delta\nu/\Delta\nu_D.$$

A typical value for α is about 0.1. Figure 6.18 shows $H(\alpha, v)$ for several values of α. It will be seen that in the center of the line the opacity is dominated by the Doppler broadening, and is unaffected by changes in α, whereas the wings of the line are determined by the value of α. This is crucial in understanding line formation in stellar spectra: the core of the line is produced by the Doppler effect; the wings are produced by damping. In the wings of the line, $v \gg \alpha$, and the Voigt function is given quite accurately by

$$H(\alpha, v) \simeq \frac{\alpha}{\sqrt{\pi}} \frac{1}{v^2}.$$

The damping wings extend far from the line center, for any significant value of α. This is most important, because it makes difficult, in practice, the exact definition of a continuum level (the brightness level outside the absorption line) in stellar spectroscopy, and because the effects of saturation (which we will discuss in the next section) are much less significant in the wings than in the core of the line. An important part of the astrophysics of line formation is the collisional damping constant.

The second approximation to the treatment of collisional effects is when the time for the collision is much longer than the lifetime of the level. That is, the atom makes its transitions in a perturbing potential field, so that its energy-level structure is slightly different from that of an unperturbed atom. The total effect will then be a statistical average over all such perturbed energy levels, which leads to statistical line broadening. It is important for hydrogen and helium lines in A stars, where the perturbation is the Stark effect. We shall discuss this later also.

6.7. Line Intensities in Stellar Spectra

We have, in principle, expressions for the line opacity and the continuous opacity in the vicinity of an absorption line. The total opacity in the line is thus $k_c(\nu) + k_l(\nu)$, and in the continuum it is $k_c(\nu)$. We now relate this to the observed intensity of the line in the spectrum of the star. Formally, we already have the result. The stellar atmosphere has some temperature structure, $T(\tau)$, where τ is some standard (perhaps Rosseland mean) optical depth. In order to evaluate the emergent flux from a star at some frequency, ν, we must evaluate the flux integral

$$F_\nu(0) = \int S(\tau_\nu)\, E_2(\tau_\nu)\, d\tau_\nu, \tag{6.153}$$

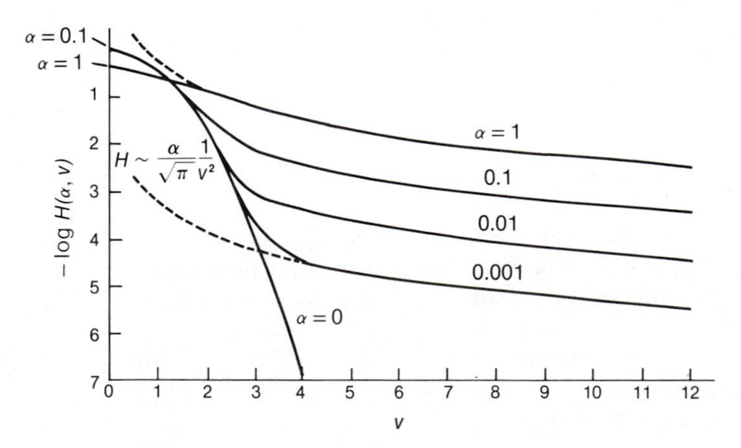

Figure 6.18. Voigt function $H(\alpha, v)$. SOURCE: A. Unsöld, *Physik der Sternatmosphären,* 2nd ed. (Berlin: Springer-Verlag, 1955).

where the optical depth at frequency ν is given by

$$\tau_\nu = \int_0^\tau \frac{k_\nu}{\bar{k}} d\tau \qquad (6.154)$$

and where the source function is equal to the Planck function if LTE is assumed. Equation (6.154) relates τ_ν to τ at the same geometrical depth s. This is the correct way to handle the problem, and is how it is done in numerical programs for calculating stellar spectra from model atmospheres. It is not immediately clear what to expect, however; so it is useful to try to get some physical insight by examining simplified models.

Image that the lines are actually formed in a layer of the plane, parallel atmosphere of uniform physical properties (T, ρ, P) called the *reversing layer,* which overlies a background source of continuous radiation. The optical thickness of this layer, whose physical thickness is h, is then (see Figure 5.4)

$$\tau_\nu = \int_0^h k_\nu \, dh = k_\nu h. \qquad (6.155)$$

The radiation intensity emerging from this layer is

$$I_\nu(h) = I_\nu(0) \, e^{-k_\nu h} = I_\nu(0) \, e^{-\tau_\nu}. \qquad (6.156)$$

We note that $I_\nu(h) = I_\nu(\tau = 0)$, and $I_\nu(h = 0) = I_\nu(\tau)$ and that $I_\nu(0)$ is essentially constant across the line. We see that passage through the region of line formation produces two effects; an attenuation $e^{-k_c(\nu)h}$, which varies smoothly over a large frequency range; and an additional attenuation $e^{-k_l(\nu)h}$, which is strong in the vicinity of the line center ν_0, but vanishes rapidly for $\nu > \nu_0$ or $\nu < \nu_0$. It is the difference between these two effects that is actually observed. Therefore we define the *rectified line profile* or residual intensity as the ratio of the line plus continuum intensity to continuum intensity:

$$r(\nu) \equiv \frac{I_\nu(0)\, e^{-[k_l(\nu)+k_c(\nu)]h}}{I_\nu(0) e^{-k_c(\nu)h}} = e^{-k_l(\nu)h} \qquad (6.157)$$

(see Figure 6.19). The spectral line absorption contains a great deal of information about conditions in the atmosphere. This information, especially in strong lines, applies to a considerable range of depths in the atmosphere as we move from the wings of the line (corresponding to deeper layers) to the center of the line (shallow layers). Recalling (6.64) and the assumptions leading to (6.156), we find that each absorbing atomic species present will contribute a term to $k_l(\nu)$ of the form

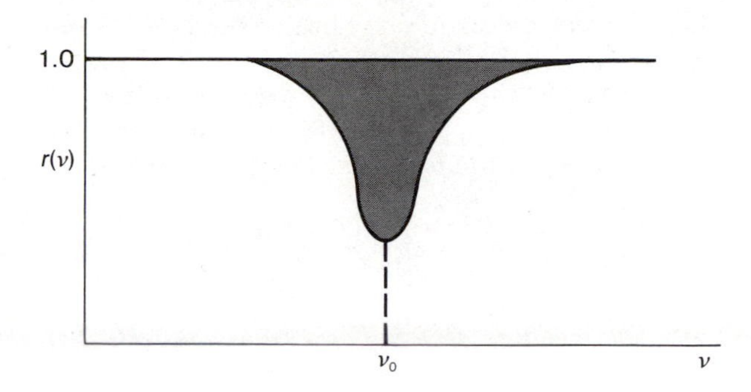

Figure 6.19. Rectified line profile (6.157).

$$\begin{aligned} k_i(\nu) &= n_i a_i(\nu) \\ &= (1 - e^{-h\nu/kT})\, n_i(T, P_e) \frac{\pi e^2}{mc} f_i \, \phi_\nu{}^i. \end{aligned} \qquad (6.158)$$

The first two factors reflect the thermodynamic conditions in the gas, which determine the overall strength of the line, as we will see. The last factor brings in the effects of turbulence or rotation as modifications to the natural and thermal Doppler profiles.

Often detailed data on line spectra are not available, and only the total amount of energy removed by the line is known. This is measured by the shaded area between the two curves in Figure 6.19, and is called the *equivalent width W* of the line:

$$W_\nu = \int_0^\infty [1 - r(\nu)] \, d\nu \qquad (6.159)$$

in frequency units.

Problem 6.28. It is more usual to define the equivalent width in wavelength units of milli-Angströms,

$$W_\lambda = \int_0^\infty [1 - r(\lambda)] d\lambda.$$

What is the relation between W_λ and W_ν?

Now, suppose that the line is weak; i.e., the line optical thickness $k_l(\nu)h$ is small. Then the residual intensity is

$$r(\nu) \simeq 1 - k_l(\nu)h. \qquad (6.160)$$

Therefore for a weak line, the line profile parallels the shape of the absorption coefficient. The quantity $k_l(\nu)$ is given by (6.158), where n_i is the number density of atoms in the level appropriate for absorbing the line. The equivalent width of the line in this case is

$$W_\nu = \int_0^\infty k_l(\nu)h\,d\nu = n_i h \int_0^\infty a(\nu)\,d\nu$$

$$= n_i h \frac{\pi e^2}{mc} f_i (1 - e^{-h\nu/kT}), \qquad (6.161)$$

where we have assumed that a_ν is sharply peaked about the line center ν_0, so that the stimulated emission factor could be treated as a constant across the line. The equivalent width of a weak line is then proportional to the total number of atoms along the line of sight capable of absorbing the line, $n_i h$. It is also proportional to f, which is a measure of the intrinsic strength of the line.

As the number of absorbing atoms per cm^2, $n_i h$, increases, the weak line approximation will break down, and the equivalent width of the line will be less than given by the simple proportionality. When n_i becomes sufficiently large, most of the photons will have been removed from the beam by absorption, and even a large increase in their number will cause very little decrease in the intensity. This is called *saturation.* Thus, we expect a behavior like that shown in Figure 6.20. Calculation of the exact shape of this curve is somewhat complicated, and depends on the form assumed for $a(\nu)$. For example, suppose we assume the line profile is pure Doppler broadening. Then the equivalent width of the line is

$$W_\nu = \int_0^\infty \left\{1 - \exp\left[-Nhf\frac{\pi e^2}{mc} \times \frac{1}{\Delta\nu_D\sqrt{\pi}} e^{-(\nu-\nu_D)^2/\Delta\nu_D{}^2}\right]\right\} d\nu. \qquad (6.162)$$

A series expression for this can be developed by writing

$$e^{-x} = 1 + \sum_{n=1}^{\infty} \frac{(-x)^n}{n!}. \qquad (6.163)$$

This gives

$$W_\nu = \sum_{n=1}^{\infty} \frac{1}{n!}\left(-\frac{Nhf}{\Delta\nu_D\sqrt{\pi}}\frac{\pi e^2}{mc}\right)^n \times \int_0^\infty e^{-n(\nu-\nu_D)^2/\Delta\nu_D{}^2}\,d\nu. \qquad (6.164)$$

The explicit evaluation of the terms in this series is possible, and yields

$$W_\nu = \Delta\nu_D \sqrt{\pi}\, C \times \left[1 - \frac{C}{2!\sqrt{2}} + \frac{C^2}{3!\sqrt{3}} - \frac{C^4}{4!\sqrt{4}} + \ldots\right], \qquad (6.165)$$

where

$$C \equiv \frac{\pi e^2}{mc}\frac{n_i f}{\Delta\nu_D} = n_i h f\, a(\nu_0) = \tau(\nu_0).$$

It is possible to prove that the asymptotic form for large C is

$$W_\nu \simeq \Delta\nu_D \sqrt{\ln C}.$$

Thus the equivalent width grows very slowly with the number of absorbing atoms; it is almost constant.

In the opposite extreme, of pure damping broadening, when the number of absorbing atoms is large we imagine that the equivalent width is dominated by the damping wings. If we make the approximation

$$H(\alpha, v) \simeq \frac{\alpha}{\sqrt{\pi}}\frac{1}{v^2},$$

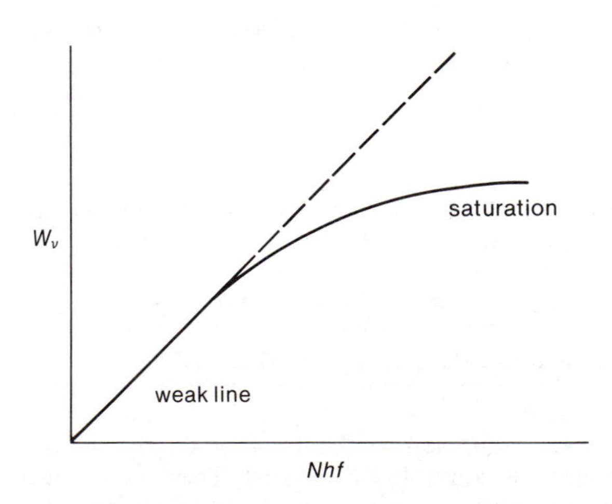

Figure 6.20. Equivalent line width W_ν versus the product of line strength f and column density of absorbers Nh.

which is valid when points far from the line center are dominating the equivalent width, we have

$$W_\nu = \int [1 - e^{(-a/\sqrt{\pi})(C/v^2)}]\, dv\, \Delta\nu_D. \quad (6.166)$$

By substituting $z = (\alpha C/\sqrt{\pi}v^2)$ into this expression, one has finally

$$W_\nu = \sqrt{\alpha\pi^{1/2}}\, \sqrt{C}\, \Delta\nu_D. \quad (6.167)$$

Thus the equivalent width, when dominated by the damping wings, grows as the square root of the number of absorbing atoms.

Curves of Growth

Now, suppose we take account of both Doppler effect and damping. For weak lines, the equivalent width is always proportional to the number of absorbing atoms; we have a linear relation. Since the Doppler broadening is intrinsically greater than the damping broadening, as the number of atoms increase, saturation begins to set in, because the Doppler core is exhausting the available photons. After a while, however, the damping wings, which are extensive, will dominate the equivalent width of the line, and so W_ν will grow again as $C^{1/2}$. The total effect is summarized in Figure 6.21, which is known as a *curve of growth*.

The curve of growth is an important tool in a rough investigation of the parameters of a stellar atmosphere. For example, suppose we consider all the lines that originate from the same lower level. An example might be the hydrogen Balmer lines (not too good an example, because the broadening mechanism for hydrogen lines is more complicated than we have discussed so far). These lines will have different wavelengths, but, for a given value of C, the equivalent widths are proportional to the Doppler widths. Therefore, at a fixed C, lines of different wavelength should have a constant value of W_λ/λ or W_ν/ν. A graph of $\log (W/\lambda)$ against $\log C$ will then be a single curve. The different lines will have different values of C, because the oscillator strengths will be different, generally lower for the higher members of the series. Therefore, for lines from a given lower level, we may plot $\log (W/\lambda)$ against $\log f$ to find the general shape of the curve of growth. Now, imagine that another line of the same series is observed, but its oscillator strength is unknown. By placing it on the curve of growth at the observed value of $\log (W/\lambda)$, we can find the oscillator strength for the line. Thus we have in principle an observational technique for finding oscillator strengths in a series of lines, provided that some of the oscillator strengths are already known. Now, consider the lines from a lower level in the atom; its excitation energy will be different from that of the ground state, and so the number of atoms in the two states are proportional

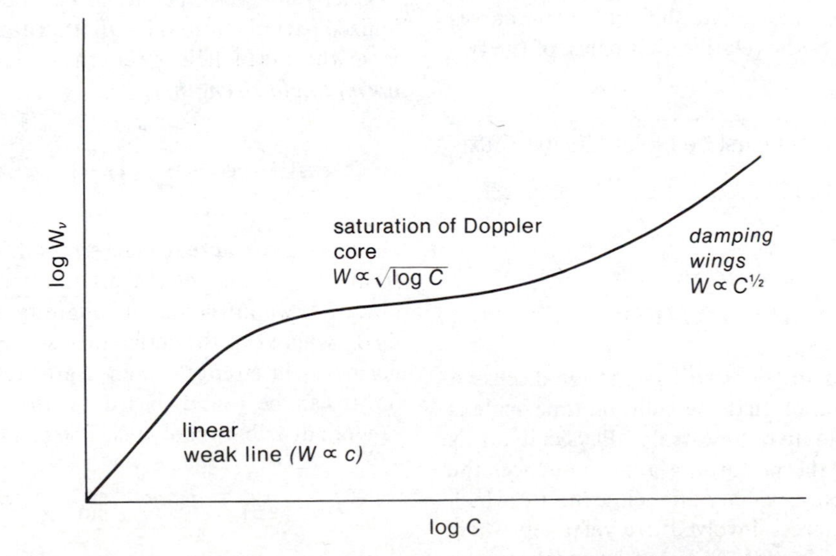

Figure 6.21. Equivalent line width (curve of growth) including saturation of the core and wing damping, versus optical depth at the line center (6.164).

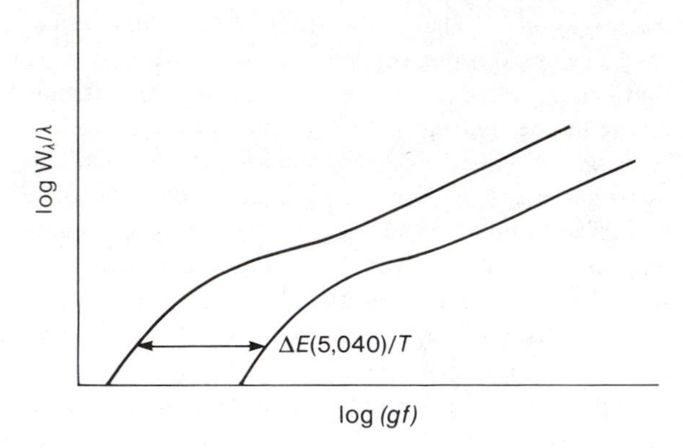

Figure 6.22. Curves of growth arising from two different series of lines. The shift in the two curves is related to the relative abundance of the absorbers.

to $g_1 e^{-E_1/kT}$ and $g_2 e^{-E_2/kT}$. The quantity C for the two levels is thus proportional to $gfe^{-\Delta E/kT}$. Suppose we plot $\log (W/\lambda)$ against $\log (gf)$ for the two series of lines (Figure 6.22). The two curves of growth will be displaced in the direction of the C axis by an amount $\log (e^{-\Delta E/kT}) = \Delta E\ (5{,}040/T)$, where T is in degrees K. By measuring this shift, and knowing the energy separation of the two levels, we can calculate $(5{,}040/T)$, or T, which is the excitation temperature for this atomic species. Finally, with a known excitation temperature, we can compare the curves of growth of two different atomic species, using all the observable lines of each element. The shift in the two curves of growth then gives the relative abundance of the two elements.

Problem 6.29. What must be plotted as the C axis in this case?

6.8. Line Broadening in Hydrogen and Helium

The impact approximation we have so far discussed was the approximation that the collision time-scale is less than the radiative time-scale. Physically, this would mean that the perturbing atom influences the radiating atom only when quite close to it, which implies that the forces involved are relatively short range. For example, in the solar atmosphere the sodium D lines are broadened by impact with neutral hydrogen atoms. The interatomic potential between two such atoms has something like an r^{-6} dependence, and hence is of short range. This is essentially why the impact approximation can be used.

In hotter stellar atmospheres, however, another phenomenon enters. A substantial number of atoms are ionized, including perhaps some hydrogen in the hotter stars, and the Coulomb field of a charged particle is of much longer range (r^{-2}) than the van der Waals force between neutral atoms. Thus collision times are rather longer, and the opposite approximation can be used, namely, that the collision time is much longer than the radiating time. Because the ion sets up an electric field in the vicinity of the radiating atom, this process is generally called the *statistical Stark effect,* since the Stark effect describes the effect of electric fields on spectral lines. Statistical Stark broadening depends on two things: (1) the electric field $\mathcal{E}$ at a radiating atom produced by a given density of ions, n (this will have some probability distribution, arising from the random distribution of ions in the material); and (2) the effect of a given field strength, $\mathcal{E}$, on the spectral line of interest.

We begin with the statistics of the electric field. First we define a mean distance between ions, $\bar{r}$, according to the formula

$$\frac{4\pi\bar{r}^3}{3} n = 1 \qquad \text{or} \qquad \bar{r} = \left(\frac{3}{4\pi n}\right)^{1/3}. \tag{6.168}$$

We now introduce one radiating atom into this assembly of ionized particles. The mean distance of an ionized particle from the radiating atom is very close to $\bar{r}$; so the mean field strength, sometimes called the *normal field strength,* is

$$\mathcal{E}_0 = \frac{e}{\bar{r}^2} = \left(\frac{4\pi}{3}\right)^{2/3} n^{2/3} e. \tag{6.169}$$

Of course, the actual field strength will be different from this because of the distribution of particle distances. If we introduce the quantity β defined by $\beta = \mathcal{E}/\mathcal{E}_0$, where $\mathcal{E}$ is the actual field strength, and $\mathcal{E}_0$ is the normal field strength, then a probability distribution for β can be found, based on the assumption of a random distribution of ions. The result is

$$w(\beta) = \frac{2}{\pi\beta} \int_0^\infty v \sin v\, e^{-(v/\beta)^{3/2}}\, dv, \tag{6.170}$$

which is known as the *Holtsmark distribution,* and $w(\beta)\, d\beta$ is the probability that β will lie in the range (β,

$\beta + d\beta$). The integral cannot be evaluated analytically, but may be approximated as follows:

$$w(\beta) \simeq \begin{cases} \dfrac{1.496}{\beta^{5/2}}\left(1 + \dfrac{5.107}{\beta^{3/2}} + \dfrac{14.43}{\beta^3} + \ldots\right) \\ \beta \gg 1 \\ \dfrac{4}{3\pi\beta^2}(1 - 0.463\,\beta^2 + 0.1227\,\beta^4 + \ldots) \\ \beta \ll 1 \end{cases} \tag{6.171}$$

The probability $w(\beta)$ is shown in Figure 6.23. Thus we have the probability of the field strengths for a given density of ionized particles.

We next consider what effect an electric field has on the atomic energy levels. This is quite complicated. Figure 6.24 shows the pattern of lines found in some members of the Balmer series. The shift in atomic energy level ΔE due to the field $\vec{\mathscr{E}}$ may be estimated semiclassically. In an atom whose dipole moment is $e\vec{\ell}$, the associated energy is approximately

$$\Delta E = h\Delta\nu = \frac{h\Delta\lambda}{\lambda^2} = e\vec{\mathscr{E}} \cdot \vec{\ell} \sim e\mathscr{E}\ell.$$

If the field is not too strong, then $\vec{\ell}$ is independent of $\vec{\mathscr{E}}$, and it is comparable to the Bohr radius ($\ell \sim a_0 = h^2/m_e e^2$), so that $\Delta\lambda \sim \lambda^2\mathscr{E}$. Exact calculations show that the lines are separated by a wavelength interval

$$\Delta\lambda = \frac{3h}{8\pi^2 me}\frac{\lambda^2}{c}\,n_k\mathscr{E} \equiv C_k\mathscr{E}. \tag{6.172}$$

Here n_k is an integer that depends on the quantum numbers of the initial and final states. Because this expression depends directly on the field strength, $\mathscr{E}$, the broadening of hydrogen lines by this process is called the *linear Stark effect*. The relative line intensities must be calculated quantum-mechanically. For any given one of these component lines, the probability that it will contribute to the line profile at a wavelength displaced by $\Delta\lambda$ from the line center is just $w(\beta)d\beta$, where

$$\beta = \frac{\mathscr{E}}{\mathscr{E}_0} = \frac{\Delta\lambda}{C_k\mathscr{E}_0}.$$

The contribution of this component at this wavelength is weighted by its intensity, I_k. Thus the total opacity at wavelength $\Delta\lambda$ is given by

$$\sum_k I_k w\left(\frac{\Delta\lambda}{C_k\mathscr{E}_0}\right) d\left(\frac{\Delta\lambda}{C_k\mathscr{E}_0}\right). \tag{6.173}$$

It is usual to introduce the variable $\alpha = \Delta\lambda/\mathscr{E}_0$, so that this becomes

$$S(\alpha)\,d\alpha = \sum_k I_k\, w\left(\frac{\alpha}{C_k}\right)\frac{d\alpha}{C_k} \tag{6.174}$$

It may be verified that

$$\int_{-\infty}^{\infty} S(\alpha)\,d\alpha = 1. \tag{6.175}$$

Since the total line opacity must obey the Ladenburg relation (6.56), we have finally the absorption cross section per atom (6.57)

$$\sigma\,(\Delta\lambda) = \frac{\pi e^2}{mc} f\lambda^2 \frac{S(\Delta\lambda/\mathscr{E}_0)}{c\mathscr{E}_0}. \tag{6.176}$$

It is of interest to develop an approximate expression for the line shape in the wings of the line, for the Stark broadening produces extensive wings. The wings of the line correspond to large field strengths, i.e., large values of β, and for these the approximation $w(\beta) = 1.496\,\beta^{-5/2}$ can be used. In this case, the line opacity in the wings is given by

$$\sigma(\Delta\lambda) = \frac{\pi e^2}{mc^2} f\lambda^2\,(\Delta\lambda)^{-5/2} \sum_k \frac{I_k}{C_k\mathscr{E}_0}\left(\frac{1.496}{C_k}\right)^{-5/2} \mathscr{E}_0^{5/2}$$

$$= \left[\frac{\pi e^2}{mc^2}\lambda^2 f \frac{\sum_k I_k C_k^{3/2}}{(1.496)^{5/2}}\right]\frac{\mathscr{E}_0^{3/2}}{(\Delta\lambda)^{5/2}}. \tag{6.177}$$

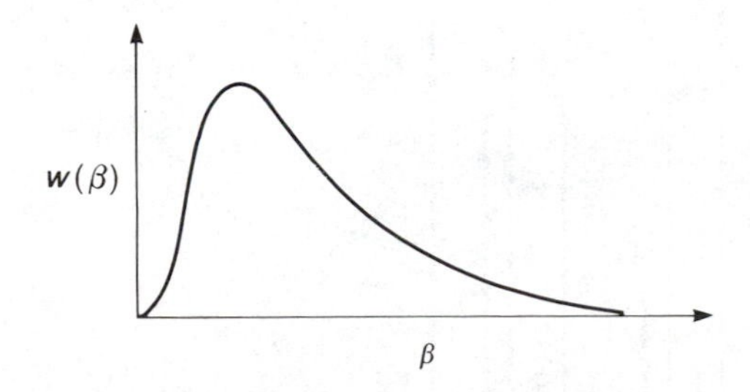

Figure 6.23. Holtsmark distribution (6.170) versus $\beta = \mathscr{E}/\mathscr{E}_0$.

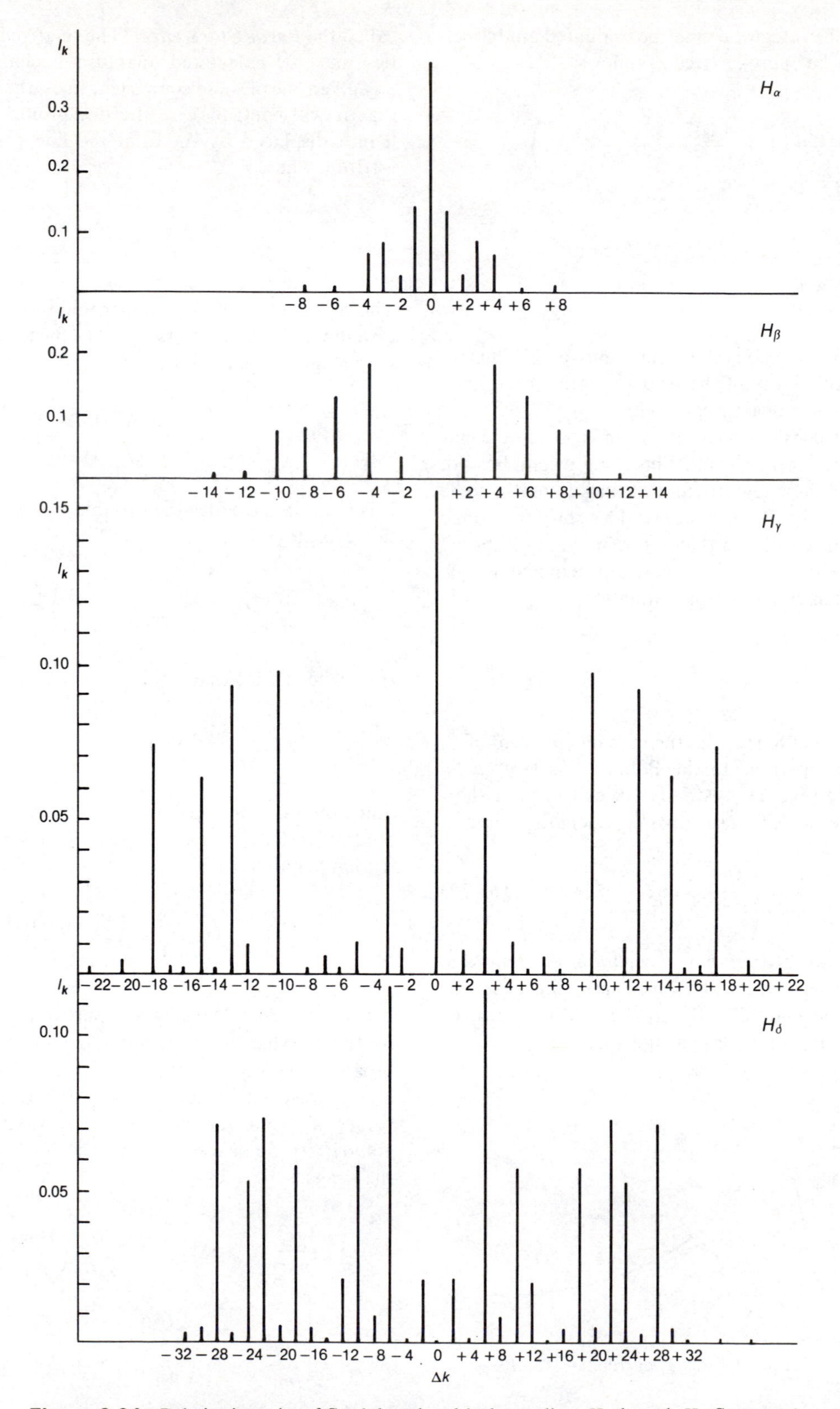

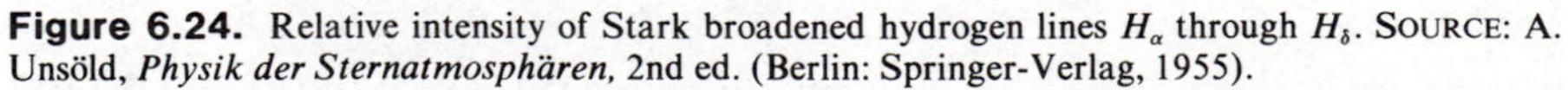

Figure 6.24. Relative intensity of Stark broadened hydrogen lines H_α through H_δ. SOURCE: A. Unsöld, *Physik der Sternatmosphären*, 2nd ed. (Berlin: Springer-Verlag, 1955).

Thus the line opacity in the wings depends on $(\Delta\lambda)^{-5/2}$, clearly giving quite extensive wings, and also on the normal field strength to the 3/2 power, that is, directly on the density of the ions, n. The numerical value of the constant in square brackets for the first few Balmer lines is as follows.

H_α: 3.13×10^{-16}
H_β: 0.885×10^{-16}
H_γ: 0.442×10^{-16}
H_δ: 0.309×10^{-16}

Problem 6.30. In a typical A star, such as Sirius, the atmospheric temperatures are around 10,000 K and the electron pressure is around 630 dyne cm^{-2}. What is the normal field strength, $\mathcal{E}_0$? Give the answer both in cgs units and in volts cm^{-1}. What is the value of α at 5 Å from the line center for (H_α)? Does this justify the approximation $\beta \gg 1$ at this distance from the line center? How does the line opacity at 5 Å displacement compare with that at, say, 2 Å? At what displacement from the line center does the approximation that $\beta \gg 1$ break down?

The hydrogen atom has a relatively simple structure; in particular, the various principal quantum states have essentially constant dipole moment, which is why they interact with an external electric field to produce a linear Stark effect; the energy shift in each level depends linearly on the field strength. In more complicated atoms, this is not necessarily the case, and the effective dipole moment may depend on the field; that is, the energy shift caused by an external field may depend on a higher power of $\mathcal{E}$, for example, on $\mathcal{E}^2$, giving rise to the *quadratic Stark effect.* This is a special complication in the helium spectrum, where the dependence of the energy shifts on the field strength is neither linear nor quadratic, but something in between. This makes the theory of Stark broadening of helium quite complicated. A further complication occurs in at least one of the helium lines. The HeI line at 4471.6 Å comes from the transition $2^3P \rightarrow 4^3D$ in the triplet series. A transition $2^3P \rightarrow 4^3F$ has a very similar wavelength, 4469.9 Å, but is forbidden by the usual rules of quantum mechanics, which do not allow transitions in which the orbital angular momentum changes by 2. This forbiddenness, however, is characteristic of the spherical symmetry of the Coulomb field of the nucleus. Because of this spherical symmetry, the overlap integrals of the wave functions are zero if the orbital quantum numbers differ by more than unity. However, in the presence of an externally imposed field, this spherical symmetry is destroyed, and the wave functions overlap where they did not before. Thus the normally "forbidden" transition [HeI] 4469.9 Å has its transition probability increased by the external field. It thus appears in the wings of the permitted HeI 4471.6 Å line. This naturally complicates the appearance of the line profile, and the astrophysics of He line formation.

Problem 6.31. Noting that the base of the photosphere is 2/3 of a mean free path into the stellar surface, use (5.77) to establish (6.121). Review the assumptions implicit in the use of (5.77).

Part 3

STELLAR EVOLUTION

Chapter 7

NUCLEAR ENERGY SOURCES

During most of a star's lifetime, energy is released by the fusion of lighter nuclei into more massive ones. In some stars, the fusion process may proceed to its endpoint, which is the formation of Fe^{56} and Fe^{56}-group nuclei. However, any lighter nuclei produced by a nuclear process may become fuel for a subsequent series of reactions. The temperature, and to a lesser extent the density, of the gas determines the rate at which thermonuclear energy is released. The end products of nuclear reactions may also depend on the evolutionary timescale characteristic of a stellar core.

7.1. Thermonuclear Energy Sources

Figure 7.1 shows schematically the total potential energy of two nuclei as a function of their separation. For large r the potential energy is due entirely to the electrostatic Coulomb repulsion between the nuclei, but for $r \lesssim r_0 \sim 10^{-13}$ cm the attractive nuclear force dominates. Typically the height of the Coulomb barrier E_c is several MeV, and the depth of the nuclear well E_N is about -30 MeV. In order for two nuclei to interact, they must approach near enough for the attractive portion of the potential ($r \lesssim r_0$) to dominate.

In a star gravitational contraction compresses the gas, heating it to high temperatures. High temperature means high relative velocities of the nuclei; so they are more likely to come close enough to interact. The energy of the Coulomb barrier is roughly

$$E_c \sim \frac{e^2}{r_0} \simeq 1.65 \times 10^{-6} \text{ ergs} \simeq 1 \text{ MeV}. \qquad (7.1)$$

The gas in the solar core is in thermal equilibrium, with $T \simeq 2 \times 10^7$ K; so the velocities of the nuclei (protons) obey the Maxwellian distribution. Since the typical kinetic energy and velocity of a nucleus at these temperatures are nonrelativistic, the distribution of energies is

$$f(E) = \frac{2}{\sqrt{\pi}} \left(\frac{1}{kT}\right)^{3/2} e^{-E/kT} E^{1/2}. \qquad (7.2)$$

The average thermal energy of a proton, $\langle E \rangle = (3/2)kT$, is several keV or about 10^{-3} times E_c. From the viewpoint of classical physics, the Sun simply does not contain enough protons with energies greater than E_c to be able to produce the energy output that it does produce.

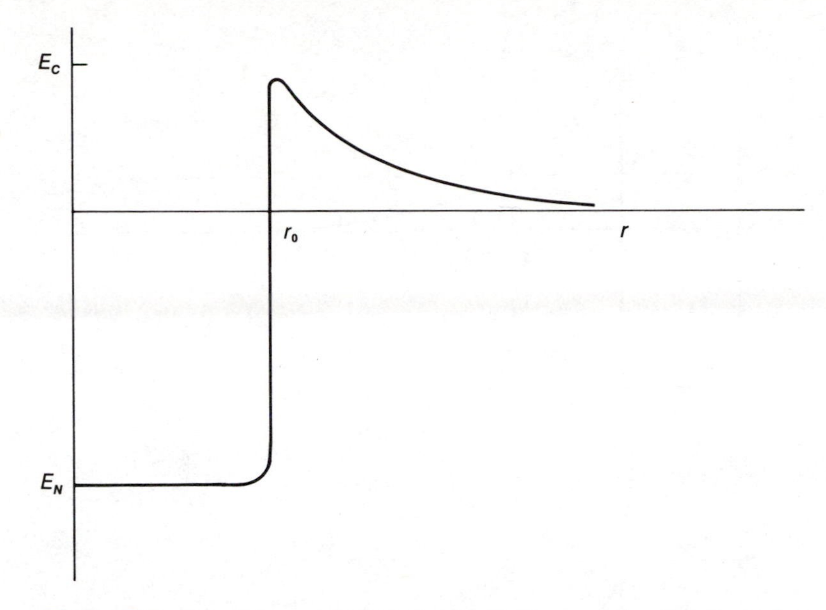

Figure 7.1. Potential energy between two nuclei. For $r < r_0$ the attractive nuclear forces dominate; for $r > r_0$ Coulomb repulsion dominates.

The solution to this dilemma involves quantum mechanics. To see the principle, let us look at the radioactive decay of $_{84}Po^{212}$, an alpha emitter whose lifetime in the laboratory is $\tau_\alpha = 3 \times 10^{-7}$ sec. In the parent nucleus, $E_c \simeq 30$ MeV, but the energy of the emitted α particle, E_α, is about 9 MeV. Classically, the α particle should therefore be confined within the nucleus by the Coulomb barrier, and $_{84}Po^{212}$ should be stable against α decay (Figure 7.2a). The particle escapes because of the quantum-mechanical phenomenon known as *barrier penetration.* In quantum theory the particles in a nucleus are described by a spatial probability distribution that is greatest where the classical theory (based, for example, on a potential such as the one shown in Figure 7.1) predicts the particles to be, but that is non-zero at other points not far away. For $_{84}Po^{212}$, the alpha particle, though confined within the Coulomb barrier most of the time, has a finite probability of being found at distances $r > r_0$ from the parent nucleus. There it is no longer subject to the nuclear forces and can escape.

In the solar core, the situation is reversed (Figure 7.2b); protons that at temperature T would be separated in classical theory by distances $r > r_0$ have in quantum theory a non-zero probability of actually being closer together, say, with $r < r_0$. Classically two such protons would not interact, but quantum mechanically there is a finite probability that they will. It is this small but finite probability that the highest-energy protons in the Maxwell distribution will interact that makes thermonuclear fusion in stars possible.

Another important feature is the height of the barrier, which increases proportionally to $Z_1Z_2e^2$, where Z_1e and Z_2e are the charges of the two reacting nuclei. Since Z increases with the mass of the nucleus, higher temperatures are required to burn more massive nuclei; hence low-mass stars can burn only nuclei of low atomic weight.

We can calculate, to an order of magnitude, the minimum mass a star must have to be able to exploit a nuclear fuel, since the core temperature T_c must exceed the ignition temperature for a given fuel. The density of a uniform distribution of matter is roughly $\rho \simeq M/4R^3$, and the central temperature for an ideal gas in hydrostatic equilibrium is

$$T \simeq (1/5)MG\mu m_H/kR.$$

At high densities matter becomes degenerate, and the thermal energy of an electron must be less than its Fermi energy,

$$kT < \epsilon_{F,e} = \frac{p_{F,e}^2}{2m_e} = \frac{\hbar^2(3\pi^2 n_e)^{3/2}}{2m_e}, \qquad (7.3)$$

where the electron's Fermi momentum is $p_{F,e} = \hbar(3\pi^2 n_e)^{1/3}$ (see Section 2.5). Charge neutrality

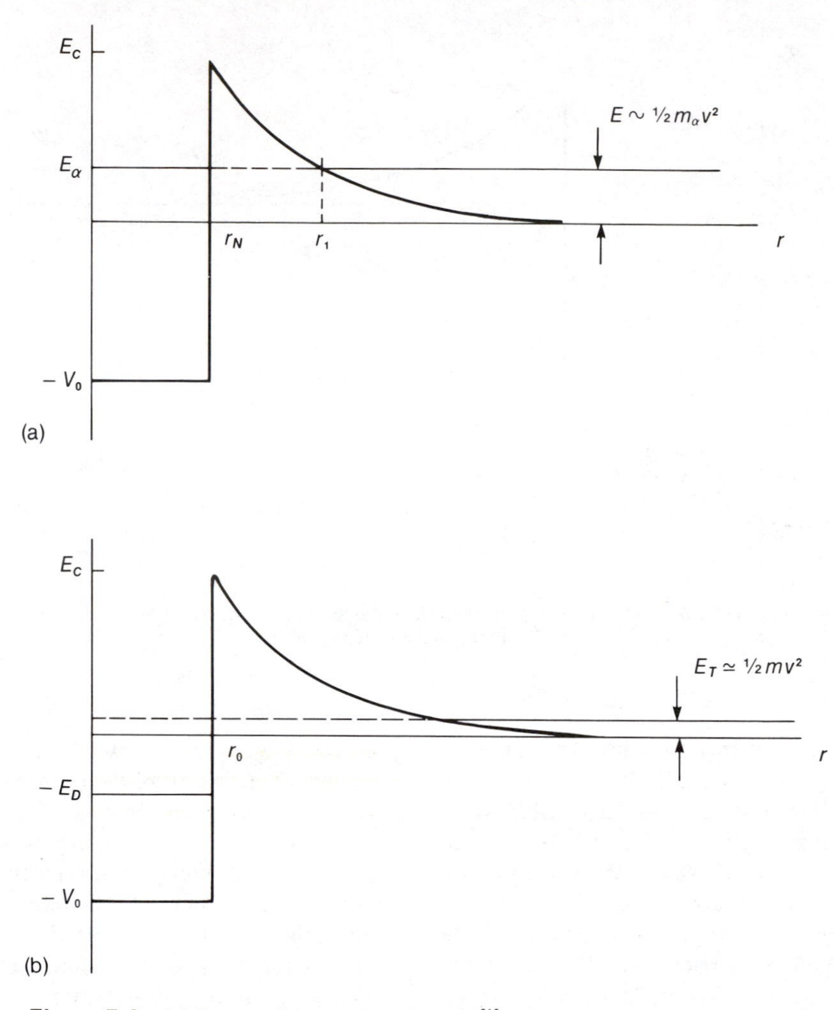

Figure 7.2. (a) Energy of α particle (E_α) in ${}_{84}Po^{212}$ nucleus. Classically the α particle can escape the nucleus only if it gains energy $\Delta E = E_C - E_\alpha$. The kinetic energy of an emitted α particle is $E_\alpha - E_C$, which is negative for $r < r_1$. Thus the α can not exist (classically) in the region $r_N < r < r_1$. Here r_N is the nuclear radius, and $-V_0$ is the nuclear potential energy.
(b) Total energy of two protons separated by a distance r. The binding energy of the deuteron is $-E_D$. Note that the thermal energy of protons in the Sun, E_T, is much less than E_C.

requires that the total electron and proton number densities be equal, $n_e = n_p = (\rho/m_H)$, which may be used to eliminate n_e in (7.3). The result may be rewritten as

$$(m_H/\rho) > \hbar^3/(2m_e kT)^{3/2}, \tag{7.4}$$

which is equivalent to the requirement that the inter-nuclear separation $(m_H/\rho)^{1/3}$ exceed the thermal wavelength of the electron $\hbar/p_e \simeq \hbar/(2m_e kT)^{1/2}$. Elimination of ρ and R from the formulas for the density and central temperature given just before (7.3) leads to the inequality

$$\mu^{3/2}(M/M_\odot) > \left[\frac{5^3\hbar^3(kT)^{3/2}}{4m_H{}^4 m_e{}^{3/2}(\mu G)^3 M_\odot{}^2}\right]^{1/2} \simeq 2.9 \times 10^{-7} T^{3/4}. \tag{7.5}$$

This may be used to estimate the minimum mass of a star that can burn a fuel whose ignition temperature is

Table 7.1
Thermonuclear ignition temperatures for nuclear fuels. Columns give: nuclear fuel; ignition temperature (K); mean molecular weight; and minimum stellar mass capable of igniting the fuel.

Fuel	T_{ignition} (K)	μ[a]	$M/M_{\odot}$[a]
H	10^7	1/2 (0.67[b])	0.14 (0.093[b])
He	10^8	4/3	0.19
C	4–7×10^8	2	0.44
O	$1.4 - 2 \times 10^9$	2	0.73
Si	3–5×10^9	2	1.43

[a]Assumes $\mu^{-1} = 2X + 3Y/4 + Z/2$, with the composition being $X = 1$, $Y = 1$, $Z = 0$.

[b]For Population I stars, μ is obtained by setting $X = 0.6$, $Y = 0.38$, $Z = 0.02$.

T. Table 7.1 summarizes the estimates for several nuclei.*

7.2. Thermonuclear Energy Release

The important aspects of thermonuclear reactions as energy sources in stars are the amount of energy released and the rate of the reactions. We will consider the amount of energy released per reaction first, and consider the rates in the next section.

For many thermonuclear burning stages, the energy-generation rate may, within a limited temperature range, be approximated by

$$\epsilon = \epsilon_0 \rho^a T^n \text{ ergs g}^{-1} \text{ sec}^{-1}, \qquad (7.6)$$

where ϵ_0 is a constant. The parameters a and n are usually slowly varying functions of ρ and T, and may be taken as constants for simple calculations. For most fuels a is near unity, and n may vary from 4 for hydrogen burning to 30 or more for carbon burning. More exact expressions for the energy-generation rate will be developed later.

The energy released by a network of nuclear reactions can be calculated from the Q-values for the reactions. Energy is released primarily in two forms: electromagnetic (photons), and as neutrinos (we will deal with neutrinos separately, in Sections 7.4 and 13.1). Only the electromagnetic form contributes to the rate of nuclear-energy generation.

Many nuclear networks can proceed by several branches. The extent to which any one branch contributes will be determined by the temperature and by the available amounts of the fuel needed to initiate the sequence. The net energy released as photons when fuel A is converted into ash B may depend on the conditions under which the process develops. For example, hydrogen burning by the pp chain is initiated by the reactions

$$\mathrm{p} + \mathrm{p} \rightarrow \mathrm{D} + \mathrm{e}^+ + \nu_e \qquad 1.18(0.26)\text{ MeV}, \qquad (7.7)$$

$$\mathrm{p} + \mathrm{D} \rightarrow \mathrm{He}^3 + \gamma \qquad 5.49\text{ MeV}. \qquad (7.8)$$

The numbers on the right give the photon energy (neutrino energy) in MeV per reaction; (7.7) includes the energy released by $\mathrm{e}^+\mathrm{e}^-$ annihilation (see Table 7.2). For temperatures below 10^7 K, the pp chain ends with He^3 production. For $T > 10^7$ K, the final step of the sequence is

$$\mathrm{He}^3 + \mathrm{He}^3 \rightarrow \mathrm{He}^4 + 2\mathrm{p} \qquad 12.86\text{ MeV}. \qquad (7.9)$$

The net photonic energy released by the formation of one He^4 is twice that released by (7.7) plus (7.8), since these reactions must occur twice for each He^4 produced, plus 12.86 MeV, for a total of 26.2 MeV. Notice that the last step releases nearly half the energy, but that the first step, because of its slowness, determines the rate at which the process occurs. Completion of the series requires that enough He^3 has built up for the terminal step (7.9) to occur with reasonable frequency.

As He^4 accumulates and the temperatures rise above 2×10^7 K, the alternative series can occur:

$$\mathrm{He}^3 + \mathrm{He}^4 \rightarrow \mathrm{Be}^7 + \gamma \qquad 1.59\text{ MeV}. \qquad (7.10)$$

The most likely branch following Be^7 production is

$$\mathrm{Be}^7 + \mathrm{p} \rightarrow \mathrm{B}^8 + \gamma \qquad 0.13\text{ MeV}, \qquad (7.11)$$

followed by

$$\mathrm{B}^8 \rightarrow \mathrm{Be}^8 + \mathrm{e}^+ + \nu_e \qquad 10.78\,(7.2)\text{ MeV} \qquad (7.12)$$

*The smallest mass of gas that can burn H is estimated in Table 7.1 to be about 0.14 $M_{\odot}$. However, more detailed arguments indicate lower values for these masses. For H burning the value is nearer 0.05 $M_{\odot}$. Furthermore, configurations that are even less massive may be able to radiate enough energy to be observable by converting gravitational energy into thermal energy.

Table 7.2
Thermonuclear energy-release parameters for the principal reaction channels that contribute to (a) the pp chain, and (b) the CNO cycle. The third column gives the energy carried off by the emitted neutrino per reaction. The last column gives the half-life for the reaction.

Reaction[a]	Q value, MeV	Average ν loss, MeV	S_0, keV barns	$\frac{dS}{dE}$, barns	B^c	τ_{12}, years[b]
The pp *chain*						
$H^1(p, \beta^+\nu)D$	1.442	0.263	3.78×10^{-22}	4.2×10^{-24}	33.81	7.9×10^9
$D(p, \gamma)He^3$	5.493		2.5×10^{-4}	7.9×10^{-6}	37.21	4.4×10^{-8}
$He^3(He^3, 2p)He^4$	12.859		5.0×10^3		122.77	2.4×10^5
$He^3(\alpha, \gamma)Be^7$	1.586		4.7×10^{-1}	-2.8×10^{-4}	122.28	9.7×10^5
$Be^7(e^-, \nu)Li^7$	0.861	0.80				3.9×10^{-1}
$Li^7(p, \alpha)He^4$	17.347		1.2×10^2		84.73	1.8×10^{-5}
$Be^7(p, \gamma)B^8$	0.135		4.0×10^{-2}		102.65	6.6×10^1
$B^8(\beta^+, \nu)Be^{8*}(\alpha)He^4$	18.074	7.2				3×10^{-8}
The CNO *cycle*						
$C^{12}(p, \gamma)N^{13}$	1.944		1.40	4.26×10^{-3}	136.93	
$N^{13}(\beta^+\nu)C^{13}$	2.221	0.710				
$C^{13}(p, \gamma)N^{14}$	7.550		5.50	1.34×10^{-2}	137.20	
$N^{14}(p, \gamma)O^{15}$	7.293		2.75		152.31	
$O^{15}(\beta^+, \nu)N^{15}$	2.761	1.00				
$N^{15}(p, \alpha)C^{12}$	4.965		5.34×10^4	8.22×10^2	152.54	
$N^{15}(p, \gamma)O^{16}$	12.126		2.74×10^1	1.86×10^{-1}	152.54	
$O^{16}(p, \gamma)F^{17}$	0.601		1.03×10^1	-2.81×10^{-2}	166.96	
$F^{17}(\beta^+\nu)O^{17}$	2.762	0.94				
$O^{17}(p, \alpha)N^{14}$	1.193		Resonant reaction		167.15	

SOURCE: Adapted from D. D. Clayton, *Principles of Stellar Evolution and Nucleosynthesis* (New York: McGraw-Hill, 1968).

[a]For an explanation of the notation in this column, see Section 7.4.

[b]Computed for $X = Y = 0.5$, $\rho = 100$, $T_6 = 15$ (Sun).

[c]See equation (7.40) with T in units of 10^6 K.

and

$$Be^8 \rightarrow 2He^4 \qquad 0.095 \text{ MeV}. \qquad (7.13)$$

The complete chain releases 19.27 MeV as photons for each He^4 produced (note that the He^4 in the first step (7.10) was assumed to be present when the reaction commenced).

Problem 7.1. Verify that the production of He^4 by the alternate process ending with (7.10) to (7.13) releases 19.27 MeV as photons. What percentage of the total energy released is in the form of neutrinos? Compare this to the fractional energy in neutrinos from the sequence (7.7) to (7.9).

Table 7.2 gives the half-life for the reactions in the pp chain under solar-core conditions. For all branches the weak production of deuterium limits the rate. Since the formation of one He^4 by (7.9) requires the formation of twice as much deuterium as (7.10) requires, the energy-generation rates for the two branches will be different. In the time required for one He^4 to be produced by (7.9), with the release of 26.2 MeV, the sequence (7.10) to (7.13) will occur twice, once for each D formed in (7.7), releasing 38.54 MeV in the process. Since the completion of either sequence (7.9) or (7.10) to (7.13) requires the same time on average, the energy-generation rate by (7.10) alone will be about 1.47 times greater than that by (7.9) alone.

Problem 7.2. Use the half-lives given in Table 7.2 for individual steps of the pp chain to show that the two branches (7.9) and (7.10) to (7.13) occur in essentially the same time.

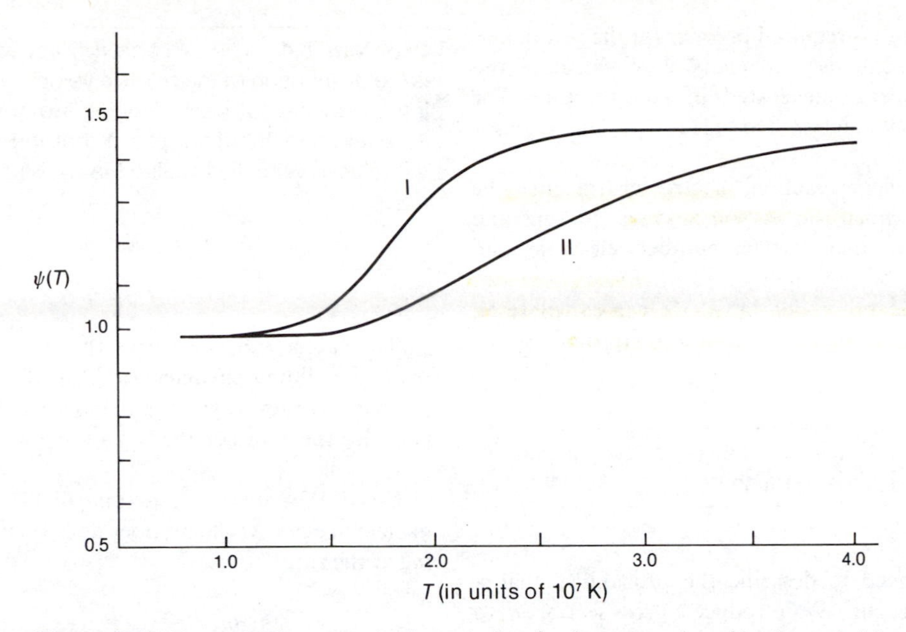

Figure 7.3. Factor $\psi(T)$ corrects simple expressions for the energy released by the pp chain's alternative reaction processes in Population I stars with solar composition (curve I), and in Population II stars with $X = 0.9$ (curve II). Temperature is in units of 10^7 K.

In practice, both branches will contribute simultaneously to He^4 production, the relative importance of each being determined by the temperature and composition. One way to allow for this behavior is to calculate the energy released by the $He^3 + He^3$ branch, use it to calculate the energy-generation rate, and multiply the result by a factor $\psi(T)$ that corrects for the admixture of other branches. The factor $\psi(T)$ for the pp chain is shown in Figure 7.3 for Population I stars (with solar composition) and Population II stars ($X = 0.9$).

The energetics of other nuclear networks may be calculated by analogy with these arguments for hydrogen burning. We will now consider the rate of energy release.

7.3. Nuclear Energy Generation Rates

The energy released by each of the nuclear reactions discussed in Section 7.2 depends only on the physics of the nuclei involved. The rate at which this energy is released depends on the relative velocity of the reacting nuclei, their average separations, and possibly on the nature of other particles that play a role in aiding or hindering the process. These factors in turn depend on the temperature, density, and composition of the gas, which are the ultimate parameters that determine ϵ. = en. gen. rate

A typical nuclear reaction may be described by

$$a + X \rightarrow b + Y. \quad (7.14)$$

If the sum of the masses of the initial particles involved in (7.14) does not equal that of the final masses ($M_a + M_x \neq M_b + M_y$), then there will be a net gain or loss of energy to the center of mass of the system, according to Einstein's formula

$$Q \equiv \Delta E = (M_x + M_a - M_y - M_b)c^2. \quad (7.15)$$

We can also write this in terms of the mass excess of each atom, defined as the difference between the actual mass of the atom M and the product Am_1, where A is the atomic weight and m_1 is a conventional mass of unit atomic weight:

$$\Delta M = M - Am_1. \quad (7.16)$$

Appendix 2 gives the mass excess ΔM in MeV for several common nuclei. By convention, m_1 is taken to be $\frac{1}{12}$ the mass of C^{12}. The energy generated (if $Q > 0$) in a nuclear reaction may now be given as

$$Q = \Delta M_a + \Delta M_x - \Delta M_b - \Delta M_y. \quad (7.17)$$

If $Q < 0$, energy is required in order for the reaction to proceed. We can use Appendix 2 to calculate the amount of energy generated in any reaction. For example, each reaction He^3 (He^3, 2p) He^4 releases 12.86 MeV.*

In any nuclear reaction, electric charge must be conserved, as must the baryon number (protons and neutrons have equal baryon number; electrons and photons have baryon number zero). From these two fundamental conservation laws, one can calculate all possible nuclear reactions that might happen.

Problem 7.3. Verify that charge and baryon number are conserved in the reaction (7.7).

We now need to describe the probability that a reaction will occur. We introduce a cross section, σ, to describe this process. Imagine one nucleus (say, X) as fixed, and the other nucleus as a projectile a approaching it with velocity v. The probability per unit path length that the reaction will take place is $n_X\sigma$, where n_X is the density of the target nuclei. In a beam of particles a moving at velocity v, the total number of reactions that take place in a unit volume and unit time is given by

$$r = n_a n_X v\sigma(v), \tag{7.18}$$

where the number densities of beam particles a and target particles X are n_a and n_X, respectively. The cross section depends explicitly on the relative velocity v of the nuclei; actually, this equation defines the cross section, $\sigma(v)$. In a stellar interior, nuclei have a range of relative velocities. If these relative velocities have a distribution $f(v)$, then the total reaction rate is obtained by summing over all velocities:

$$r = n_a n_X \int v\sigma(v) f(v)\, dv \equiv n_a n_X \langle\sigma v\rangle. \tag{7.19}$$

The total energy generated per unit mass per unit time is

$$\epsilon = \frac{Qr}{\rho} = \frac{n_a n_X}{\rho} Q\,\langle\sigma v\rangle. \tag{7.20}$$

*For an explanation of this notation, see Section 7.4.

Problem 7.4. The function $f(v)$ appearing in (7.19) is the distribution of the relative velocities of the nuclei. If the velocities of each species are governed by the Maxwellian distribution, prove that the distribution of the relative velocities is also Maxwellian, with reduced mass

$$\mu = M_a M_x/(M_a + M_x).$$

The factor $n_a n_X$ measures the number density of pairs of colliding particles, (a,X), if they are different particles. However, when X and a are the same type of particle, the number density of pairs of particles is $\frac{1}{2} n_X^2$.

Since the nuclear reaction (7.14) destroys the nuclear species X, the number density of X is decreasing at the rate

$$\frac{\partial n_X}{\partial t} = -\, n_X n_a \langle\sigma v\rangle \tag{7.21}$$

because of this reaction. If other reactions involving X are taking place, then other terms must be added, with negative sign if the reactions are destroying X, and positive sign if they are creating X. It follows from (7.21) that the characteristic time-scale for the destruction of X is of the order of

$$\tau \simeq [n_a \langle\sigma v\rangle]^{-1}. \tag{7.22}$$

Problem 7.5. What is the time-scale when the two species a and X are the same? Explain physically why the extra factor is present.

When stellar evolution is brought about by changes in composition due to thermonuclear reactions, the rate calculated for the appropriate nuclear reaction also gives the time-scale for the evolution. For example, the time-scale for evolution of the sun is essentially determined by the first reaction of the proton-proton chain

$$H(H,e^+\nu_e)D. \tag{7.23}$$

For typical densities in the solar core, the time-scale for the destruction of H by this reaction is about 7×10^9 years.

Reaction rates are determined by barrier penetration, which we began discussing in Section 7.1, and will deal with more fully here. A nucleus of charge $Z_1 e$ incident on another of charge $Z_2 e$ encounters a potential barrier of height

$$Z_1 Z_2 \frac{e^2}{r_n}, \tag{7.24}$$

where r_n is the radius of the combined nucleus. This is shown schematically in Figure 7.1. Typical nuclear diameters are a few times 10^{-13} cm. The Coulomb barrier presented to two protons is therefore $e^2/r_n \simeq$ 1.4 MeV. At the temperatures near the center of the Sun, 10^7 K, typical thermal energies are $kT \approx 861$ eV, which is about 0.1 percent of the barrier height, and far too small to exceed the Coulomb potential barrier. It was first realized in connection with studies of beta decay that one must consider penetration of the barrier by the particles. This is a purely quantum-mechanical effect, with no classical counterpart. The penetration probability can be shown to be proportional to

$$\exp\left(-\frac{4\pi^2 Z_1 Z_2 e^2}{hv}\right), \tag{7.25}$$

where $Z_1 e$, $Z_2 e$, are the charges on the nuclei, and v is their relative velocity. The argument of the exponential is (to an order of magnitude) the ratio of the Coulomb energy to the kinetic energy E_c/E_{kin} of the reacting nuclei. If we denote a typical nuclear separation by r_0, then

$$\frac{E_c}{E_{\text{kin}}} \simeq \frac{Z_1 Z_2 e^2}{r_0} \bigg/ \frac{mv^2}{2}. \tag{7.26}$$

Using the uncertainty principle in the form $pr_0 \approx \hbar$ to express $r_0 = \hbar/p = \hbar/mv$, E_c/E_{kin} reduces to $4\pi Z_1 Z_2 e^2/hv$, which differs from the argument of (7.25) by a factor π. That the probability of barrier penetration should depend on the ratio E_c/E_{kin} is plausible, since increasing E_c increases the barrier height, but increasing E_{kin} means that the particles approach more closely and penetration becomes easier. It is sometimes also useful to write the penetration factor as a function of the energy $(\frac{1}{2})mv^2$ of the particles (in this case m must be the reduced mass of the particles),

$$\exp\left[-2\pi a Z_1 Z_2 \left(\frac{mc^2}{2E}\right)^{1/2}\right] = \exp(-b/E^{1/2}), \tag{7.27}$$

where the constant b is defined as

$$b \equiv 2\pi a Z_1 Z_2 \,(mc^2/2)^{1/2} \tag{7.28}$$

and $a = (e^2/hc)$.

We will see later that most nuclear reactions take place on the low-energy end of this penetration factor, because the Maxwellian velocity distribution gives relatively few particles with high energies. A useful rule of thumb for the temperatures at which reactions take place is obtained by setting

$$kT \simeq Z_1^2 Z_2^2 mc^2 \xi, \tag{7.29}$$

where ξ is a small factor, of the order of 10^{-6}. For example, the proton mass is 931 MeV; so the temperature at which proton-proton reactions take place is of the order of

$$T \sim 10^{-6} mc^2/k \simeq 10^7 \text{ K}. \tag{7.30}$$

Notice also how the penetration factor, or equivalently the required temperature, changes with increasing particle charge and mass. As the charge increases, so does the potential barrier; so the required temperature must go up. Also, the masses of high-Z nuclei are generally greater than those of lower-Z nuclei, and more difficult to accelerate to the required velocity. Consequently, the temperature necessary to make nuclear reactions "go" in stellar interiors increases quite rapidly with nuclear charge. For example, the temperature required for the reaction

$$\text{He}^4\,(\text{He}^4, \text{b})\text{Y} \tag{7.31}$$

will be roughly $2^2 \cdot 2^2 \cdot 4 = 64$ times that for the proton-proton reaction. Barrier penetration and other quantum-mechanical effects generally occur within a de Broglie wavelength of the particle, $\lambda = h/p$. We might therefore expect the cross section for a nuclear reaction to be proportional to the "area" $\pi\lambda^2$ of the particle. Since this is proportional to $1/p^2$ (or to $1/E$ for nonrelativistic energies), we may combine these two effects (penetration probability and quantum-mechanical "area") to define the cross-section factor $S(E)$ by the expression

$$\sigma(E) = \frac{S(E)}{E} \exp\,(-b/E^{1/2}). \tag{7.32}$$

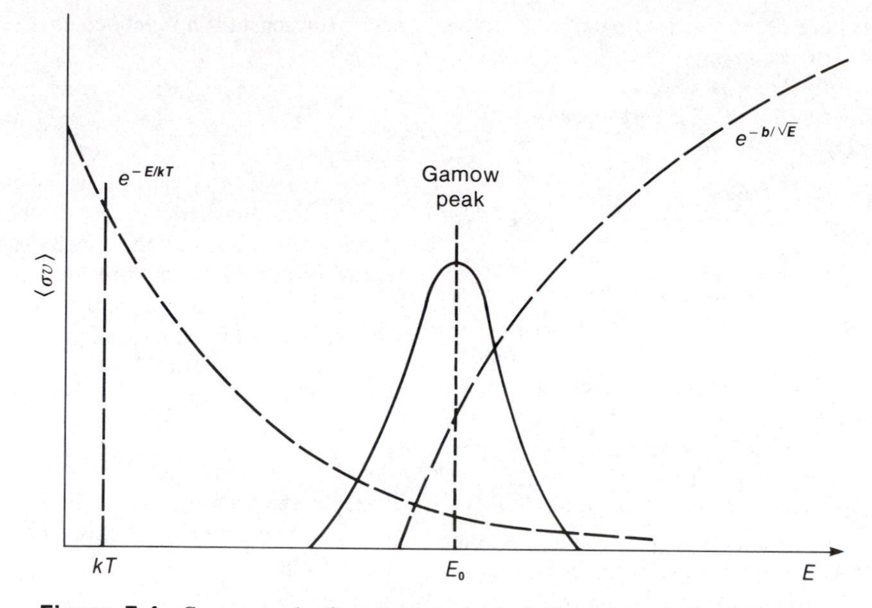

Figure 7.4. Gamow peak. Contributions to $\langle\sigma v\rangle$ from the velocity distribution [$\sim\exp(-E/kT)$] and the cross section [$\sim\exp(-b/E^{1/2})$] as a function of energy.

$S(E)$, which has units of energy $\times$ area, is usually a slowly varying function of energy over the range of significance. This definition is useful because the cross sections important in most stellar situations occur at lower energies than can be conveniently dealt with in laboratory measurements; so one must often extrapolate measurements to get the required cross sections at stellar energies. This extrapolation is easier in the slowly varying function $S(E)$ than in the cross section itself.

If we insert this expression for the cross section into the expression for the mean cross section, $\langle\sigma v\rangle$ in (7.19), and use the Maxwellian distribution in energy,

$$f(E)\,dE = \frac{2}{\sqrt{\pi}}\frac{E}{kT}e^{-E/kT}\frac{dE}{(kTE)^{1/2}}, \qquad (7.33)$$

we find

$$\langle\sigma v\rangle = \int_0^\infty \sigma(E)v(E)f(E)\,dE$$
$$= \left(\frac{8}{m\pi}\right)^{1/2}(kT)^{-3/2}$$
$$\times \int_0^\infty S(E)\exp\left(-\frac{E}{kT}-\frac{b}{\sqrt{E}}\right)dE. \qquad (7.34)$$

The exponential factor in the integrand is the product of two factors, the E/kT factor, which causes it to decrease with increasing E, and the $b/E^{1/2}$ factor, which causes it to increase with increasing E. The product of these two is shown schematically in Figure 7.4, and is significant only within a rather small range of energies in the vicinity of

$$E_0 = (bkT/2)^{2/3}. \qquad (7.35)$$

This effect is sometimes known as the Gamow peak. The energy E_0 can be thought of as the most effective energy for nuclear reactions at temperature T (for nuclear charges Z_1 and Z_2). Note that in these formulas E is the energy in the center of mass of the two colliding particles.

Problem 7.6. What is the most effective energy for nuclear reaction H(H, $e^+\nu_e$)D at the temperature of the solar center, 10^7 K?

Because of the Gamow peak effect, we can take the slowly varying factor $S(E)$ outside the integral in the

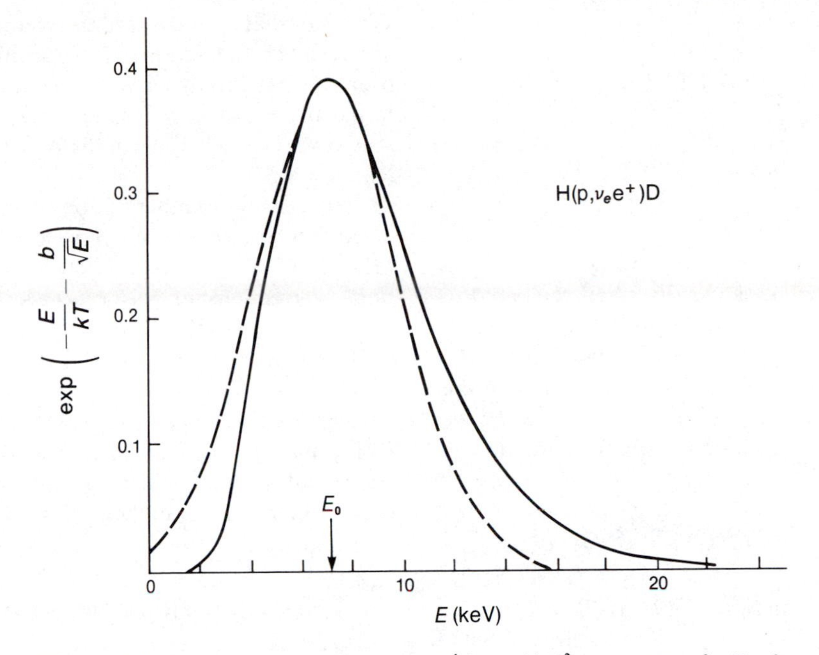

Figure 7.5. Gamow peak for the reaction H^1 (p, ν_e e^+) H^2 versus energy for $T = 2 \times 10^7$ K; $E_0 = 7.16$ keV, $\Delta = 8.11$ keV, and $b = 22.22$ keV. The dashed curve is the Gaussian fit. The ordinate is in units of 10^{-5}.

last step of (7.34) and so obtain, approximately,

$$\langle \sigma v \rangle = \left(\frac{8}{m\pi}\right)^{1/2} \frac{S(E_0)}{(kT)^{3/2}} \times \int_0^\infty \exp\left(-\frac{E}{kT} - \frac{b}{\sqrt{E}}\right) dE. \quad (7.36)$$

The integral that appears here, which is a function of b and T, is not analytically calculable, but can be approximately evaluated if the Gamow peak is represented as a Gaussian curve. For example, the Gamow peak for the H(H, ν_e e^+)D reaction at 2×10^7 K is shown in Figure 7.5. It is not far from a Gaussian curve, and one may show that it is well approximated by

$$\exp\left(-\frac{E}{kT} - \frac{b}{\sqrt{E}}\right) \simeq \exp\left(-\frac{3E_0}{kT}\right) \exp - \left(\frac{E - E_0}{\Delta/2}\right)^2, \quad (7.37)$$

where $\Delta \equiv 4\sqrt{E_0 kT/3}$. From this formula, the width of the Gamow peak, which measures the range of significant energies involved, is clearly

$$\Delta = \frac{4}{\sqrt{3}} (E_0 kT)^{1/2} = \frac{4}{\sqrt{3}} \left(\frac{b}{2}\right)^{1/3} (kT)^{5/6}. \quad (7.38)$$

Substituting this approximation for the Gamow peak into (7.36), one obtains

$$\langle \sigma v \rangle = \left(\frac{8}{\pi m}\right)^{1/2} \frac{S(E_0)}{(kT)^{3/2}} \exp\left(-\frac{3E_0}{kT}\right) \times \int_{-\infty}^{\infty} \exp\left[-\left(\frac{E - E_0}{\Delta/2}\right)^2\right] dE$$

$$= \left(\frac{8}{\pi m}\right)^{1/2} \frac{S(E_0)}{(kT)^{3/2}} \Delta \frac{\sqrt{\pi}}{2} e^{-3E_0/kT}. \quad (7.39)$$

Note that the lower limit of the integral has been taken to $-\infty$, since this introduces negligible error in the result. Introducing the dimensionless factor

$$\eta = \frac{3E_0}{kT} = 3\left(\frac{b}{2}\right)^{2/3} (kT)^{-1/3} = BT^{-1/3}, \quad (7.40)$$

we have

$$\langle \sigma v \rangle = \left(\frac{8}{\pi m}\right)^{1/2} \frac{\sqrt{\pi}}{2} \frac{8}{9\sqrt{3}} \frac{S(E_0)}{b} \eta^2 e^{-\eta}$$

$$= \frac{8\sqrt{2}}{9\sqrt{3}} \frac{S(E_0)}{mb} \eta^2 e^{-\eta}. \tag{7.41}$$

The variation in temperature throughout nuclear-burning regions in stars is usually small compared to typical temperatures in the region. It is therefore useful to develop an approximate expression for the reaction cross section (7.41), for temperatures in the vicinity of some value T_0. Assuming that $T \approx T_0$, we may set $T/T_0 = 1 + \xi$, where ξ is small, and write

$$\frac{\eta}{\eta_0} = \left(\frac{T_0}{T}\right)^{1/3} \simeq 1 - \frac{1}{3}\xi. \tag{7.42}$$

Denoting the reaction rate (7.19) at $T = T_0$ by r_0, substituting (7.42) into r/r_0, and ignoring the small variation in $S(E_0)$ with temperature, we find

$$\frac{r}{r_0} = \frac{S(E_0)_T}{S(E_0)_{T_0}} \left(\frac{\eta}{\eta_0}\right)^2 e^{-\eta - \eta_0} \simeq \left(\frac{\eta}{\eta_0}\right)^2 e^{-\eta_0(\eta/\eta_0 - 1)}$$

$$\simeq \left(1 - \frac{2}{3}\xi\right)\left(1 - \eta_0 \frac{\xi}{3}\right) \simeq 1 + \xi\left(\frac{\eta_0 - 2}{3}\right), \tag{7.43}$$

where all factors of order ξ^2 or higher have been dropped. Expanding $(T/T_0)^n$ to first order in ξ and comparing the result with (7.43) shows that

$$n = (\eta_0 - 2)/3, \qquad (T/T_0)^n \simeq r/r_0. \tag{7.44}$$

Therefore, when T/T_0 is of order unity, we may set

$$\langle \sigma v \rangle \simeq \frac{\langle \sigma v \rangle_0}{T_0^n} T^n. \tag{7.45}$$

To the same order of accuracy, the energy-generation rate per unit mass (7.20) for temperatures near T_0 is given by

$$\epsilon = \epsilon_0 \rho T^n, \tag{7.46}$$

where

$$\epsilon_0 = \frac{n_a n_X}{\rho} Q \frac{\langle \sigma v \rangle_0}{T_0^n}. \tag{7.47}$$

We recall that n_a and n_X are the abundance by number of nuclei per unit volume of stellar material. It follows from the first expression in (7.44) and the definition of h that the exponent n depends on T_0. For the Sun, we may take $T_0 \simeq 10^7$ K when the pp chain dominates; so here $n \simeq 4.5$.

Expressions of higher accuracy may be obtained by expanding the factor $S(E_0)$ about T_0, and using the result in (7.43). In accurate stellar structure calculations, corrections of this type are included, as are others for such effects as electron screening of the nuclear charge.

The preceding analysis applies to nonresonant nuclear reactions. For some nuclear processes, reaction (7.14) may proceed by intermediate formation of a compound nucleus, which can give a larger reaction cross section than would otherwise be expected.

7.4. Nuclear-Burning Stages

The nuclear-burning stages beginning with hydrogen (the most common nuclear fuel) involve a relatively slow buildup of nuclear species, at least through the formation of C^{12} and O^{16}. The most important reactions in this regime ($1 \le Z \le 8$) and ($1 \le A \le 16$) supply energy at the rates given approximately by the power law (7.6). These are discussed qualitatively below.

The Hydrogen-Burning Main Sequence

Before entering the main sequence, stars are mostly hydrogen. We therefore consider nuclear reactions involving hydrogen and possibly other light nuclei. The primary result of hydrogen burning is the conversion $4H \rightarrow He^4$, which takes place by a chain of reactions. To decide which reactions are important we must consider all possible reactions, with their cross sections, and find out which dominate in a given temperature range.

A typical nuclear reaction,

$$a + X \rightarrow b + Y, \tag{7.48}$$

where a is an impinging particle (perhaps an atomic nucleus itself), X is the target nucleus, Y is the new nucleus produced, and b is a particle (or particles) given off, may also be described by the simplified notation $X(a, b)Y$. For example, the capture of a proton H by N^{14} to produce O^{15} with the emission of a photon is N^{14} (H, γ) O^{15}. In fact, when a nucleus X captures a and emits b, the notation may be simplified to (a, b), since, given the nucleus X, the nature of Y is fixed.

There are two major chains in the H-burning main-sequence phase of stellar evolution: the proton-proton chain (pp); the carbon-nitrogen-oxygen cycle (CNO).

The proton-proton chain. The main pp chain consists of the set of reactions

$$\begin{aligned} &H(H, e^+\nu_e)\, D\, (H, \gamma)\, He^3, \\ &He^3\, (He^3, 2p)\, He^4, \end{aligned} \tag{7.49}$$

the net result of which is the conversion of 4H to He^4. In any such chain, the overall speed, and the rate at which energy is generated, is governed by the slowest reaction, which here is the formation of deuterium:

$$H(H, e^+\nu_e)D. \tag{7.50}$$

In fact, the cross section for this reaction is so small that it has not been measured in the laboratory. The reaction involves the conversion of a proton to a neutron:

$$p \rightarrow n + e^+ + \nu_e. \tag{7.51}$$

Since this is a β-decay reaction, its probability depends also on the β-decay constants, i.e., the weak interactions. For hydrogen burning via the pp chain at temperatures of about 10^7 K, as is typical for stars in the lower part of the main sequence, (7.6) may be used with $a = 1$ and $n \simeq 4$ to 5 in rough calculations.

The pp chain is believed to be the main energy-producing reaction in the Sun and other lower-main-sequence stars. In addition to the main pp chain, several subsidiary chains occur:

$$He^3(He^4, \gamma)\ Be^7(e^-, \nu_e)Li^7(H, He^4)He^4 \tag{7.52}$$

$$\downarrow$$

$$Be^7(H,\gamma)B^8(e^+,\nu_e)\,Be^8(\ \ , He^4)He^4. \tag{7.53}$$

The importance of these chains through Be^7 depends in part on how much He^4 is already present. In very cool stars, the $He^3(He^3, 2p)$ reaction rarely occurs, because of the greater Coulomb barrier. In these stars the chain stops at He^3, and He^4 is not produced. This fact, combined with the $He^3(He^4, \gamma)$ Coulomb barrier in chains (7.51) and (7.52), means that, at the lower temperatures, the net reaction rate depends on temperature in a very complicated way, being made up of three parts, $H^1(H^1, e^+\nu_e)$, $He^3(He^3, 2p)$, and $He^3(He^4, \gamma)$. At the temperature of the solar interior, however, the reaction rate is dominated by the first of these. Table 7.2 summarizes some of the results for reaction rates in the solar interior.

Some of these reactions produce neutrinos. Any energy carried by the neutrinos is lost to the star, whose neutrino opacity is effectively zero unless its density is 10^{11} g/cm^3 or more. To allow for neutrino energy loss, the net energy generation of the reaction (1.442 MeV for the pp reaction) must be reduced by averaging over the theoretically expected neutrino energy spectrum. The loss amounts to about 0.263 MeV for the pp reaction.

The CNO cycle. The pp chain suffers from a very small cross section for the first reaction (7.50). However, many stars contain an appreciable amount of the heavier elements. For example, if current ideas of stellar evolution are basically correct, the surface layers of most stars have not been changed by the nuclear reactions that occur in their interiors. The composition of many stellar atmospheres must therefore reflect the initial chemical composition of the stellar material. In fact, we find appreciable amounts of elements like C,N,O in stellar atmospheres, as well as trace amounts of Na, Fe, and heavier elements. This raises the possibility that reactions involving the nuclei C,N,O may be important. For example, the C^{12} nucleus may act as a catalyst to convert 4H to He^4, itself remaining unchanged. This leads to the CN cycle,

$$\begin{aligned} &C^{12}(H, \gamma)\, N^{13}\, (e^+, \nu_e)\, C^{13}\, (H, \gamma)\, N^{14}\, (H, \gamma) \\ &\qquad O^{15}\, (e^+, \nu_e)\, N^{15}\, (H, He^4)\, C^{12}. \end{aligned} \tag{7.54}$$

The importance of this reaction in stellar interiors will depend on the C^{12} abundance and on the temperature, since a greater Coulomb barrier ($Z_1Z_2 = 6$) has to be overcome than for the pp reaction. The first reaction in the chain is the slowest and governs the overall energy-generation rate. At temperatures of about 10^7 K, the value $n = 18$ in (7.6) is appropriate. Since it requires higher temperatures, the CN cycle will generally dominate in the upper main sequence, for stars more massive than the Sun, whereas the pp chain dominates in the lower main sequence, for stars of solar mass or less.

As in the pp chain, there are subsidiary cycles of importance, particularly one involving oxygen:

$$O^{16}\ (H, \gamma)\ F^{17}\ (e^{+}, \nu_e)\ O^{17}\ (H, He^4)\ N^{14}. \qquad (7.55)$$

This set of reactions creates N^{14} nuclei, which can serve as catalyst nuclei in the CN cycle. At first sight, this chain appears to destroy all O^{16}. However, another complication arises from branching of the N^{15} (H, α) reaction. It does not always produce C^{12}; occasionally (about once every 2,500 times) the reaction

$$N^{15}\ (H, \gamma)\ O^{16} \qquad (7.56)$$

regenerates O^{16}. Thus there are two coupled cycles going on simultaneously. The entire sequence is sometimes called the CNO bi-cycle.

Representation of the rate of nuclear energy generation as a power law (7.6) is clearly only a rough approximation. Nevertheless, it permits qualitative insights into stellar structure, as we will discuss in Section 7.5 in some detail. The main point is that the energy-generation rate is very temperature-sensitive (more so for CNO than pp); so the luminosity of a star might be expected to be very sensitive to the interior temperature. Conversely, because of the high temperature sensitivity, the effects of small changes in the chemical composition (say, represented by the constant ϵ_0) can be compensated for by rather small adjustments in temperature, which for a star of given mass will mean a rather small adjustment in its radius. Such considerations help explain the basic features of stellar interior structure and, to some extent, of stellar evolution.

In real stars both the pp and the CNO cycles will generally occur simultaneously, and accurate calculations must take account of both.

Other Thermonuclear Reactions in Stellar Interiors

Before main-sequence nuclear burning. Although not significant as energy-generating sources, some reactions involving light nuclei can occur at temperatures of around 10^6 K. Because of these reactions, Li, B, Be, and D are broken down by reactions such as

$$\begin{array}{ll} Li^6(H, He^4)He^3, & \\ Li^7(H, He^4)He^4, & Be^9(2H, 2He^4)He^3, \\ B^{10}(2H, He^4)2He^4, & D(H, \gamma)He^3. \end{array} \qquad (7.57)$$

The destruction of these light elements by relatively moderate temperatures makes it difficult to explain their observed abundances, since all the nuclei should have been destroyed if the stellar material were processed through stellar interiors. We must suppose that low-temperature nuclear reactions are taking place elsewhere than in the centers of stars in order to account for the observed abundances of these elements.

Helium burning. Hydrogen burning dominates the life of stars. Nevertheless, when hydrogen exhaustion in stellar interiors reduces core pressure (because the molecular weight has increased), the resulting contraction of the stellar core may heat the residual helium-rich material enough to ignite helium burning, *if* the star is massive enough. Stars less massive than about 0.5 $M_\odot$ do not have sufficient potential energy stored as gravitation to proceed to helium burning, which requires a core temperature of about 10^8 K.

The helium-burning reaction is also known as the triple-alpha process, and may be summarized by

$$3He^4 \rightarrow C^{12} + \gamma, \qquad (7.58)$$

with a Q of 7.27 MeV. The reaction actually takes place in two stages:

$$He^4\ (He^4),\ Be^8\ (He^4, \gamma)\ C^{12}. \qquad (7.59)$$

The first of these reactions is endothermic, requiring a supply of energy to make it go; so the third alpha particle must be immediately available to ensure reaction with the Be^8 nucleus, and so obtain a net output of energy. The reaction is therefore effectively a three-body collision, and the reaction rate is given approximately by

$$\epsilon_{3\alpha} = \epsilon_0\, \rho^2\, T^{30}. \qquad (7.60)$$

During He burning, the reaction

$$C^{12}\ (He^4, \gamma)\ O^{16} \qquad (7.61)$$

also occurs.

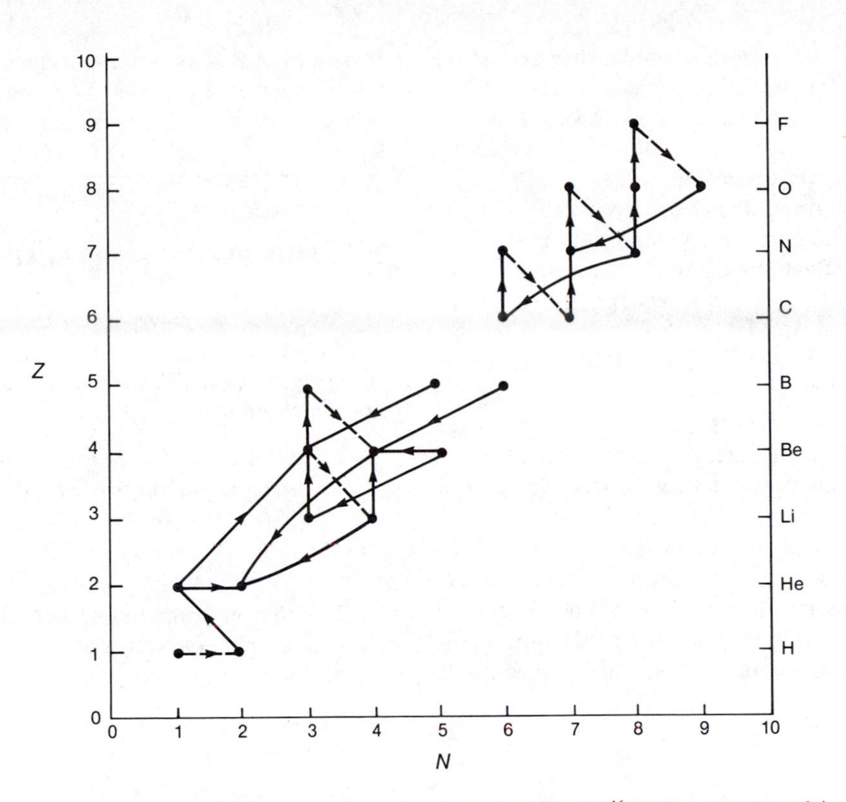

Figure 7.6. Nuclear burning network through formation of O^{16} (major processes only). These include strong interactions (solid lines), which are rapid; and weak interactions (dashed lines), which are slow.

Because of this reaction, C^{12} and O^{16} are the two most abundant nuclei after He burning, and are the nuclei that engage in the next two phases of nuclear-energy generation. In addition, any N^{14} present will be converted to O^{18} and Ne^{22} by the reactions

$$N^{14}\,(He^4, e^+\nu_e)\,O^{18}\,(He^4, \gamma)\,Ne^{22}. \tag{7.62}$$

Carbon burning. At temperatures greater than about $T = 6 \times 10^8$ K carbon burning can take place. The most important reactions are

$$C^{12} + C^{12} \longrightarrow \begin{cases} Na^{23} + H, \\ Ne^{20} + He^4, \\ Mg^{23} + n, \\ Mg^{24} + \gamma, \end{cases} \tag{7.63}$$

and the energy generation rate is approximately

$$\epsilon_c = \epsilon_0\, \rho\, T^{32}. \tag{7.64}$$

Oxygen burning. Oxygen, left from the C-burning stage (7.61) is the next contributor to energy generation, and undergoes the reactions

$$O^{16} + O^{16} \longrightarrow \begin{cases} S^{32} + \gamma, \\ P^{31} + H, \\ S^{31} + n, \\ Si^{28} + He^4, \\ Mg^{24} + 2He^4. \end{cases} \tag{7.65}$$

The reactions following oxygen burning do not follow a pattern like that seen so far. The reasons for this, and the consequences of it, will be explored in Chapter 12.

The major nuclear-burning stages through the formation of O^{16} are shown in Figure 7.6.

Neutrino Energy Loss

The neutrino has a very small interaction cross section ($\sigma \approx 10^{-44}$ cm^2) with other particles; so essentially all the neutrinos produced inside a typical star will escape.

For example, if n_b is the baryon density, and R the stellar radius, the probability that a neutrino will undergo an interaction with a baryon is roughly $\sigma\, n_b\, R$, which is approximately 10^{-9} for the Sun. This fact has several potential consequences.

First, if enough neutrinos are produced in stellar interiors, they may cause an important energy loss from the star without contributing to the pressure, an effect that must be taken into account during the late stages of stellar evolution.

Second, if such neutrinos could be observed on Earth, we would have some direct information on the processes going on in the centers of stars. However, the fact that the neutrino interaction cross section is so small means that they are very hard to detect (see Chapter 13).

Although neutrinos are produced in many thermonuclear reactions in stellar interiors, these do not represent the major source of neutrinos. The most important mechanisms include pair production by the mutual annihilation of an electron and a positron,

$$e^+ + e^- \rightarrow \nu_e + \bar{\nu}_e, \tag{7.66}$$

and photoneutrinos created when a photon decays into two neutrinos in matter of densities above 10^{10} g/cm^3,

$$\gamma \rightarrow \nu + \bar{\nu}. \tag{7.67}$$

This process requires a third body in order to conserve total energy and momentum. The third body may be a particle of the gas, or even a plasma oscillation in the ionized medium (plasma neutrinos). A third process, neutrino *Bremsstrahlung* (the free-free emission of a neutrino-antineutrino pair by an electron in the field of a proton, rather than the normal emission of a photon), is also important at high densities. This energy loss will be observable as a speeding up of the expected evolutionary process in a star. In the late evolution of a star, it may even be the dominant process. In addition, the cores of some stars may become so dense that they become at least partially opaque to neutrinos. This may happen in the collapsing core of supernovae.

Problem 7.7. What is the opacity of a neutron star to neutrinos? (What is the probability that a neutrino will escape?) Assume an average density and radius of $\bar{\rho} \simeq 10^{15}$ g/cm^3 and $R \simeq 10^6$ cm.

Problem 7.8. Show that a star collapsing without losing mass will eventually become opaque to neutrinos (assuming other processes do not halt the collapse).

7.5. Homologous Stellar Models

We can use the power-law expression for nuclear-energy generation (7.6) to construct an important class of approximate stellar models, if we make three additional assumptions:

(1) the opacity of the stellar material is given by the Kramers opacity formula,

$$\kappa = \kappa_0\, \rho T^{-3.5};$$

(2) either gas pressure or radiation pressure operates, but not both; and
(3) convective energy transport is ignored.

With these assumptions the equations of stellar structure become

$$\frac{dP}{dr} = -\frac{mG\rho}{r^2}, \tag{7.68}$$

$$\frac{dm}{dr} = 4\pi r^2 \rho, \tag{7.69}$$

$$\frac{dT}{dr} = -\frac{3\kappa_0 \rho^2 T^{-3.5}}{16\pi a c r^2 T^3} L, \tag{7.70}$$

$$\frac{dL}{dr} = 4\pi r^2 \epsilon_0 \rho^2 T^n, \tag{7.71}$$

and the equation of state is either

$$P = \rho \frac{kT}{\mu m_H} \tag{7.72}$$

or

$$P = \frac{1}{3} a T^4. \tag{7.73}$$

In these expressions, the chemical composition enters through the parameters μ, κ_0, and ϵ_0.

Such a set of equations is subject to a homology

transformation. That is, given a solution to the equations, with the quantities (P, T, L, ρ) stated as functions of r, for a given total mass M and chemical composition, then we can find a new solution for a new total mass, M', simply by multiplying the other physical variables by appropriate scaling factors.

Physically, two stars related by a homology transformation have the same relative mass distribution. If the two stars have masses M_1 and M_2, and radii R_1 and R_2, we can define

$$x = \frac{r_1}{R_1} = \frac{r_2}{R_2}. \tag{7.74}$$

Then

$$q \equiv \frac{m_1(r_1)}{M_1} = \frac{m_2(r_2)}{M_2} \tag{7.75}$$

expresses the fact that each star contains the same fraction, q, of its mass within the same fraction, x, of its radius. Equations (7.68) to (7.73) can be used to describe how the physical variables, such as ρ and P, scale when (7.74) and (7.75) hold. For example, (7.69) for star 2 may be transformed as

$$\frac{dm_2}{dr_2} = \frac{M_2}{M_1}\frac{dm_1}{dr_2} = \frac{M_2}{M_1}\frac{R_1}{R_2}\frac{dm_1}{dr_1}.$$

But

$$\frac{dm_2}{dr_2} = 4\pi r_2^{\,2}\rho_2 = 4\pi\left(\frac{R_2}{R_1}\right)^2 r_1^{\,2}\rho_2.$$

Equating the right-hand sides, and rearranging constant factors, gives

$$\frac{dm_1}{dr_1} = 4\pi r_1^{\,2}\frac{M_1}{M_2}\left(\frac{R_2}{R_1}\right)^3\rho_2.$$

Since the left-hand side must equal $4\pi r_1^{\,2}\rho_1$, it follows immediately that the density at x in star 2 is obtained by scaling the density at x in star 1 by $(M_2/M_1)(R_1/R_2)^3$:

$$\rho_2(x) = \left(\frac{M_2}{M_1}\right)\left(\frac{R_1}{R_2}\right)^3\rho_1(x). \tag{7.76}$$

The pressure transformation follows from the equation of hydrostatic equilibrium; making liberal use of (7.68) and (7.74) through (7.76), we can easily show that

$$\frac{dP_2(x)}{dx} = \left(\frac{M_2}{M_1}\right)^2\left(\frac{R_1}{R_2}\right)^4\frac{dP_1(x)}{dx}.$$

Since both sides are a function of x only, this equation can be integrated from the surface $x = 1$, where $P_2 = P_1 = 0$, to give

$$P_2(x) = \left(\frac{M_2}{M_1}\right)^2\left(\frac{R_1}{R_2}\right)^4 P_1(x). \tag{7.77}$$

Given a solution to the equations of hydrostatic equilibrium for one star, a family of homologous solutions can be found by using (7.74) through (7.77), which are also in hydrostatic equilibrium.

The temperatures in homologous stars can be related by the equation of state for the stellar material. In a star supported by ideal gas pressure, (7.72) and (7.76) to (7.77) imply that

$$T_2(x) = \frac{\mu_2}{\mu_1}\frac{M_2}{M_1}\frac{R_1}{R_2}T_1(x). \tag{7.78}$$

This is an homology transformation only if (μ_2/μ_1) is independent of x, as will be true for stars of uniform chemical composition, or for stars whose composition varies in the same manner from $x = 0$ to $x = 1$.

If star 1 is in hydrostatic and thermal equilibrium, then (7.74) to (7.78) guarantee only that all homologous stars will be in hydrostatic equilibrium; they will not, in general, be in thermal equilibrium. If they are not, then their structure will change over a Kelvin time-scale, and the stars will no longer be homologous. The luminosity and temperature gradients (7.70) and (7.71) must also admit homology transforms if the family of stars is to be in thermal equilibrium. Proceeding in analogy with the discussion above, it can be shown from (7.71) that the luminosity transforms according to

$$L_2(x) = \frac{\epsilon_{02}}{\epsilon_{01}}\left(\frac{\mu_2}{\mu_1}\right)^n\left(\frac{M_2}{M_1}\right)^{n+2}\left(\frac{R_1}{R_2}\right)^{n+3}L_1(x). \tag{7.79}$$

This form holds only when the ratios $(\epsilon_{02}/\epsilon_{01})$ and (μ_2/μ_1) are constant in x.

Finally, the requirement that the temperature gradient (7.70) retain its form under the transforms

Box 7.1
Stellar-structure equations: constitutive relations (left column), and implied relations (right column) between dependent variables. The energy-generation rates are appropriate for H burning, and Kramers opacity is assumed.

Constitutive relations	Implied relations
$\frac{dP}{dr} = -\frac{m(r)G}{r^2}\rho;$	$\frac{P}{R} \sim \frac{MG}{R^2}\cdot\frac{M}{R^3}$
$\frac{dm}{dr} = 4\pi r^2 \rho$	
$\frac{dL}{dr} = 4\pi r^2 \rho\epsilon;$	$\frac{L}{R} \sim R^2\frac{M}{R^3}\epsilon \sim R^2\frac{M}{R^3}\epsilon_0\frac{MT^n}{R^3}$
$L = \frac{4\pi r^2 c}{\rho}\frac{d}{\kappa\, dr}\left(\frac{1}{3}aT^4\right);$	$L \sim R^2\frac{1}{\rho\kappa}\frac{T^4}{R} \sim R^2\left(\frac{R^3}{M}\right)^2\frac{T^{7/2}}{\kappa_0}\frac{T^4}{R}$
$P = \frac{\rho kT}{\mu m_H};$	$P \sim \frac{M}{R^3}\frac{T}{\mu}$ (gas pressure)
$P = \frac{1}{3}aT^4;$	$P \sim T^4$ (radiation pressure)

Energy generation rates:
$\epsilon = \epsilon_0 \rho T^n$
Opacity:
$\kappa = \kappa_0 \rho T^{-7/2}$

(7.76) to (7.79) leads to the following constraint:

$$\frac{\epsilon_{02}}{\epsilon_{01}}\frac{\kappa_{02}}{\kappa_{01}}\left(\frac{\mu_2}{\mu_1}\right)^{n-7.5}\left(\frac{M_2}{M_1}\right)^{n-3.5}\left(\frac{R_1}{R_2}\right)^{n+2.5} = 1. \quad (7.80)$$

If the structure of star 2 is related to that of star 1 by the complete set of relations (7.76) to (7.80) and the material is described by the perfect gas equation of state, then it will be in hydrostatic and thermal equilibrium.

Problem 7.9. Show that the luminosity in two homologous stars transforms according to (7.79), and verify the constraint (7.80). Discuss the implications of (7.80).

The analysis leading to (7.79) and (7.80) can be generalized to stars supported by radiation pressure, or stars in which the energy generation rates and opacities are proportional to arbitrary powers of ρ and T. However, homology transformations do not exist for arbitrary forms of the constitutive relations.

By applying these scaling arguments to the complete set of differential equations (7.68) to (7.71), one can eliminate uninteresting variables, and obtain scaling relationships between, for example, the luminosity of a star and its mass, or its radius. This is how we discover mass-luminosity relationships for stars, as well as radius-mass relationships and luminosity-temperature relations. These are summarized in Boxes 7.1 to 7.3. Box 7.1 gives the equations of stellar structure. Box 7.2 gives the deduced relationships for gas pressure (GP) and radiation pressure (RP). Box 7.3 anticipates some later results, namely, that in the upper part of the main sequence (UMS) radiation pressure dominates the structure and the important nuclear reactions yield a value of n of about 18, whereas in the lower part of the main sequence (LMS) gas pressure is more important and other nuclear reactions yield n about 5.

We can learn several interesting things from these formulas. First, we have a mass-luminosity relation $L \sim M^{5.4}$ for the upper main sequence, and $L \sim M^{1.6}$ for the lower main sequence, where we have assumed GP in LMS and RP in UMS.

Box 7.2

Relations between M, R, L, and T for stars supported by gas pressure (GP) and by radiation pressure (RP). The index n is the temperature exponent in the energy-generation rate.

Radius-mass	$R \sim (\kappa_0\epsilon_0)^{2/(2n+5)}\,(G\mu)^{(2n-15)/(2n+5)}\,M^{(2n-7)/(2n+5)}$	GP
	$R \sim (\kappa_0\epsilon_0)^{2/(2n+5)}\,G^{(2n-15)/(4(2n+5))}\,M^{(2n+1)/(4(2n+5))}$	RP
Temperature-mass	$T \sim (\kappa_0\epsilon_0)^{-2/(2n+5)}\,(G\mu)^{20/(2n+5)}\,M^{12/(2n+5)}$	GP
	$T \sim (\kappa_0\epsilon_0)^{-2/(2n+5)}\,G^{5/(2n+5)}\,M^{2/(2n+5)}$	RP
Luminosity-temperature	$L \sim \kappa_0^{-1/6}\,\epsilon_0^{5/6}\,(G\mu)^{-4/3}\,T^{(10n+31)/12}$	GP
	$L \sim \kappa_0^{1/2}\,\epsilon_0^{3/2}\,G^{-2}\,T^{(6n+17)/4}$	RP
Mass-luminosity	$L \sim \kappa_0^{-(2n+6)/(2n+5)}\,\epsilon_0^{-1/(2n+5)}\,(G\mu)^{(14n+45)/(2n+5)}\,M^{(10n+31)/(2n+5)}$	GP
	$L \sim \kappa_0^{-(2n+6)/(2n+5)}\,\epsilon_0^{-1/(2n+5)}\,G^{(14n+45)/(2n+5))}\,M^{(6n+17)/(2(2n+5))}$	RP

Problem 7.10. Verify the results in Boxes 7.1–7.3. How different would the M-L relation be if gas pressure dominated on the upper part of the main sequence?

The observed mass-luminosity relation is about $L \sim M^4$. We see that the index of the mass-luminosity relation is primarily an index of the relative importance of gas pressure and radiation pressure, and the observed index of 4 suggests that in real stars both are important.

The position of the main sequence in the Hertzsprung-Russell diagram (Chapter 3) can be estimated. In the LMS we have $L \sim T^7$, whereas in the UMS, $L \sim T^{31}$. Certainly the observed upper main sequence bends upward, as suggested by these results.

The luminosity of stars on the lower main sequence is very sensitive to the value of G. This is of potential interest for some theories of cosmology that demand the gravitational "constant" be different in the past from what it is now. If it was different, the luminosity of the Sun would have been significantly different in the past, and this could be detected in principle from fossil records on the Earth. Such investigations have set limits to the possible time-variation of G.

The luminosity is rather insensitive to the constant ϵ_0, because the temperature sensitivity is so great that a star can easily make up for deficiencies in ϵ_0 by small adjustments in temperature. Changes in chemical composition will affect μ, κ_0 and ϵ_0. However, L is not only insensitive to ϵ_0 but also inversely proportional to κ_0, whereas it is proportional to $\mu^{23/3}$ for the LMS. The temperature of a star of given M is, on the other hand, nearly proportional to $\mu^{4/3}$. To first order, then, a change in the chemical composition of a star is felt first through μ, and increases T by a factor $\mu^{4/3}$ and L by a factor $\mu^{23/3}$. It therefore moves a star along a line $L \sim T^6$ in the Hertzsprung-Russell diagram. This is fairly close to the $L \sim T^7$ of the main sequence. Hence position of the main sequence in the HR diagram is not very sensitive to chemical composition, although of course the luminosity of a star of given mass is quite sensitive to chemical composition.

Problem 7.11. When a star's structure changes homologously, it remains in hydrostatic equilibrium. Show that for an ideal gas, T, ρ, and p change with R according to

$$\frac{\Delta T}{T} = \frac{1}{3}\frac{\Delta\rho}{\rho} = \frac{1}{4}\frac{\Delta P}{P} = -\frac{\Delta R}{R}.$$

Box 7.3

Same as Box 7.2, but assumes gas pressure and the pp chain ($n = 5$) for the lower main sequence (LMS), and radiation pressure and CNO cycle ($n = 18$) for the upper main sequence (UMS).

$R \sim (\kappa_0\epsilon_0)^{2/15}\,(G\mu)^{-1/3}\,M^{1/5}$	$L \sim \kappa_0^{-1/6}\,\epsilon_0^{5/6}\,(G\mu)^{-4/3}\,T^{81/12}$	(LMS)
$R \sim (\kappa_0\epsilon_0)^{3/4}\,G^{21/164}\,M^{37/82}$	$L \sim \kappa_0^{1/2}\,\epsilon_0^{3/2}\,G^{-2}\,T^{125/4}$	(UMS)
$T \sim (\kappa_0\epsilon_0)^{-2/15}\,(G\mu)^{4/3}\,M^{12/15}$	$L \sim \kappa_0^{-16/15}\,\epsilon_0^{1/15}\,(G\mu)^{23/3}\,M^{27/5}$	(LMS)
$T \sim (\kappa_0\epsilon_0)^{-2/111}\,G^{5/41}\,M^{2/41}$	$L \sim \kappa_0^{-42/41}\,\epsilon_0^{-1/41}\,G^{297/164}\,M^{125/82}$	(UMS)

7.6. Electron Screening in Nuclear Reactions

When a nuclear fuel is very dense, as in the cores of red giants, in hot white dwarfs, and in the central regions of highly evolved stars, the energy-generation rates are somewhat different. At low density, the reacting nuclei must overcome or tunnel through their mutual Coulomb barrier, which is proportional to $Z_1Z_2e^2$, in order to react. When the density is high, the electron gas tends to screen the ion charges, reducing the effective Coulomb barrier. Screening is more effective in hot, dense systems than in degenerate ones.

Electron screening can be approximated when the screening is weak. First, rewrite (7.34) in the equivalent form

$$\langle \sigma v \rangle = \int_0^\infty \sigma(E)v(E)f(v)\,dv. \tag{7.81}$$

Without screening, the ion's motion is determined by the energy difference

$$E - V(r) = E - \frac{Z_1Z_2e^2}{r}, \tag{7.82}$$

where Z_1e and Z_2e are the charges of the two reacting nuclei. When screening is important, the Coulomb potential energy $V(r)$ must be replaced by the screened potential energy, which we may write as

$$V_s(r) = V(r) + U_s(r). \tag{7.83}$$

The ion motion is then determined by

$$E - V_s(r) = E - U_s(r) - V(r). \tag{7.84}$$

Examination of (7.81), and the derivation leading to it, shows that introducing screening replaces the particle energy E by $E - U_s(r)$, and that the screened reaction cross section is

$$\langle \sigma v \rangle_s = \int_0^\infty \sigma[E - U_s(r)] \times V[E - U_s(r)]\,f(v)\,dv. \tag{7.85}$$

Screening is a collective effect, in which electrons tend to clump around ions, producing a slight increase in the electron-charge density at the ion sites. The effectiveness of screening is measured by the size of the charge cloud around the ion, or equivalently by the distance (screening length) from the ion within which the charge density drops to a certain fraction of its peak value. As shown in what follows, this is roughly the Debye length, given by

$$\lambda_D^{-1} \simeq \left(\frac{4\pi n_e e^2}{kT}\right)^{1/2}, \tag{7.86}$$

where n_e is the average electron-number density. Strong screening will result when λ_D is small relative to the ion-ion separation. According to the discussion of (7.35), nuclear reactions occur when the particle energies E_0 are in the keV range. The classical distance of closest approach is then given (for un unscreened potential) by

$$r_0 \sim \frac{Z_1Z_2e^2}{E_0}. \tag{7.87}$$

If $r_0 > \lambda_D$, the screening is strong, and the fuel behaves more like a liquid or solid than a gas. We consider only the case $r_0 \ll \lambda_D$, where screening is weak. The nuclei are then well within the screening cloud. Approximating this cloud by a smoothly varying distribution without point sources, we may set $U_s(r)$, the screening correction to the potential, approximately equal to $U_s(0)$,

$$U_s(r) \simeq U_s(0). \tag{7.88}$$

Problem 7.12. Show that electron screening of the pp chain for the Sun is weak. Assume $n_e \simeq 10^{26}$ cm^{-3} and $T \simeq 10^7$ K.

Furthermore, since the material is assumed to be gaslike, we may safely assume that the energies E and E_0 are large compared to $U_s(0)$. This simply requires that electrostatic effects be small. Using (7.88) in (7.85), and noting that $U_s(0)$ is also small,

$$\langle \sigma v \rangle_s \simeq \langle \sigma v \rangle \exp\,[-U_s(0)/kT]. \tag{7.89}$$

That is, the screened energy-generation rates will be given by the product of the unscreened rate and the factor $e^{-U_0/kT}$ when screening is weak.

Next the factor $U_s(0)$ must be found, by deriving the Poisson equation for the screening-charge distribution created by the ions and electrons that are present.

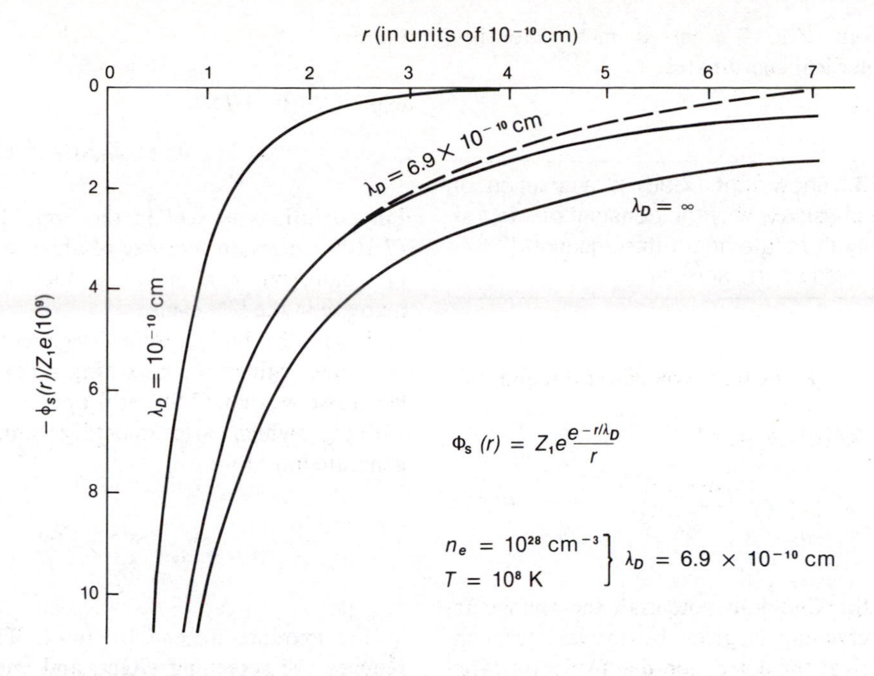

Figure 7.7. Screened Coulomb potential $\phi_s(r)$, for three values of the screening length λ_D (see text). The dashed curve is the weak-screening approximation (7.93) for $\lambda_D = 6.9 \times 10^{-10}$ cm.

This analysis is complex; so we will consider the simpler problem of a test charge $Z_1 e$ immersed in an initially uniform distribution of electrons of number density n_e. The screened potential due to the electron distribution is given by the solution to Poisson's equation

$$\nabla^2 \phi_s(r) = 4\pi e\,[n_e(r) - n_e], \tag{7.90}$$

where $n_e(r)$ is the electron distribution when screening develops. It is not constant, but peaks near the test charge $Z_1 e$. The chemical potential of an electron in a uniform electron gas at high temperature ($kT \gg \mu_0$) is

$$\mu_0 = kT \ln\left[\frac{n_e}{2}\left(\frac{2\pi\hbar^2}{m_e kT}\right)^{3/2}\right]. \tag{7.91}$$

In the presence of the external potential $\phi_s(r)$, the electron chemical potential becomes

$$\mu_e = \mu_0 + e\phi_s(r), \tag{7.92}$$

$$\mu_e = kT \ln\left[\frac{n_e(r)}{2}\left(\frac{2\pi\hbar^2}{m_e kT}\right)^{3/2}\right]. \tag{7.93}$$

Substituting (7.93) into (7.92), and using (7.91), we find, for $|e\phi_s(r)| \ll kT$,

$$n_e(r) - n_e \simeq e\phi_s(r)/kT. \tag{7.94}$$

Substituting this into (7.90) yields the differential equation

$$\nabla^2 \phi_s(r) = \lambda_D^{-2}\,\phi_s(r), \tag{7.95}$$

where λ_D is given by (7.86).

If the ion being screened is of charge $Z_1 e$ in vacuum, then the solution to (7.95) is

$$\phi_s(r) = Z_1 e \frac{e^{-r/\lambda_D}}{r} \tag{7.96}$$

The screened potential $\phi_s(r)$ is shown for several values of the Debye length in Figure 7.7. The case $\lambda_D = \infty$ gives the unscreened Coulomb potential, and the case $\lambda_D = 10^{-10}$ cm is representative of the strong screening limit. The intermediate curve results when $n_e = 10^{28}$ cm^{-3} and $T = 10^8$ K, as in the core of highly evolved stars. Since the screening charge will be spherically symmetric about $Z_1 e$, $\nabla^2 \phi_s(r)$ is most naturally expressed in spherical coordinates.

Problem 7.13. Show that (7.96) is a solution of (7.95). Explain physically why the constant of integration is Z_1. Justify the statement after equation (7.87) that small λ_D implies strong screening.

In the limit $r_0 \ll \lambda_D$ we may expand $\phi_s(r)$ to obtain

$$\phi_s(r) \simeq \phi(r) - Z_1 e \lambda_D^{-1}$$

$$= \phi(r) - Z_1 e \left(\frac{4\pi n_e e^2}{kT}\right)^{1/2}. \qquad (7.97)$$

Since $\phi(r)$ is the Coulomb potential, the change in $\phi_s(r)$ due to screening is given by the last term in (7.97), which gives the correction due to electrostatic screening in the potential of the ion of charge $Z_1 e$. If the ion with which it must react has charge $Z_2 e$, then the screened potential energy is

$$U_s(0) = -Z_1 Z_2 e^2 \left(\frac{4\pi e^2 n_e}{kT}\right)^{1/2}. \qquad (7.98)$$

Therefore

$$-\frac{U(0)}{kT} = \frac{Z_1 Z_2 e^2}{(kT)^{3/2}} (4\pi n_e e^2)^{1/2}$$

$$= 0.188\, Z_1 Z_2 \zeta \rho^{1/2} T_{(6)}{}^{-3/2}, \qquad (7.99)$$

where ζ is a factor of order unity that depends on composition. When the analysis is carried out for a mixture of ions (A_Z, Z) with mass fractions X_Z, it is found that

$$\zeta^2 = \sum_Z \frac{Z(Z+1)}{A_Z} X_Z. \qquad (7.100)$$

In stars with low metal content ($X_Z \approx 0$ for $Z > 2$), $\zeta^2 \simeq (X + 3)/2$. As expected, (7.99) shows that electron screening increases the rate of nuclear-energy generation. When (ρ/T^3) is small, the electron-shielding factor

$$\exp(-U_s(0)/kT)$$

$$\simeq 1 + 0.188\, Z_1 Z_2 \zeta \rho^{1/2}\, T_{(6)}{}^{-3/2}. \qquad (7.101)$$

For conditions typical in the Sun, the correction in (7.101) causes an increase of about 8 percent for the pp chain, but can yield as much as a 50 percent increase in the CN cycle.

When the fuel becomes degenerate in the weak screening regime, the preceding analysis can be modified to show that (7.99) still applies if ζ^2 is replaced with $\zeta^2{}_{\text{deg}}$, which varies smoothly from ζ^2 in the nondegenerate limit to

$$\zeta^2{}_{\text{deg}} = \sum_Z Z^2 \frac{X_Z}{A_Z}, \qquad (7.102)$$

in the extreme degenerate limit. Thus degeneracy reduces the screening effect, and energy is released more slowly than under nondegenerate conditions. For low metal abundance, $\zeta^2{}_{\text{deg}} \simeq 1$. The temperature when ζ is replaced by ζ_{deg} in (7.101) is that of the ions that satisfy Boltzmann statistics, even though the electrons are described by a Fermi-Dirac distribution.

Pycnonuclear Reactions

The number of nuclei in the high-energy tail of the Boltzmann distribution decreases rapidly with decreasing temperature. Below about 10^5 K, the reactions described in the preceding become ineffective as energy sources, since too few nuclei are energetic enough to tunnel through the Coulomb barrier. At very high densities, however, two effects combine to make nuclear burning possible, even at low temperatures. First, electron screening, which is important at high density and low temperatures, reduces the Coulomb barrier (Figure 7.7). Second, as the density increases, even at $T = 0$ the ions are squeezed into a smaller volume. Eventually the uncertainty principle becomes important, requiring that the uncertainty in their momentum Δp and position Δx satisfy $\Delta p\, \Delta x \gtrsim \hbar$. Since $\Delta x \simeq a$, the interparticle spacing, and $p \approx \Delta p$, this implies a zero-point energy

$$E_0 \simeq \frac{\Delta p^2}{2m_H} \simeq \frac{\hbar^2}{2m_H a^2} \approx (\hbar^2/2m_H{}^{5/3})\rho^{2/3}. \qquad (7.103)$$

For high densities (about 5×10^4 g/cm^3 for H; 8×10^8 g/cm^3 for He^4; and 6×10^9 g/cm^3 for C^{12}) and $T = 0$, the zero-point energy may suffice to initiate nuclear reactions between neighboring nuclei. The process, called pycnonuclear burning, is most effective when the nuclei have formed a crystal lattice, in which case the energy E_0 contributes to their vibrational energy about the equilibrium position. For light nuclei at high densities, the zero-point motion may be large enough that the nuclei spend much of their time far from the equilibrium point. The lattice then represents a quantum crystal.

Chapter 8

INTRODUCTION TO STELLAR EVOLUTION

8.1. Phases of Stellar Evolution

Stellar evolution may be divided into three important phases: before, during, and after the main sequence. The first phase follows the formation of a star from a cloud of interstellar material, through its gravitational contraction and subsequent heating, to the main sequence, defined as the point at which hydrogen burning begins. The second phase follows the development of a star on the main sequence. A star spends most of its lifetime on the main sequence; so this is in a sense the most important, though the least spectacular, phase. The third phase follows the increasingly more rapid evolution that begins when hydrogen burning has been exhausted in the stellar core, through the often spectacular but brief events that foreshadow the star's ultimate death.

Understanding the details of these phases requires lengthy calculation on large computers, and it is not always easy to pick out precisely which physical process is responsible for what phenomenon in the evolution. Although many phenomena of stellar evolution can be understood in terms of a quasistatic model, this is not true for evolution before the main sequence, or for some advanced phases of stellar evolution, since these phases involve nonequilibrium processes, and are often characterized by free-fall time-scales that are comparable to the evolutionary time-scale.

Once a star has separated gravitationally from the rest of the universe and is a bound system, its future will be a continual battle against the force of gravitation, which will try to make the star collapse. During various stages the star can halt this collapse temporarily by burning nuclear fuel, and so maintaining a high internal temperature gradient that will give the pressure gradient necessary for support. However, eventually this fuel supply must run out, and the star will collapse further until a new fuel supply is found, or the collapse is halted by degeneracy, or some violent event rearranges its structure. When none of these options is available to the star, continuing dynamic collapse appears inevitable. The result is a black hole.

A star is formed in the region of large radius and low temperature in the HR diagram. If the subsequent collapse is quasistatic, then, according to the virial theorem, half the energy of collapse goes to heating the material, and the other half is radiated away. If the collapse is not quasistatic, some of the energy of collapse goes into rotational energy of the whole star. A more complete statement of the virial theorem for

stars not in equilibrium is

$$\frac{1}{2}\frac{d^2I}{dt^2} = 2U + \Omega, \qquad (8.1)$$

where $I = \int r^2\, dm$ is the generalized moment of inertia for the star. Initially the gas is transparent, and collapse occurs at nearly uniform temperature, so that most of the change in Ω goes into I; the collapse is therefore rapid. When the density is great enough, the gas becomes opaque, and some of the gravitational energy goes into heating. This reduces the rate of collapse, and a state of quasistatic equilibrium is approached. The interior of the star may now be hot enough for hydrogen burning to begin, at which point the collapse is effectively halted, and the star adjusts itself to equilibrium on the zero age main sequence. Notice that for this to happen there must be enough gravitational potential energy available to raise the temperature to the hydrogen ignition point. For a gas cloud of very low mass, this will not happen, and the collapse is halted by electron degeneracy pressure. The lowest-mass star that can burn hydrogen is about 0.05 $M_\odot$. The planet Jupiter, and possibly some other giant planets, may be such objects. We do not yet have a complete theory of the distribution function by mass of objects that condense from interstellar material, but there may be many such stellar "failures" around the universe, shining only at the expense of internal energy built up during collapse, and with very weak gravitational fields. Presumably they would be rather hard to detect. Theory suggests that they may be infrared "stars" whose surface temperatures are less than 2,000 K.

Problem 8.1. Suppose collapse is halted when the temperature reaches a critical value T_c required for H burning. Suppose also that all the initial gravitational energy can go into thermal energy. Show that the greater the mass of the star, the smaller the density of the star at the point where T_c is reached:

$$\rho_c = \frac{3}{4\pi M^2}\left(\frac{kT_c}{m_H G}\right)^3. \qquad (8.2)$$

Note the criterion for electron degeneracy, and estimate the critical mass below which collapse is halted by electron degeneracy, not by H burning. Show that this critical mass is related to the Chandrasekhar mass limit by the approximate relation

$$\frac{M_{\rm crit}}{M_{\rm ch}} \simeq \mu_e{}^2 \frac{3}{2}\frac{1}{\pi 5^{3/2}}\left(\frac{kT_c}{m_e c^2}\right)^{3/4} \qquad (8.3)$$

The major contributor to this expression is the factor $1/5^{3/2}$, which is about 1/11. This relation provides an estimate of the critical mass as about 0.1 $M_\odot$, which is not too different from the 0.05 $M_\odot$ stated in the preceding paragraph.

Evolution on the main sequence begins with the burning of hydrogen in the stellar core. Roughly speaking, the core is radiative if the pp chain dominates the energy production, and convective if the CNO cycle dominates. One of the important features of a convection zone is that it mixes the stellar material within the convective core. For stars with $M \lesssim 1.3\ M_\odot$, there is no convection zone at the center. For stars with $M \geq 3\ M_\odot$, the energy generation is entirely contained within the convective core. For intermediate cases, energy generation goes on in both a convection zone at the center and the surrounding radiative region.

Evolution off the main sequence involves a complicated series of interrelated effects. As the evolution advances, nuclei of increasing Z are burned, leading to the development of a highly inhomogeneous structure, with energy generation taking place in various more or less discrete zones in the star. Nevertheless, a few general principles can be established.

(a) The initial effect of nuclear burning is to increase the mean molecular weight. Thus, as $4H^1 \rightarrow He^4$, the molecular weight is increased by a factor of 8/3. This increase lowers the pressure in the energy-generation zone (other things being equal). The pressure then no longer supports the star, and the core begins to contract, which will increase T, and hence P, and may restore hydrostatic equilibrium.

(b) This rise in temperature increases the temperature gradient, which would tend to increase the flow of energy from the star. To restore this balance, so that the stellar luminosity balances the energy generation, the star must increase its radius. Concurrently, the energy generation increases somewhat because of the rise in temperature. The result is that the total luminosity of the star tends to increase. (The sensitivity of L to molecular weight μ is shown in Box 7.2 for homology equations.) These two general principles appear to govern the evolution of a star, although

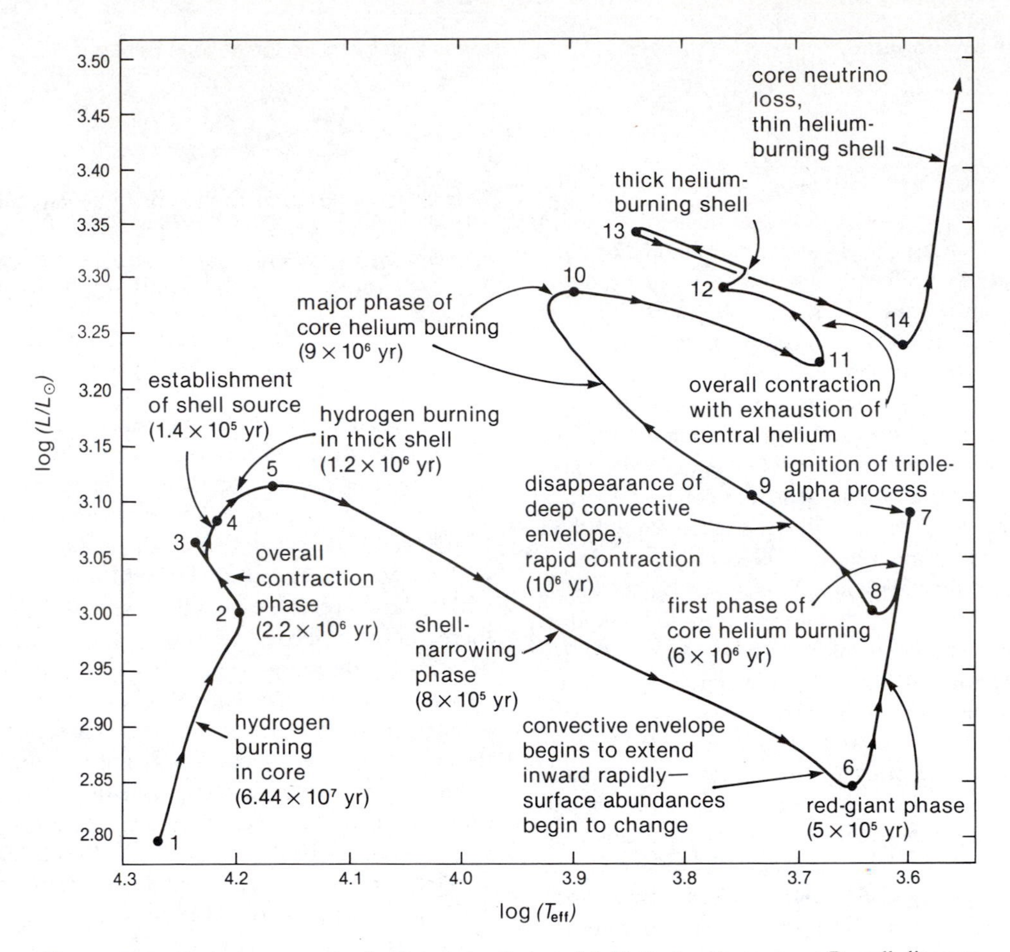

Figure 8.1. Evolutionary path of a Population I star of 5 $M_\odot$ in the Hertzsprung-Russell diagram, showing processes characterizing each stage. Time in parentheses is the duration of the stage between the numbered points. Luminosity is in $L_\odot$ and temperature in K. The elapsed time between points 10 and 11 is 10^6 years.

many details are specific to various phases. In general, the star will tend to develop a high-density core of thermonuclear-processed stellar material, surrounded by a large, low-density envelope. This is more or less the picture of a red giant star.

(c) In addition to these factors, the rising temperature in the core of more massive stars will eventually ignite higher-mass nuclear fuel. Surrounding the core is a zone of material processed from the previous burning stage, but not yet ignited; surrounding that is a shell zone in which the previous burning stage continues. This in turn is surrounded by unignited material, and so on until the hydrogen envelope is reached. This "onion-skin" model has important consequences for theories of stellar nucleosynthesis.

In Figure 8.1, the evolution of a main-sequence star of 5 $M_\odot$ is summarized. From points 1 to 2, the molecular weight is increasing, and the core contracting, as indicated above. From 2 to 3, the star has significantly depleted its H at the center, and contracts overall until the H is totally exhausted at the center. Continuing contraction of the core proceeds between points 3 and 4, until a shell source of H burning is initiated at 4, continuing to point 5. From 5 to 6, the H-burning shell narrows, and the inner He core contracts still further, increasing the central temperature. The radius expands greatly, attempting to reduce the temperature gradient. In the process, the outer layers cool, becoming relatively opaque as they pass the H ionization zone. This increase in opacity generates a surface convection zone, which rapidly deepens. The star continues to expand, but the surface temperature stays nearly constant, with the core still contracting (points 6–7). This surface convection zone may extend down to the region that has been enriched by H-He burning and the reactions $O, C \rightarrow N^{14}$. Hence, because of convective mixing, the surface abundances of O, C, and N change during the phases 6–7. At point 7 the core temperatures are high enough to ignite He, and the contraction stops. The star readjusts itself through points 8, 9, 10 until He is established as the main energy source for the star; the star has contracted overall, so that the outer layers are now reasonably transparent, and the surface convection zone has disappeared. The star has now reestablished a fairly stable structure, similar to what it had on the main sequence, although now with He burning as the energy source rather than H burning. The subsequent phases of evolution are very similar to the previous ones in their general characteristics: core contraction, expanding envelope, exhaustion of central He, establishment of shell source of He, convective envelope, ignition of C, and so on.

It is assumed here (by choosing a star of 5 $M_{\odot}$ as the example) that the core will in fact get hot enough to ignite helium, and further that it does so at reasonably low densities. Two alternatives are possible for stars of less mass.

(a) The core density may reach the value required for electron degeneracy before the temperature for He ignition is reached. If this happens, the core ceases to contract and remains isothermal (because of the high thermal conductivity of degenerate electrons). The outer, nondegenerate layers may still contract, however, and their increasing temperature is communicated to the core because of its high conductivity. If the star is massive enough, the whole of the core is ignited at once, because of its isothermal character. Also, because the degeneracy pressure is not very sensitive to T, this rise in temperature does not do much initially to remove the degeneracy. As a result, the ignition of He in this way is explosive, and the whole star undergoes a very rapid (i.e., with a time-scale of seconds) adjustment of its structure. This process is known as the *He flash*. What happens to the star after this is not certain, although it is known that this explosive ignition of He does not have really drastic (i.e., disruptive) effects on the star.

(b) The other alternative, for stars of mass less than about 1 $M_{\odot}$, is that the temperature will never rise enough to ignite the He. The star will then sink into oblivion as a white dwarf, composed primarily of He.

A similar situation occurs at the time of C ignition, certain stars undergoing quiet C ignition, others undergoing a C flash, and others not making it, becoming white dwarfs composed primarily of C.

The most advanced stages of stellar evolution are quite complicated, and are discussed in Chapter. 12.

The continuing production of a very dense stellar core and a very tenuous outer envelope suggests that the envelope might relatively easily become so detached from the star that it is no longer gravitationally bound. Although the mechanisms that might accomplish this are by no means settled, this gives an attractive model for the production of planetary nebulae, objects that consist of a tenuous shell of gas surrounding the now-visible stellar core. The surrounding gas would then be excited by absorption of photoionizing radiation. Section 24.3 will explore these evolutionary features in greater detail.

Evolutionary Sequences

The problem of stellar evolution is much more difficult that that of stellar structure, for two primary reasons. First, some stages involve structural changes on dynamic time-scales, so that the full hydrodynamic equations must be solved. Second, the structure, and its rate of change at a given instant, necessarily depend on the structure during the previous instant. This latter complication implies that stellar evolution must ultimately be treated by constructing a sequence of models beginning with some known state that either is static, or has a rate of structural change that is known at each point throughout the star. Given such a state, the usual procedure is to choose a time-step large enough to be practical, yet small relative to the system's time-scale, and to obtain a new, evolved model. The new model then serves as input for the next step, and the procedure is repeated. The sequence of models, separated by finite (and usually variable) time-steps, approximates the evolution of the star.

The stellar-structure equations, which have been developed in earlier sections, are

$$\frac{\partial P}{\partial r} = -\frac{m(r)G\rho}{r^2} - \rho\frac{d^2r}{dt^2}, \tag{8.4}$$

$$\frac{\partial m}{\partial r} = 4\pi r^2 \rho, \tag{8.5}$$

$$\frac{\partial T}{\partial r} = -\frac{3\kappa\rho}{16\pi ac}\frac{L(r)}{r^2T^3}, \tag{8.6}$$

$$\frac{\partial L}{\partial r} = 4\pi r^2 \rho \left[\epsilon(r) - T\frac{ds}{dt}\right]. \quad (8.7)$$

Unless the system is changing on a time-scale comparable to the free-fall time, the last term in (8.4) may usually be ignored. Equations (8.4) to (8.7) differ from the static definition in that all quantities will be functions of time. For truly static situations, d^2r/dt^2 and ds/dt vanish. When the system changes quasistatically, the term ds/dt must usually be retained.

8.2. Evolution of a Protostar

The birth of a protostar begins when local regions of interstellar gas become gravitationally bound and begin to contract. We will postpone discussion of this phase until Chapter 22. Here we will consider the evolution immediately before main-sequence hydrogen burning. The initial model for this stage of evolution may be obtained by examining the transition from rapid to quasistatic contraction.

The principal constituents of protostellar clouds are atomic and molecular hydrogen, atomic helium, and dust. The dissociation and ionization of the molecular and atomic species reduces γ below 4/3; so collapse must continue until these processes have been completed. The temperatures and ionization or dissociation energies are shown in Table 8.1. When these processes are complete, γ rises to 5/3, and the collapse becomes quasistatic contraction. We may use the virial theorem and the observation that collapse halts when ionization and dissociation are complete to relate the maximum radius of the stable configuration to its mass and to the energies in Table 8.1.

As the cloud contracts, half the change in gravitational energy goes into radiation, and half goes into internal energy. The internal energy goes into ionization and dissociation until these processes are complete. The total energy per gram E_I associated with these processes may be expressed as follows.

The total energy absorbed to ionize He completely is given by the product (number of He atoms/g) × (mass function of He) × (ionization energy of He), and is $(N_0/4)(Y)(E_{He})$ where N_0 is Avagadro's number. Continuing in this way for the discussion of H_2 and the ionization of the resulting H, we find

$$E_I = N_0XE_H + 1/2\, N_0XE_D + 1/4\, N_0YE_{He}$$
$$= 1.9 \times 10^{13}\,[1 - 0.2\,X], \quad (8.8)$$

Table 8.1
Ionization or dissociation energy, and temperature, for principal constituents in stars before the main sequence.

Constituent	Dissociation or ionization energy (eV)	T(K)
H_2	$4.48 = E_D$	5.2×10^4
H	$13.6 = E_H$	1.6×10^5
HeI	$24.58 = E_{HeI}$	2.9×10^5
HeII	$54.4 = E_{HeII}$	6.3×10^5

using $X + Y = 1$, and denoting $E_{HeI} + E_{HeII} = E_{He}$. From the virial theorem,

$$\frac{1}{2}\alpha\frac{M^2G}{R} \simeq ME_I, \quad (8.9)$$

with α of order unity (for $n = 1.5$ polytrope, $\alpha = 6/7$). Using (8.8) and solving for the radius, we find

$$\frac{R}{R_\odot} = \frac{43.2(M/M_\odot)}{1-0.2X}. \quad (8.10)$$

This represents the maximum radius for a stable star as it begins its evolution. For larger radii, ionization and dissociation are incomplete, and gravitational potential energy is not converted into heat, and thus pressure, needed to stop the collapse. Once the radius (8.10) is reached, the internal temperature rises, and a state of quasiequilibrium is possible.

Problem 8.2. Find the average density and the central temperature of the stable protostar having a radius given by (8.10) if its structure is approximated by an $n = 1.5$ polytrope, and $X = 0.7$.

The central temperature of the protostar ($n = 1.5$) is given by

$$T_c \simeq 3 \times 10^5 \mu\,(1 - 0.2X)\text{ K}, \quad (8.11)$$

which is well below the point at which thermonuclear reactions become efficient. Therefore the system's only source of energy is the conversion of gravitational potential energy into thermal energy and radiation. Denoting the total energy of the protostar by E_T, it follows that the luminosity is

$$L = dE_T/dt = \frac{1}{2}\alpha\frac{d}{dt}\frac{M^2G}{R}$$

$$= -\frac{\alpha}{2}\frac{M^2G}{R}\frac{\dot{R}}{R}. \qquad (8.12)$$

Since L is positive, $\dot{R}$ must be negative, and, as expected, the star must contract to supply the needed luminosity. The time-scale for the evolution is set by the ratio of energy lost, ΔE, to the luminosity:

$$\Delta t \simeq \frac{\Delta E}{L} \sim -\frac{1}{2}\frac{\Delta\Omega}{L} \sim \frac{M^2G}{2RL}$$

$$= 1.6 \times 10^7 \left(\frac{M}{M_\odot}\right)^2 \left(\frac{R_\odot}{R}\right)\left(\frac{L_\odot}{L}\right) \text{yrs.} \qquad (8.13)$$

Since L may be large, Δt is relatively short.

Hayashi Track

The central temperatures of newly formed protostars are still sufficiently low that ionization contributes to the opacities, which will be very large throughout. This means that the limiting radiative luminosity (5.135) will be small, and energy transport by convection dominates. Recall that for stars in hydrostatic equilibrium with an ideal-gas equation of state, the quality (T^3/ρ) varies only slowly through the star. The initial stellar model, with radius (8.10) will therefore be fully convective, except for the outermost layer (photosphere). The models developed in Section 6.5 are therefore applicable to this stage of evolution. In general, however, the evolution must be followed numerically. To do so, one chooses M and T_c, and adjusts the remaining parameters to satisfy the stellar structure equations. Nevertheless, these simplified models give a reasonable order-of-magnitude estimate of the star's early evolution, and illustrate the qualitative nature of its evolution. For example, (6.131) shows that Population I stars of low surface density undergo a reduction in luminosity and a slight increase in T_{eff} as they contract. Their evolutionary track in the HR diagram is given approximately by

$$\log L = 10 \log M - 7.24 \log T_{\text{eff}} + \text{constant}. \qquad (8.14)$$

This track (known as the Hayashi track) is steeply descending, and shifts upward with increasing initial mass.

As the star contracts, its surface density increases. In low-mass stars, the approximations (6.135) eventually become appropriate. The high surface density increases the dependence of L on R, but changes the qualitative dependence of T_{eff} on R. Now L drops more rapidly, and T_{eff} decreases very slightly as contraction continues. The approximate evolutionary track, when (6.135) holds, is

$$\log L = 38.7 \log T_{\text{eff}} - 6.74 \log M + \text{constant}. \qquad (8.15)$$

Therefore, as the star contracts, its surface temperature grows slightly, and it continues to grow fainter.

Problem 8.3. Obtain the evolutionary track $L(T_{\text{eff}})$ for stars of high surface density that contain H ionization zones. Compare the result with (8.15).

As the interior temperatures increase, Kramers' opacity becomes important in the core, and κ is reduced. The limiting luminosity (5.135) is thus increased, and energy transport by radiation becomes dominant. At this stage the star developes a radiative core, which, with continued contraction, grows at the expense of the convective envelope. The star is now less sensitive to surface conditions, and its luminosity may be described approximately by the linear model (6.112). Using (6.112) and (1.3) to eliminate the radius, we can express the evolutionary track as

$$\log L = 0.8 \log T_{\text{eff}} + 4.4 \log M + \text{constant}. \qquad (8.16)$$

Continued contraction increases T_{eff} and the luminosity is also increased, though to a lesser extent. The star now moves sharply to the left and slightly upward in the HR diagram. Eventually the convective envelope shrinks to a relatively thin layer, and radiation dominates the interior.

In more massive stars, electron scattering dominates and, as a result, the leftward movement is essentially horizontal, since the luminosity depends primarily on mass.

Problem 8.4. Obtain the evolutionary track $L(T_{\text{eff}})$ for massive stars in which electron scattering domi-

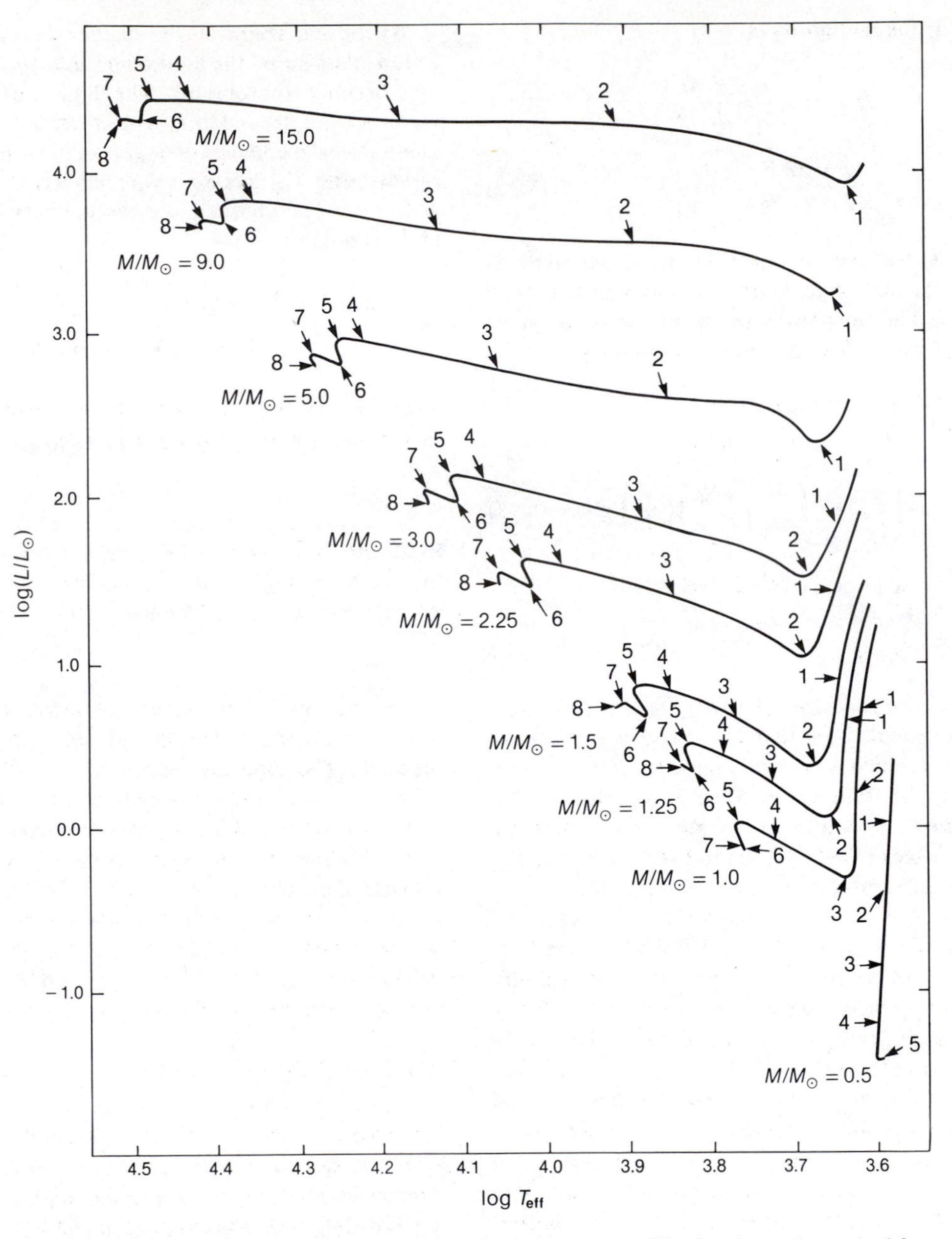

Figure 8.2. Evolutionary tracks for stars before the main sequence. The time intervals required for stars to reach the numbered points are given in Table 8.2. The locus of the terminal points of the paths defines the zero age main sequence.

nates. Explain why these stars do not develop high surface densities during their convective stages. Compare the evolutionary tracks for high-mass and low-mass stars during their convection and radiative stages.

The development of a radiative core and subsequent evolution away from the Hayashi track marks the last stage in the star's approach to the main sequence. Evolutionary tracks based on realistic models of protostars are shown in Figure 8.2, for initial masses in the range $0.5 \leq M/M_{\odot} \leq 15.0$. The numbered arrows

Table 8.2
Elapsed time in years of selected points along evolutionary tracks leading to the main sequence, as shown in Figure 8.2.

	$M/M_\odot$								
Point	15.0	9.0	5.0	3.0	2.25	1.5	1.25	1.0	0.5
1	6.740×10^2	1.443×10^3	2.936×10^4	3.420×10^4	7.862×10^4	2.347×10^5	4.508×10^5	1.189×10^5	3.195×10^5
2	3.766×10^3	1.473×10^4	1.069×10^5	2.078×10^5	5.940×10^5	2.363×10^6	3.957×10^6	1.058×10^6	1.786×10^6
3	9.350×10^3	3.645×10^4	2.001×10^5	7.633×10^5	1.883×10^6	5.801×10^6	8.800×10^6	8.910×10^6	8.711×10^6
4	2.203×10^4	6.987×10^4	2.860×10^5	1.135×10^6	2.505×10^6	7.584×10^6	1.155×10^7	1.821×10^7	3.092×10^7
5	2.657×10^4	7.922×10^4	3.137×10^5	1.250×10^6	2.818×10^6	8.620×10^6	1.404×10^7	2.529×10^7	1.550×10^8
6	3.984×10^4	1.019×10^5	3.880×10^5	1.465×10^6	3.319×10^6	1.043×10^7	1.755×10^7	3.418×10^7	
7	4.585×10^4	1.195×10^5	4.559×10^5	1.741×10^6	3.993×10^6	1.339×10^7	2.796×10^7	5.016×10^7	
8	6.170×10^4	1.505×10^5	5.759×10^5	2.514×10^6	5.855×10^6	1.821×10^7	2.954×10^7		

SOURCE: I. Iben, Jr., *Astrophys. J.*, **141**:993 (1965). Reprinted by permission of The University of Chicago Press. Copyright 1965 by The University of Chicago.

indicate the models whose evolutionary lifetimes, as measured from formation, are given in Table 8.2. These are in order-of-magnitude agreement with the rough estimate (8.13) if main-sequence values of M, L, and R are used. Figure 8.3 shows the variation in several parameters with time for a star of one solar mass as it evolves down the Hayashi track and over to the main sequence. We see that the star actually moves away from the Hayashi track only after the radiative core has taken over most of the interior, leaving very little mass in the convective envelope.

The preceding simple model misses the final decline in luminosity as the star settles onto the main sequence. This decline is due to thermonuclear burning of C^{12} at temperatures above about 1.2×10^6 K, and will be considered in Chapter 9.

In their convective stages, protostars will evolve down the Hayashi track in a relatively short time (10^7

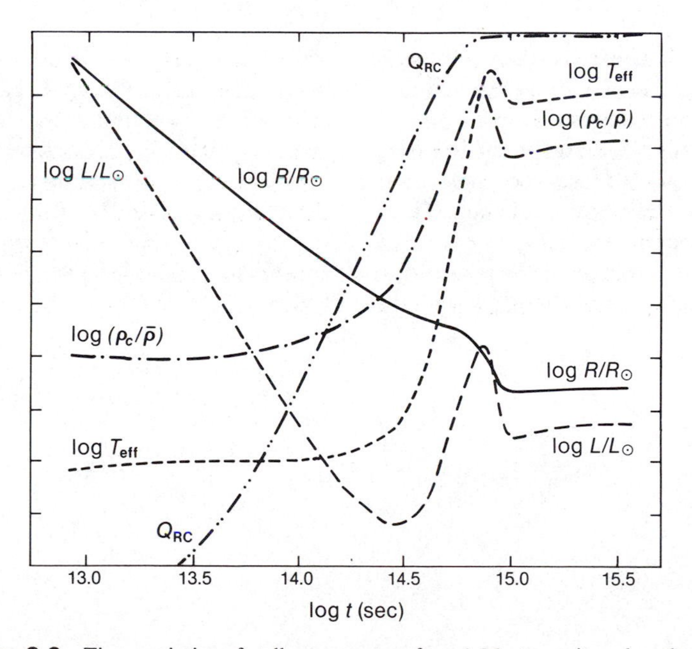

Figure 8.3. Time-variation of stellar parameters for a 1 $M_\odot$ star as it evolves down the Hayashi track; the time is in seconds. The curve Q_{RC} gives the mass fraction in the radiative core, with the ordinate scaled such that $0 \le Q_{RC} \le 1.0$. For other curves, $3.58 \le \log T_{eff} \le 3.78$; $0 \le \log(\rho_c/\bar{\rho}) \le 2.0$; $-0.4 \le \log(L/L_\odot) \le +0.6$; and $-0.4 \le \log(R/R_\odot) \le +0.6$.

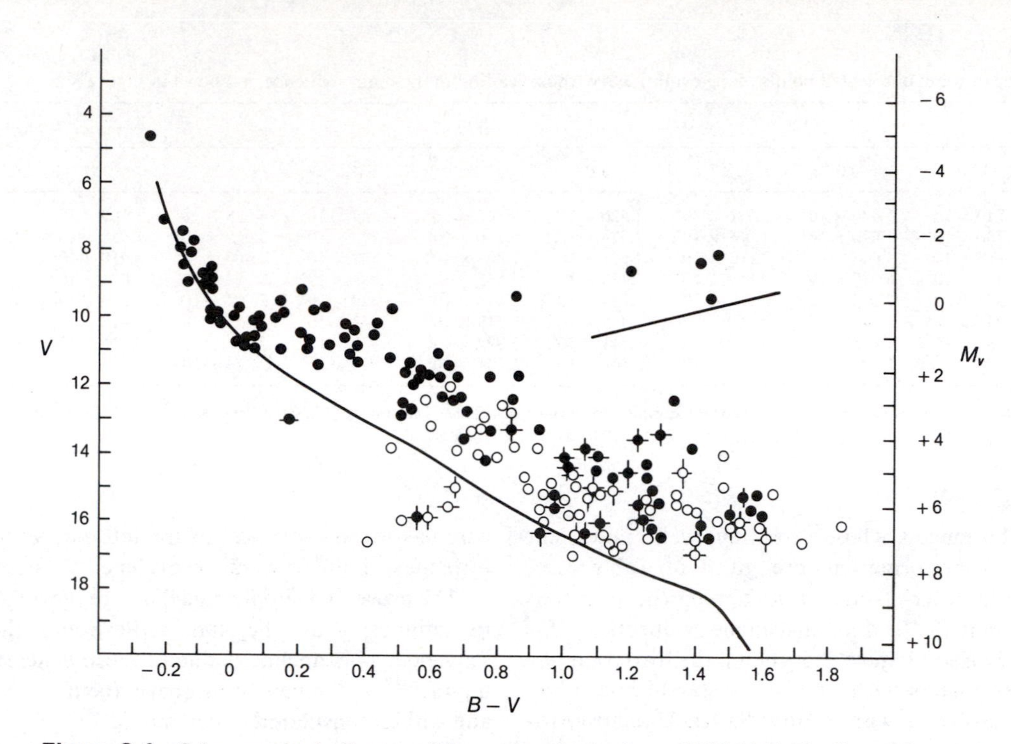

Figure 8.4. Color-magnitude diagram of the young galactic cluster NGC 2264. The lines represent the main sequence and the giant branch, corrected for uniform reddening. Points with horizontal bars denote stars with H_α line emission; those with vertical bars denote variable stars.

years or less for stars more massive than the Sun). Their luminosities during this period may be substantial, but their surface temperatures are only 3 or 4 × 10^3 K; so most of their light is radiated in the infrared. Finally, their association with dense, cool interstellar gas clouds is a source of additional reddening. These factors make observation of the early protostellar evolution difficult. Once a radiative core develops, however, the surface temperature increases and the evolutionary rate decreases, making observation more likely. The red irregular T Tauri variables, found in associations containing hot young stars and dense clouds of hydrogen, are believed to be stars in the final approach to the main sequence. In fact, the location of these stars in the HR diagram (Figure 8.4), lying slightly above the main sequence for field stars, appears to terminate on the zero age main sequence, to which we turn in Chapter 9.

Chapter 9

THE MAIN SEQUENCE

9.1. The Zero Age Main Sequence

The curve defined by values of (L, T_{eff}) corresponding to static stars that are homogeneous and have just commenced hydrogen burning forms the *zero age main sequence* (ZAMS). Central temperatures of about 2×10^7 K separate stars whose primary mode of hydrogen burning is the pp chain from those in which the CN and CNO bi-cycle dominate. These two classes differ in structure primarily because of the temperature dependence of the two nuclear processes. Consider the CNO cycle; for $1.2 \times 10^7 < T < 5 \times 10^7$ K, the energy generation rate varies roughly as T^n with $20 \gtrsim n \gtrsim 13$. Rates that are so temperature-sensitive lead to the development of convective cores, because temperature-sensitive sources tend to be centrally concentrated, which in turn leads to large radiative temperature gradients. In order to reduce these temperature gradients, the core becomes convective. Reaction rates for the pp chain are far less sensitive, with $6 \gtrsim n \gtrsim 3.5$ for $4 \times 10^6 \lesssim T \lesssim 2.4 \times 10^7$ K. Radiative transport is therefore possible for reasonable temperature gradients.

Problem 9.1. Approximate a star's luminosity by $L(r) = L =$ constant (point source) when the nuclear-energy generation rates are highly temperature-sensitive. Show that a radiative core is impossible under these circumstances.

The envelope structure of stars on the upper main sequence is also different from that of stars on the lower main sequence. In the latter stars, which are cool, hydrogen ionization zones in the envelope reduce the adiabatic index, and convective regions develop; in stars on the upper main sequence, which are hotter, ionization zones are very shallow, and the envelope is radiatively stable.

The star's final approach to the zero age main sequence is marked by a drop in luminosity, which occurs just as hydrogen burning begins in the core, as C^{12}, though present only in small quantities, burns to form N^{14}. The energy-generation rate is $\sim T^{19}$, and therefore leads to the development of a convective core. Convective regions are well-approximated by polytropes of index $n = 1.5$, whereas radiative ones correspond to larger values of n. Examination of the ratio of central to average density $\rho_c/\bar{\rho}$ for polytropes as a function of n shows that larger values of n imply greater degree of central condensation. When C^{12} burning begins, the star thus must adjust its density distribution from the relatively more condensed configurations typical of radiative cores to the less condensed cores typical of convective equilibrium, that is, must move some matter outward, against the pull of gravity. Since this represents work done on the star, not all the energy released is available for radiation. Calculations show that only about 80 percent of the thermonuclear energy released is available, the remainder going into work. Since gravitational contraction effectively ceases at this point, there is a net reduction in luminosity and T_{eff}, and the star's evolutionary track moves down and slightly to the right in the HR diagram. Since very little C^{12} is available to the star at this stage of its life, this phase is short-lived.

Hydrogen burning, which has been going on along with C^{12} burning, remains as the star's primary energy source once the initial C^{12} supply has been exhausted. Since low-mass stars favor the pp chain, whose temperature dependence is weak, a radiative core again develops. In more massive stars, which operate on the CN and CNO bi-cycle, convection is again favored because of the extreme temperature-sensitivity of the fuel.

For the first time in the formation process, the star is essentially static, supplying its energy losses by nuclear burning.

9.2. Evolution on the Main Sequence

Main-sequence stars evolve on the extremely slow nuclear time-scale $t_n \sim Mc^2/L$; so at each instant their properties are well represented by solutions to the static structure equations. As time passes, the star's composition changes as a result of thermonuclear processes. It is these composition changes that lead to new static models, and the succession of static models so obtained represents the star's slow evolution on the main sequence. The static structure equations are therefore augmented by specifying the time-rate of change of the star's composition. For example, suppose that only the hydrogen and helium abundances X and Y are needed to specify a static model at time t. We then relate their time-rates of change to the energy-generation rates and the energy released per gram of matter. For hydrogen, X is reduced by the pp chain, the CN and CNO bi-cycle, or both; so the time-rate of change dX/dt is described as follows.

For the pp chain, $Q_{\text{pp}} \approx 26.73$ MeV of energy is liberated each time four H are converted into one He^4 nucleus. The energy released per gram of transmitted matter is therefore $Q_{\text{pp}}/4m_{\text{H}} = E_{\text{pp}}$. The contribution to the time-rate of change of X due to the pp chain is thus $4\epsilon_{\text{pp}}\, m_{\text{H}}/Q_{\text{pp}} = \epsilon_{\text{pp}}/E_{\text{pp}}$, where the energy-generation rate has been written ϵ_{pp}. Similar expressions for the CN cycle and CNO bi-cycle are easily constructed. Then it follows that

$$\frac{dX}{dt} = -\frac{\epsilon_{\text{pp}}}{E_{\text{pp}}} - \frac{\epsilon_{\text{CN}}}{E_{\text{CN}}}, \tag{9.1}$$

where the last term includes the contributions from the CN cycle and the CNO bicycle. For low-mass stars with $T < 2 \times 10^7$ K, the first term dominates, whereas for $T > 2 \times 10^7$ K the last term dominates. The terms E_{pp} and E_{CN} are constants, but ϵ_{pp} and ϵ_{CN} depend on the structure of the star.

Helium burning may be analyzed similarly, except for the complication that any remaining hydrogen burning supplies fuel for helium burning. When only the triple-alpha process depletes the helium supply, then

$$\frac{dY}{dt} = -\frac{\epsilon_{3\alpha}}{E_{3\alpha}} - \frac{dX}{dt}. \tag{9.2}$$

The last term, which is positive, represents helium produced by concurrent hydrogen burning. This is an important feature of nuclear burning: the ashes of one burning stage become potential fuel for a subsequent stage. Whether or not this fuel ignites depends on the star's mass. The term $\epsilon_{3\alpha}/E_{3\alpha}$ in (9.2) represents the rate of buildup of heavy elements.

The preceding discussion must be modified where nuclear burning occurs in convective regions. The time-scale for convective motion across typical zones is in months, which is essentially instantaneous when compared to the time-scales of nonviolent nuclear processes. Consequently, an average composition, X_c, Y_c, and Z_c, may be associated with such regions, and its rate of change will depend on the rates at which energy is generated and released. Although we will not discuss them further here, one must also include corrections that take into account movement of the boundaries of the convective zone into radiative regions. If this is neglected, then the rate of change of X_c is given by ϵ/E for the relevant process, averaged through the zone. For hydrogen burning by the pp chain,

$$\frac{dX_c}{dt} = -\int_{M_1}^{M_2} \frac{\epsilon_{pp}}{E_{pp}} \frac{dM}{\Delta M}, \tag{9.3}$$

where $\Delta M = M_2 - M_1$ is the mass of the convective zone.

Calculations of evolution on the main sequence (or during any quasistatic phase) therefore begin by assuming an initial model, in this case, one that is chemically homogeneous, whose initial values include X and Y at each point in the star. The structure is calculated and used to estimate the changes $\Delta X = (dX/dt)\, \Delta t$ and $\Delta Y = (dY/dt)\, \Delta t$ during a small time interval Δt at each point in the star. The new composition $X' = X + \Delta X$ and $Y' = Y + \Delta Y$ serves as input for the next static model, and the process is repeated. It is assumed in this process that the total mass is constant to within the small fraction carried away by radiation. Some evolutionary processes exist in which mass loss plays an important role. The most obvious of these is mass exchange in close binary systems. Another, whose significance is not clear observationally, may be associated with mass loss by, for example, stellar winds. Mass exchange will be discussed in Chapter 17.

Problem 9.2. For a star beginning to consume hydrogen on the zero age main sequence via the CNO cycle, with Kramers' opacity dominating, the linear (Chapter 4) model gives (for Population I)

$$\frac{R}{R_\odot} = 0.451\mu^{0.395}(M/M_\odot)^{0.697}, \tag{9.4}$$

$$L/L_\odot = 43.5\mu^{7.3}(M/M_\odot)^{5.18}, \tag{9.5}$$

$$T_{\text{eff}} = 2.34 \times 10^4 \mu^{1.63}(M/M_\odot)^{0.871}, \tag{9.6}$$

$$T_c = 1.98 \times 10^7 \mu^{0.606}(M/M_\odot)^{0.364}, \tag{9.7}$$

$$\rho_c = 65.8\mu^{-0.455}(M/M_\odot)^{-0.909}, \tag{9.8}$$

where the subscript c denotes central values. Discuss the evolution of such a star during the first stage of nuclear burning qualitatively. For simplicity, assume $Z = 0$.

9.3. Lower Main Sequence

Stars in which the pp chain dominates, and which therefore have radiative cores, are usually referred to as being on the lower main sequence. Typically their central temperatures are below about 2×10^7 K, and their masses below about 2 $M_\odot$. The structure and evolution of lower-main-sequence stars is believed to be typified by the Sun, at least for compositions corresponding to Population I. We will focus primarily on Population I objects, though some aspects of Population II will also be mentioned and developed in the problems.

When they reach the main sequence, most low-mass stars develop radiative cores where hydrogen burns predominantly via the pp chain. These cores are surrounded by convective envelopes, which extend to the base of the photosphere. The size of the core becomes smaller as we move down the main sequence. Initially the core is chemically homogeneous.

The star's energy losses are supplied by nuclear reactions. The latter occur on the nuclear time-scale t_n, which is as much as 10^{16} times the dynamic time-scale. Thus the star is easily able to rearrange its structure quasistatically to accommodate the slow depletion of H in the core. As fuel is used up, however, the energy-generation rate, which is crudely proportional to $\rho X^2 T^n$, would decrease with X unless ρ, T, or both increase. A reduction in energy-generation rate would lower the pressure; so the star would contract, increasing X. By the virial theorem, part of the released potential energy goes into raising the temperature, and part goes into radiation. Therefore both ρ and T must increase. The net effect is an increase in energy output, and a slight increase in the radius of both the core and the envelope. The higher energy-generation rate in turn increases L; the increase in T_{eff} is smaller because of the small change in radius. Therefore the evolutionary track is one of movement to slightly higher T_{eff} and proportionately higher L. The curve marked 1 $M_\odot$ in Figure 9.1 shows the evolution of a star like the Sun. The portion of the track between points marked 1 and 2 corresponds to the evolution just described, which requires about half the main-sequence lifetime, or about 4.3×10^9 yrs. At the end of this period, the star's structure is as shown in Figure 9.2, which shows X, $L(r)$, $T(r)$, $\rho(r)$, and r as functions of the mass fraction $m(r)/M$. Recall that the stellar surface corresponds to $m(r)/M = 1$. Roughly 90 percent of the mass lies within $r = 0.5R$. The reduction in H, represented by $X < 1$, occurs primarily within $r = 0.3R$

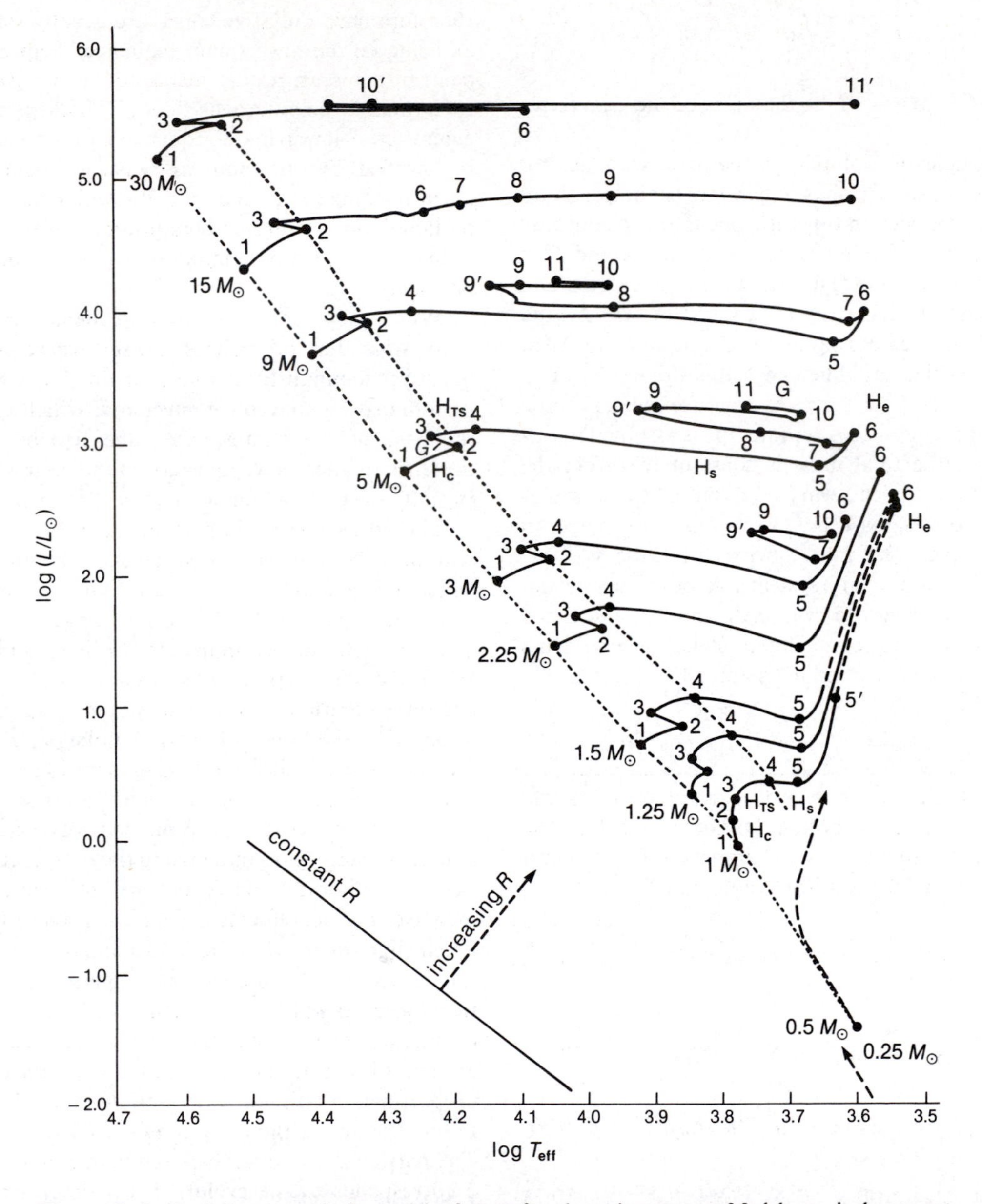

Figure 9.1. Evolutionary tracks for models of stars after the main sequence. Model mass is shown next to the initial point on zero age main sequence. Dotted lines indicate boundaries of the main sequence. Lines of constant radius and increasing radius as shown in lower left. Elapse times between points are shown in Table 9.1. The stages are labeled as: H_c, hydrogen core burning; H_{TS}, thick hydrogen shell burning; H_s, shell hydrogen burning; He, helium core burning; and *G*, gravitational energy release. The 15 $M_\odot$ track does not reverse in the giant region, because the semiconvective region was treated as fully convective in this model.

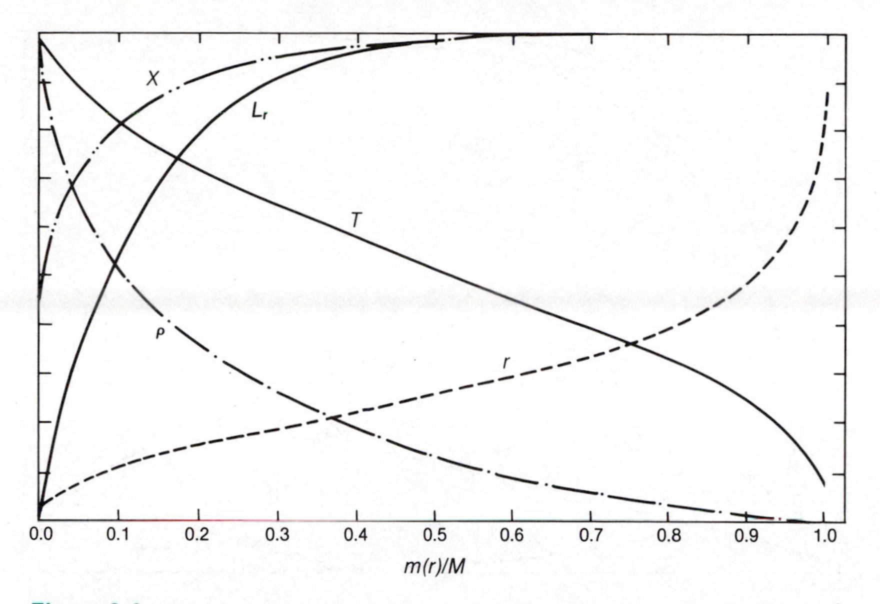

Figure 9.2. A 1 $M_\odot$ model during main-sequence hydrogen burning at time 4.2699×10^9 years (between points 1 and 2 in Figure 9.1), showing radius, density, temperature, total luminosity, and hydrogen abundance versus mass fraction. The lower limits of the ordinate are zero. The upper limit for each curve is: $r = 0.9683\ R_\odot$, $\rho_c = 159.93$ g cm^{-3}, $T_c = 1.591 \times 10^7$ K, $L = 1.0575\ L_\odot$, and $X_c = 0.708$; $P_c = 2.5186 \times 10^{17}$ dynes cm^{-2}. The elapsed time is measured from the initial model for the phase before the main sequence.

Table 9.1
Elapsed time, in years, between numbered points on the evolutionary tracks in Figure 9.1. Numbers in parentheses are powers of 10 by which entries are multiplied.

	Interval				
Mass ($M_\odot$)	1–2	2–3	3–4	4–5	5–6
30	4.80 (6)	8.64 (4)	←	~1.0 (4)	→
15	1.010 (7)	2.270 (5)	←	7.55 (4)	→
9	2.144 (7)	6.053 (5)	9.113 (4)	1.477 (5)	6.552 (4)
5	6.547 (7)	2.173 (6)	1.372 (6)	7.532 (5)	4.857 (5)
3	2.212 (8)	1.042 (7)	1.033 (7)	4.505 (6)	4.238 (6)
2.25	4.802 (8)	1.647 (7)	3.696 (7)	1.310 (7)	3.829 (7)
1.5	1.553 (9)	8.10 (7)	3.490 (8)	1.049 (8)	≥2 (8)
1.25	2.803 (9)	1.824 (8)	1.045 (9)	1.463 (8)	≥4 (8)
1.0	7 (9)	2 (9)	1.20 (9)	1.57 (8)	≥1 (9)

	Interval				
Mass ($M_\odot$)	6–7	7–8	8–9	9–10$^{(r)}$	10$^{(r)}$–11$^{(r)}$
30		53.1 (4)			1.3 (4)
15	7.17 (5)	6.20 (5)	1.9 (5)	3.5 (4)	
9	4.90 (5)	9.50 (4)	3.28 (6)	1.55 (5)	2.86 (4)
5	6.05 (6)	1.02 (6)	9.00 (6)	9.30 (5)	7.69 (4)
3	2.51 (7)	4.08 (7)		6.00 (6)	

Source: Adapted from I. Iben, Jr. 1967. *Annual Reviews of Astronomy and Astrophysics, 5*, p. 571

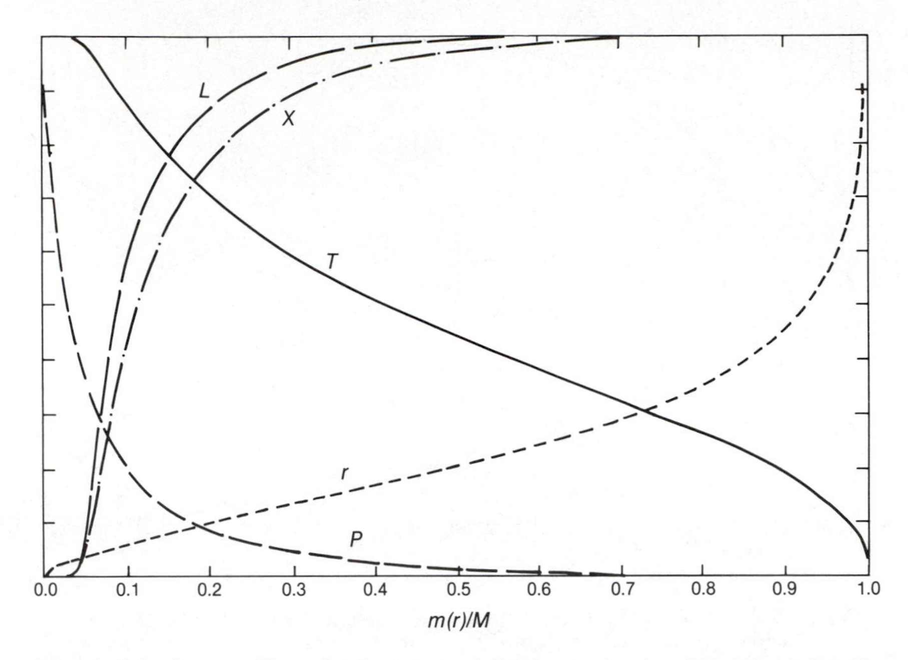

Figure 9.3. Same as Figure 9.2 for 1 $M_{\odot}$ model between points 2 and 3 of Figure 9.1. The elapsed time since the initial model for the phase before the main sequence is 9.2015×10^9 years. The lower limits of the ordinate are zero. The upper limit of the ordinate for each curve is $r = 1.268\ R_{\odot}$; $P_c = 1.315 \times 10^{18}$ dynes cm^{-2}; $T_c = 1.91 \times 10^7$ K; $L = 2.13\ L_{\odot}$; $X_c = 0.708$. The actual stellar radius is $R + 1.353\ R_{\odot}$, and the central density is 1026.0 g cm^{-3}.

and $L(r) = L$ for $r > 0.3R$. Hydrogen burning therefore occurs throughout a core of radius about one-third of the stellar radius, corresponding to about 10 percent of the mass.

The early, homogeneous structure of the lower main sequence can be qualitatively approximated by the linear model. Assuming that the sole energy source is a pp chain with $n = 4$ and opacity of the Kramers' type, then for Population I stars, it can be shown that

$$R/R_{\odot} = 0.312\mu^{-0.538}(M/M_{\odot})^{0.0769}, \tag{9.9}$$

$$L/L_{\odot} = 49.1\mu^{7.77}(M/M_{\odot})^{5.46}, \tag{9.10}$$

$$T_{\text{eff}} = 2.73 \times 10^4\mu^{2.21}(M/M_{\odot})^{1.327}, \tag{9.11}$$

$$T_c = 3.05 \times 10^7\mu^{1.54}(M/M_{\odot})^{0.923}, \tag{9.12}$$

$$\rho_c = 186\mu^{-1.615}(M/M_{\odot})^{0.769}. \tag{9.13}$$

Problem 9.3. Use dimensional arguments and the stellar structure equations, or the homology models of Section 7.3, to show that the quantities on the left in (9.9) to (9.13) have the correct dependence on μ and M.

According to the model, an increase in mean molecular weight produces an increase in all quantities except the radius. The major drawback to the linear model is its underestimate of the degree of condensation, yielding a value of ρ_c for the Sun about one-third that obtained from accurate models. The surface temperature is therefore higher by a factor of about three. The main sequence is shifted nearly parallel to the observed main sequence and to higher temperatures. The general trend of the expressions is nevertheless born out by numerical calculations (except for the μ dependence of R). Notice in particular that increasing mass yields increased ρ_c, T_c, and L.

Problem 9.4. Discuss qualitatively the expected relation of Population II stars on the lower main sequence stars to similar Population I stars. You may assume that the primary difference is in the mean molecular weight μ, since the details are insensitive to the composition-dependence of the opacity coefficient κ_0. Suggest an observational way of checking these results.

Problem 9.5. Estimate the mass of a star, in hydrostatic equilibrium, consisting of an ideal gas plus radiation for which the gas pressure equals the radiation pressure. Assume $\mu \simeq 1/2$.

As the star evolves on the main sequence, its hydrogen content is reduced, and helium builds up in the core. Eventually the core becomes dominantly He. Most of the H burning is now confined to a relatively thick shell surrounding a small but growing He core. For a star of one solar mass, the structure after 9.2×10^9 yrs in the main sequence is shown in Figure 9.3, which corresponds to point 3 in the evolutionary track of Figure 9.1. The region $r \lesssim 0.03\ R$ is now essentially pure He; hydrogen burning is occurring in the intermediate zone $0.03\ R \lesssim r \lesssim 0.3\ R$. The rate of burning, measured by $\epsilon = dL/dm$, is the proportional to the slope of the curve $L(m)$. Comparison with Figure 9.2 shows that it is considerably higher, and that the He core is now essentially isothermal. The development of a chemically inhomogeneous structure marks the beginning of the end of the main-sequence life of the star. It is now evolving more rapidly, and is moving toward lower surface temperatures. Comparison of $m(r)$ for various points shows that as the star evolves, it is becoming more centrally condensed.

In stars of $1.1\ M_\odot$ and less, the core remains stable against convection during the following evolutionary phase, so that there is a gradual change in composition from the newly formed isothermal He core, through the thick H-burning zone, into the envelope; see curve for $X(m)$ in Figure 9.3. Because the core is radiative, core H burning continuously gives way to thick-shell H burning.

The final evolutionary phase on the main sequence, points 3–4 will be described in Section 9.6, after the behavior of the isothermal core has been discussed.

9.4. Upper Main Sequence

For temperatures in excess of 2×10^7 K, H burning proceeds via the CN cycle and CNO bi-cycle in convective core. Although radiation pressure becomes increasingly important for more massive stars, it will be considered only where necessary. Its major effect is to reduce adiabatic indices, which reinforces the tendency toward convection.

The cores of stars on the upper main sequence are convective; consequently the composition throughout is uniform, even though H burning proceeds much more rapidly at the center than near the outer edge of the zone. In the intermediate mass range, Kramers' opacity dominates; for higher masses, electron scattering is more important. Energy-generation rates for upper-main-sequence stars can be approximated by the product $\rho X Z_{\mathrm{CNO}} T^n$, where Z_{CNO} denotes the mass fraction of C, N, and O necessary for the specific reactions. This differs most significantly from the corresponding pp-chain rate in being linear in the hydrogen content X rather than quadratic, and in that n is typically ~ 18. A decrease in hydrogen concentration therefore has less effect on energy rates in stars on the upper main sequence, and their cores do not need to contract as much as those of stars on the lower main sequence in order to maintain their energy output. As the mass of the model increases, so do R, L, T_{eff}, and T_c, but the central density becomes smaller. When electron scattering dominates, the luminosity and T_{eff} show weaker dependence on mass.

The zero age main sequence, as described by the linear model for Population I stars, is given by the following relations. For Kramers' opacity, and CNO cycle with $n = 18$,

$$R/R_\odot = 0.451\mu^{0.395}(M/M_\odot)^{0.697}, \tag{9.14}$$

$$L/L_\odot = 43.5\mu^{7.3}(M/M_\odot)^{5.18}, \tag{9.15}$$

$$T_{\mathrm{eff}} = 2.34 \times 10^4\mu^{1.63}(M/M_\odot)^{0.871}, \tag{9.16}$$

$$T_c = 1.98 \times 10^7\mu^{0.606}(M/M_\odot)^{0.364}, \tag{9.17}$$

$$\rho_c = 65.8\mu^{-0.455}(M/M_\odot)^{-0.909}. \tag{9.18}$$

For election scattering,

$$R/R_\odot = 0.454\mu^{0.588}(M/M_\odot)^{0.765}, \tag{9.19}$$

$$L/L_\odot = 112\mu^4(M/M_\odot)^3, \tag{9.20}$$

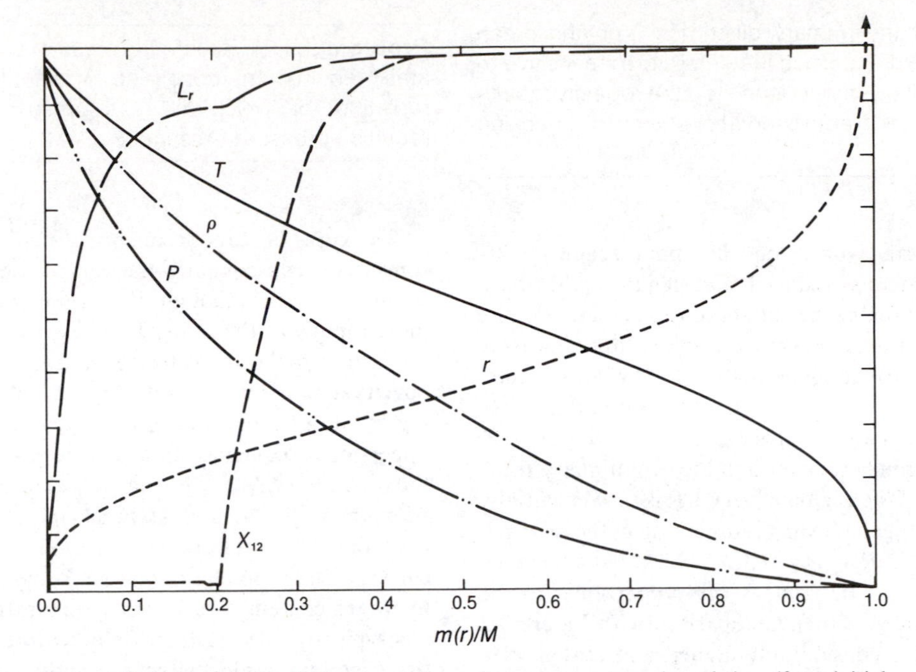

Figure 9.4. Model of a 5 $M_\odot$ star on the main sequence at an elapsed time (from initial pre-main-sequence model) of 6.099×10^5 years. Lower ordinate is zero, and upper values such that $r = 1.8466\ R_\odot$; $P_c = 8.0274 \times 10^{16}$ dynes cm^{-2}; $\rho_c = 21.43$ g cm^{-3}; $T_c = 2.7339 \times 10^7$ K; $L = 631.3\ L_\odot$; initial abundance of C^{12} is $X_{12} = 3.61 \times 10^{-3}$. The total radius of the model is $R = 2.397\ R_\odot$.

$$T_{\text{eff}} = 2.77 \times 10^4 \mu^{0.706}(M/M_\odot)^{0.368}, \quad (9.21)$$

$$T_c = 2.12 \times 10^7 \mu^{0.412}(M/M_\odot)^{0.235}, \quad (9.22)$$

$$\rho_c = 60.3\mu^{-1.765}(M/M_\odot)^{-1.294}. \quad (9.23)$$

The switchover from Kramers' to electron-scattering opacity occurs around $M \simeq 3\ M_\odot$ for Population I. The linear model underestimates radii for stars on the zero age main sequence by about as much as it did on the lower main sequence. For $M = 5M_\odot$, accurate models give $R = 2.4\ R_\odot$ but (9.19) gives only $R = 1.19\ R_\odot$. The estimates of T_c and ρ_c are reasonably good, however. The error is due in part to the model's underestimate of the degree of central condensation. Figure 9.4 shows the initial structure of a 5 $M_\odot$ star shortly after it arrives on the main sequence. The additional contribution to the luminosity near $m(r)/M \simeq 0.2$ is due to C^{12} burning outside the core. In the more massive stars, the initial C^{12} overabundance, which burns completely in low-mass stars as they approach the zero age main sequence, takes longer to disappear. Since the mass contained in the outer 23 percent of the star is negligible, the figure portrays only the inner 77 percent.

The evolution initially proceeds as in a 1 $M_\odot$ star. When nuclear burning has reduced X to about 0.05, the rate of energy release is no longer adequate to support the core, which begins to contract. This corresponds to point 2 on the 5 $M_\odot$ curve of Figure 9.1, and marks the end of the star's main-sequence life. An important difference between the 1 $M_\odot$ and 5 $M_\odot$ stars is that the latter lack a thick H-burning shell, partly because of convection, which thoroughly mixes elements in the core, reducing X throughout uniformly. This in turn reduces the amount of fuel available. Therefore, for all but the low-mass stars, the thick-shell burning stage begins well after the main-sequence stage.

9.5. Isothermal Cores

The cores of low-mass stars become isothermal when nuclear burning ceases there. This follows since $L(0) = 0$ and $dL/dr \simeq \epsilon(r)$; thus if $\epsilon = 0$ throughout the core, then $L = 0$ as well. But $dT/dr \sim L(r)$; so $T(r) = T_c$ throughout the core (see Figure 9.3). The density must therefore rise rapidly enough to supply the needed pressure. When it does, the isothermal core can support itself, and will not be forced to contract by the weight of overlying material. There is a limit, however, to how much mass may be supported in this way. The limiting value, called the Schönberg–Chandrasekhar limit, may be estimated by the virial theorem (4.10) and the linear model.

Suppose the isothermal core is of radius r_1, its temperature is T_c, and it is surrounded by the remainder of the stellar mass. Denote core and total masses by M_c and M, respectively. We will assume that the core is nondegenerate, and use the virial theorem (4.10) to relate the internal energy U_1, gravitational potential energy Ω_1, and volume V_1 of the core. The first term in (4.10) is simply

$$4\pi r_1^{\,3} P_1 = 3V_1 P_1,$$

where P_1 is the pressure at the outer boundary of the core. Next, we use (2.6) to write the second term in the form

$$-3\int_0^{M_1} \frac{P}{\rho}\,dm = -\,3(\gamma^1 - 1)\int_0^{M_1} \frac{u\,dm}{\rho}$$

$$= -\,3(\gamma^1 - 1)U_1.$$

We have assumed that γ^1 is constant throughout the core mass M_1, and have denoted the internal energy by U_1. Substituting these expressions into (4.10) yields the virial theorem for the isothermal core,

$$3(\gamma^1 - 1)\,U_1 - 3P_1V_1 + \Omega_1 = 0. \qquad (9.24)$$

For an isothermal ideal-gas sphere, $U_1 = 3M_ckT_1/m_H$, and (9.24) gives the maximum pressure capable of supporting the material above the core. For $\gamma^1 = 5/3$,

$$P_{\max} = (\pi/G^3M_c^{\,2})(kT_c/\mu_c)^4. \qquad (9.25)$$

Notice that $P_{\max}$ decreases as M_c increases, and that the core mean molecular weight is denoted by μ_c. Stellar evolution will determine M_c, μ_c, and T_c, and therefore set $P_{\max}$. To relate $P_{\max}$ to the total mass of the star, one must use a model giving P_1, the pressure at the core-envelope interface, and T_c. Since the linear model gives reasonable estimates of T_c and ρ_c, we use it. For T_c we have

$$T_c = \frac{5}{12}\frac{\mu MG}{kR} \simeq T_1 \qquad (9.26)$$

for an isothermal core. The central pressure is given by (4.32); for P_1 we can use the average pressure, which, from (4.33) is easily shown to be $0.294\,P_c = P_1$.

Problem 9.6. Verify (9.26) by extremizing P. Assume that

$$\Omega = -\frac{3}{5}\frac{M^2G}{R}.$$

What is the critical radius of the core? Take $\gamma = 5/3$ for an ideal gas.

In order for the isothermal core to support the overlying material, P_1 must be less than $P_{\max}$, which leads to the restrictions

$$M_c/M \lesssim 0.3\,(\mu/\mu_c)^2, \qquad (9.27)$$

where μ denotes the mean molecular weight in the envelope. More accurate calculations based on realistic stellar models show that the coefficient on the right of the inequality should be 0.39. These models show that, as stars move off the main sequence toward the red-giant region of the HR diagram,

$$M_c \lesssim M_{SC} = 0.12\,M. \qquad (9.28)$$

When M_c exceeds the Schönberg–Chandrasekhar limit M_{SC}, the pressure exerted by the core is insufficient to support the star, and gravitational contraction must ensue.

The size of the isothermal core is determined initially by the stellar mass, the type of nuclear burning, and the internal structure of the star. Two cases are of interest when the core is initially nondegenerate. First, when $M_c < M_{SC}$, the stable core is in equilibrium as long as (in low-mass stars) the shell-burning phase continues. As H is converted into He ash by the shell, mass is gradually added to the core, which readjusts itself to support the added mass isothermally. However, if the shell-burning stage adds enough matter to the core that $M_c > M_{SC}$, then gravitational contraction ensues. Thus, depending on the difference $M_{SC} - M_c$ when shell burning begins, the star may exist for a long time with an isothermal core, as stars like the sun do. When $M_c > M_{SC}$, the core begins to contract, immediately raising its temperature and density. Any mass added by shell burning simply augments the contraction. The fate of the contracting core depends on its mass. If it is less than the Chandrasekhar mass limit, electron degeneracy ultimately stops the collapse. If it exceeds M_{Ch}, then contraction continues until the next nuclear fuel is ignited by rising central temperatures.

9.6. Termination of the Main Sequence

The final main-sequence evolution of low-mass stars is exemplified by the 1 $M_\odot$ model discussed earlier (points 3–4 of Figure 9.1). Section 9.3 developed this model up to the formation of the isothermal core surrounded by a thick H-burning shell. During the final phase, the shell becomes progressively thinner as it adds ash to the core. For 1 $M_\odot$, the Schönberg–Chandrasekhar mass limit is roughly 0.12 $M_\odot$. The initial core (Figure 9.3) is well below this limit; so it can support itself by increasing its density gradient. Therefore the core remains stable as H is converted into He by the shell. On the zero age main sequence, the temperature in the region $0.1 \lesssim m(r)/M \lesssim 0.3$ was low enough that very little hydrogen burning occurred there, and what did burn was characterized by low rates (slope dL/dm in Figure 9.2). By the time an isothermal core has developed, however, the temperature in the $0.1 \leq m(r)/M \lesssim 0.3$ zone is high enough to support vigorous H burning. The energy released is now greater than when core burning dominated, and some of it increases the core temperature. The rest leads to a gradual expansion of the outer portions of the star. Because of the increased luminosity, the evolutionary track moves slightly upward. The added energy released goes mostly into work and a gradual expansion of the envelope as a whole. The increase in L tends to raise T_{eff}, but the increase in R, which lowers T_{eff}, is more pronounced; so T_{eff} actually drops. The star therefore moves toward the right in the HR diagram with only a slight increase in L (points 3–4 in Figure 9.1). The time required for this stage of evolution is about 12 percent of the entire main-sequence lifetime, the increased rate of evolution being attributable to the increased energy-generation rate in the shell.

As shell burning proceeds, the core mass increases with added He ash, and the width of the shell decreases. In all but stars with very low mass, eventually $M_c \simeq M_{SC}$, and the core begins to contract. This marks the termination of the main-sequence phase, and the star begins to move toward the red-giant region in the HR diagram. The 1 $M_\odot$ star in this model spends about 10×10^9 yrs on the main sequence, and has transmuted about 12 percent of its hydrogen into helium. During this period its radius has nearly doubled, its luminosity has tripled, and its surface temperature has dropped by about 16 percent.

In stars more massive than about 1.25 $M_\odot$, the evolution is somewhat different. The growing importance of the CN and CNO cycles produces convective cores; so isothermal conditions do not prevail. The core supplies energy until X has been reduced to a few percent. At this point energy production ceases, and the core, which continues to radiate energy, begins to contract. In the absence of burning shells, the entire star will contract and heat up. The source of energy during this phase (points 2–3 in Figure 9.1) is gravitation. For a short time, the star behaves somewhat as it did during its final approach to the main sequence, increasing its luminosity slightly and its temperature T_{eff} more. The contraction phase is stopped, however, not by further core burning (the conversion of He to C^{12} requires temperatures above 10^8 K, but the core is initially at $T_c \simeq 3 \times 10^7$ K), but by the ignition of a H-burning shell surrounding the core.

The 5 $M_\odot$ model has spent about 7.2×10^7 yrs on the main sequence. At the end of the gravitational-contraction stage, its radius is about half again as large as on the zero age main sequence, its central temperature is up by about 30 percent, and its luminosity has nearly doubled.

Chapter 10

EVOLUTION AWAY FROM THE MAIN SEQUENCE

10.1. Post-Main-Sequence Evolution

The time required for a star to evolve through stages subsequent to the main sequence is small compared to its total lifetime. However, some of the most spectacular phenomena are associated with this part of its life. The details of the evolution are more complicated, and the interplay between various parts of the star more important. Unfortunately, these aspects of the analysis lead at times to much uncertainty, and the results are more sensitive to the details of the model. Therefore, as we move further away from the main sequence, we are less able to make firm statements about the system's behavior. Three major sources of uncertainty arise: (1) dynamical effects become more important; (2) observational data to guide and check theories become scarcer; and (3) there appears to be a point beyond which the qualitative features of evolution must be fundamentally different. For the moment, we will consider the stages immediately following the main sequence, through He burning in the core. The most important new ingredient for this stage is the development of composition inhomogeneities, and the associated nuclear shell sources. We refer to thin shells of matter that are releasing energy by thermonuclear fusion as *active shell sources.* If the temperature at a shell is reduced sufficiently (by expansion, for example), nuclear-energy release will cease; we call such shells *inactive shell sources.* Shell sources usually occur at composition inhomogeneities, and inactive shell sources may reignite if the temperature becomes high enough.

10.2. Composition Inhomogeneities

The most dramatic result of main-sequence evolution is the formation of He^4 nuclei at temperatures well below their ignition point, surrounded by a hydrogen-rich envelope. We will develop the basic phenomena associated with these composition inhomogeneities, and explore their effect on stellar evolution. Three basic phenomena will be considered in detail in the rest of this chapter:

(1) inhomogeneous composition leads to greater central condensation in a star;
(2) the position of a shell-burning source tends to remain nearly constant as the star evolves;
(3) volume changes (contraction, expansion) reverse at each active nuclear-burning shell, but

remain unaffected by the presence of inactive shells.

In nearly all stages of nuclear burning, the mean molecular weight of the ash exceeds that of the fuel. Thus $\mu(\mathrm{He}) = 4/3$ and $\mu(\mathrm{H}) = 1/2$. Consequently, composition inhomogeneities mean greater central core density; that is, more of the mass will be found near the center. Since an increase in central condensation translates roughly into a decreased adiabatic exponent γ, the central regions may often be approximated by polytropes of index $n \simeq 3$. In some evolutionary stages, the fact that γ is near $4/3$ means that a slight modification in internal structure (the ignition of energy-absorbing, or endothermic, reactions, for example) could trigger collapse. Such effects may be related to explosive processes associated with supernovae, and to local instabilities associated with novae.

Point (2) is related to the thermostatic nature of shell-burning sources. Movement of a shell source toward the center, and therefore into regions of higher temperature, would lead to increased energy generation and raise the pressure. This in turn would tend to counter the original movement, causing the shell source to return to its original position.

The last point has a dramatic effect on evolution. Because volume changes reverse at active shell sources, core contraction may be accompanied by envelope expansion, and vice-versa. When there is more than one active shell source, with intermediate regions of matter that are not burning, then reversal occurs at each shell. Therefore the radius of a star with one active shell source will expand when the core contracts, but the radius of a star with two active shells will contract. Finally, the presence of inactive sources has little effect on the propagation of volume changes.

10.3. Central Condensation

A useful measure of the extent to which matter in a star is concentrated toward the center is given by the ratio of the density at r to the average density $\bar{\rho}(r)$ of matter contained within a sphere of radius r. The degree of central condensation may be defined by

$$U \equiv 3\rho(r)/\bar{\rho}(r), \tag{10.1}$$

where $\bar{\rho}(r) = 3m(r)/4\pi r^3$. It is straightforward to show that the following useful relations follow from (10.1):

$$U = \frac{d \ln m(r)}{d \ln r} = \frac{d \ln q}{d \ln r}, \tag{10.2}$$

where $q = m(r)/M$ is the mass fraction. It follows from (10.1) that $0 \leq U \leq 3$ as r varies from the surface to the center.

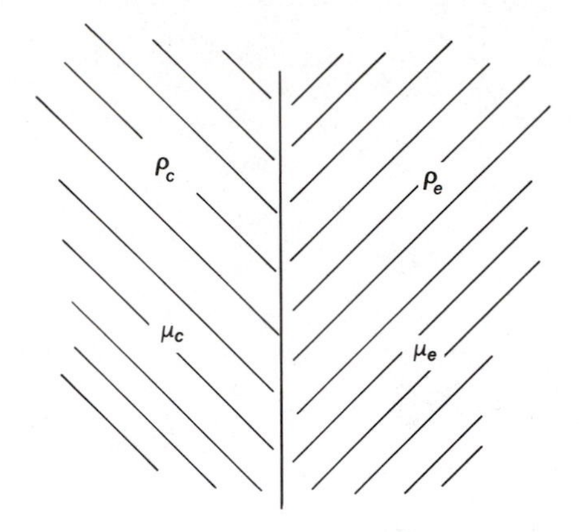

Figure 10.1. Composition discontinuity between He-rich core (left) and H-rich envelope (right).

Now consider the behavior of ρ, P, and T across a composition discontinuity, such as separates the He-rich core from the H-rich envelope in a star without active energy sources (Figure 10.1). If the star is to be hydrostatically and thermally stable, both T and P must be continuous across the interface. Assuming that the gas satisfies the ideal gas law, it follows that $\rho_e/\mu_e = \rho_c/\mu_c$. Subscripts c and e denote quantities evaluated in the core or envelope, respectively. Since nuclear burning increases μ_c relative to μ_e, it immediately follows that $\rho_c > \rho_e$. To the extent that the boundary between the two regions is sharp, $\bar{\rho}(r)$ will be constant across it, so that a discontinuity in $\rho(r)$ implies one in $U(r)$. Thus, evaluating the following ratios just to the inside and outside of the core-envelope interface, we find

$$\rho_c/\mu_c = \rho_e/\mu_e \quad \text{or} \quad U_c/\mu_c = U_e/\mu_e. \tag{10.3}$$

As defined by (10.2), U is dimensionless. Two other dimensionless variables are also useful in analyzing stellar structure:

$$V \equiv -\,d \ln P/d \ln r, \tag{10.4}$$

$$N + 1 \equiv d \ln P / d \ln T. \tag{10.5}$$

The last, if evaluated adiabatically, is related to Γ_2.

10.4. Characteristics of Shell-Burning Sources

We have already noted that the thermostatic nature of active shell sources tends to keep their position within a star fixed. To see why, let us consider an active shell source that separates the core from an envelope with most of the envelope's mass lying near the core-envelope interface; in Figure 10.2, the thin shell source is located at r_s. Now suppose that the gas obeys the ideal gas law; then, to the extent that the envelope may be characterized by a polytropic relation of the form $\rho = KT^n$, it follows that

$$\frac{dP}{dr} = \frac{dP}{dT}\frac{dT}{dr} = \frac{dT}{dr}\left(\frac{d}{dT}\frac{\rho k T}{\mu m_H}\right)$$

$$= K\frac{k\rho(n+1)}{\mu m_H}\frac{dT}{dr}. \tag{10.6}$$

Using the equation of hydrostatic equilibrium on the left, and noting that since most of the envelope's mass lies near r_s, $m(r) \simeq M_c$ for $r \gg r_s$, we find immediately from (10.6) that

$$T_s = \frac{\mu m_H G M_c}{k(n+1)}\frac{1}{r_s} + \text{constant}. \tag{10.7}$$

The integration constant is unimportant for our purposes; T_s is the temperature at the shell source. We see that $T_s \sim r_s^{-1}$. Suppose the location of the shell were to move inward; then the rate of energy generation, which goes as some power of T_s, will increase, increasing the pressure on the overlying envelope. In response the envelope will tend to move back out, lowering T_s until equilibrium is again attained. Similar arguments may be made for expansion, and again show that r_s tends to remain constant. Although several restrictive assumptions underlie the derivation of (10.7), numerical results confirm its validity under more general conditions. For example, T_s is expected to rise gradually as the star evolves. But at the same time, mass is added to the core, increasing M_c, and tending to hold r_s fixed. In 1 $M_\odot$ models with shell sources and inert cores, the location of the shell remains very nearly at $r_s = 0.03\ R_\odot$ during evolution away from the main sequence and up the red-giant branch; during this time the star's radius has nearly tripled.

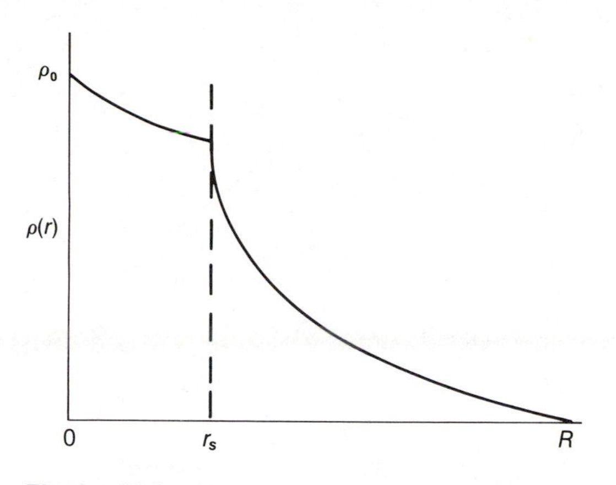

Figure 10.2. Density profile in a star containing an active thermonuclear shell source located at r_s.

Problem 10.1. A thin, active shell source separates the stellar core and envelope. Show that the shell position tends to remain constant by supplying the details to the following local analysis. Suppose that an initial radial displacement $(\Delta R/R_1)$ in the shell's position is made, and that the energy-generation rate in the shell is approximated by (7.6). The star makes a homologous change in its structure [see Section 7.5] to regain hydrostatic equilibrium. Show that the fractional change in energy generation is given by

$$\frac{\Delta\epsilon}{\epsilon} \simeq -(3+n)\left(\frac{\Delta R}{R}\right)_1. \tag{10.8}$$

The star will not be in thermal equilibrium, however, since the rate of energy generation exceeds the luminosity. Approximate the opacity by (6.86), and show that, if the star is to establish thermal equilibrium (energy loss = energy released locally), then an additional temperature change is necessary, which is of order

$$\left(\frac{\Delta T}{T}\right)_{\text{T.E.}} \simeq -\frac{3a+n+3}{4-b}\left(\frac{\Delta R}{R}\right)_1. \tag{10.9}$$

In deriving (10.9) assume radiation transport. The change in temperature in (10.9) produces an additional change in pressure at constant density $\Delta P/P = \Delta T/T$, which forces the star out of hydrostatic equilib-

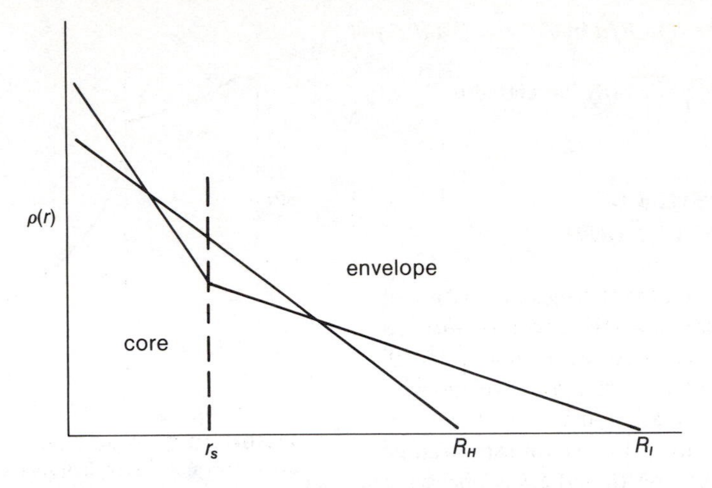

Figure 10.3. Effect of core contraction and envelope expansion associated with an active thermonuclear shell source. R_H is the radius in the homogeneous state.

rium. It therefore makes a final homologous change in R: show that this is given by

$$\left(\frac{\Delta R}{R}\right)_2 = -\frac{3a + n + 3}{4(4 - b)}\left(\frac{\Delta R}{R}\right)_1 \qquad (10.10)$$

(note the sign). Discuss the range of values of the coefficient of $(\Delta R/R)_1$, for typical energy sources and opacities. Note that $(\Delta R/R)_2$ is always opposite in sign to $(\Delta R/R)_1$.

Volume Changes and Shell Sources

Suppose the core-envelope structure is as shown in Figure 10.2, and consider the envelope as the core becomes more centrally condensed. According to (10.2),

$$\ln R = \ln r_s + \int_{q_s}^{1} \frac{d \ln q}{U}, \qquad (10.11)$$

with R the stellar radius, and $q_s = M_c/M$ the mass fraction in the core. Since r_s remains fixed, the behavior of R is described by the integral in (10.11). Imagine now that the core gradually increases its degree of central condensation. If r_s remains fixed during the process, and the central density is increasing, $\rho(r)$ at the inside edge of the shell will decrease. Since $\overline{\rho}(r_s)$ is fixed to first approximation, U_c will also decrease. However, this implies that U_e will also decrease as indicated in (10.3). This will reduce $U(r)$ in the lower portions of the envelope where $1/q$ is largest; so the integral in (10.11) is increased, as is the stellar radius. In other words, as the degree of central condensation grows, the density near the shell drops, and the stellar radius grows. The preceding argument is easily modified to show that the radius R decreases as the degree of central condensation drops. Figure 10.3 shows how $\rho(r)$ behaves when ρ_c changes in a star with a single active shell source at r_s. Also shown is $\rho(r)$ before the increase in ρ_c.

In the absence of an active shell source, the radius would tend to shrink as the central density increased. In fact, it is relatively easy to show that volume changes are unaffected by inactive shells. This fact has already been used in our discussion of the evolution of stars on the upper main sequence, where exhaustion of H burning in the core results in contraction, not only of the core, but of the star as a whole (stage 2–3 in Figure 9.1 for $M = 5\ M_{\odot}$).

Problem 10.2. *Inactive shell sources.* Suppose a star of radius R has an inactive shell source at r_s, with envelope mass M_e. Denote its average density by $\overline{\rho} = (\rho_0 + \rho_s)/2$, where ρ_0 is the density at the base of the photosphere. Assuming $r_s \ll R$, show that a homologous change throughout the core (expansion or contraction) produces the same type of change in the envelope. Specifically, show that if $\Delta r_s/r_s \lessgtr 0$, then $\Delta R/R \lessgtr 0$ also. Note that r_s need not remain fixed when the source is inactive.

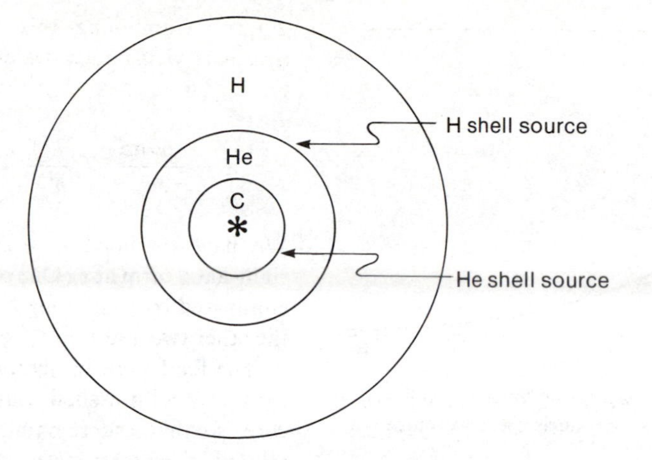

Figure 10.4. Series of active thermonuclear shell sources, separated by inactive regions containing their ashes. C burning has started at the center.

Problem 10.3. *Active shell sources.* Suppose a star contains a single active shell source at r_s, and that its envelope density is approximated by

$$\rho(r) = \rho_s(r_s/r)^n, \quad (10.12)$$

with photospheric density $\rho(R) = \rho_s(r_s/R)^3$. If the mass in the core remains constant, show that

$$\Delta R/\Delta\rho < 0.$$

Set $n = 3$.

Problem 10.4. A star contains an inert C^{12} core, surrounded by an inert He^4 inner envelope and finally an outer H envelope. The C^{12}/He^4 interface is inactive, but the He/H boundary is an active shell source. (a) Suppose the core begins to contract. Describe the response of each zone, and how the radius changes. (b) Eventually the temperature of the C^{12}/He^4 boundary exceeds 10^8 K, and an He shell source ignites, heating the core and releasing additional energy to the entire star. The central density in the core decreases. Describe the response of each zone and of the radius, assuming that the H shell source also remains active.

10.5. Evolution of Shell Sources

A feature characteristic of advanced evolution is the formation of nuclear-energy sources in thin shells outside the core. In massive stars, more than one shell source, separated by inert zones, may be active at one time. The core itself may or may not also be active. Figure 10.4 shows the nature of these zones schematically for a star massive enough to have reached core C^{12} burning; here there may be two active shells (H and He burning), and inert H and He zones. During some evolutionary stages, the shell sources may be responsible for most of the star's luminosity. Let us consider the major factors that cause thin shell sources to form, and develop some of their properties. Their principal features are their location at composition discontinuities, and their limited thickness.

First consider the general behavior of $T(r)$ and $\rho(r)$ in the envelope. When most of the mass lies within the core, $m(r) \simeq M$, and the temperature goes as r^{-1}, according to (10.7). This is characteristic of extreme central condensation. It follows from the perfect gas law that if $\rho \sim r^{-N}$ then $P \sim r^{-N-1}$. Finally, we can fix N by noting that, since $m(r) \simeq M$ for $r > r_s$, then ρr^3 is essentially constant, and $N \simeq 3$. In actuality, some mass lies in the outer regions and this effectively reduces N below 3. Figure 10.5 shows ρ, T, and X schematically for a star that has exhausted H burning in the core. Since the energy-generation rates go as $\rho X T^n$, these will be extremely small in the core, where $X \simeq 0$. Beyond r_s, X is large ($X \simeq 0.6$ for Population I), but the temperature and density are dropping rapidly. If $T(r_s)$ is high enough for ignition, then a small increase in r will produce a large reduction in the product ρT^n, and energy production will cease. These two effects—a near-discontinuity in X, and the rapid

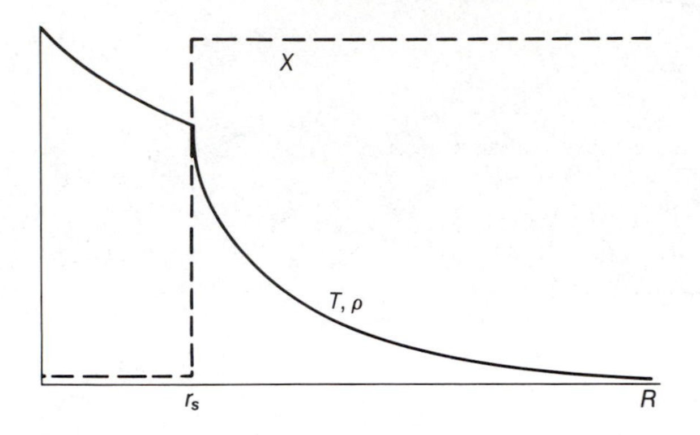

Figure 10.5. Schematic profiles of T and ρ, and mass fraction of H for a star whose only nuclear-energy source is a H-burning shell at r_s.

drop in both T and ρ in the envelope—combine to limit energy generation to a narrow region (see Problem 10.5).

Problem 10.5. A star has a single shell source at $r_s = 0.0321\ R_\odot$. If its envelope density and temperature are $\rho(r) = \rho_s(r_s/r)^3$ and $T(r) = T_s(r_s/r)$, find the shell's luminosity. Suppose

$$\epsilon(r) = 3.98 \times 10^{-26}\, \rho\, X^2 T^{3.5}\ \mathrm{erg\ g^{-1}\ sec^{-1}} \qquad (10.13)$$

for $r \geq r_s$, and that $X \simeq 0$ in the core. Take $T_s = 2.13 \times 10^7$ K and $\rho_s = 930\ \mathrm{g/cm^3}$. What is the thickness of the region from which 99 percent of the shell's energy originates? What is the mass in the shell? How long can the star supply its luminosity in this way?

For example, numerical models of 1 $M_\odot$ stars indicate that, just prior to evolution up the red-giant branch, these stars have shell sources at about $r_s = 0.03\ R_\odot$, that $L(r)$ varies from $0.1L$ to $0.9L$ for $0.04 < r < 0.05\ R_\odot$, and that the width narrows as the star evolves.

Assuming density and temperature profiles as in Problem 10.5, we find that the luminosity from the shell source is

$$L = \frac{4\pi r_s^3}{n+3}\, \epsilon_0\, X^2\, \rho_s^2\, T_s^n, \qquad (10.14)$$

if $\epsilon \sim \rho\, X^2 T^n$. Since the temperature is reasonably approximated by $T \sim 1/r$, the temperature gradient is $T_s r_s / r^2$. Combining this with (5.54) for radiative transport yields a second expression for the luminosity:

$$L = \frac{16\pi ac}{3\kappa_0}\, r_s T_s \frac{T^{3-b}}{\rho^{1+a}} \simeq \frac{16\pi ac}{3\kappa_0} \frac{r_s T_s^{4-b}}{\rho_s^{1+a}}. \qquad (10.15)$$

We have assumed $\kappa = \kappa_0 \rho^a T^b$, and evaluated the right-hand term at r_s. One of the shell variables may be eliminated by combining (10.14) and (10.15), leaving the other two and L.

The final steps in obtaining the evolutionary track for a star with a shell source require a model of the core, so that the remaining shell variables may be related to the total mass and the central density. If we approximate the shell source as being of essentially zero thickness, then a relatively simple analytic model may be constructed to show the principal evolutionary features of stars having a single shell and no significant energy generation in the core. Though straightforward, the model becomes algebraically complex; so we make the following additional assumptions at the outset:

(1) the core radius $r_s \ll R$; and
(2) $\rho_c \gg \rho_s$.

The first assumption is quite good, though the second may often not be true. The core is supposed to be described by the linear model, with $\rho(r)$ given by (4.22) and R replaced by r_s, and the envelope is described by (10.14). The analysis will next be outlined, with algebraic details supplied by the problems.

The basic behavior of the star's evolutionary track (L, T_{eff}) depends on the opacity in the envelope. At the higher temperatures characteristic of the initial evolution away from the main sequence, the envelope is radiative, and electron-scattering opacity dominates. Later, as the outer envelope cools (in lower-mass stars), convection becomes significant, and the surface conditions become critical. Broadly speaking, the star moves back toward the region of the HR diagram from which it evolved toward the main sequence.

First consider the core, whose density is given by (4.22). The core mass is easily obtained from (4.27) and is

$$M_c \simeq \frac{\pi r_s^3}{3}\, \rho_c. \qquad (10.16)$$

The pressure is found from the equation of hydrostatic equilibrium, and gives a relation between the central and shell pressure:

$$P_c = P_s + \frac{5\pi}{36} G\rho_c^2 r_s^2. \qquad (10.17)$$

Finally, using the ideal-gas relation, we find that the core temperature is given by

$$T_c = \frac{5\pi G}{36} \frac{\mu_c}{k} m_H r_s^2 \rho_c. \qquad (10.18)$$

Problem 10.6. Derive (10.16)–(10.18), assuming that $\rho_c \gg \rho_s$ and the ideal-gas equation of state. Show that the core, or shell radius, is given approximately by

$$\frac{r_s}{R_\odot} = 0.178 \left(\frac{M_c}{M_\odot}\right)^{1/3} \left(\frac{\rho_c}{10^3}\right)^{-1/3} \qquad (10.19)$$

and that

$$T_c = 5.4 \times 10^7 \mu_c \left(\frac{M_c}{M_\odot}\right)^{2/3} \left(\frac{\rho_c}{10^3}\right)^{1/3}. \qquad (10.20)$$

The expressions for T_c apply only when the core is nondegenerate. When degeneracy is important, the core will be isothermal, so that a reasonable approximation is

$$T_c \simeq T_s. \qquad (10.21)$$

Now construct the envelope, using (10.12) and assuming that $1.5 \leq n < 3.0$ (the case $n = 3$ must be handled separately). The mass of the star is given by the integral of $(4\pi r^2 \rho)$, with $\rho(r)$ described by (10.12):

$$M = M_c + \frac{4\pi}{3-n} \rho_s r_s^n R^{3-n}, \qquad (10.22)$$

again assuming $r_s/R \ll 1$. This may be solved for the radius of the star,

$$R = \frac{r_s}{3-n} + \left[\frac{(3-n)M(1-q_c)}{4\pi\rho_s r_s^n}\right]^{1/(3-n)}. \qquad (10.23)$$

To complete the description of the envelope structure, we need $T(r)$ and from it the value at r_s. To obtain this, construct the pressure from the equation of hydrostatic equilibrium, the density (10.12), and the mass profile $m(r)$, which gave (10.22). For $r_s/R \ll 1$, it follows that

$$P_s = P(r_s) = \frac{M_s G\rho_s}{r_s(n+1)} + \frac{2\pi G\rho_s^2 r_s^2}{n^2-1}. \qquad (10.24)$$

This assumes that $P(R) = 0$. The temperature is then given by the ideal gas law: at $r = r_s$ the temperature is

$$T_s = \frac{2.3 \times 10^7}{n+1} \mu_e (M_s/M_\odot)(R_\odot/r_s) + \frac{2.46 \times 10^7}{n^2-1} \mu_e (r_s/R_\odot)^2 \rho_s, \qquad (10.25)$$

where μ_e is the mean molecular weight in the envelope, and $1.5 \leq n \leq 3$.*

Problem 10.7. Estimate the shell temperature for a 1 $M_\odot$ star of Population I.

Problem 10.8. Derive (10.25), assuming that $r_s/R \ll 1$. Note that P_s is obtained by requiring that $P(R) = 0$.

The core equations (10.16)–(10.20) and the envelope equations (10.23)–(10.25), plus the ideal-gas relation, allow us to find ρ_s, T_c, T_s, r_s, and R, given the model parameters ρ_c, M, and M_c (or $q_c = M_c/M$). In principle, ρ_c is obtained from earlier calculations at the instant the shell source is ignited. Similarly, q_c is determined by the conditions when the shell becomes effective.

Since the core is inert, ρ_c must increase (at least until electron degeneracy results, or a sufficiently large T_c develops that the next stage of nuclear burning sets in). Therefore, we specify M and M_c, and consider the star's behavior, that is, L and T_{eff}, as ρ_c increases. Eventually ρ_c will become constant, or the next stage of core burning will begin, and the analysis will become more involved. The model here will take the star up the red-giant branch; so we will limit ourselves to this stage of evolution.

*The value $n = 3$ may be used in (10.24)–(10.25), though not in the expression for the radius (10.23). For $n = 3$, the radius depends exponentially on the quantity $(\rho_c/\rho_1)(1/q_1 - 1)$.

Problem 10.9. Show how a value for $q_c = M_c/M$ could in principle be found from the previous stage of stellar structure, and thus demonstrate that the preceding model depends only on M and ρ_c. Recall the discussion of the core structure in Section 9.5.

The next step is to find expressions for ρ_s and T_s in terms of the parameters ρ_c, q, and M. The shell temperature is obtained by substituting (10.19) into (10.25):

$$T_s = \frac{1.29 \times 10^8}{n+1} \mu_e (M_c/M_\odot)^{2/3} (\rho_c/10^3)^{1/3}. \quad (10.26)$$

The second term in (10.25) has been dropped, since it is small. For low-mass stars, say, $M \lesssim 3$ or $4\ M_\odot$, it is probably reasonable to replace (10.26) with (10.21), since the core rapidly develops degeneracy pressure and is essentially isothermal. The ratio of (10.20) to (10.26) gives

$$T_c/T_s = 0.419\,(n+1)\,(\mu_c/\mu_e), \quad (10.27)$$

which shows that, as T_c increases, so does T_s. For a pure He core and extreme central condensation ($n = 3$), (10.27) gives $T_c \simeq 3.4\ T_s$.

So far, the opacity of the envelope and its dominant mode of energy transport have not been used. These factors enter as we calculate the shell density ρ_s, and are critical for finding the star's track in the HR diagram. In the model considered here, there are several primary situations. The envelope structure may be: (1) radiative with electron scattering dominant; or (2) convective with low surface density. The first possibility occurs during evolution away from the main sequence in all stellar models, surviving to lower and lower T_{eff} as the star's total mass increases. In low-mass and intermediate-mass stars, case (2) takes over near the red-giant branch. Case (3) exists in the final portions of the red-giant branch, particularly in low-mass stars (the details will be given shortly for Population I stars of about 1 $M_\odot$; the analysis is easily extended to other cases as well).

The condition fixing ρ_s is that the shell luminosity L equal the energy-generation rates, which is just energy conservation applied at the core-envelope boundary. In all cases, one expression for L is given by (10.14). A second, which depends on the modes of energy transport and opacity, is (10.15), and is applicable to radiative envelopes. Equating these two expressions and using (10.26) and (10.19) to eliminate T_s and r_s yields an expression for ρ_s:

$$\rho_s = 217\,(M_c/M_\odot)^{2(3-n)/9} \times (\rho_c/10^3)^{(6-n)/9} \text{g cm}^{-3}, \quad (10.28)$$

which assumes Population I composition, and the CNO cycle as the principal energy source, with $\epsilon_0 = 8.99 \times 10^{-3}$ for T in units of 10^7 K. The index n, which appears in (10.28) and the following equations, is the temperature exponent in $\epsilon(r)$, and may be assumed to lie in the vicinity of $n = 16$. For Population II stars, the coefficient 217 in (10.28) is replaced by 583. The stellar radius is given by (10.23). Using (10.28) and (10.19) to eliminate ρ and r_s, and rewriting in solar units, we find

$$R/R_\odot = 1.54 \times 10^{-2}(M/M_\odot)^{(10n-21)/27} \times (1-q)^{5/3} q^{(10n-66)/27} \times (\rho_c/10^3)^{(5n+6)/27}. \quad (10.29)$$

For Population II, the coefficient is 2.97×10^{-3}. An average polytropic index $n \simeq 2.4$ has been used in obtaining some quantities above, such as the exponent of the mass fraction in the envelope $1 - q_c$. The stellar luminosity is given by (10.14) if r_s, T_s, and ρ_s are eliminated. The result is

$$\frac{L}{L_\odot} = 39.5 \left(\frac{M_c}{M_\odot}\right)^{(2n+21)/9} \left(\frac{\rho_c}{10^3}\right)^{(n+3)/9}. \quad (10.30)$$

For Population II the coefficient 39.5 is replaced by 5.01. Finally, the effective temperature is given by substituting (10.30) and (10.31) into (1.3):

$$T_{\text{eff}} = 1.16 \times 10^5 (M/M_\odot)^{(15-2n)(7/108)} \times (1-q_c)^{-5/6} \times q_c^{(195-14n)/108} (\rho_c/10^3)^{-(7n+3)/108} \quad (10.31)$$

with a coefficient of 1.58×10^5 for Population II stars. The slope of the evolutionary track given by (10.30) and (10.31) for radiative envelopes is $d \log L / d \log T_{\text{eff}} = -12\,(n+3)/(7n+3)$, which is -1.98 for $n \simeq 16$. The star therefore moves to lower values of T_{eff}, as seen from (10.31), with increased ρ_c and to somewhat higher luminosities. As the evolution proceeds (increasing ρ_c), the core becomes more dense and the outer envelope cools.

Problem 10.10. Derive the dependence of T_{eff} and L on ρ_c as given in (10.30)–(10.31). Do not worry about dependence on the parameters M or M_c, or the numerical coefficients.

Eventually the opacity increases, and convective instabilities develop. At this stage the model corresponding to convective stars with high surface density applies. This marks the beginning of the red-giant branch in low-mass stars. In intermediate-mass and high-mass stars, core He burning begins at about this stage. As expansion continues, the surface density drops, and the model for stars with convective envelopes and low surface density applies. Since this model describes most of the region of the HR track, we will give the results. They are obtained as those for a radiative envelope are, except that equations (6.131) and (6.135) are used for L rather than (10.15). The results in (10.32)–(10.35) assume $n = 16$ and a surface opacity typified by (6.120) for Population I stars:

$$\rho_s = 31.6(M/M_\odot)^{2.64} \times (1 - q_1)^{0.51} q_1^{-3.24} (\rho_c/10^3)^{-0.64}; \tag{10.32}$$

$$R/R_\odot = 0.383(M/M_\odot)^{4.73} \times (1 - q_1)^{0.81} q_1^{4.07} (\rho_c/10^3)^{2.4}; \tag{10.33}$$

$$L/L_\odot = 0.835(M/M_\odot)^{6.39} \times (1 - q_1)^{1.03} q_1^{5.18} (\rho_c/10^3)^{3.05}; \tag{10.34}$$

$$T_{eff} = 8.89 \times 10^3 (M/M_\odot)^{-0.77} \times (1 - q_1)^{-0.15} q_1^{-0.74} (\rho_c/10^3)^{-0.44}. \tag{10.35}$$

Problem 10.11. At what central density will a 1 $M_\odot$ star begin core He burning, which sets in at 10^8K? What is the condition of matter under these conditions? Your answer should justify setting $T_c = T_s$.

Assuming q_1 to be constant during evolution, we can easily show that (10.34) and (10.35) give

$$\log L = -6.93 \log T_{eff} + \text{constant}.$$

The star therefore moves to the upper right-hand corner of the red-giant branch for low-mass stars. Although they are not shown explicitly in (10.32)–(10.35), the exponents of M, ρ_c, and q_1 are sensitive to the surface opacity via the constants a and b that appear in (6.120). For example, the track in the HR diagram has slope

$$d \log L / d \log T_{eff} = 0.5 - b - 3.5a, \tag{10.36}$$

if we assume q_1 is constant. We see that uncertainties in the surface opacity may strongly influence the results.

The model developed in the preceding is far too simple to give good numeric estimates of many quantities. Nevertheless, it does reflect the overall qualitative aspects of evolution away from the main sequence. In order to better understand the quantitative details, we will discuss the evolution of a $1 M_\odot$ star based on numerical models in Section 10.6. Some aspects of the evolution for $M = 5\, M_\odot$ are also reviewed.

Before we do so, let us estimate the evolutionary time-scale for core contraction in the presence of a H shell source. The contracting core converts gravitational potential energy into radiation, which heats the shell. As H burns in the shell, He ash is added to the core, increasing its mass. The shell luminosity L_s governs the rate of mass increase of the core, which in turn gives the core luminosity L_c. The gravitational potential energy

$$\Omega_c = -\alpha M_c^2 G / R_c$$

for the core. According to the virial theorem, half the change in Ω_c goes into radiation. The corresponding energy loss is

$$\Delta E_c = -\frac{\alpha}{2} \frac{M_c^2 G}{R_c} \left\{ 2 \frac{\Delta M_c}{M_c} - \frac{\Delta R_c}{R_c} \right\}, \tag{10.37}$$

where ΔM_c and ΔR_c give the change in core mass and radius, respectively. The factor α depends on the specific core model. For a linear density law, $\alpha = 6/7$; for a uniform density core, it is 3/5. The core luminosity is the product of the mass infall rate, dM_c/dt, and the gravitational potential, $\Phi_c = M_c G / R_c$, at the core surface:

$$L_c = \frac{M_c G}{R_c} \frac{dM_c}{dt}. \tag{10.38}$$

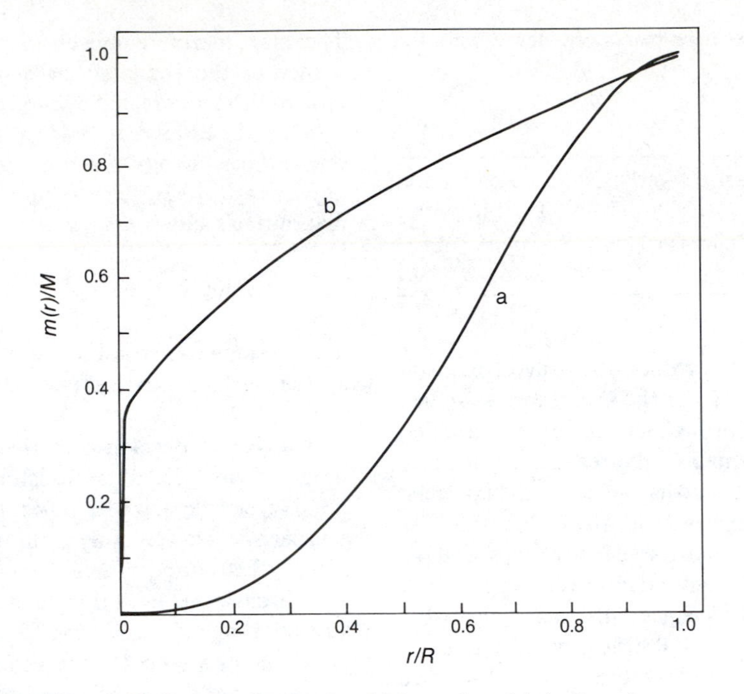

Figure 10.6. Mass fraction in a 5 $M_\odot$ star from (a) the linear homogeneous model; (b) an inhomogeneous model with core mass $M_c = 0.256\ M$.

The H-burning shell luminosity is given by the product of the mass infall rate, the hydrogen abundance in the envelope X_e, and the thermonuclear energy released per gram of matter when four H atoms fuse to form He, $E_H \simeq 6.4 \times 10^{18}$ ergs/sec:

$$L_s = E_H X_e \frac{dM_c}{dt}. \tag{10.39}$$

Eliminating dM_c/dt from (10.38)–(10.39) gives the core luminosity in terms of the shell luminosity:

$$\begin{aligned} L_c &= \frac{M_c G}{R_c} \frac{L_s}{E_H X_e} \\ &= \frac{2.98 \times 10^{-4}}{X_e} \left(\frac{M_c}{M_\odot}\right)\left(\frac{R_\odot}{R_c}\right)\left(\frac{L_s}{L_\odot}\right). \end{aligned} \tag{10.40}$$

This may be used, with estimates of L_s and M_c/R_c, to show that in general $L_c \ll L_s$. Therefore it is sometimes useful to replace L_s by L, the total stellar luminosity, on the right-hand side.

The time-scale for core contraction, which governs the evolution of the overlying matter, is given by the ratio of (10.37) to (10.40):

$$\begin{aligned} \Delta t_c &= \alpha \frac{M_c E_H X_e}{2L_s}\left(2\frac{\Delta M_c}{M_c} - \frac{\Delta R_c}{R_c}\right) \\ &= 5.1 \times 10^{10} \alpha X_e (M_c/M_\odot)(L_\odot/L_s) \\ &\quad \times (2\Delta M_c/M_c - \Delta R_c/R_c) \text{ yrs.} \end{aligned} \tag{10.41}$$

For Population I stars, assuming a 5 $M_\odot$ star with $M_c \simeq 0.2\ M$ and $L_s \simeq 10^3\ L_\odot$, as is typical for stages near the red-giant branch, a 10 percent change in M_c or R_c requires about 10^6 yrs, which is roughly 10^{-2} times the corresponding main-sequence lifetime. The increase in mass of the core during the time Δt_c is given by (10.39):

$$\begin{aligned} \Delta q_c &= \frac{\Delta M_c}{M} = \frac{L_s \Delta t}{M E_H X_e} \\ &= 1.6 \times 10^{-11} (L_s/L_\odot)(M_\odot/M)\Delta t_c, \end{aligned} \tag{10.42}$$

assuming $X_e = 0.6$, $E_H = 6.4 \times 10^{18}$ ergs/g, and Δt_c is in years. For a 5 $M_\odot$ star with typical values for L_s and Δt_c, Δq_c is about 0.03.

The preceding simplified analysis shows that evolution away from the main sequence, which is dominated by core contraction after exhaustion of H burning in the center, but with a coexisting H-burning shell, is more rapid than core H burning, but is still quasistatic in nature. This justifies the continued use of quasistatic models, at least in gaining a qualitative understanding of the subsequent evolution. Figure 10.6 shows the mass distribution in a 5 $M_\odot$ star with a linear density relation, and in an inhomogeneous model.

10.6. RED GIANTS

Once a star has developed a thin shell source, its core is inert, though it may supply a small fraction of the luminosity because of gravitational contraction, which sets in once $M_c > 0.13\ M$. Because of the shell source, the envelope expands as the core contracts. During this process, $L \simeq L_s$ remains nearly constant, and the surface temperature drops; the star moves nearly horizontally to the right in the HR diagram, as described qualitatively by (10.30)–(10.31). The structure of a 1 $M_\odot$ star during this phase of its evolution is shown in Figure 10.7, and would lie midway between points 4 and 5 of Figure 9.1. The H-burning shell is quite thin, and is located very close to the star's center ($r_c/R \simeq 0.03$). The distinction between core and envelope is now quite evident in $T(r)$ and in the luminosity $L(r)$. The energy-generation rate, dL/dm, has evidently increased over its values during the main-sequence stage. The slight decrease in $L(r)$ in the outer envelope represents energy spent in expansion.

Problem 10.12. Show that the variation in luminosity with mass fraction in a star is

$$\frac{\partial L}{\partial m} = \epsilon - \frac{3}{2}\rho^{2/3}\frac{d}{dt}(p/\rho^{5/3})$$

for an ideal gas with $\gamma = 5/3$. Use this expression to establish:

(a) nuclear-energy release ($\epsilon > 0$) adds to the star's luminosity;

(b) expansion reduces the luminosity.

Shell burning adds mass gradually to the isothermal core, which contracts, raising T_s slightly. In response,

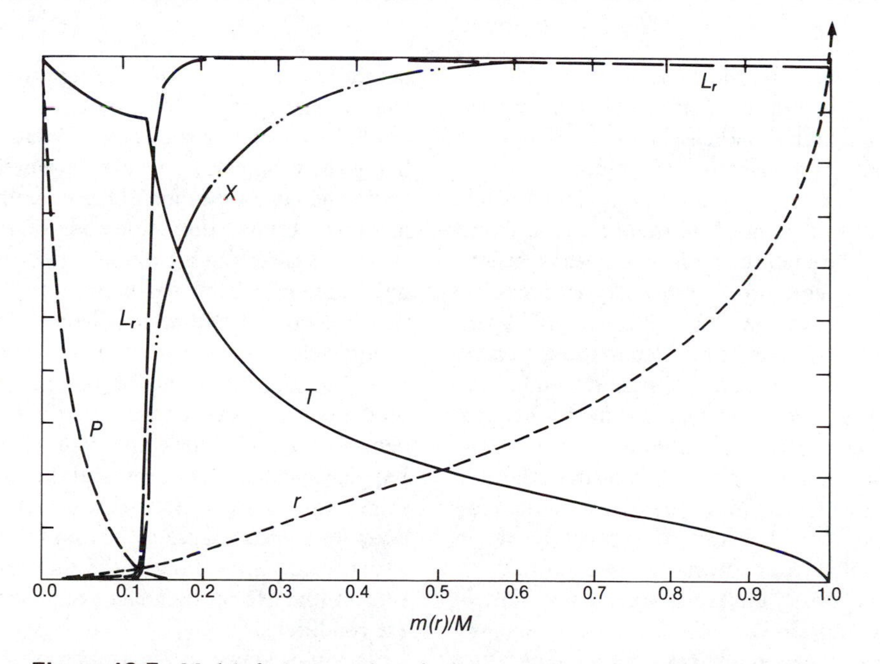

Figure 10.7. Model of a 1 $M_\odot$ star just after it leaves the main sequence, at time 10.31 × 10^9 years. The maximum value of the ordinate for each curve is: r = 2.13 $R_\odot$; P_c = 5.15 × 10^{19} dynes cm^{-2}; T_c = 2.39 × 10^7 K; L = 2.82 $L_\odot$; X = 0.708. The total radius is 2.2179 $R_\odot$, and the central density is 1.52 × 10^4 g cm^{-3}. Note the gradual increase in L for $m(r)M \gtrsim$ 0.4.

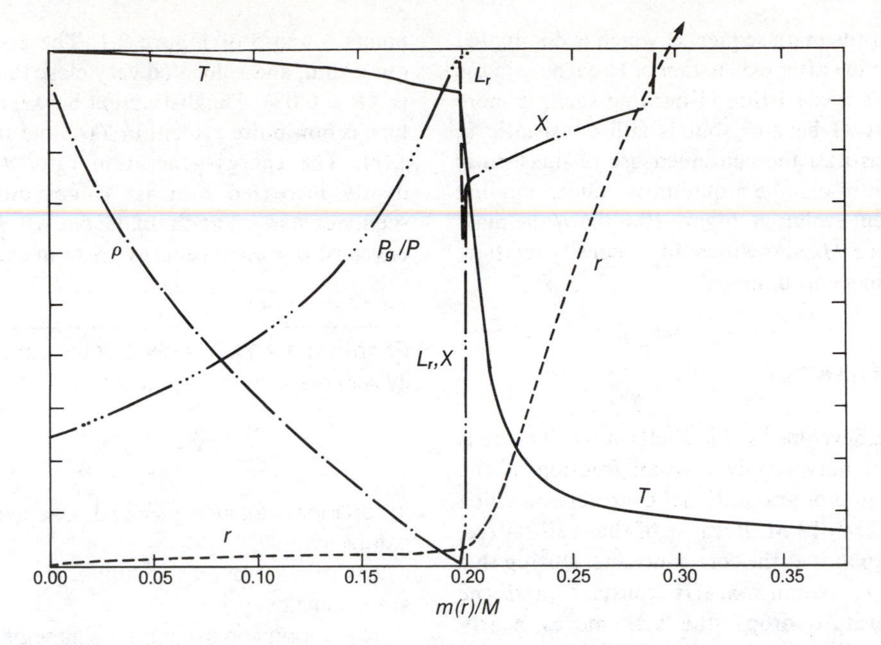

Figure 10.8. Model of the core of a 1 $M_\odot$ star during the subgiant stage, at a time of 10.8747×10^9 years. P_g/P is the ratio of gas pressure given by the ideal gas law to the total pressure including degeneracy. The luminosity rises steeply from zero to $L = 11.422\ L_\odot$ across the inner He core boundary. The maximum value of the ordinate for each curve is: $r = R_\odot$; $P_g/P = 1$; $\rho_c = 9.117 \times 10^4$ g cm^{-3}; $T_c = 2.7351 \times 10^7$ K; and $X = 0.693$. Note that only the inner 0.38 M is shown.

the surface is driven outward. Some of the shell energy heats the core enough to flatten the temperature profile, but not enough to increase T to the He ignition point. Eventually the envelope becomes convectively unstable, and the star moves up the red-giant branch. The structure at this point (5′ in Figure 9.1) is shown in Figure 10.8. The shell source is now a near discontinuity in $L(r)$. The core pressure-density relation is no longer that of an ideal gas, but reflects a significant and growing contribution from electron degeneracy, since $\rho_c \simeq 10^5$ g/cm^3. Convection in the region beyond $m(r)/M = 0.29$ gives the envelope a uniform chemical composition. Core contraction continues, driving the envelope outward more rapidly. The near-isothermal nature of the core means that the density must increase rapidly in order for the core to support itself. In particular, the increase is now more rapid than T^3.

In low-mass stars, electron degeneracy sets in before core He ignition is reached. Figure 10.9 shows the central density plotted against central temperature schematically for a series of models. Solid lines are obtained from the linear model for the core. For stars of about $M \lesssim 3\ M_\odot$, T_s remains nearly constant while core contraction proceeds, and degeneracy sets in before He burning. When He ignition does occur, the energy release raises the core temperature. In normal matter expansion would follow, reducing T_c and stabilizing the rate of He burning. However, because matter in the core is degenerate, no significant change in pressure results; the increased T_c leads to further energy release, and a thermal runaway develops. The phenomenon is known as ***helium flash***, and the evolution is dynamic. Numerical models indicate that the energy-release rate at the peak of helium flash reaches $L \simeq 10^{11}\ L_\odot$, but lasts for only a short time. This prodigious energy release rapidly lifts the electron degeneracy and allows the core to expand. As this happens, He burning ceases and the envelope contracts somewhat. Most of the increased luminosity of the cores goes into expansions, so that the star's luminosity actually decreases. Finally the core readjusts its structure so that He burning can occur under nondegenerate conditions.

In higher-mass stars (such as the 5 $M_\odot$ model shown in Figure 10.9), the increasing importance of gravitational energy release by the core forces it to develop a temperature gradient. Therefore the onset of He burning occurs before electron degeneracy be-

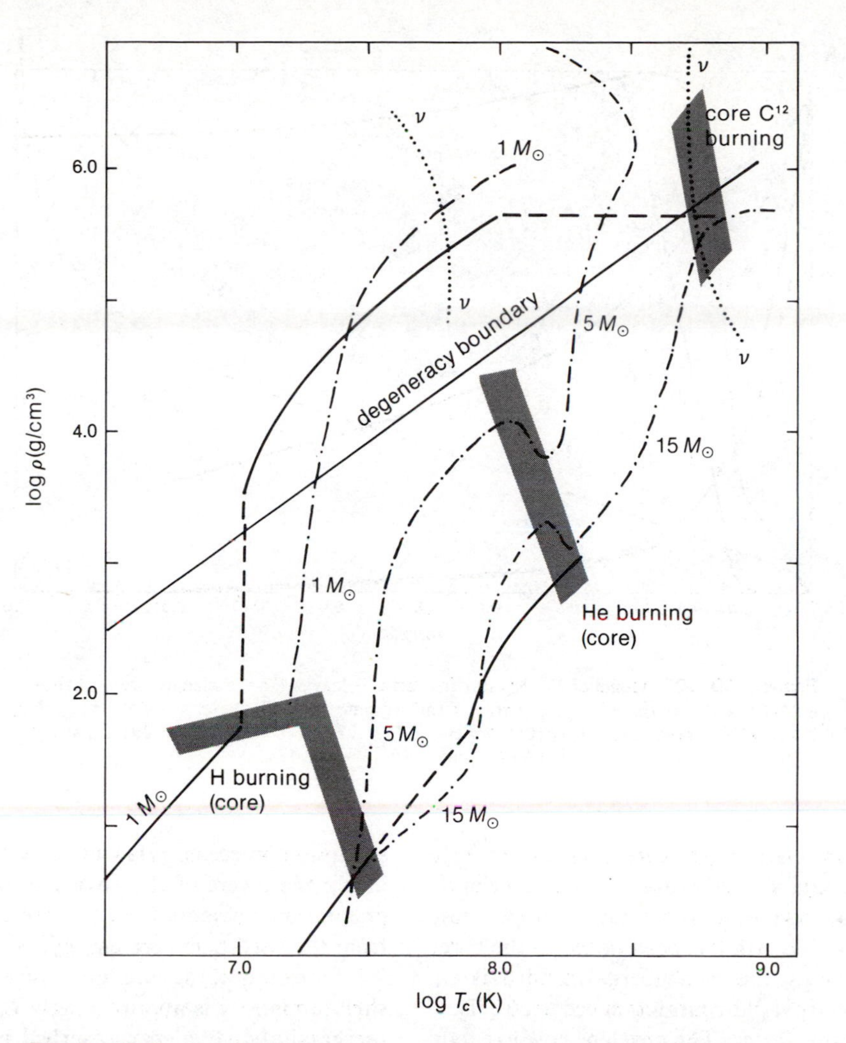

Figure 10.9. Central density versus central temperature during stellar evolution. Results are from the linear model discussed in the text (solid), and from numerical models (dash-dot). Dashed trajectories are interpolations between linear models. Above the dotted lines marked ν, the neutrino losses may be comparable to energy generation from core burning.

comes dominant. In fact, as M increases, ρ_c decreases; so the problem of helium flash is restricted to lower-mass stars. By analogy, C^{12} flash can also occur in higher-mass stars when ρ_c becomes great enough for electron degeneracy to again dominate. The situation here is complicated by the importance of neutrino processes, which can cool the star rapidly enough to produce a positive temperature gradient near the core.

Now consider the evolution of a 5 $M_\odot$ star as obtained from numerical models. At point 3 in Figure 9.1 the star is near the end of core contraction and is leaving the main sequence. Its structure at this point is shown in Figure 10.10. The initial increase in $L(r)$ through the inner 10 percent of the mass is due primarily to gravitational contraction, although a small contribution from H burning persists. The H shell source is relatively thick, covering about 10 percent of the mass. Because of the high rate of energy production in the shell, the envelope expands greatly. Some of the released energy is used up in doing work against gravity during this process; so $L(r)$ drops somewhat throughout the envelope.

By the time point 4 in Figure 9.1 has been reached,

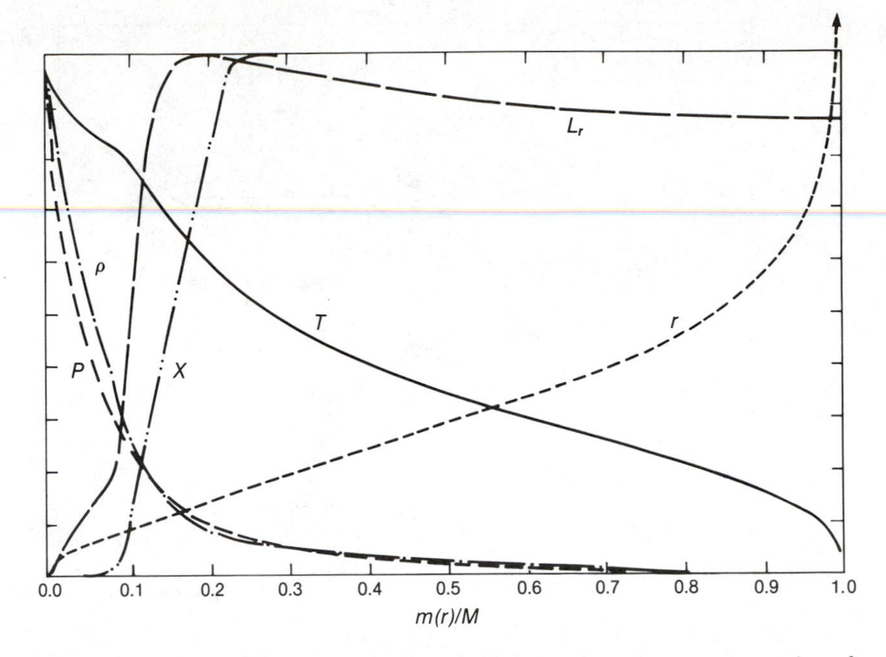

Figure 10.10. Model of a 5 $M_\odot$ star just after it leaves the main sequence at a time of 6.84461×10^7 years. Maximum value of the ordinate for each curve; $r = 2.9198\ R_\odot$; $P_c = 2.427 \times 10^{17}$ dynes cm^{-2}; $\rho_c = 106.59$ g cm^{-3}; $T_c = 3.6164 \times 10^7$ K; $L = 1.291\ L_\odot$; and $X = 0.708$. The total radius is $R = 3.943\ R_\odot$.

the rate of energy release in the core from H burning is essentially zero, and an isothermal core has developed (see Figure 10.9). As the shell continues to supply most of the luminosity, He ash has been added to the core, which now exceeds the Schönberg–Chandrasekhar limit. Consequently rapid contraction occurs, T_c rises, and the shell burns faster. The envelope now expands rapidly, with most of the increased energy release going into work as mass is moved to regions of lower gravitational potential by expansion. The star's track in the HR diagram is to the right (reducing T_{eff}) and slightly downward. From Table 9.1 it follows that the time required to reach point 5, the base of the red-giant branch, is less than a million years, or about 1 percent of the star's main-sequence lifetime. This region is known as the Hertzsprung gap, and the time spent here is considerably less than the time spent in the corresponding region by lower-mass stars. The Hertzsprung gap separates the main sequence from the red-giant region. Examination of the HR diagram for clusters (Figure 3.2) shows that this region contains relatively few stars, as expected.

The star's motion across the Herzsprung gap sets the stage for its subsequent motion up the red-giant branch, as may be seen from the envelope's behavior as expansion proceeds. Work must be done by the star in lifting the layers of the envelope to regions of lower gravitational potential. Part of the energy used comes from the shell plus core energy. At point 4 in Figure 9.1, for example, the core luminosity is $L_c \simeq 0.6\ L_\odot$, the shell luminosity is approximately $L_s \simeq 1{,}300\ L_\odot$, and the gravitational energy absorbed per second is $L_g \simeq 45\ L_\odot$. When point 5 is reached, the core and shell luminosity are roughly $L_c \simeq 6\ L_\odot$ and $L_s \simeq 950\ L_\odot$, respectively, and $L_g \simeq -250\ L_\odot$. Roughly 36 percent of the total energy released at point 5 therefore goes into expansion.

Problem 10.13. Use the virial theorem to show that the internal energy of the expanding envelope decreases as described. Assuming that 5 percent of the luminosity goes into increasing the gravitational potential energy, estimate the loss in internal energy.

According to the virial theorem, radiation is absorbed, and the matter cools, as it expands. Consequently, the envelope temperature drops, which

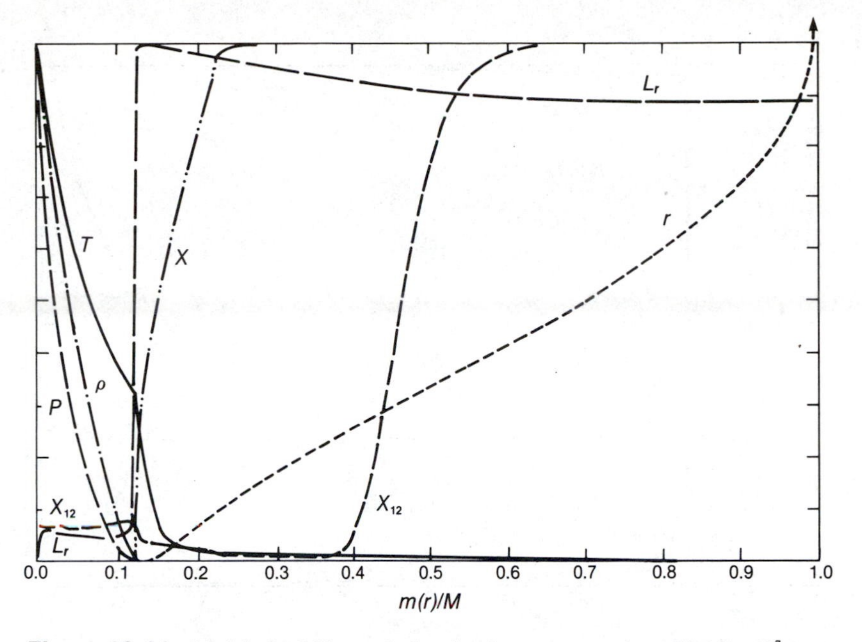

Figure 10.11. Model of a 5 $M_\odot$ star during the giant stage at a time of 7.04×10^7 years. The maximum value of the ordinate for each curve is: $r = 47.088\ R_\odot$ (total radius $R = 51.328\ R_\odot$); $P_c = 5.18 \times 10^{19}$ dynes cm^{-2}; $\rho_c = 7.7 \times 10^3$ g cm^{-3}; $T_c = 1.0323 \times 10^8$ K; $L = 8.76 \times 10^2\ L_\odot$; $X = 0.708$; and $X_{12} = 3.61 \times 10^{-3}$.

increases the opacity. When point 5 is reached, the luminosity exceeds the radiation limit, and convective transport becomes increasingly more important in the outer layers. The outer convective zone extends rapidly toward the core, and ultimately includes just over half the stellar mass. At about this point, the surface temperature has dropped so low that there are not enough free electrons to form H^- ions. The consequent reduction in surface opacity increases the radiation loss from the stellar surface, and the luminosity rises sharply, but the effective temperature remains very nearly constant. As one might expect from the discussion of a star's evolutionary track up to the main sequence, the steady rise L is due in part to the formation of a convective envelope that expands adiabatically. The work of expansion comes entirely from decreased internal energy (first law of thermodynamics, with $dQ = 0$). Therefore the radiation from the shell source and core passes through the envelope unimpeded.

The star's structure just after it begins to ascend the red-giant branch is shown in Figure 10.11 (just after point 5 in Figure 9.1). There is a slight reduction in L in the inner portions of the envelope where energy transport is radiative. But $L(m)$ becomes constant in the outer convective region $[m(r)/M \gtrsim 0.6]$. The shell source is now extremely thin, and there is a small contribution to L in the center, where N^{14}, which is produced as ash by the CNO bi-cycle, burns to form O^{18}. This accounts for part of the rapid increase in luminosity at the center; it also leads to the development of a small convective core, and brings about a slight expansion of the core as well. The N^{14} burning stage is short, since there is very little fuel available; so eventually the core begins to contract again, and T_c increases. The star moves rapidly up the red-giant branch, taking slightly under 5×10^5 years to reach point 6 in Figure 9.1. This represents less than 1 percent of the star's main-sequence lifetime.

Problem 10.14. Explain why the luminosity drops (see Figure 10.11) in the core region $m(r)/M < 0.1$ as the star begins its ascent up the red-giant branch.

As in 1 $M_\odot$ models, the onset of He burning via the triple-alpha process marks the tip of the red-giant branch, with $T_c \simeq 2 \times 10^8$ K. A rapid rate of energy generation results. In a 5 $M_\odot$ star (unlike 1 to 3 $M_\odot$

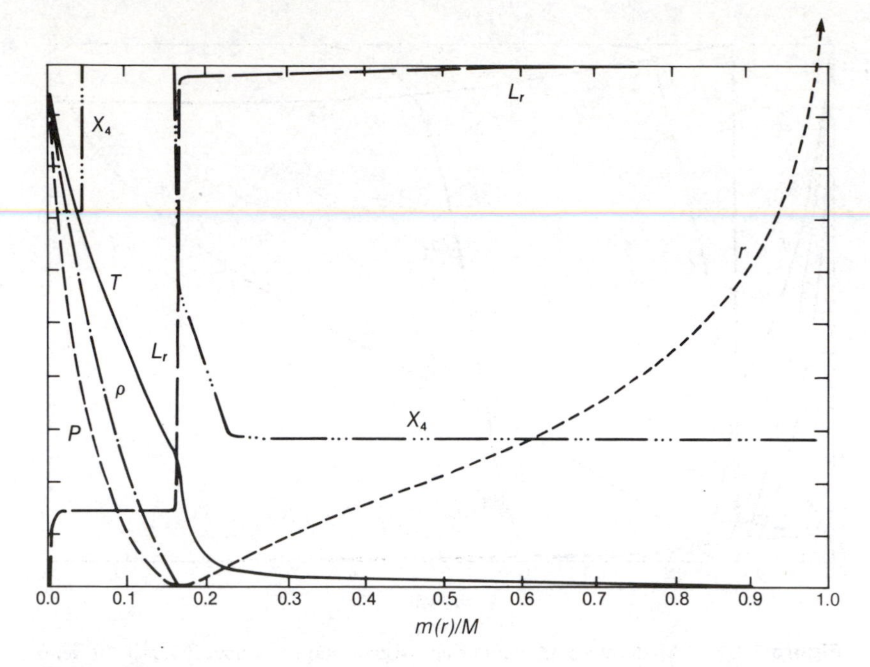

Figure 10.12. The 5 $M_\odot$ during the giant stage at time 7.7×10^7 years. The maximum values of the ordinate for each curve are: $r = 35.85\ R_\odot$; $P_c = 6.21 \times 10^{19}$ dynes cm^{-2}; $\rho_c = 7.7$ g cm^{-3}; $T_c = 1.33 \times 10^8$ K; $L = 1.14\ L_\odot$; and helium abundance $X_4 = 0.9763$. Total radius $R = 50.61\ R_\odot$.

stars) the core is nondegenerate when He burning begins; so expansion follows, and both thermal runaway and the helium flash are avoided. Core expansion is reversed at the active H-burning shell, and the star contracts with a subsequent reduction in luminosity. The evolutionary track is downward from point 6 toward point 7 of Figure 9.1. Since the energy-generation rates are highly temperature sensitive, a convective core forms. The temperature of the envelope rises as it contracts, which has two primary effects; a reduction in opacity, which causes the thickness of the convective zone to shrink toward the surface; and an increase in luminosity because of the reduced opacity. The surface temperature therefore increases, as does the luminosity.

The star has two major energy sources at this point: a highly concentrated He-burning core, and a H shell source. The star can be thought of as on the He-burning main sequence. The analogy is only partially valid, because the shell source, between points 7 and 8, supplies more than six times as much luminosity as the core. When the star reaches the vicinity of point 7, the convective zone is narrow, and is confined to the surface. The evolution from point 6 to point 7 is much slower than during previous stages since leaving the main sequence. In fact, as is easily shown from Table 9.1, the star takes about five times longer to descend the red-giant branch than to ascend to the tip from the main sequence (points 4 to 6). By the time point 7 has been reached, the gradual contraction of the envelope has raised the temperature and density near the shell; so the rate of energy generation rises, causing L to increase and reversing the downward motion of the evolutionary track. The star's structure between points 7 and 8, where both L and $T_{\rm eff}$ increase, is shown in Figure 10.12. The small He core is converting He into C^{12} and O^{16}, and the ashes are mixed convectively. Central expansion is gradual, so the core luminosity remains nearly constant between the highly concentrated He source and the H shell. The latter is extremely narrow and, as noted above, supplies most of the energy. A small amount of energy is also produced by gravitational energy release in the contracting envelope. The increase in shell density and temperature accelerates the star's evolution, moving it more rapidly toward point 8.

When point 8 is reached, core expansion, which has persisted since He burning began at point 6, ends and

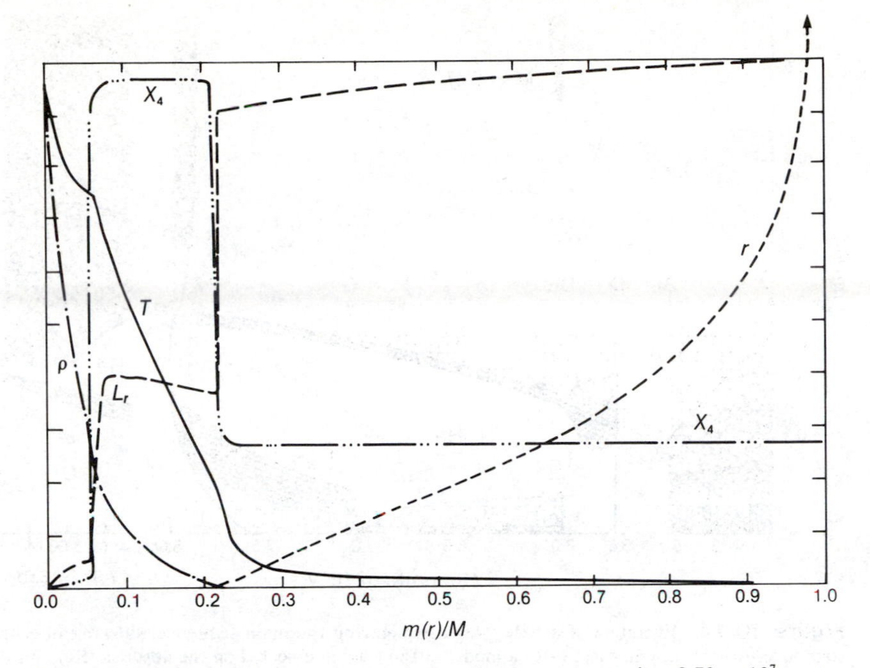

Figure 10.13. Model of a 5 $M_\odot$ star during the giant stage at time 8.79×10^7 years. Maximum ordinate for each curve: $r = 23.77\ R_\odot$ (total radius $R = 44.14\ R_\odot$); $\rho_c = 2.16 \times 10^4$ g cm^{-3}; $T_c = 1.84 \times 10^8$ K; $L = 1.94\ L_\odot$; and $X_4 = 1.0$.

is replaced by contraction. Since only one active shell lies between the core and the surface, the envelope contraction is halted, and gradual expansion begins again following point 9. The He content of the core is now low, and after point 9′ the central energy-generation rate is supplemented by gravitational contraction. Because of the active shell source, the envelope begins to expand, reducing $T_{\rm eff}$ but keeping the luminosity essentially constant. The evolutionary rate increases as the star moves back toward the giant region. The total time required for the star to reach point 10 is less than a quarter of the main-sequence lifetime.

The evolution from this point on is of increasing rapidity. Point 10 marks the termination of significant core He burning. The primary core-energy source becomes gravitational, though a small contribution from thermonuclear reactions is also present. The major sources are He and H, burning in shells that are separated by an inactive He zone. The core continues to contract. The He shell reverses this trend, and the H shell reverses it again, so that the stellar surface contracts. The increase in density and temperature at the edge of the core raises the rate of He shell burning; but the reduction in these quantities at the H shell lowers its rate of energy production. As a result, the luminosity remains nearly constant. This trend continues as the star moves to the left.

The structure at point 11 is shown in Figure 10.13. The core is now mostly C^{12} and O^{16}, with gravitational contraction supplying most of the luminosity for $m(r)/M < 0.07$. The luminosity increases rapidly at the He shell, which is thin. There is a temporary reduction in the expanding He zone as some of the energy released goes into work against gravity. At point $m(r)/M \simeq 0.22$, the thin H shell adds to the luminosity. Beyond this shell, gravitational energy release also adds to the luminosity. The density undergoes an extremely rapid drop across the He shell. In the core and envelope the temperature drops rapidly with increasing mass fraction, though in the intermediate He zone it decreases almost linearly with mass fraction.

The star's subsequent evolution depends on several factors that are similar to those of earlier stages. Eventually, with increasing central temperature, C^{12} and O^{16} will burn in the core. Although a stage of degeneracy develops (see Figure 10.9), some calculations indicate that it is lifted before carbon burning begins. In these models, neutrino processes lead to rapid core cooling, as the neutrinos carry off a signifi-

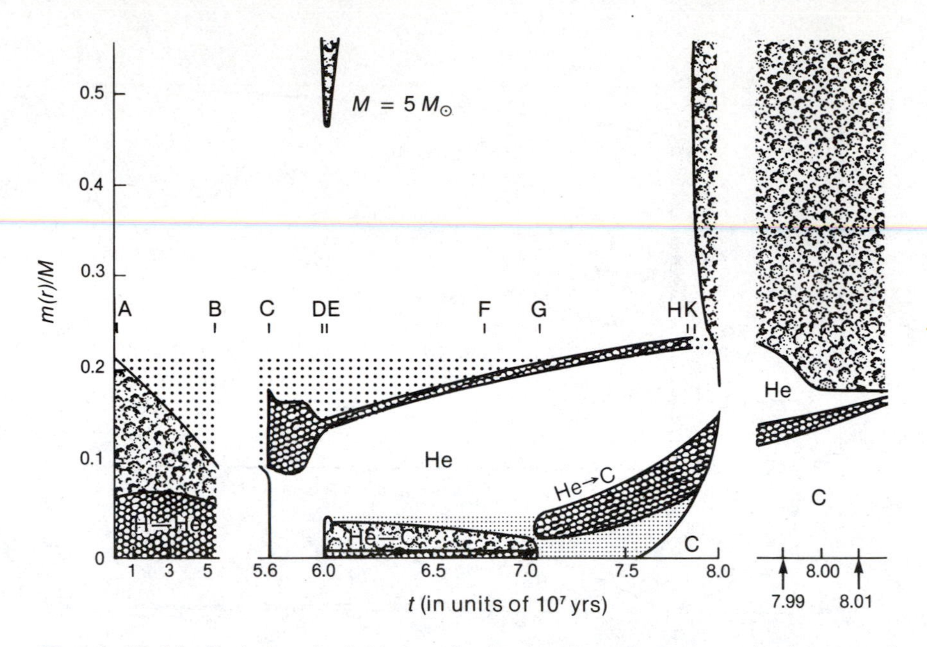

Figure 10.14. Evolution of a 5 $M_\odot$ star after leaving the main sequence, showing internal structure. A vertical line constitutes a model at the time intersected on the abscissa. Regions of convective transport shown by clouds; regions where nuclear burning exceeds 10^3 erg g^{-1} sec^{-1} are shown striped. Dots mark regions in which composition varies with mass fraction, with wide spacing for transition from initial H to He; and close spacing for transition from He to C. Note nonuniform abscissa.

cant amount of thermal energy. Energy then flows from the He shell inward because of the rapid central heat losses. In this case carbon burning may be delayed. The absence of these neutrino processes would seem to lead to carbon flash, a process qualitatively similar to He flash in stars of about 1 to 3 $M_\odot$. In any case, the subsequent evolution is still not certainly known. There are indications that stars less massive than about 4 $M_\odot$ pass through advanced stages of nuclear burning in a nonviolent manner. It is likely

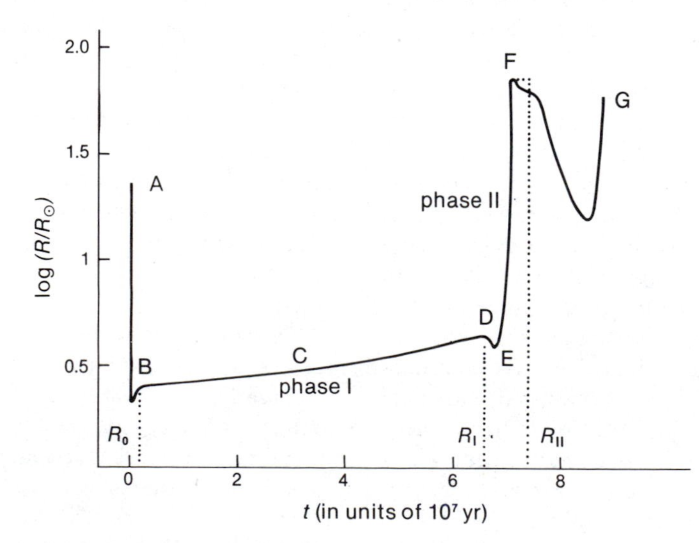

Figure 10.15. Radius versus time for a 5 $M_\odot$ star.

that these stars reach the supergiant stage, and then lose enough mass to become stable white dwarfs. These topics, and the possible fate of stars more massive than 5 $M_\odot$, will be considered in Chapters 11 to 15.

The evolution of a 5 $M_\odot$ star after it leaves the main sequence may be summarized by Figure 10.14, which shows regions of nuclear burning, and regions where convective transport dominates, as a function of time. Figure 10.15 gives the stellar radius versus time, showing most dramatically the increased rate of evolution as the star moves up the red-giant branch.

10.7. Modifications: Composition and Mass Loss

The preceding discussion dealt primarily with Population I stars, of specified chemical composition. The effect of changes in initial composition can be understood by several simple arguments. First, the reduction in He content on the main sequence can be shown to increase the lifetime, as might be expected. For stars of low metal abundance ($Z \simeq 0$) in which electron scattering dominates, the lifetime for H burning is roughly

$$t \simeq 10^{2X}. \tag{10.43}$$

The relation shows that helium-rich stars tend to leave the main sequence more rapidly than those with solar-type abundances.

Problem 10.15. Use elementary arguments to show that if the core opacity is dominantly Kramers' opacity with $Z = 0$, then the main-sequence lifetime is given roughly by

$$\log t \sim \log X + \frac{1}{2} \log \kappa_0 - 1.4 \log \mu. \tag{10.44}$$

Use (10.44) to show that $\partial \log t / \partial X$ is roughly 2 to 3, as X varies between 0.5 and 1.0. In this way establish (10.43) as a lower bound to the dependence of t on hydrogen abundance X.

Problem 10.16. Using the preceding results, find the fractional change in main-sequence lifetime given a change ΔX in hydrogen abundance in the stellar core. Evaluate this, assuming a 10 percent increase in H abundance (10 percent reduction in He).

Problem 10.17. Use the linear model for main-sequence stars to relate the central temperature to chemical composition, and explain physically why (in addition to increasing the star's fuel supply) increasing X increases the star's lifetime, as described by (10.43). Do this for upper-main-sequence stars.

Several additional factors may modify quantitatively the star's evolution on and away from the main sequence. These include uncertainties in chemical composition (as has been discussed); and uncertainties in the mixing length approach to convection in stellar envelopes. Besides changing the lifetime, changes in chemical composition can lead to large shifts in position on the main sequence. For example, a reduction in Y from 0.3 to 0.2 for a Population I star ($Z = 0.03$) of 5 $M_\odot$ reduces $T_{\rm eff}$ by about 10 percent, but reduces L by nearly a factor of two. Change in heavy-metal abundance Z has nearly the opposite effect, though the magnitude of the shift is not as large.

As the star moves away from the main sequence, the situation becomes more complex. The evolutionary tracks for three 5 $M_\odot$ stars, differing in initial chemical compositions but with all other factors the same, are shown in Figure 10.16. Uncertainties in Y are generally more important than those in Z for fixing the luminosity. The importance of Z increases as the star evolves, in part because the later stages of nuclear burning are quite sensitive to Z, particularly when the CNO bi-cycle dominates, as it does here. Notice that the position of the red-giant branch is insensitive to small changes in either Y or Z, but that the location of the onset of He burning (tip of the red-giant branch) varies by nearly a factor of four in luminosity.

Problem 10.18. Show qualitatively that reducing Y with Z fixed for main-sequence stars reduces both L and $T_{\rm eff}$, but L more than $T_{\rm eff}$. The dependence of opacity on X may be neglected for simplicity, since its effect is small.

Another source of uncertainty lies in the choice of mixing length employed in treating convection in the envelopes of stars near or on the red-giant branch. These problems are not significant until point 5 in Figure 9.1 is reached, and have little effect on the luminosity, but are found to shift $T_{\rm eff}$ to higher values

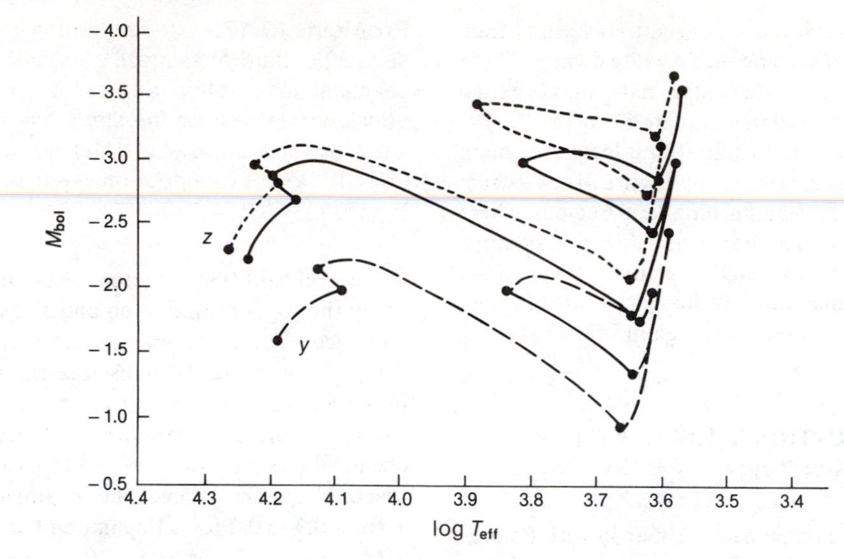

Figure 10.16. Effect of chemical composition on the evolutionary track for a Population I star of mass 5 $M_\odot$. The compositions are: solid curve, $X = 0.67$, $Y = 0.30$, and $Z = 0.03$; (dashed curve, $Y = 0.20$ and $Z = 0.03$; dotted curve, $Y = 0.30$ and $Z = 0.015$).

when the mixing length is increased relative to the pressure scale height.

Throughout the preceding discussions, it has been assumed that the star's mass remains constant. There is in fact a gradual reduction in M, since nuclear burning converts a fraction of the rest mass into energy that is carried off by photons and neutrinos. The effect is small, and does not play a role in stellar evolution. Other processes can lead to a more significant mass-loss rate. For example, the Sun is ejecting baryons (protons, primarily) from its photosphere at such a rate that their velocity and density at the Earth's orbit are about 400 km/sec and 10 m_H/cm^3, respectively. This implies a mass-loss rate of about 4×10^{-14} $M_\odot$/yr, which is extremely small. Nevertheless, mass-loss rates observed in other types of stars may reach values of 10^7 or 10^9 times as great. Such phenomena may be important for massive main-sequence stars, whose surface temperatures are high, and in extended red giants, whose surface opacity is extremely low.

Problem 10.19. Calculate an upper limit to the mass lost because of nuclear burning in the Sun. Express your answer as a mass-loss rate in units of $M_\odot$/yr.

These observations are consistent with the idea that stars more massive than the Chandrasekhar limit for white dwarfs, but not massive enough to become supernovae, must somehow lose their excess mass. It is presently estimated that stars less massive than about 4 $M_\odot$ somehow end their evolution as white dwarfs; so they must shed up to 2.8 $M_\odot$ during their lifetime. If we assume that much of this loss occurs gradually during the star's lifetime, conventional models with M constant would yield dM/dt as large as 10^{-8} $M_\odot$/yr. Lower values would be possible if significant mass loss occurred during the relatively short red-giant stages, as will be discussed shortly. Nevertheless, it is not obvious that stellar-wind effects are always negligible.

Problem 10.20. Verify that the solar mass-loss rate because of the stellar wind is about 4×10^{-14} $M_\odot$/yr.

To illustrate this last point, let us consider a simple model in which two stars with similar compositions are compared at some time when they have identical masses M_2. One star is assumed to have evolved with constant mass, but the other loses mass at a constant

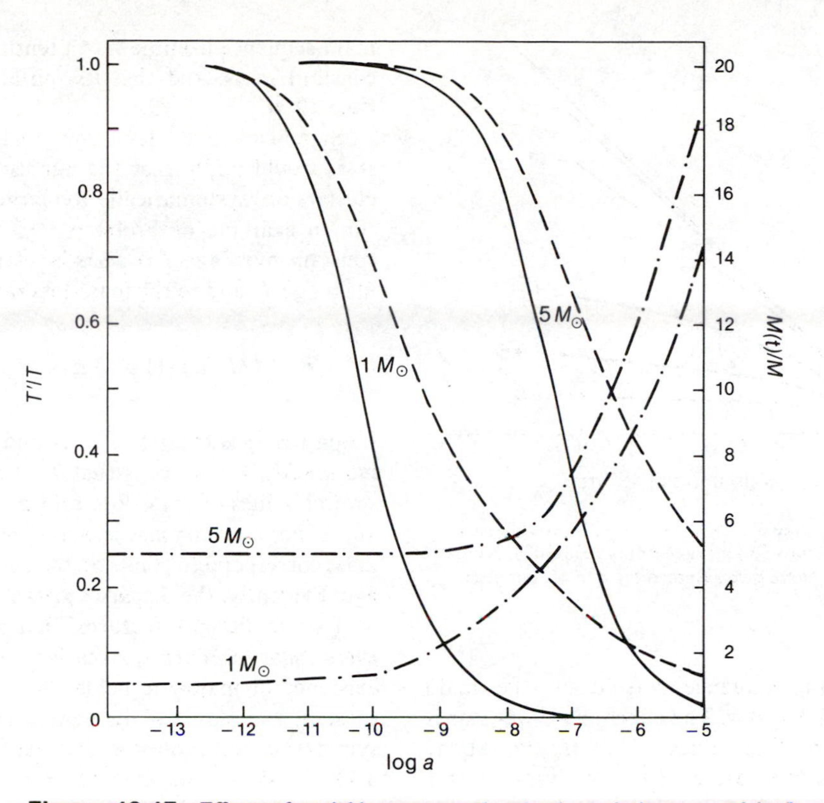

Figure 10.17. Effects of variable mass on the simple evolutionary model of Problem 10.21. Solid curve is T'/T (left scale, from 0.0 to 1.0); dashed curve is $M(t)/M$ (right scale, from 0.0 to 1.0); dash-dot curve gives $M/M_\odot$ (right scale, from 1.0 to 20.0).

rate $dM/dt = -aM_\odot/\text{yr}$. Since the star with variable mass must have started its evolution with $M(0) > M_2$, it clearly must have evolved faster than did the one with fixed mass. If both stars have converted equal amounts of H into He by the time $M(T') = M_2$, then they will have radiated equal amounts of energy. Assuming a mass-luminosity relation of the form $L \sim M^\alpha$ to hold for each star, it follows that

$$M_2{}^\alpha T = \int_0^{T'} M(t)^\alpha \, dt, \qquad (10.45)$$

where T is the age of the constant-mass star, and T' the age of the star with mass-loss rate $dM(t)/dt$. Assuming that $M(t)$ is known, we may then solve (10.45) to obtain T' as a function of M_2 and α, and T (as given by stellar evolution with no mass loss).

Problem 10.21. Suppose that $\alpha = 4$ in (10.45), that

$$\frac{dM}{dt} = -a = \text{constant}, \qquad (10.46)$$

and that $M(0) = M_1$, $M(T') = M_2$. Find the initial mass M_1 that would equal 5 $M_\odot$ after $T = 6.77 \times 10^7$ years, which is the main-sequence lifetime of a 5 $M_\odot$ star of constant mass. Then find the lifetime T' required for the variable-mass star to reach this point. Assume $a = 10^{-7}\ M_\odot/\text{yr}$.

Figure 10.17 shows the effects of variable $M(t)$ as compared to the main-sequence lifetime for a constant-mass star with $M = 5\ M_\odot$ and $M = 1\ M_\odot$, as obtained for several values of the mass-loss rate described in Problem 10.21. Clearly, values of $a \lesssim 10^{-12}\ M_\odot/\text{yr}$ have little or no effect, and values $\sim 10^{-14}\ M_\odot/\text{yr}$ leave T' unchanged to one part in 10^4. However, mass-loss rates $a \sim 10^{-10}\ M_\odot/\text{yr}$ in a 1 $M_\odot$ star would halve the main-sequence lifetime in this simple model, and would mean that a star of 1 $M_\odot$ when

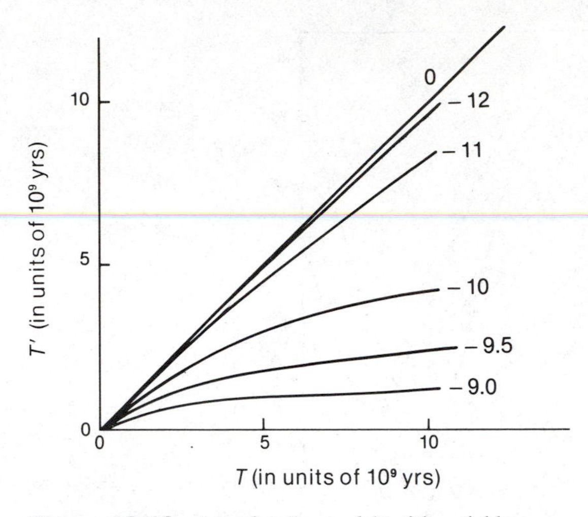

Figure 10.18. Age of stellar models with variable-mass (T') versus fixed-mass (T) age according to (10.47). Numbers next to each curve give assumed log a in $M_\odot$ per year.

leaving the main sequence started on the main sequence as a 1.4 $M_\odot$ star. The effect on a 5 $M_\odot$ star is reduced until a reaches values $\sim 10^{-9}$ $M_\odot$/yr. At the extreme, a mass-loss rate of 10^{-6} $M_\odot$/yr in a star whose final mass was 5 $M_\odot$ would imply that the star's main-sequence lifetime was a tenth of that of a star of constant mass, and that its initial mass would have been 12 $M_\odot$.

Significant mass loss from typical main-sequence stars would imply that the age estimates of globular clusters are systematically too large. Returning to the simple example of Problem 10.21, we see that the constant-mass age T of stars is related to the variable-mass age T' and to the mass-loss rate a by

$$T' = (M_2/a)\,[(1 + 5\,aT/M_2)^{1/5} - 1]. \quad (10.47)$$

Using the ages from Table 9.1 and the corresponding masses M_2, we can construct T' as a function of T for several values of mass-loss rate a as shown in Figure 10.18. For a given mass-loss rate, models of increasing mass correspond to points on the curve with decreasing age. Evidently, the disparity grows with cluster age.

Two additional features that could change the evolutionary scenario qualitatively are rotation and the existence of magnetic fields. When these effects are present, the star will in general not be spherically symmetric. Rotation was discussed briefly in Section 5.11. It should be evident from the discussion of evolution away from the main sequence that rotational

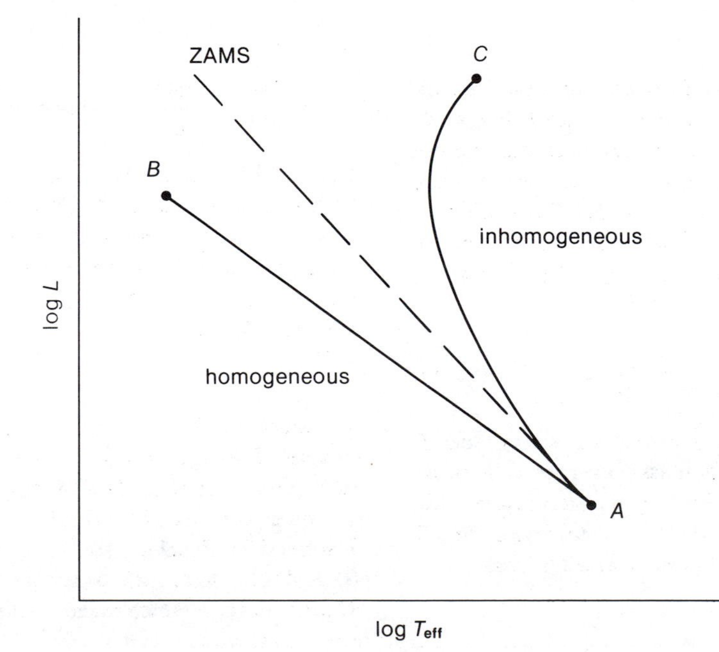

Figure 10.19. Effect of inhomogeneities in composition on evolution from zero-age main sequence.

mixing could dramatically alter the results. For example, if rotation mixed matter in the core and envelope on a time-scale small compared with the time-scale for nuclear reactions, then the evolution would be homogeneous. In this case, the linear model shows the qualitative behavior of L and T_{eff} as H is burned in the star. Figure 10.19 shows schematically the evolution of a star with and without rotational mixing. The dashed line is the zero-age main sequence from which both stars start their evolution at point A. When the composition remains homogeneous, the star evolves toward point B, whereas the development of inhomogeneities causes the star to swing to the right, toward point C. We note that rotation, acting in conjunction with stellar magnetic fields, may set up stellar winds, and may thus act as the underlying mechanism of mass loss from some stars. Strong stellar winds are believed to play an important role in x-ray emission in close binary systems (Chapter 17).

Strong magnetic fields ($B \gtrsim 10^3$ gauss) have been detected in some main-sequence-type stars (especially spectral type Ap), in some white dwarfs (fields up to 10^7 gauss), and in pulsars (up to 10^{12} gauss). Observational data on such magnetic stars suggest that stellar magnetic fields are not strong enough to directly affect their hydrostatic structure. Nevertheless, they may indirectly influence the star's structure, particularly if they occur in the outer convective regions. Stellar magnetic fields may be fossil in nature (galactic magnetic fields contained within the original interstellar material from which the star formed), or may arise from large-scale relative motion of electrons and ions (electric currents) within stellar envelopes. It is probable that magnetic and rotational phenomena occur together, and that their effects are closely interrelated. For example, Ap stars have strong magnetic fields, but unlike their nonmagnetic counterparts, type A, they are very slow rotators. It is believed that Alfvén waves associated with the magnetic fields in Ap stars carry off angular momentum, reducing their rotation rates.

Chapter 11

DEVIATIONS FROM QUASISTATIC EVOLUTION

Stellar models based on equations (8.4) to (8.7) are in hydrostatic or quasistatic equilibrium. However, not all such models are stable against small perturbations. We will consider here the preliminary steps that determine whether or not an equilibrium state is stable. Unstable equilibrium states have important applications to variable stars, to stellar mass limits, and to the final and initial stages of stellar evolution.

11.1. DEVIATIONS FROM HYDROSTATIC EQUILIBRIUM

One way to find out whether a system is stable or not is to subject all quantities to small perturbations and see if the amplitude of the perturbations grows in time. If some or all of the amplitudes grow, then instabilities will result. Two cautions are in order.

First, if small perturbations are found to grow, then one needs to do further analysis to find out whether they will continue to grow, eventually dominating or altering the system's structure, or whether nonlinear terms that are usually not retained in iinearized theory will limit their growth. When perturbations grow to a large but finite size, we obtain a new equilibrium state that will vary with time. For example, pulsational stability results whenever finite oscillations can be set up and maintained for extended periods of time. Pulsational modes are characteristic of many variable stars, such as Cepheid and RR Lyrae variables.

Second, there may be other nonlinear effects. Many physical systems are stable against small perturbations, but become unstable if they undergo a disturbance of more than a certain amplitude. Systems of this type are termed *metastable.*

For these reasons, stability analysis in linearized theory (small perturbations) may not be sufficient to establish stability in general, although it can usually prove instability. Unfortunately, nonlinear problems are extremely difficult to solve, and must usually be carried out numerically. The following discussion is therefore restricted to stability analysis using linearized theory. Furthermore, we will concentrate on the stability of spherically symmetric systems.

In order to carry out the proposed analyses, we need two equations of hydrodynamics. The first has, in fact, already been introduced in (4.1), which we rewrite in the form

$$\rho \frac{d\mathbf{v}}{dt} = -\nabla P - \rho \nabla \phi, \qquad (11.1)$$

where ϕ is the gravitational potential. The second equation is the mass continuity equation

$$\frac{\partial \rho}{\partial t} + \nabla \cdot \rho \mathbf{v} = 0, \qquad (11.2)$$

which states that the rate of change in mass Δm of a volume element ΔV of fluid, $\Delta m = \int \rho \, dV$, is equal to the net mass flux into the volume element.

Problem 11.1. Show that the statement about (11.2) is correct for a fluid element of arbitrary shape.

Equation (11.1) is equivalent to Newton's second law, and (11.2) is a statement of mass conservation. The variables in these equations are time-dependent; as we will show, equations (11.1) and (11.2) describe the time-development of perturbations imposed on equilibrium configurations.

The perturbation approach is developed in greater detail in Section 20.1 for the hydrodynamic equation. Basically, if only first-order quantities are required, the procedure is as follows. Express all variables, such as $\rho(\mathbf{r}, t)$, as a sum of the unperturbed, time-independent quantity $\rho_0(\mathbf{r})$ and a small, time-dependent perturbation denoted by a prime, $\rho'(\mathbf{r}, t)$:

$$\rho(\mathbf{r}, t) = \rho_0(\mathbf{r}) + \rho'(\mathbf{r}, t). \qquad (11.3)$$

Expressing P, ϕ and $\mathbf{v}$ in similar fashion, substitute the results into (11.1)–(11.2) and require that the zero-order equations be satisfied identically (in this case, that $\nabla P_0 = -\rho_0 \nabla \phi_0$). Finally, dropping all terms that contain products of two or more small quantities (denoted by primes) and assuming that $\mathbf{v}_0(\mathbf{r}) = 0$, (11.1) and (11.2) reduce to

$$\frac{\partial \rho'}{\partial t} + \nabla \cdot \rho_0 \mathbf{v}' = 0, \qquad (11.4)$$

$$\rho_0 \frac{\partial \mathbf{v}'}{\partial t} = -\nabla P' - \rho_0 \nabla \phi' + (\rho'/\rho_0) \nabla P_0. \qquad (11.5)$$

Problem 11.2. Carry out the analysis leading to (11.4) and (11.5). The last term is equivalent to $\rho' \nabla \phi_0$.

Continuing in this way, Poisson's equation, which relates the gravitational potential ϕ to the mass distribution, $\nabla \phi = 4\pi \rho G$, takes the form

$$\nabla \phi' = 4\pi \rho' G. \qquad (11.6)$$

The radiative energy flux in the system is given by the vector radiative equation

$$\mathcal{F} = -K_c \nabla T, \qquad (11.7)$$

where the thermal conductivity $K_c = 4\,acT^3/3\kappa\rho$. This reduces to the linearized form

$$\mathcal{F}' = -(K_{c,0} \nabla T' + K_c' \nabla T_0), \qquad (11.8)$$

where a prime signifies that all quantities are to be evaluated so that the net result is first order in perturbations. Thus K_c' will be given by

$$K_{c,0}[1 + 3T'/T_0 - \upsilon'/u_0 - \rho'/\rho_0]$$

to lowest order.

Next, we need an expression for the perturbation caused by a net heat flux into an element of matter. This may be obtained from the first law of thermodynamics. Starting with (2.12) and defining $u = Nmu$ and $S = Nms$, where N is the particle number and m is the particle mass, we find for an ideal gas

$$\rho \frac{du}{dt} = \rho T \frac{ds}{dt} + \frac{P}{\rho} \frac{d\rho}{dt}. \qquad (11.9)$$

The last step uses $V = mN/\rho$. Solving this for the time-rate of change in the specific entropy yields

$$\rho T \frac{ds}{dt} = \frac{1}{\gamma - 1} \left[\frac{dP}{dt} - \frac{\gamma P}{\rho} \frac{d\rho}{dt} \right]. \qquad (11.10)$$

The right-hand side gives the change in compressional and internal energy where heat flows into the matter element. The quantity $T\,ds/dt$ gives the heat transfer per gram of matter per second. But by conservation of energy, this must equal the difference between the heat generated per gram per second in the element itself by thermonuclear or other particle processes, ϵ_N, and the net energy flux out of the element per gram,

$\nabla \cdot \mathcal{F}/\rho$. Therefore,

$$T\frac{ds}{dt} = \epsilon_N - \frac{1}{\rho}\nabla \cdot \mathcal{F}. \tag{11.11}$$

Combining (11.10) and (11.11), and noting that if the adiabatic exponents are not equal, then $\gamma - 1 \rightarrow \Gamma_3 - 1$ and $\gamma \rightarrow \Gamma_1$ in (11.10), we have

$$\frac{dP}{dt} - \frac{\Gamma_1 P}{\rho}\frac{d\rho}{dt} = (\Gamma_3 - 1)(\rho\epsilon_N - \nabla \cdot \mathcal{F}). \tag{11.12}$$

When this is linearized, one finds

$$\frac{\partial P'}{\partial t} + \mathbf{v} \cdot \nabla P_0 - (\Gamma_1 P_0/\rho_0)\left(\frac{\partial \rho'}{\partial t} + \mathbf{v} \cdot \nabla \rho_0\right) = (\Gamma_3 - 1)\,\rho_0\left(\epsilon_n - \frac{1}{\rho}\nabla \cdot \mathcal{F}\right)', \tag{11.13}$$

where the prime on the last factor on the right is used as in (11.8). Notice that when the net heat transfer to the element vanishes ($T\,ds = 0$), then so does the left-hand side of (11.12), leaving the familiar equilibrium equation

$$\rho_0\epsilon_{N,0} = \nabla \cdot \mathcal{F}_0.$$

The five equations (11.4) to (11.6), (11.8), and (11.13) describe the behavior of the perturbations. They may be combined to eliminate the spatial derivatives of P', ρ' and ϕ', yielding a third-order (in time) differential equation for the radial displacement $\delta r/r$ caused at point r by the perturbations. If it is assumed that

$$\delta r/r = \xi(r)\,e^{\omega t} \tag{11.14}$$

with $\xi(r)$ dimensionless, then the differential equation is equivalent to the algebraic cubic equation

$$\omega^3 + \omega(A - B) + C = 0. \tag{11.15}$$

The quantities A, B, and C are integrals involving various adiabatic indices, the displacement $\xi(r)$, and the zero-order structure of the system. The solutions of (11.15) give critical information about the stability of radial perturbations (11.14), as will be seen below. The analysis, however, requires that something be known about the quantity $\xi(r)$. A simplified approach to this problem will be considered next.

11.2. Adiabatic Stellar Pulsations

Pulsating variable stars form an important stellar class, of which the Cepheids and RR Lyraes are relatively simple and straightforward examples, and long-period variables, dwarf Cepheids, β Cephei stars, and δ Scuti stars are more complicated examples. Here we need to make some general comments about stellar pulsation. To keep matters simple, we will make two important assumptions.

First, we assume that the pulsations are adiabatic; that is, the pulsation periods are short compared to the time-scale needed for thermal equilibrium to be established, so that there is no gain or loss of energy for any moving element of the star. This assumption enables us to relate the temperature and pressure fluctuations of any element of gas to the density fluctuations by the equations of adiabatic change. Using $P \sim \rho^\gamma$ and $T \sim \rho^{\gamma-1}$, we find

$$\frac{\delta P}{P} = \gamma\frac{\delta\rho}{\rho}, \qquad \frac{\delta T}{T} = (\gamma - 1)\frac{\delta\rho}{\rho}. \tag{11.16}$$

Second, we assume that the oscillations in the star are purely radial, so that the star maintains spherical symmetry. This is probably a good approximation for Cepheids and perhaps RR Lyraes, but not so good for β Cephei or δ Scuti stars, in which nonradial oscillations may be dominant.

We first do a very simple, volume-averaged analysis of the pulsations to develop a feeling for the physics involved. This yields a simple description of the time behavior of perturbations about equilibrium at a given point in the star. In equilibrium, the star is in balance between the pressure gradient and the gravitational force. Let $\overline{P}$ and $\overline{\rho}$ be the mean internal pressure and mass density, respectively. Then the equation of hydrostatic equilibrium can be approximated as

$$\frac{\overline{P}}{R} = \frac{\overline{\rho}MG}{R^2} = \frac{4\pi G}{3}\overline{\rho}^2 R. \tag{11.17}$$

We now suppose that a star changes its radius by an amount δR, and consider the effect this change has on the equilibrium, assuming that the motion is adiabatic.

The changes in $\overline{P}$ and $\overline{\rho}$ are therefore

$$\delta\overline{\rho} = \delta(3M/4\pi R^3) = -3\overline{\rho}\delta R/R \quad (11.18)$$

and

$$\delta\overline{P} = \gamma\frac{\overline{P}}{\overline{\rho}}\delta\overline{\rho} = -3\gamma\overline{P}\delta R/R. \quad (11.19)$$

Dividing the left-hand side of (11.17) by $\overline{\rho}$ gives the buoyancy force per gram of stellar material f_b, and its variation is

$$\delta\left(\frac{f_b}{\overline{\rho}}\right) = \frac{\delta\overline{P}}{R\overline{\rho}} - \frac{\overline{P}\delta R}{\overline{\rho}R^2} - \frac{\overline{P}\delta\overline{\rho}}{\overline{\rho}^2 R}. \quad (11.20)$$

Dividing the right-hand side of (11.17) by $\overline{\rho}$ gives the gravitational force per unit mass f_g, and its variation is

$$\delta\left(\frac{f_g}{\overline{\rho}}\right) = \frac{4\pi G}{3}(\overline{\rho}\delta R + R\delta\overline{\rho}). \quad (11.21)$$

The net force per gram acting in the positive radial direction is the acceleration $\delta\ddot{R}$ and is given by

$$\delta(f_b/\overline{\rho}) - \delta(f_g/\overline{\rho}) = \delta\ddot{R}$$
$$= -(3\gamma - 4)\frac{4\pi G}{3}\overline{\rho}\delta R. \quad (11.22)$$

This will be recognized as the equation of simple harmonic motion

$$\frac{d^2}{dt^2}\delta R = -\omega^2\delta R, \quad (11.23)$$

where

$$\omega^2 = (3\gamma - 4)(4\pi G/3)\overline{\rho}$$

is the square of the angular frequency of oscillation. We see that stars will be able to undergo stable oscillations if $\gamma > 4/3$. We had already found this limiting value earlier when discussing the virial theorem. If $\gamma < 4/3$, the star is unstable, and an initial displacement δR produces a force of disequilibrium that causes it to grow in time ($\delta R \approx e^{\omega t}$). Note also that the period T of the oscillation, which is $2\pi/\omega$, is given by

$$T = \frac{2\pi}{(4\pi(\gamma - 4/3)\overline{\rho}G)^{1/2}}$$
$$= \frac{\sqrt{\pi}}{\sqrt{(\gamma - 4/3)\overline{\rho}G}}. \quad (11.24)$$

Problem 11.3. Solve (11.22) assuming that δR was finite in the past, and show that $\delta R \longrightarrow \infty$ as $t \longrightarrow \infty$ if $\gamma < 4/3$. What happens if $\gamma > 4/3$?

Problem 11.4. Plot the period-density law (11.24) for stars of mass 0.5, 1.0, and 10 $M_\odot$, and average density between 10^{-8} g/cm^3 (supergiant) and 10^{15} g/cm^3 (neutron stars). Locate the regions of supergiant, giant, main-sequence, white dwarf, and neutron stars.

In conjunction with radius-luminosity relations from hydrostatic stellar models, (11.24) leads to the period-luminosity law discovered empirically for Cepheid variables.

Equation (11.22) is a simplified description of the evolution in time of small disturbances at a given point inside a star. Now consider how those perturbations change with distance through the star at a fixed time. Rewrite the hydrodynamic equation (11.1) for an element of matter at radial distance r from the center of the star as

$$\frac{d^2r}{dt^2} = -\frac{1}{\rho}\frac{\partial P}{\partial r} - g, \quad (11.25)$$

where $g = m(r)G/r^2$ is the gravitational acceleration at r, and d^2r/dt^2 is the acceleration of the volume element at r in a frame of reference moving with the matter. In equilibrium, the acceleration is zero. Now, suppose a radial displacement of the material occurs, and focus on a specific piece of material, denoting the equilibrium values by subscript zero,

$$\begin{aligned} r &= r_0 + \delta r, \\ \rho &= \rho_0 + \delta\rho, \\ P &= P_0 + \delta P, \\ g &= g_0 + \delta g, \end{aligned} \quad (11.26)$$

where δr, δP, $\delta\rho$, and δg are all assumed to be small. We now obtain a differential equation for the perturbations, assuming that they are small, and that the changes are all adiabatic. It is convenient to express them all in terms of δr. A shell of the star that previously lay between r and $r + dr$, with density ρ, now lies between $r + \delta r$, and $r + \delta r + d(r + \delta r)$, with density $\rho + \delta\rho$. Since we are following a piece of material,* the mass between these limits is by assumption the same. Hence,

$$\rho r^2 dr = (\rho_0 + \delta\rho)(r_0 + \delta r)^2 d(r_0 + \delta r) = \rho_0 r_0^2 dr_0.$$

Expanding, and dropping terms containing products of two or more small quantities, we find

$$\frac{\delta\rho}{\rho_0} = -\frac{2\delta r}{r} - \left(\frac{d\delta r}{dr}\right)_{r=r_0}. \tag{11.27}$$

Perturbations $\delta\rho$ and δr that satisfy (11.27) will conserve mass during the oscillations. In addition, the adiabatic assumption (11.19) becomes

$$\begin{aligned}\frac{\delta P}{P_0} = \gamma\frac{\delta\rho}{\rho_0} &= -2\gamma\frac{\delta r}{r_0} - \delta\left(\frac{d}{dr}\delta r\right)_{r=r_0} \\ &= -3\gamma\frac{\delta r}{r_0} - \gamma r_0\left(\frac{d}{dr}\frac{\delta r}{r}\right)_{r=r_0}\end{aligned} \tag{11.28}$$

Since the mass of a fluid element is constant, the change in g following the motion is

$$\delta g = \delta\left(\frac{m(r)G}{r^2}\right) = -\frac{2m(r)G}{r_0^2}\frac{\delta r}{r_0} = -2g_0\frac{\delta r}{r_0}. \tag{11.29}$$

We may now substitute these expressions into the equation of motion (11.25). The pressure gradient must be handled with care, since the quantities that enter into it take their instantaneous values ($P_0 + \delta P$, $r_0 + \delta r$, and so on). We rewrite (11.25), dividing each term by r^2 to obtain

$$\begin{aligned}\frac{1}{r^2}\frac{d^2r}{dt^2} &= -\frac{1}{\rho r^2}\frac{dP}{dr} - \frac{g}{r^2} \\ &= -\frac{1}{\rho_0 r_0^2}\frac{\partial P}{\partial r_0} - \frac{g}{r^2},\end{aligned} \tag{11.30}$$

where we use conservation of mass of the moving fluid element to replace $\rho r^2\,dr$ by $\rho_0 r_0^2\,dr_0$. Now write $r = r_0(1 + \delta r/r_0)$, and expand to first order in terms like r/r_0. We get

$$\begin{aligned}\frac{1}{r^2}\frac{d^2r}{dt^2} &= -\frac{1}{\rho_0 r_0^2}\left(\frac{dP_0}{dr_0} + \frac{d\delta P}{dr_0}\right) \\ &\quad - \frac{g_0}{r_0^2}\frac{(1 + \delta g/g_0)}{(1 + 2\delta r/r_0)} \\ &\simeq -\frac{1}{\rho_0 r_0^2}\frac{dP_0}{dr_0} - \frac{1}{\rho_0 r_0^2}\frac{d\delta P}{dr_0} \\ &\quad - \frac{g_0}{r_0^2}\left(1 - \frac{2\delta r}{r_0} - \frac{2\delta r}{r_0}\right) \\ &= -\frac{1}{\rho_0 r_0^2}\frac{d\delta P}{dr_0} + \frac{g_0}{r_0^2}4\frac{\delta r}{r_0},\end{aligned} \tag{11.31}$$

where we have canceled the equilibrium terms. This can be written more simply as

$$\frac{d\delta P}{dr_0} = 4\rho_0 g_0\frac{\delta r}{r_0} - \rho_0\frac{r_0^2}{r^2}\frac{d^2r}{dt^2}. \tag{11.32}$$

Now, from equation (11.12),

$$\begin{aligned}\frac{d\delta P}{dr_0} &= \frac{d}{dr_0}\left\{P_0\left[-3\gamma\left(\frac{\delta r}{r_0}\right) - \gamma r_0\frac{d}{dr_0}\left(\frac{\delta r}{r_0}\right)\right]\right\} \\ &= \frac{dP_0}{dr_0}\left[-3\gamma\frac{\delta r}{r_0} - \gamma r_0\frac{d}{dr_0}\left(\frac{\delta r}{r_0}\right)\right] \\ &\quad + P_0\left[-3\gamma\frac{d}{dr_0}\left(\frac{\delta r}{r_0}\right)\right. \\ &\quad \left. - \gamma\frac{d}{dr_0}\left(\frac{\delta r}{r_0}\right) - \gamma r_0\frac{d^2}{dr_0^2}\left(\frac{\delta r}{r_0}\right)\right] \\ &= -g_0\rho_0\left[-3\gamma\frac{\delta r}{r_0} - \gamma r_0\frac{d}{dr_0}\left(\frac{\delta r}{r_0}\right)\right] \\ &\quad + P_0\left[-4\gamma\frac{d}{dr_0}\left(\frac{\delta r}{r_0}\right) - \gamma r_0\frac{d^2}{dr_0^2}\left(\frac{\delta r}{r_0}\right)\right] \\ &= -\gamma P_0 r_0\frac{d^2}{dr_0^2}\left(\frac{\delta r}{r_0}\right) \\ &\quad + (\gamma r_0 g_0\rho_0 - 4\gamma P_0)\frac{d}{dr_0}\left(\frac{\delta r}{r_0}\right) \\ &\quad + 3\gamma g_0\rho_0\left(\frac{\delta r}{r_0}\right).\end{aligned} \tag{11.33}$$

*This will be discussed in Chapter 21.

Putting this into equation (11.32) gives us

$$\frac{d^2}{dt^2}\left(\frac{\delta r}{r_0}\right) + \left(\frac{4}{r_0} - \frac{g_0\rho_0}{P_0}\right)\frac{d}{dr_0}\left(\frac{\delta r}{r_0}\right)$$

$$- (3 - 4/\gamma)\frac{g_0\rho_0}{P_0 r_0}\left(\frac{\delta r}{r_0}\right) = \frac{r_0\rho_0}{\gamma P_0 r^2}\frac{d^2r}{dt^2}. \quad (11.34)$$

A regularly pulsating star, such as a Cepheid or RR Lyrae variable, must pulsate with the same period at all levels. This fact enables us to simplify the right-hand side of (11.34), for if the angular frequency of pulsation is ω, then we must have, from (11.22),

$$\frac{d^2r}{dt^2} = \frac{d^2}{dt^2}\delta r = -\omega^2\delta r. \quad (11.35)$$

Hence

$$\frac{r_0\rho_0}{\gamma P_0}\frac{1}{r^2}\frac{d^2r}{dt^2} = -\frac{\omega^2 r_0\rho_0}{\gamma P_0}\frac{\delta r}{r^2}$$

$$= -\frac{\omega^2\rho_0 r_0^2}{\gamma P_0 r^2}\frac{\delta r}{r_0} = -\frac{\omega^2\rho_0}{\gamma P_0}\frac{\delta r}{r_0} \quad (11.36)$$

to first order in $\delta r/r$. Hence equation (11.34) can be written in the final linearized form that describes small adiabatic pulsations about a mean state of hydrostatic equilibrium:

$$\frac{d^2}{dr_0^2}\left(\frac{\delta r}{r_0}\right) + \left(\frac{4}{r_0} - \frac{g_0\rho_0}{P_0}\right)\frac{d}{dr_0}\left(\frac{\delta r}{r_0}\right)$$

$$+ \left[\frac{\omega^2\rho_0}{\gamma P_0} - (3 - 4/\gamma)\frac{g_0\rho_0}{P_0 r_0}\right]\left(\frac{\delta r}{r_0}\right) = 0. \quad (11.37)$$

This basic equation describes the amplitude of the pulsation, $\delta r/r_0$, as a function of r_0, subject, of course, to the relevant boundary conditions, which are that the perturbations in δr vanish at the origin, and that the perturbations in the pressure δP vanish at the surface,

$$\delta r = 0, \qquad r = 0,$$

$$\delta P = -\Gamma_1 P\left(3\xi + r\frac{d\xi}{dr}\right) = 0, \qquad r = R, \quad (11.38)$$

where $\xi \equiv r/r_0$.

Remember that the actual radius is a function of time when pulsations occur, and that the quantities entering into the middle expression for δP corresponds to adiabatic motion. Although δr is small compared to r_0 at each point, the net displacement δR may be a substantial fraction of the equilibrium radius.

Problem 11.5. The mass and mean radius of a typical Cepheid variable are given by $\log M/M_\odot = 0.8$ and $\log R/R_\odot = 1.4$. Show that $\delta r/r_0$ = constant satisfies (11.37)–(11.38), and find the period and surface velocity of the star. What is the range in effective temperature T_{eff} assuming $\delta r/r_0 = 0.1$?

11.3. Stellar Stability

The simple model developed in the beginning of Section 11.2 produced the period-density relation (11.24) and stability criterion $\gamma > 4/3$ in terms of average quantities $\bar{P}$, $\bar{\rho}$, and γ. In real stars, the adiabatic index will vary throughout, and the way in which it varies with position may have important consequences for the star's behavior. We therefore return to the issue of stability from a more general standpoint.

The stability of a system against small radial perturbations described by (11.14) depends on the nature of the roots of (11.15). In general, $\omega = i\omega_1 + \omega_2$, where ω_1 and ω_2 are real. The physical displacement will then be given by the real part of $\delta r/r$. When ω_1 is nonzero, the perturbation has an oscillatory part for either sign of ω_1. Whether the system is stable depends in part on the sign of ω_2. If $\omega_2 = 0$, then higher-order terms must be retained in the analysis leading to (11.15). Stability in linear theory results when $\omega_2 < 0$; for finite $\xi(r)$, $\delta r/r_0 \simeq e^{-|\omega_2|t}$ will then decay with time and small perturbations damp out. Instability develops when $\omega_2 > 0$, since initial disturbances then grow with time. However, as we remarked earlier, $\omega_2 > 0$ need not necessarily lead to unbounded motion, since nonlinear effects may ultimately limit their growth. Therefore, when $\omega_2 > 0$, we might expect either unbounded growth or growth to a finite though possibly large amplitude.

The stability analysis using the perturbations $\delta r/r_0 = \xi(r)$ obtained from (11.37) is equivalent to an eigenvalue problem, and the corresponding modes are similar to those familiar from elementary physics. The fundamental mode is the one in which the perturbations vanish at the center and reach maximum values at or near the surface (compare a standing sound wave in a pipe with one open end). The wavelength of the disturbance is therefore $\sim 4R$. Overtones correspond to the existence of one or more surfaces within the star where the perturbations vanish.

A general analysis of the roots of (11.29) is complicated. But some headway is possible if the motion is restricted. The simplest example is given by adiabatic motion in the fundamental mode. In this case $\delta r/r$ grows gradually from the core to the surface of the star. As a first approximation to this situation, we may assume homologous motion with $\xi(r)$ constant. We then find that (11.15) reduces to the integral for the frequency

$$\omega_1^2 = \int_0^R (3\Gamma_1 - 4)3P\,dV \Big/ \int_0^M r^2\,dm, \qquad (11.39)$$

where P is the unperturbed pressure, and Γ_1 is a function of position in the star. When $\omega_1^2 > 0$, the system is stable, since $\delta r/r$ does not grow with time (the incorporation of nonadiabatic effects would presumably lead to decay in the oscillatory motion). Now, define the average of Γ_1 over the star so that

$$\omega_1^2 = \langle 3\Gamma_1 - 4 \rangle \int_0^R 3P\,dV \Big/ \int_0^M r^2\,dm. \qquad (11.40)$$

The denominator is the generalized moment of inertia I; the numerator may be rewritten by means of the equation of hydrostatic equilibrium—since all quantities appearing in the integrals in (11.40) are zero order—and an integration by parts,

$$\begin{aligned} \int_0^R 3p\,dV &= 3\int_0^R 4\pi r^2 P\,dr \\ &= -\int_0^R 4\pi r^3 \frac{dP}{dr}\,dr \\ &= \int_0^R \frac{mG}{r^2} 4\pi r^3 \rho\,dr \\ &= \int_0^M \frac{mG}{r}\,dm = |\Omega|. \qquad (11.41) \end{aligned}$$

Therefore, the frequency of the fundamental mode is given by

$$\omega_1^2 = \langle \Gamma_1 - 4/3 \rangle \frac{3|\Omega|}{I}. \qquad (11.42)$$

Since ω_1^{-1} gives the time-scale for these oscillations, it follows that

$$\tau \sim \omega_1^{-1} \sim (\bar{\rho}G)^{-1/2}, \qquad (11.43)$$

which is a dynamical time-scale. In effect (11.43) follows from, and is equivalent to, the period-density relation (11.24).

It is important to note that stability depends on the volume-averaged value of Γ_1. Whenever Γ_1 falls below 4/3, as happens in extensive ionization zones in some stellar envelopes, instability may result. Conversely, a narrow region of ionization (occurring, for example, when the pressure is low) may not be sufficient to cause instability. We will see shortly that the instabilities believed to be responsible for Cepheids and RR Lyrae, and their restriction to a narrow region in the Hertzsprung-Russell diagram, are related to this point. Instabilities are particularly important in red giants and supergiants, where both matter and radiation contribute to the equation of state. Therefore, the adiabatic indices Γ_1 will be between 5/3 and 4/3. Although the presence of a radiation component will reduce Γ_1, it can at most move it toward a value $\gtrsim 4/3$, and will not in itself produce instability. When ionization occurs, however, Γ_1 will be reduced well below 4/3 (Chapter 2), and instabilities for the star as a whole will become more likely.

Other sources of instability are also important. For example, the dissociation of molecular hydrogen and the ionization of atomic hydrogen and helium probably play an important role in reducing the value of Γ_1 below 4/3 and initiating rapid collapse in protostars. The onset of convection in the tenuous, cool regions of some red giants and supergiants may also reduce Γ_1. In addition to their role in pulsational stars, processes such as these that reduce Γ_1 also contribute to gradual mass loss and possibly to planetary nebula formation. During advanced stages of stellar evolution, matter in the stellar core may undergo nuclear photodisintegration. In this type of process, elements in the Fe group dissociate to form He^4, which in turn may disintegrate to form nucleons. Each such process can effectively lower Γ_1 below 4/3. The resulting instability, it has been suggested, may lead to supernovae and the formation of neutron stars. In massive stars, temperatures may be high enough to induce electron-positron pair production. Subsequent decay of pairs into electron neutrinos will lower Γ_1 and initiate instability. At higher densities typical of white dwarfs, electron capture (inverse β-decay) can reduce Γ_1 throughout a significant volume, reducing the maximum allowed mass of a white dwarf at zero temperature. The decay of less massive hyperons to produce more massive ones with less Fermi energy brings about the same effect in the core of neutron stars, and contributes to their upper mass limit. Finally, the instabilities associated with $\Gamma_1 < 4/3$ discussed in the next section set upper bounds to the masses of stable main-sequence stars.

11.4. Pulsational Stability

A system in which the volume-averaged Γ_1 is greater than 4/3 is dynamically stable. If the system is given a small perturbation, the resulting motion will either decay (dissipation) or grow in time (negative dissipation). This raises the issue of pulsational stability. In a star that is pulsationally stable, Γ_1 is greater than 4/3 throughout enough of its volume that $\omega_1{}^2 > 0$, but the star may have regions where perturbations can grow to a large but finite amplitude. The star's gross properties will therefore be functions of time, but its time-averaged state will be one of hydrostatic equilibrium.

Bearing in mind that the linearized theory does not guarantee stability, but can uncover instability unless it sets in only above a certain amplitude, we can attack the issue of pulsational stability by using (11.15), which gives the frequency of radial perturbations $\delta r/r$. Denote the solutions to the cubic equation for ω by

$$\omega = i\omega_1 + \omega_2. \tag{11.44}$$

According to Section 11.3, ω_1 determines the stability of oscillatory motion (dynamic stability). When the volume-averaged value of Γ_1 exceeds 4/3, ω_1 is real. If ω_2 is nonzero, it causes changes in $\delta r/r$, which decrease the amplitude with time when ω_2 is negative, and increase it when ω_2 is positive. The physical processes that accomplish this are dissipative in nature, and normally act against the motion set up by perturbations. Sometimes, as is particularly important for variable stars, the dissipation is out of step with the perturbations, and acts effectively to amplify the motion. Phase delay is found to be associated with the energy-generating and energy-transferring properties of the medium, as will be seen shortly.

Dissipation is associated with entropy production, and therefore involves nonadiabatic processes. When nonadiabatic effects are included in the analysis of (11.15), it can be shown that

$$\omega_2 \simeq \frac{\int_0^R \left(\frac{\delta T}{T}\right)_0 \left(\epsilon_n - \frac{1}{\rho}\nabla \cdot \mathcal{F}\right)' dm}{2\omega_1{}^2 \int_0^M |\xi|^2 r^2\, dm} \tag{11.45}$$

as long as the nonadiabatic effects are small. The analysis leading to (11.45) is involved, but can be clarified by the following physical arguments. In effect, we wish to see how a gas element in a pulsating star can act as a heat engine. Let us assume adiabatic motion, and note that if the pulsations are to be steady, the element's internal energy must remain constant from one pulsational cycle to the next cycle. Then the internal energy U and entropy S of a given mass element to be followed during pulsations satisfy

$$\oint dU = 0, \qquad \oint dS = 0. \tag{11.46}$$

The integrals are over a complete dynamic cycle of the mass-element motion. The first law of thermodynamics applied to a gas element can be written as

$$dU = dQ - dW_s = T\,dS - dW_s, \tag{11.47}$$

where dW_s is the work done by the gas element on its surroundings, and would equal $+\ p\,dV$. Using (11.46)–(11.47), we immediately find that the total work done by the star in a single cycle is

$$\oint dt \int dm \frac{dW_s}{dt} = \oint dt \int dm \frac{dQ}{dt} \equiv \overline{W}, \tag{11.48}$$

with the mass integral running over the entire system. Next consider the entropy. From (11.46) we obtain, for the mass element during a complete cycle,

$$\begin{aligned} 0 = \oint ds &= \oint \frac{dS}{dt}\,dt \\ &= \int dt \frac{1}{T}\frac{dQ}{dt} \simeq \oint dt \frac{1}{T_0}\left(T - \frac{\delta T}{T_0}\right)\frac{dQ}{dt}. \end{aligned} \tag{11.49}$$

The last step supposes that the temperature, which varies periodically with time, can be written as $T(t) \simeq T_0 + \delta T(t)$, with T_0 characterized by the equilibrium configuration, and $\delta T/T_0$ small. Since (11.49) vanishes,

$$\oint dt \frac{dQ}{dt} \simeq \oint dt \frac{\delta T}{T_0} \frac{dQ}{dt}.$$

Integrating over the star, and comparing with (11.48), we find

$$\overline{W} = \oint dt \int dm \frac{\delta T}{T_0} \frac{dQ}{dt}. \qquad (11.50)$$

For a pulsating star, both $\delta T/T_0$ and dQ/dt will vary periodically with time. If the mass element is to act as a heat engine and perform work on its surroundings, then $\overline{W}$ must be positive. Therefore dQ/dt, which represents the rate of heat transfer to the element, must be positive when δT is largest, and negative when δT is smallest. In other words, heat energy must be transferred to the element during compression, and lost during expansion. According to (11.11), the rate of heat transfer per element is

$$\frac{dQ}{dt} = T\frac{ds}{dt} = \left(\epsilon_N - \frac{1}{\rho}\nabla \cdot \mathcal{F}\right)_0 + \left(\epsilon_N - \frac{1}{\rho}\nabla \cdot \mathcal{F}\right)'.$$

The first term, which corresponds to the equilibrium state, vanishes; so dQ/dt depends entirely on the last term. Substitution into (11.50) yields $\overline{W}$ in terms of the perturbed heat flux, and its time derivative yields

$$\frac{d\overline{W}}{dt} = \oint dm \frac{\delta T}{T_0}\left(\epsilon_N - \frac{1}{\rho}\nabla \cdot \mathcal{F}\right)'. \qquad (11.51)$$

Next we consider the average kinetic energy associated with the pulsations. The velocity of a mass element at point r undergoing pulsations of magnitude δr with period P is

$$v \simeq 2\pi\delta r/P. \qquad (11.52)$$

If we assume that P is constant throughout the star, the pulsational kinetic energy must be proportional to

$$\begin{aligned}\frac{1}{2}\int \rho v^2\, dV &= \frac{1}{2}\int_0^R 4\pi r^2 \rho\, dr \left(\frac{2\pi}{P}\right)^2 \delta r^2 \\ &= \frac{1}{2}\left(\frac{2\pi}{P}\right)^2 \int_0^M \left(\frac{\delta r}{r}\right)^2 r^2\, dm \\ &= \frac{\Omega^2}{2}\int_0^M \left(\frac{\delta r}{r}\right)^2 r^2\, dm, \end{aligned} \qquad (11.53)$$

where $(2\pi/P) = \Omega$ is the pulsation frequency. The rate at which pulsational work is done is given by (11.51). The ratio of $d\overline{W}/dt$ to the kinetic energy stored in pulsation is proportional to the inverse of the time required for a significant change in pulsational energy to occur. Denoting the inverse time by ω_2,

$$\omega_2 \sim \frac{\int_0^R dm\,(\delta T/T_0)\left(\epsilon_N - \frac{1}{\rho}\nabla \cdot \mathcal{F}\right)'}{\omega_1^2 \int_0^M r^2 |\xi|^2\, dm}$$

with $(\delta r/r)^2 = |\xi|^2$ and $\Omega = \omega_1$. This is just (11.45) to within a constant. Note that all quantities appearing in ω_2 are evaluated adiabatically in the equilibrium state, and ω_1^2 is positive if the system is to be dynamically stable.

Problem 11.6. Write the energy-generation rate as

$$\epsilon_N = \epsilon_0 \rho^\alpha T^\beta = \epsilon_N(\rho, T),$$

with α and $\beta > 0$. Denoting the variation $\epsilon_N' = \delta\epsilon_N$, use this equation to show that nuclear-energy generation always tends to destabilize a star pulsationally. Note that the integrand in (11.45) is evaluated for adiabatic motion.

Return now to the stability of the pulsations. When $\Gamma_1 > 4/3$, ω_1^2 is positive, and the sign of ω_2 depends entirely on the sign of the integral in the numerator. As shown by Problem 11.6, nuclear-energy generation contributes a term to (11.45) that is always negative. Equations (11.14) and (11.44) indicate then that this leads to a growth in the amplitude of pulsations. The second term in (11.45) is more complicated, but the physical effect can be illustrated as follows.

The contribution of the last term depends essentially on the behavior of the opacity during the pulsations. It may be approximated as

$$\kappa = \kappa_0 \rho^a T^b, \qquad (11.54)$$

where $a \simeq 1$, and b may be either positive (H^- opacity in the outer, cool layers of a star) or negative (Kramers' opacity). During adiabatic pulsations, the contributions from the luminosity increase as the opacity decreases, and vice versa. The variation in κ is given by

$$\delta\kappa = \kappa[a\delta\rho/\rho + b\delta T/T]$$

$$= \kappa[a + b(\gamma - 1)]\frac{\delta\rho}{\rho}, \qquad (11.55)$$

if we assume adiabatic motion. When $\gamma \geq 4/3$ in the stellar interior and Kramers' opacity applies, then $a + b(\gamma - 1)$ is negative: $\delta\kappa/\kappa \sim -\delta\rho/\rho$. It usually follows that compression and an increase in density reduce the opacity. The heat flux from the element increases, and it tends to cool; this thermal dissipation causes the pulsations to decay. However, when γ approaches unity, $\delta\kappa/\kappa \simeq \delta\rho/\rho$. Here an increase in density during compression will increase the opacity, reducing the heat flux from the element. Energy is then stored thermally in the matter, to be released during expansion stages. The result is negative dissipation, which causes the amplitude of the pulsations to grow. The mechanism that results when $\gamma \simeq 1$ is known as the *valve mechanism,* and represents one way in which a star drives pulsations like a heat engine.

Problem 11.7. Show that $\tau \sim 1/\omega_2$ associated with pulsations is the Kelvin time-scale given by $L/|\Omega|$, where L is the star's luminosity and Ω is its gravitational potential energy.

It follows from Problem 11.7 and the discussion of ω_1 that $\tau_D \sim 1/\omega_1 \ll \tau_p \simeq 1/\omega_2$. This usually guarantees that the system can remain dynamically stable while exhibiting steady pulsations. In effect, the star makes necessary small adjustments on the time-scale τ_D in order to remain stable and accommodate the pulsations that result from processes taking place on the longer time-scale τ_p.

Problem 11.8. Consider a thin spherical shell of radius r and thickness Δr in the envelope of a star. Assuming that the mass density of an element of volume $(\Delta r)^3 = \rho/\Delta m$, where Δm is the element's mass, and that the fractional change in temperature across the shell is large, show that the shell luminosity can be written as

$$L = \frac{4ac}{3}\frac{m_s}{\Delta m^{2/3}}\frac{T^4}{\kappa\rho^{4/3}} = \frac{T^4}{\kappa\rho^{4/3}} \cdot \text{constant}, \qquad (11.56)$$

where $m_s = 4\pi r^2 \Delta r$ is the shell mass. Assume radiative transport.

Problem 11.9. Denote the radiative flux gradient by

$$\nabla \cdot \mathcal{F} = \frac{1}{4\pi r^2}\frac{dL}{dr}. \qquad (11.57)$$

Consider the mass shell in a star as described in Problem 11.8. Assume that the shell and element masses are constant during adiabatic motion, and that the fractional change in luminosity across the shell is large. Show that the second term in the integrand of (11.45) stabilizes the star against pulsations when $\gamma > 4/3$, and destabilizes the star when $1 \lesssim \gamma < 4/3$. The opacity may be taken to be Kramers', with $a \simeq 1$ and $b \simeq -3.5$. Note that the integrand is to be evaluated for adiabatic motion, with $T \sim \rho^{\gamma-1}$.

Two mechanisms limit the effectiveness of thermonuclear reactions and opacity in driving pulsations. As noted in the discussion following (11.50), pulsations may result if the integrand appearing in (11.51) is positive over enough of the mass distribution to make $d\overline{W}/dt > 0$. Thermonuclear reactions tend to contribute in this direction (in linear theory), as does opacity when it is approximately like Kramers' opacity, and Γ_1 approaches unity.

The situation is not so simple, however. First, the effectiveness of nuclear processes may be limited by thermal dissipation, which converts mechanical energy into heat. Since nuclear reactions occur only in the relatively dense high-temperature stellar core (or shells surrounding the core), where the characteristic length of thermal diffusion is small, pulsations tend to be highly adiabatic and dissipative. Therefore, the first term in (11.51) tends to be small in general. This leaves the opacity (valve mechanism) as the most likely source of pulsations. In order for it to be effective, however, Γ_1 must approach unity throughout enough of the star to make $d\overline{W}/dt > 0$. As has already been noted, this happens in red giants, which have relatively deep H and He ionization zones. The characteristic properties of Cepheid and RR Lyrae variables are in part related to the location of these ionization zones in the star.

11.5. Classical Cepheid and RR Lyrae Variables

The classical Cepheids, and the RR Lyrae or cluster variables, are the best understood theoretically of the regularly pulsating stars; so we will limit our discussion to these objects. The most interesting properties of these stars are (1) the origin of the instability, and the maintenance of regular pulsations; (2) the observed period-luminosity relation; and (3) the restriction of these objects in the HR diagram to a narrow, nearly vertical band that terminates near the main sequence. The first of these properties has already been considered in some detail in the preceding sections; so we deal here with only those aspects that have not been covered. Investigations indicate that the deep stellar interior contributes little to the pulsations of Cepheids and RR Lyrae variables; so most models assume a core structure, and then follow the dynamics in the envelope.

In some stars where ionization zones are near the surface, or in stars that are extremely cool, convection may be the dominant means of energy transport, which would require serious modifications in the approach we have been describing. Pulsations in convective regions are not understood as well as those in radiative ones; so we will consider radiation to be the dominant means of energy transport.

Period-Luminosity Relation

A simple period-luminosity law for homologously pulsating stars may be derived from the period-density relation (11.24). Comparing this with our empirically derived relation yields valuable information about Cepheids and RR Lyrae variables. For homologous pulsations, the period $P \sim \rho^{1/2}$ and the density is $\rho \sim M/R^3$. The radius may be eliminated by using the definition of the effective temperature

$$T_{\text{eff}} = (L/4\pi\sigma R^2)^{1/4}$$

to show that

$$\log P = -\frac{1}{2}\log M + \frac{3}{4}\log L - 3\log T_{\text{eff}} + C, \qquad (11.58)$$

where C is a constant. The theory of stellar structure for homologous stars may then be used to express M and T_{eff} in terms of the luminosity. The result gives $P(L)$, which may be compared with observations. In this way one can in principle check the results of models of variable stars.

Problem 11.10. Adopt a homologous model of a star in which radiation pressure is negligible, and in which the CNO cycle supplies the energy. Use it to eliminate M and T_{eff} from (11.58), and obtain a period-luminosity relation. Indicate your assumptions, and specify a procedure to find a value for the constant term.

Next we consider observationally derived relations $P(M_V)$, where M_V is the absolute visual magnitude of the variable. Return to (11.58) and reexpress it in solar units:

$$\log P = -\frac{1}{2}\log(M/M_\odot) - 3\log(T/T_\odot) - 0.3(M_b - M_{b,\odot}) + \log Q, \qquad (11.59)$$

where temperatures are effective surface temperatures, M_b denotes bolometric magnitude, and Q is a constant. For bright main-sequence stars ($-8 < M_b < +1$) the mass-luminosity relation is well fit by

$$M_b = 3.96 - 8.22\log(M/M_\odot). \qquad (11.60)$$

This may be modified to apply to Cepheids as follows. Assuming that their evolutionary tracks are similar to those of other (nonvariable) stars in the giant region of the HR diagram, we can argue that for the same mass a Cepheid will be about one magnitude brighter than the corresponding main-sequence star. Then (11.60) becomes for Cepheids

$$M_b \simeq 2.96 - 8.22\log(M/M_\odot). \qquad (11.61)$$

Similar arguments can be used to show that the bolometric magnitude M_b and the visual magnitude M_V are related for Cepheids by

$$M_b = M_V + 0.145 - 0.322(B - V), \qquad (11.62)$$

and that the effective temperature T is given approximately by

$$\log T_{\text{eff}} = 3.886 - 0.175\,(B - V). \qquad (11.63)$$

In (11.63), $0.4 < B - V < 1.0$ is assumed. When (11.61)–(11.63) are used in conjunction with $T_{\text{eff},\odot} = 5{,}754\,K$, the period (11.59) may be written as

$$\log P + 0.239\,M_V = 0.602\,(B - V) - 0.456. \qquad (11.64)$$

The constant has been eliminated by fitting the result to Cepheids of known color and luminosity ($\log Q = -1.294$). The relation (11.64) shows that the period does depend on the luminosity (through the magnitude), but also varies with color. Fitting observed periods and magnitudes (averaged over color), we find that

$$\log P + 0.394\,M_V = -0.657. \qquad (11.65)$$

This describes the central region of the Cepheid strip in the HR diagram. The simple homologous model leading to $P(L)$ in Problem 11.10, and the empirical result (11.64), show dependence on color, which indicates that not all the scatter in the variable-star strip is due to observational uncertainty. Comparing the observed relation (11.51) with (11.50), which contains the term $\log Q$ [see (11.59)], shows that

$$\log Q = 0.241 \log P - 1.50. \qquad (11.66)$$

This result demonstrates that Q is, in fact, not a constant, as was assumed in arriving at (11.59). The source of the difficulty may be traced to the assumption of homologous structure. The fact that Q depends only weakly on period shows, however, that deviations from homologous motion are probably not large.

This simple analysis demonstrates three important facts about Cepheids: (1) whether or not the star pulsates homologously, the period depends on luminosity and weakly on the color index $B - V$; (2) the dependence on color gives a spread in the HR diagram for variables having the same period; (3) pulsations are not strictly homologous.

Pulsations in Cepheids and RR Lyrae Stars

The most dramatic properties of regular variables are similar to those of δ Cephei, shown in Figure 11.1. The brightness varies regularly (through not sinusoidally), with a period of 5.37 days; the variations relative to mean values are about 10 percent in magnitude, 18 percent in temperature, and 7 percent in radius. Cepheid variables typically show a rapid rise in light, followed by a gradual decline; maximum brightness and temperature coincide but are slightly out of phase with minimum radius. As shown in Figure 11.1, δ Cephei reaches maximum brightness slightly after minimum radius during the expansion phase. RR Lyrae light curves contain a secondary maximum, and have shorter periods than those of classical Cepheids.

Problem 11.11. The ratio of maximum to minimum radius $\eta = 1.14$ for δ Cephei, and the amplitude of the radial pulsational velocity is 19 km/s. Assume that $v(t) = v_0 \sin 2\pi t/P$ and find the average radius of the star. Suggest a way that η could be found observationally.

Extensive numerical studies of RR Lyrae variables indicate that deep HeII and HI ionization zones supply the driving mechanism in these stars, and that nuclear-energy generation plays no role. The stellar envelope is divided into mass zones, and the mechanical work done per cycle by each mass shell,

$$W_i = \oint P\,dV_i = \oint_{\Delta m_i} dt \left(\frac{dT}{T_0}\right)\left(\epsilon_N - \frac{1}{\rho}\nabla \cdot \mathcal{F}\right)', \qquad (11.67)$$

is evaluated. As discussed in Section 11.3, the system is unstable to pulsations when the sum of W_i over mass zones is positive. The ratio of mechanical work to pulsational kinetic energy for a typical RR Lyrae model is shown in Figure 11.2. The envelope consists of a neutral-hydrogen surface zone below the atmosphere, and below it a region of singly ionized helium HeII. Below the HeII layer, all matter is essentially ionized. The integrals for W_i for several representative zones are shown in Figure 11.3 as the area enclosed by each curve for $P(V)$. The sign of W_i is given by the sense of motion in the P,V-plane during each cycle, positive when the motion is clockwise, and negative when it is counterclockwise. The numbers next to each curve label the mass zones, increasing outward. The innermost zones, such as zone 22, are nearly adiabatic,

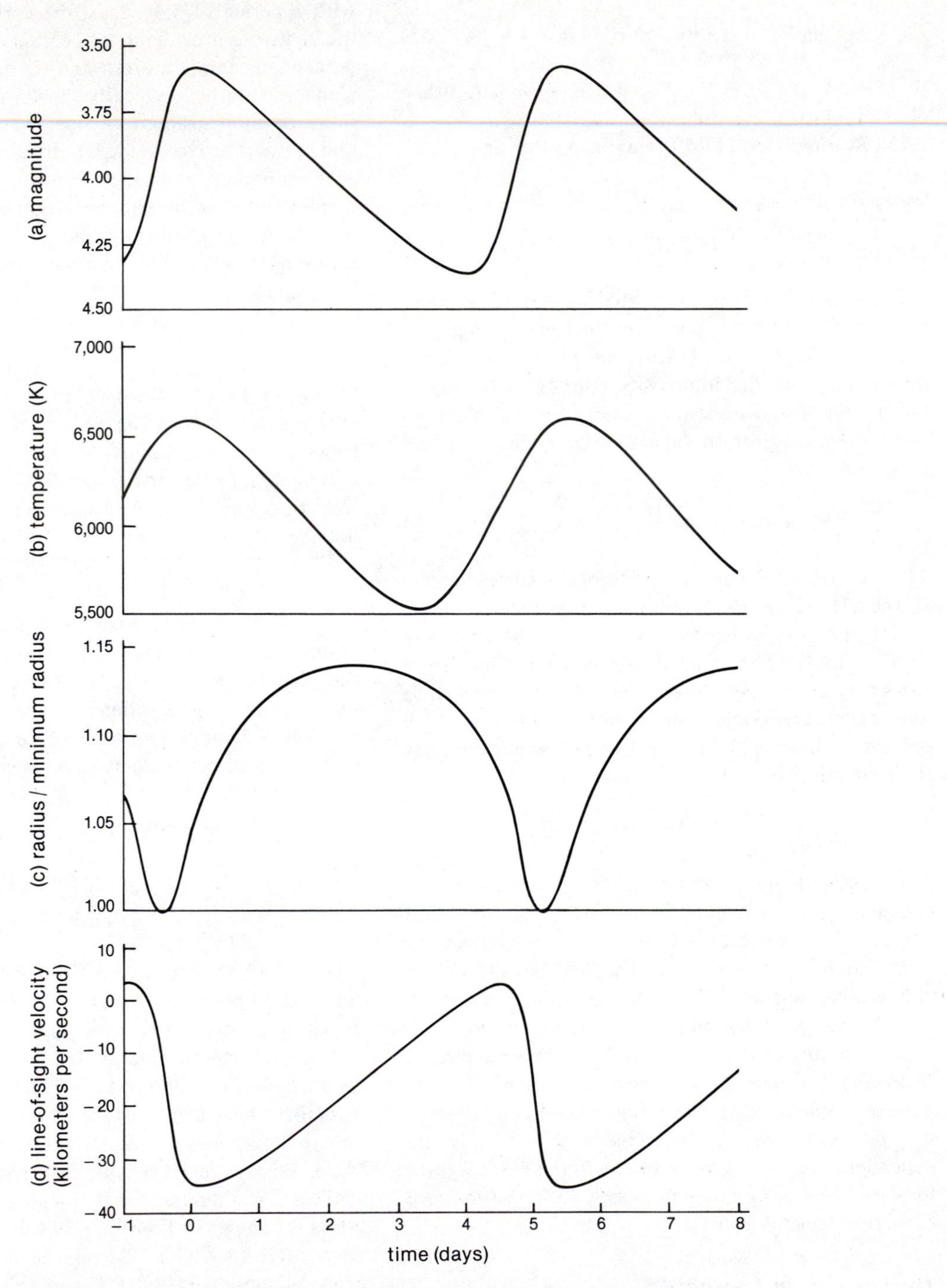

Figure 11.1. Periodic variations in δ Cephei versus time in days: (a) magnitude; (b) effective temperature in K; (c) ratio of instantaneous radius to minimum radius; (d) line-of-sight velocity (km/s) obtained from line spectra.

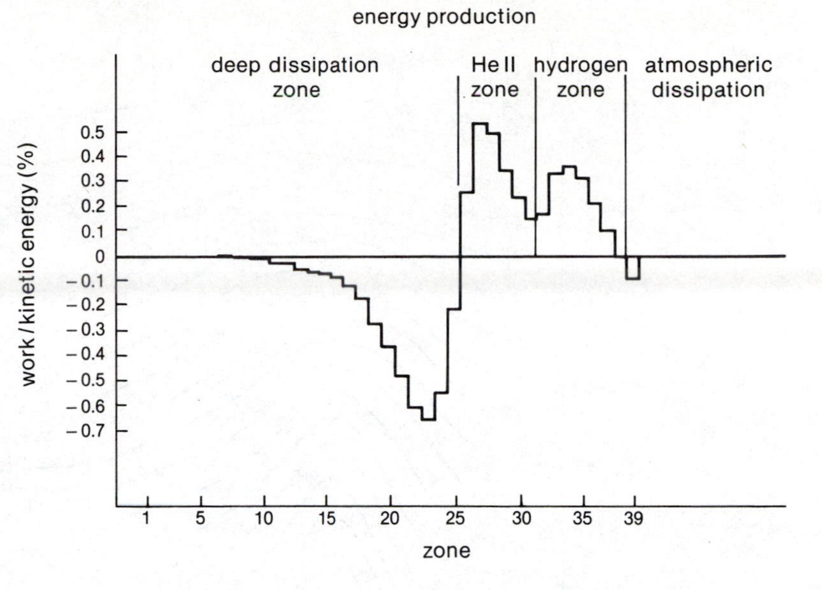

Figure 11.2 Ratio of mechanical work done per cycle to kinetic energy of pulsations in a numerical model of RR Lyrae variables. Dissipation occurs in the inner envelope (core not shown) and in the atmosphere.

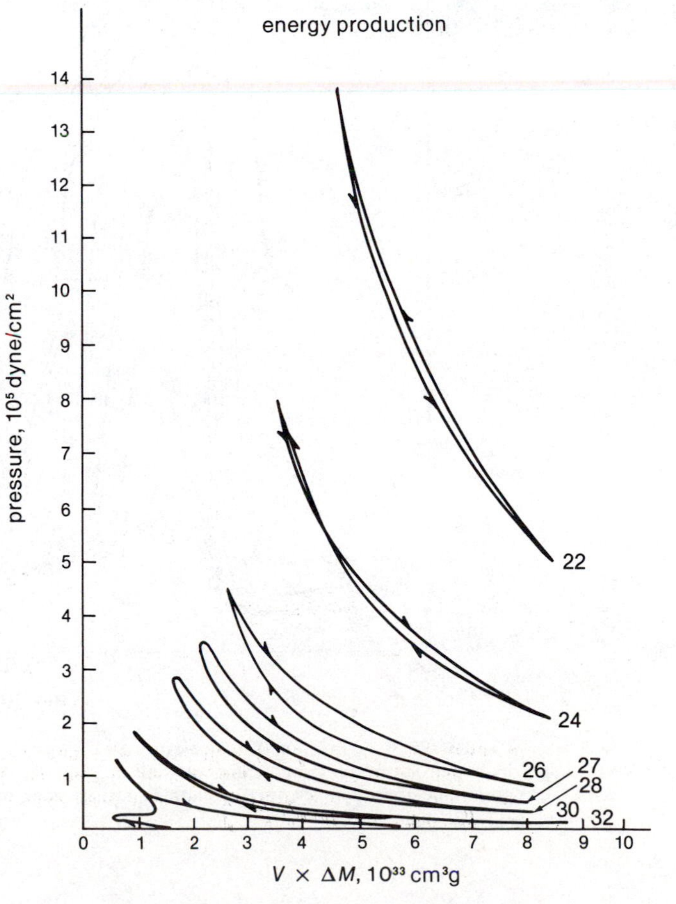

Figure 11.3. Pressure versus volume cycles for selected mass shells within a model of a RR Lyrae variable. Clockwise cycle performs positive mechanical work equal to the enclosed area.

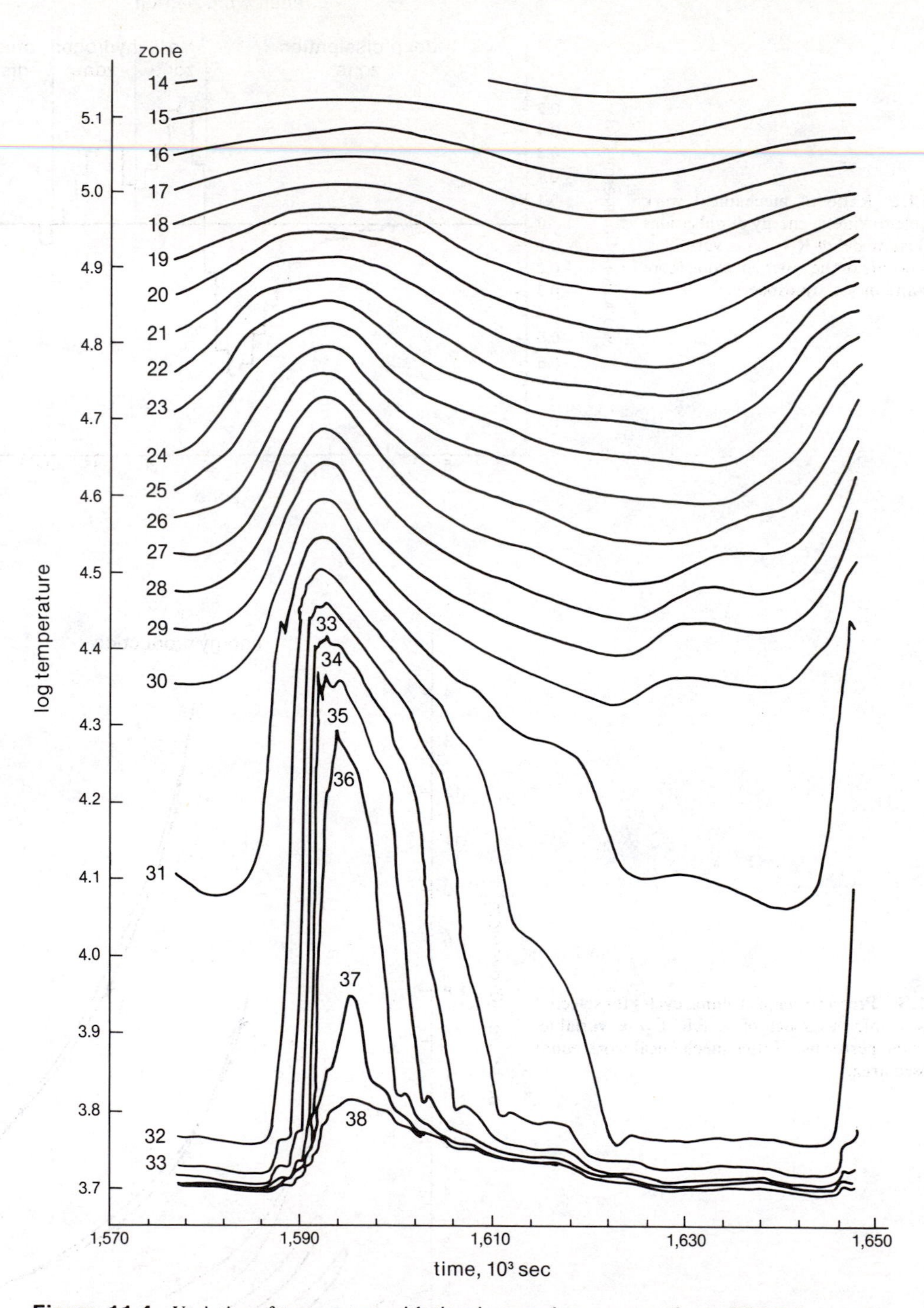

Figure 11.4. Variation of temperature with time in several mass zones of a model RR Lyrae. The hydrogen ionization zone centers at shell 33, and the HeII ionization zone centers at shell 27. The amplitude of the temperature oscillations in these zones is larger than the quiescent temperature in the static model.

and are dissipative. Nonlinear effects become quite important as the HeII and HI zones are approached. In these regions, $\gamma \rightarrow 1$, and the opacity drives pulsations as described in Section 11.4.

Although the inner zones ($i \lesssim 25$) contribute negative values to the net work, the sum over all mass zones is positive, with HI and HeII contributing about one-third and two-thirds, respectively, of the positive work done during each cycle. The work done is about 6.3×10^{38} ergs/cycle, which amounts to about 7 percent of the star's net luminosity. A finite amplitude is attained when the net thermal dissipation equals the net mechanical work done. The temperature of each zone varies by a factor of about five in the HI zone, but by only a factor of 1.6 in the HeII zone. The slight delay between maximum light and minimum radius appears to develop in the HI region, as is seen in the plot of temperature versus time in Figure 11.4.

Instability Region

Observations show that Cepheids and RR Lyrae variables lie within a narrow band ($B - V$ width about three or four magnitudes), starting above the main sequence at absolute magnitude M_V about zero and spectral type about A5, and running up to $M_V \simeq -6$ and spectral type about G5. The existence of a narrow instability band may be explained in part by the following observations.

First, consider stars of the same mass and luminosity that fall on a horizontal line in the HR diagram. Those lying to the left are hot, and the H and He ionization zones lie in the outermost part of the envelope, where there is very little mass. Consequently they can not contribute significantly to the mass integral for ω_2 in (11.45). Moving to the right, the ionization zones move toward the inside. For low temperatures, they all lie deep within the adiabatic core, where stable pulsations do not arise when energy transport is radiative. For a narrow range of intermediate temperatures, one or both of the ionization zones will lie midway in the envelope, where opacity can drive pulsations. Repeating the arguments for a larger mass and hence larger luminosity, we obtain another band above the previous one. In this way the instability strip is developed.

11.6. Unstable Shell Sources

It is unlikely that stars less massive than about 1.4 $M_\odot$ evolve beyond He burning; so their final stages consist of H and He shell-burning sources, surrounding a predominantly C^{12} and O^{16} core. The behavior of stable shell sources was considered in Sections 10.4 and 10.5. Unstable shell sources may lead to gradual mass loss from stars, and may play a key role in models of recurrent novae. Theoretical studies show that shell sources in highly evolved stars of low mass may become thermally unstable if the thermal diffusion time is less than the rate of nuclear-energy release, and if, as the shell expands, its pressure remains essentially constant. This will happen if the shell width l is less than the distance over which the pressure gradient changes significantly. The conditions under which an instability will develop are described as follows.

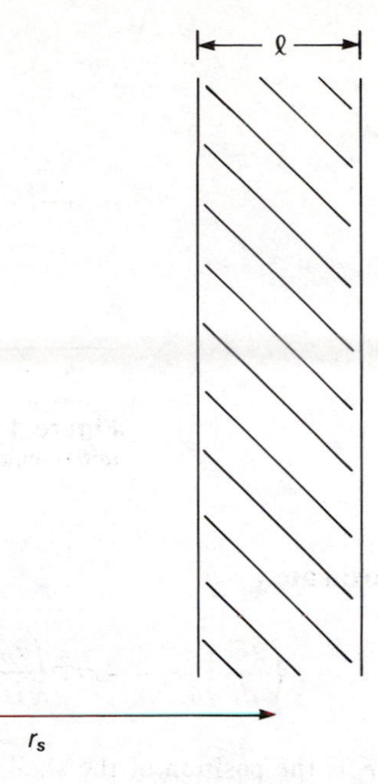

Figure 11.5. Nuclear-burning shell of thickness $l \ll r_s$.

We start by writing the energy-transport equation in the form of a diffusion equation,

$$\frac{L(r)}{4\pi r^2} = -\frac{4acT^3}{3\kappa\rho}\frac{dT}{dr} \equiv -\sigma_D\frac{dT}{dr}, \qquad (11.68)$$

where σ_D is the diffusion coefficient. Noting that energy may diffuse from the shell in either direction (Figure 11.5), we may approximate its rate of energy diffusion by $2\sigma_D T/l$. The total energy diffusing out of the shell per second is the shell's energy-loss rate, and

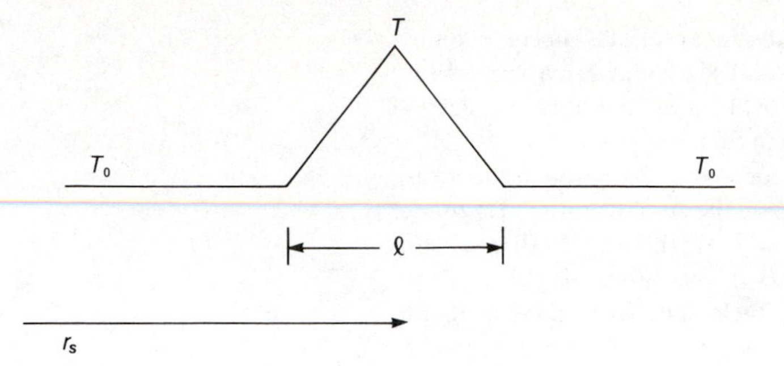

Figure 11.6. $T(r)$ near nuclear shell shown in Figure 11.5. T_0 is the initial temperature in the absence of the shell source.

is approximately

$$\left(\frac{dE_s}{dt}\right)_D \simeq -4\pi r_s^2 \left(\frac{\sigma_D T}{l/2}\right), \tag{11.69}$$

where r_s is the position of the shell, and $r_s \gg l$. If the energy-generation rate and density in the shell are ϵ and ρ, respectively, the net energy released to the shell per second by nuclear processes is of order

$$\left(\frac{dE_s}{dt}\right)_N \simeq 4\pi r_s^2 \left(\frac{l}{2}\right) \rho\epsilon. \tag{11.70}$$

Thermal instability requires that energy gains exceed losses. This leads immediately to the condition

$$(dE_s/dt)_D + (dE_s/dt)_N < 0,$$

which implies

$$\frac{\rho\epsilon l^2}{4\sigma_D T} > 1. \tag{11.71}$$

In itself (11.71) will not guarantee instability. We have shown earlier that a homogeneous star in hydrostatic equilibrium consisting of an ideal gas and nuclear-energy sources is thermally stable because

$$\frac{\delta\rho}{\rho} \simeq \frac{3}{4}\frac{\delta p}{p} \simeq 3\frac{\delta T}{T}. \tag{11.72}$$

This guarantees that heating leads to expansion, which is followed by cooling and a reduction in energy generation. Evolved stars are chemically inhomogeneous, so (11.72) need not apply. In fact, if a shell source is thin and highly temperature-sensitive (fuels with $\epsilon \sim T^n$, where $n \gg 1$, are good examples), expansion can occur with little or no change in pressure, $\delta P/P \approx 0$. Then

$$\frac{\delta\rho}{\rho} \simeq -\frac{\delta T}{T} \tag{11.73}$$

for an ideal gas, and a thermal instability will result. The simple model developed in Problem 11.12 illustrates these points.

Problem 11.12. A thin shell source of width l is located at r_s. The temperature variation across the shell is shown in Figure 11.6. Show that the luminosity at $r_s + l/2$ is given approximately by

$$L_2 = \frac{8\pi r_s^2}{l}\sigma_D(T - T_0), \tag{11.74}$$

and the luminosity at $r_s - l/2$ is $L_1 = -L_2$. Plot schematically the total luminosity $L(r)$ across the shell source, and explain the result physically.

The results of Problem 11.12 imply that the change in L with mass across the shell is roughly

$$\frac{dL}{dm} \simeq \frac{4\pi r_s^2 \sigma_D}{\Delta M/2}\frac{(T - T_0)}{l/2}, \tag{11.75}$$

where ΔM is the shell mass. Denoting the shell luminosity L_s by

$$L_s = L_2 - L_1 = 4\pi r_s^2 \sigma_D \frac{(T - T_0)}{l} \equiv \epsilon_s \Delta M, \tag{11.76}$$

we find that

$$\delta\frac{dL}{dm} = 4\frac{L_N}{\Delta M}\frac{\delta T}{T - T_0}. \qquad (11.77)$$

The dynamic implications of thermal instability follow from (5.65), which for an ideal gas is

$$\epsilon_N - \frac{dL}{dm} = \frac{3}{2}\frac{P}{\rho}\frac{d}{dt}\ln(P/\rho^{5/3}). \qquad (11.78)$$

The variation in the left-hand side due to the shell source, assuming $\epsilon_N \lesssim \epsilon_S = \epsilon_0 T^n$ and $\delta P = 0$, is

$$\delta\epsilon_N - \delta\frac{dL}{dm} = \epsilon_N n\frac{\delta T}{T} - \frac{L_s}{\Delta M}\frac{4\delta T}{T - T_0}$$

$$\simeq \epsilon_s\left(n - \frac{4T}{T - T_0}\right)\frac{dT}{T},$$

where we have used (11.76) and (11.77). Taking the variation of the right-hand side of (11.78), and noting that as the instability develops $\epsilon_N \simeq dL/dm$, we find

$$\left(n - \frac{4T}{T - T_0)}\right)\frac{\delta T}{T} \simeq t_K\frac{d}{dt}\delta\ln(P/\rho^{5/3}), \qquad (11.79)$$

$$t_K = 3P/2\rho\epsilon_N, \qquad (11.80)$$

where t_K is the local Kelvin time-scale. The change in entropy across the shell is

$$\delta S = (3Nk/2)\delta\ln(P/\rho^{5/3}).$$

The right-hand side of (11.79) will be positive when the shell expands, and the sign of the temperature perturbation δT will be determined by the sign of the coeficient $(n - 4T/\Delta T)$. For a highly temperature-sensitive fuel ($n \gg 1$), the shell will heat up ($\delta T > 0$) as it expands. The condition $n > 4T/\Delta T$ may be interpreted as a constraint on the shell's thickness. In this form it implies that the shell be thick enough that the temperature drop by at least $\Delta T = nT/4$ across the shell. Therefore, in order for a nuclear-shell source to become thermally unstable and lead to a thermal runaway, the shell must be thick enough to permit a large temperature change, but thin enough that the change in pressure on expansion $\delta P/P$ will be small.

The instability develops on the local Kelvin time-scale t_K and may, according to (11.80), be relatively short if the energy-generation rates are large.

The arguments above illustrate the conditions necessary for the development of a thermal instability. In practice, the quantities $L(r)$, $P(r)$, $T(r)$, and r, which appear in (11.78), (11.68), and the equations of hydrostatic equilibrium, are subjected to time-dependent perturbations. The resulting equations must then be solved numerically to find out if the perturbations grow in time, and on what time-scale. Such studies indicate that $l/r_s \lesssim 0.1$ and $\Delta M/M_{\text{core}} \lesssim 0.1$ for He shells. Hydrogen shells near the surface of hot white dwarfs may become unstable when $\Delta M \approx 10^{-4}\ M_\odot$. They also demonstrate that a substantial release of nuclear energy, equivalent to shell luminosities between 10^7 and $10^{10}\ L_\odot$, are common. Typically $L \simeq 100\ L_\odot$ for a thermally stable shell. Nevertheless, the total energy input into the star is probably less than 10^{49} ergs (see Problem 11.14).

Problem 11.13. Energy is transported radiatively through the He-rich region separating active, stable He and H shell sources. When the He source becomes thermally unstable, the shell luminosity rises greatly. Show that if the thermal instability is to persist, energy must be transported convectively between the shells.

Problem 11.14. Show that the luminosity of an unstable shell source is described approximately by

$$\frac{dL_s}{dt} = C L_s^2$$

during the initial stages, where

$$C \equiv 2\rho_s/3P_s\Delta M_s$$

is a constant depending on parameters in the shell. Find the luminosity as a function of time, and show that the total nuclear-energy input during shell flash depends logarithmically on the peak luminosity.

11.7. Mass Loss from Red Giants

Mass loss from the red giants and blue supergiants represents one way that a star, initially more massive than M_{Ch}, can reduce its mass and end its evolution as a white dwarf. Observations suggest that mass loss from the cool, extended envelopes of giants may be common. Planetary nebulae may also be formed by

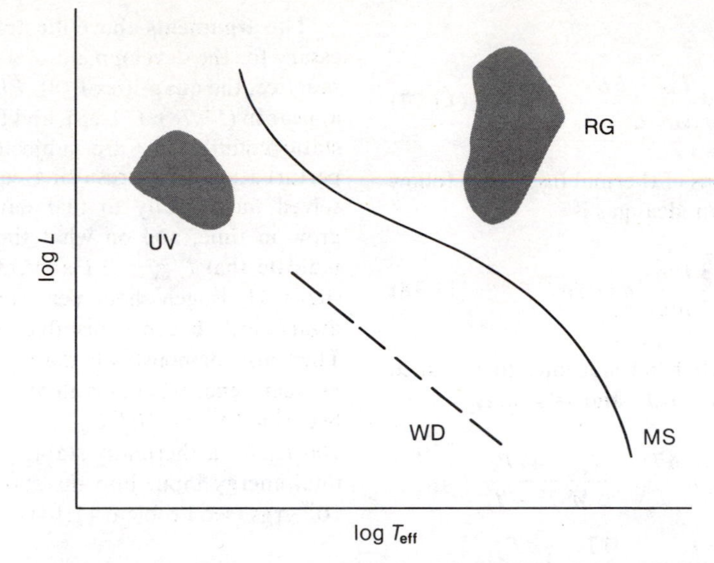

Figure 11.7. HR diagram for advanced evolutionary stages of low-mass stars (schematic), showing the main sequence (MS), the regions occupied by red giants (RG), ultraviolet dwarfs (UV) and white dwarfs (WD).

envelope ejection during a star's second ascent up the red-giant branch (Section 17.7). If a red giant's envelope is removed, the stellar core that remains will be a very hot star whose radiation peaks in the ultraviolet. Such UV stars are commonly observed at the apparent centers of young planetary nebulae.

An evolutionary scenario consistent with the preceding remarks is shown in Figure 11.7. The red giant region and the region of hot, ultraviolet central stars of planetary nebulae are labeled RG and UV, respectively. The track of a typical white dwarf (to be discussed in Section 14.5) is labeled WD, and is separated from the region UV by a gap. Consider a typical red giant as it begins shedding its envelope. Initially the optical depth in the envelope is small, and the core is shrouded from view. As the envelope expands and mass is lost, it becomes transparent, revealing the red-giant core, whose outer temperature may exceed 10^5 K. The star is now a hot object of about the same luminosity as the original red giant, and lies to the left of the main sequence.

Three principal mechanisms probably contribute to mass loss from red giants. First, repeated flashes caused by thermally unstable He shell sources set up relaxation oscillations in the envelope. Then, as the envelope expands, the gas cools and becomes transparent, and an underpressure develops, causing the envelope to fall back onto the core. If the amplitude of the shell flashes is great enough, a compression wave will form at the core-envelope interface, and may develop into a shock wave as it propagates out into the envelope. If the shock is strong enough, it will eject mass. Finally, if the radiation pressure on matter in the envelope is great enough—i.e., if the star's luminosity is greater than the Eddington limit (5.59)—it will further accelerate the ejected mass.

Unstable He Shell Burning

The evolution of a star massive enough to reach shell He burning is shown schematically in Figure 11.8. The ratio of the Kelvin time to free-fall time in the envelope

$$t_K/t_{ff} \sim (M^2G/RL)/(R^3/MG)^{1/2} \sim R^{-5/2}L^{-1}$$

increases as the star moves up the red-giant branch toward A, or as it descends during He shell burning toward E, and may exceed unity. If a region of the envelope becomes thermally unstable while $t_K \approx t_{ff}$, then dynamic motion results. The principal evolution-

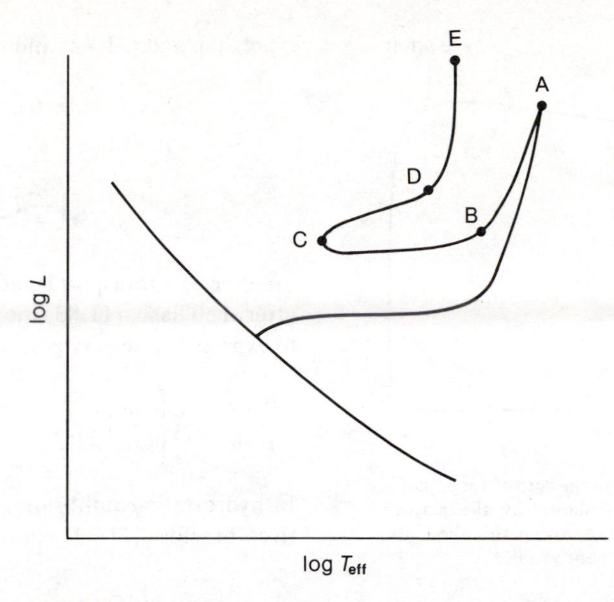

Figure 11.8. Schematic evolution of a star with a He shell source.

ary stages between B and E are the following:

- (B–D) main stage and exhaustion of He core burning;
- (D) thick He shell forms and contributes 50 percent or more of the energy release;
- (D–E) second ascent of the red-giant branch; the energy output from the H shell increases in importance and the He shell begins relaxation oscillations, each of which culminates in He shell flash;
- (E) the shell luminosity L_s becomes great enough, and the evolutionary time small enough, that mass ejection leads to the formation of a planetary nebula surrounding a hot UV star of mass $M \lesssim M_{ch}$.

Numerical calculations support this scenario, and suggest that near-degenerate conditions are needed in the He shell if envelope ejection is to occur. As the star moves from D to E, the He shell density increases, reaching values as high as 4 to 6 $\times$ 10^4 g/cm^3. For a star of about 1 $M_\odot$ at point E, $L \lesssim 10^4\ L_\odot$ and $T_{eff} \simeq$ 3,200 K. Convection extends about halfway from the He-burning shell into the He-rich region between the He and H shells during the early stages of instability. The instability, with time-scales $\sim 10^9$ sec. or more, occurs roughly in cycles with the following characteristics: (1) the He shell flashes with $L_s \lesssim 10^7\ L_\odot$, leading to (2) expansion of the He-rich zone between the He shells and the H shell, and (3) subsequent cooling of the H-burning shell. Since the H shell source contributes nonnegligibly to the luminosity, the total luminosity remains nearly constant, $L \simeq 10^4\ L_\odot$. The envelope is ejected during the relaxation oscillations (see the following) on a time-scale of order 10^5 yrs. During these cycles, the core gradually contracts until it becomes sufficiently dense and degenerate that neutrino cooling becomes important, reducing the core temperature $T_c \lesssim 10^8$ K. The temperature of the He shell remains near $T_s \simeq 2 \times 10^8$ K. Since C^{12} burning requires $T_c > 4$ to 6×10^8 K, degeneracy pressure halts the contraction before the next stage of nuclear burning begins, and a hot UV star results, which will evolve to a white dwarf. Figure 11.9 shows $T(r)$ versus $m(r)/M$ for a typical example. The total nuclear-energy release during He shell flash amounts to about $10^{48} - 10^{49}$ ergs. The gravitational binding energies of the core and envelope are of order 10^{49} ergs and 10^{45} ergs, respectively; so enough nuclear energy is released to eject the envelope without disrupting the core.

We now consider how thermal shell instabilities lead to envelope ejection. Imagine the interface between an active shell and the surrounding inert envelope, and let the radius of the interface be r_1. Normally energy is transported by convection in a

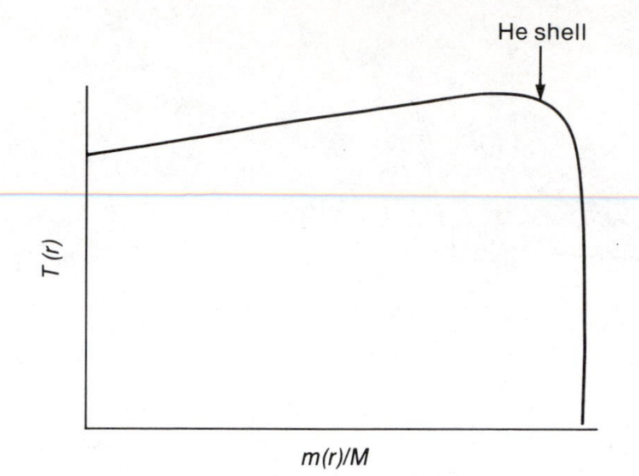

Figure 11.9. Temperature in the core of a red giant. Location of the He shell is shown by the arrow. Although most of the mass lies within this shell, its radius is about 10^{-3} times the giant's radius.

red-giant envelope. However, there is a limit to the rate at which energy may be carried convectively, obtained as follows.

An element of matter with internal energy density $u \approx \rho kT/m_H$ can move no faster than the local speed of sound $v_s \simeq (kT/\mu m_H)^{1/2}$ without forming a shock wave. Therefore the maximum energy flux due to convection is uv_s, and the maximum convective luminosity is

$$L_{\max,c} \simeq (4\pi r_1^2)uv_s \simeq 4\pi r_1^2 \rho (kT/\mu m_H)^{3/2}. \quad (11.81)$$

The right-hand side is fixed by the nature of the inert matter in the envelope. If a nuclear shell flash produces $L > L_{\max,c}$, then convective transport cannot carry the energy released, radiative transport takes over at the base of the envelope, and the temperature gradient is given by (5.54). At the temperatures typical of a shell-envelope interface ($T > 10^5$ K), the pressure $P = P_g + P_R$, and ionization is complete ($\mu \simeq$ constant). Therefore

$$\frac{dP}{dr} = \frac{dP_g}{dr} + \frac{dP_R}{dr} = \frac{4a}{3}T^3\frac{dT}{dr} + \frac{kT}{\mu m_H}\frac{d\rho}{dr} + \frac{k\rho}{\mu m_H}\frac{dT}{dr}. \quad (11.82)$$

The radial acceleration of a mass element of density ρ is given by (8.4); eliminating dP/dr between this expression and (11.82) and rewriting yields

$$\frac{P_g}{\rho}\frac{d\rho}{dr} = -\frac{mG\rho}{r^2} - \rho\ddot{r} - \left(\frac{4a}{3}T^3 + \frac{k\rho}{\mu m_H}\right)\frac{dT}{dr}. \quad (11.83)$$

Since energy transport is radiative at the envelope base after shell flash, (11.83) may be rewritten using (5.54) to express the density gradient as

$$\frac{P_g}{\rho}\frac{d\rho}{dr} = \frac{1}{r^2}\left\{\frac{\kappa\rho L}{4\pi c}\left(\frac{P_g}{4P_R} + 1\right) - mG\rho - \rho r^2\ddot{r}\right\}. \quad (11.84)$$

In hydrostatic equilibrium, $\ddot{r} = 0$ and $d\rho/dr$ is negative; therefore (11.84) requires that

$$\frac{\kappa L(r_1)}{4\pi c}\left(\frac{P_g}{4P_R} + 1\right) \leq m(r_1)G \quad (11.85)$$

at the base of the envelope. This will be satisfied as long as $L(r_1)$ is not too large [in fact, $L(r_1)$ must be less than the Eddington limit]. If $L(r_1)$ becomes too large to satisfy (11.85), then the term $\rho r^2\ddot{r}$ must be retained in (11.84) and it must be positive. Therefore $\ddot{r} > 0$, and the matter at the base of the envelope must be accelerating outward. If the shell flash is strong enough, some of the envelope will be accelerated to escape velocities. Furthermore, if a series of strong flashes occurs, then mass ejection may accompany each, producing a series of more or less concentric shells of expanding gas. If the shell flashes are relatively weak, then little matter will escape in this way. However, as the envelope expands, its mean free path increases, and radiative energy losses may lead to significant underpressure. The envelope will then collapse to be recompressed when it reaches the core. A compression wave then travels outward with increasing amplitude and develops into a shock wave* that ejects the entire envelope.

Numerical calculations indicate that such processes can eject red-giant envelopes over periods of 10^2 to 10^3 years, which corresponds to the time a star takes to evolve from the red-giant branch to the region of central stars of planetary nebulae. Numerical models of stars whose mass before He shell flash were 0.8 $M_\odot$, 1.5 $M_\odot$, and 3.0 $M_\odot$ have been shown to eject 0.2 $M_\odot$, 0.7 $M_\odot$, and 1.8 $M_\odot$ during the double-shell stage.

*We are working in the so-called Lagrangian frame.

Chapter 12

FINAL STAGES OF STELLAR EVOLUTION

Evolving stars build up nuclear ash of increasing atomic weight in their cores. If the temperature becomes high enough, the ash formed by the last burning stage becomes the fuel for a subsequent stage of evolution. The maximum attainable temperature depends, in part, on the star's total mass, with the more massive stars reaching the highest temperatures. In low-mass stars, the nature of nuclear burning and the total mass determine the core structure. For higher masses, neutrino processes become increasingly important, and may play an essential role in the evolution, since they remove energy at a rate that is sensitive to temperature.

Unfortunately, our understanding of these advanced stages is incomplete. The behavior of stars on or near the main sequence is well established, at least when rotation, magnetic fields, and mass loss can be neglected. Also well established is the hypothesis that cold configurations of stellar mass that have evolved to high densities will be stable against continued gravitational contraction if their masses are not too great. For densities and masses $10^6 \lesssim \rho \lesssim 10^9$ g/cm^3 and masses $M \lesssim 1.4\ M_\odot$, these configurations are identified with white dwarfs. For densities $10^{13} \lesssim \rho \lesssim 10^{15}$ g/cm^3 and masses probably in the range $0.5 \lesssim M/M_\odot \lesssim 2.5$, they are identified with neutron stars. The exact scenario connecting the main sequence to the probable end point of stellar evolution is still under investigation; so the discussion here focuses on general properties, and on those aspects of current theory that are likely to be contained in the final explanations of advanced stellar evolution.

12.1. Stellar Mass and the Final Stage

The existence of white dwarfs in the Hyades and Pleiades clusters is indirect evidence that stars more massive than 1.41 $M_\odot$ may evolve to this state. The main-sequence turn-off point for the Pleiades corresponds to stellar masses $\approx 6\ M_\odot$; all stars less massive than this are still on the main sequence. For an isolated star with $M \gtrsim 6\ M_\odot$ to have completed its evolution and become a white dwarf, at least 4.6 $M_\odot$ must have been lost since the main sequence. We may also argue that stars which will become supernovae are likely to be more massive than 6 $M_\odot$ while they are on the main sequence. In the Hyades, the turnoff corresponds to $M \approx 3$ to 4 $M_\odot$. Repeating the preceding argument would suggest that the lower-main-sequence mass for progenitors of supernovae could be as low as 3 or 4 $M_\odot$. Bearing these points in mind, we will find it convenient for purposes of discussion to sort stars into four categories (Table 12.1) based on their main-sequence mass. The last column estimates the fraction of the stellar mass of the Galactic disk (Population I stars) found in that mass range. Stars in Class a will become white dwarfs; those in Class b will have to lose enough matter (up to 2.5 $M_\odot$) that their final mass is less than the Chandrasekhar limit if they are to become white dwarfs. The exact fate of Class c stars is uncertain, but they seem likely to be supernova progenitors. Stars in Class d are massive enough to reach advanced stages of nuclear burning that form presupernova cores. The working hypothesis for our discussion is that all stars less massive than about 4 $M_\odot$ evolve nonviolently, and lose enough mass during their final stages to become white dwarfs. Those with $M \gtrsim 5\ M_\odot$ are likely to involve violent evolutionary stages. If an explosion does not completely disrupt the star, the compact remnant will be a neutron star if its mass is less than about 2.5 $M_\odot$. A more massive remnant is likely to collapse, forming a black hole.

Problem 12.1. Estimate the minimum gradual mass-loss rate for a star of 5 $M_\odot$ if it is to become a cold white dwarf. Assume that there is negligible loss during the main-sequence lifetime. Discuss all assumptions that go into this estimate.

The preceding discussion will not in general apply to stars in close binary or multiple systems, or to stars whose structure may be affected by rotation or magnetic fields.

Advanced stages of evolution are characterized by a high degree of central mass concentration. This has two important consequences for stellar models. First, the dynamic time-scale for the core will be greatly reduced. In a 5 $M_\odot$ star prior to C^{12} burning, the central density is 10^6 times greater than its main-sequence value, and its free-fall time is reduced by a factor 10^{-3}, to about two seconds. Dynamic changes in core structure are therefore rapid. Second, because the density changes significantly from the core to the envelope, the core and envelope may have dynamic time-scales that differ by several orders of magnitude. That is, the envelope may not have time to adjust to rapidly changing core conditions and, on the core's

Table 12.1
Stellar masses and the end points of stellar evolution.

Class	Mass range	Fractions
a	$M < 1.5\ M_\odot$	.60
b	$1.5 < M/M_\odot \lesssim 4$	.20
c	$4 < M/M_\odot \lesssim 8$	.06
d	$8 < M/M_\odot$	.14

time-scale, may resemble a static system. When regions of a star change dynamically, we can often treat the system adiabatically, as we did in the discussion of stellar pulsations. The evolution is then driven by the acceleration term in (8.4), and $\dot{s} \simeq 0$ in (8.7), so that the change in luminosity depends on local energy-generation rates. When the evolution is rapid but nonadiabatic, the full set of equations (8.4) to (8.7) must be used.

12.2. Advanced Stages of Nuclear Burning and Stellar Nucleosynthesis

A star develops a highly inhomogeneous structure during its advanced evolution, particularly if it is massive enough to reach carbon burning. A typical example is shown in Figure 12.1. The regions containing unburned fuel are labeled by the primary species present, and are separated from one another by thin (and usually active) shell regions. This structure is crucial not only to the star's structure, but to theories of element formation as well.

An important consequence of theories of stellar evolution (and a constraint on them) is the way that chemical elements form. Figure 12.2 shows the relative abundances in the solar system plotted against atomic number Z. In order to decide whether or not current scenarios of evolution are consistent with observed element abundances, we must consider nuclear reactions during the most advanced evolutionary stages.

The energy-generating mechanisms in stellar interiors give a satisfactory general picture for the formation of most of the light elements, starting from hydrogen. The most notable features are as follows.

(1) He^4 is formed by hydrogen burning, and some He^3 is formed by the incomplete proton-proton chain;

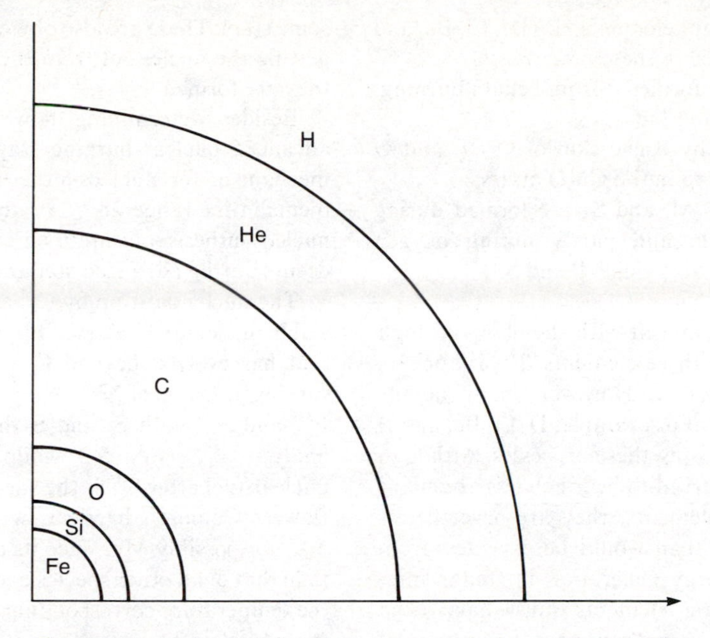

Figure 12.1. Composition shells in an evolved, massive star. The principal constituent is shown in each shell. "Fe" represents both iron and iron-group elements.

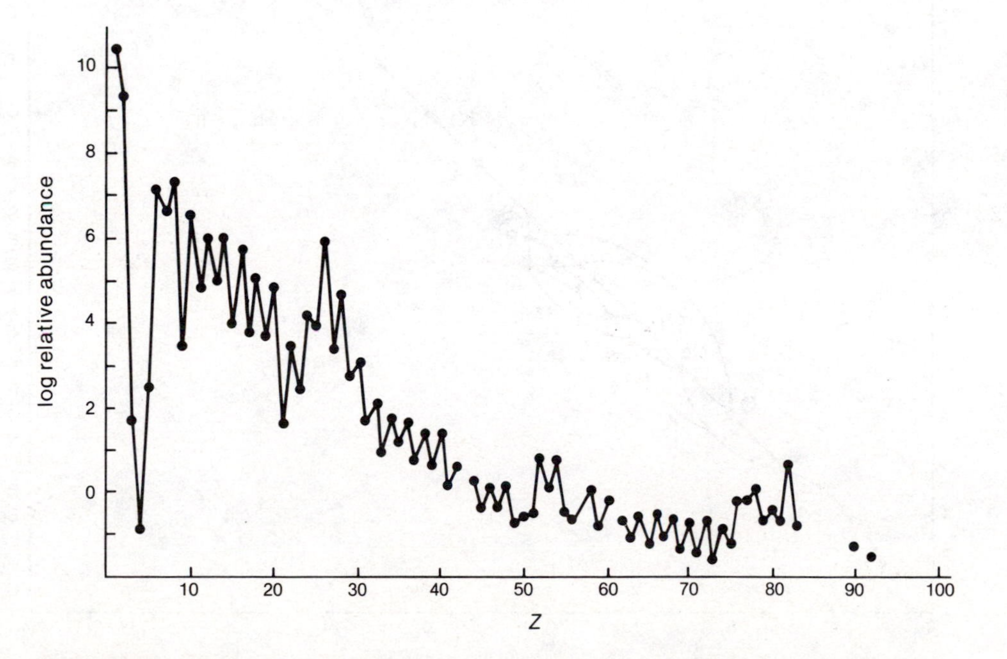

Figure 12.2. Relative cosmic abundances versus atomic number, normalized to a silicon abundance of 10^6.

(2) however, the light elements H^2 (D), Li, Be, and B are not formed in these processes;
(3) C^{12} and O^{16} are formed during helium burning, as is some O^{18} and Ne^{22};
(4) N^{14} is formed by conversion of C, N, and O during H burning in the CNO cycles;
(5) Ne^{20}, Na, Mg, Al, and Si are formed during carbon burning, and partly during oxygen burning, which also yields P and S.

This picture agrees so well with the observed high relative abundances of these elements that it appears to be substantially correct. However, many details remain to be sorted out. For example, D, Li, Be, and B are not formed at all during these processes. Although these elements are observed to be much less abundant than the other light elements, they are nevertheless much more abundant than would be expected from these processes of energy generation in stellar interiors. Presumably these elements must have been formed in some other way, perhaps near the surfaces of some stars. There are also obvious problems in actually getting the nuclei out from the stellar interiors where they are formed.

Besides determining how the star evolves, the advanced nuclear-burning stages are important as a mechanism for nucleosynthesis, particularly for elements in the range $28 \lesssim A \lesssim 60$. In fact, much of the nucleosynthesis of intermediate- and high-A nuclei occurs during advanced stellar evolution.

The nuclei built up by C^{12} and O^{16} burning include stable nuclei up to S^{32} (see Figure 12.3). A stellar core that has evolved beyond C and O burning contains varying amounts of Ne^{20}, Na^{23}, $Mg^{24,25,26}$, Al^{27}, $Si^{28,29,30}$, P^{31}, and S^{32}, with Si and S the most abundant. The analysis of Section 5.2 would suggest that the next nuclear fuel after O^{16} is the species with the smallest Z (lowest Coulomb barrier), which would be Ne^{20} or Na^{23}, or possibly Mg, since its concentration is greater than that of all other species except Si and S. However, the temperature corresponding to the Gamow peak for these nuclei, $T \simeq 2$ or 3×10^9 K, is high enough to

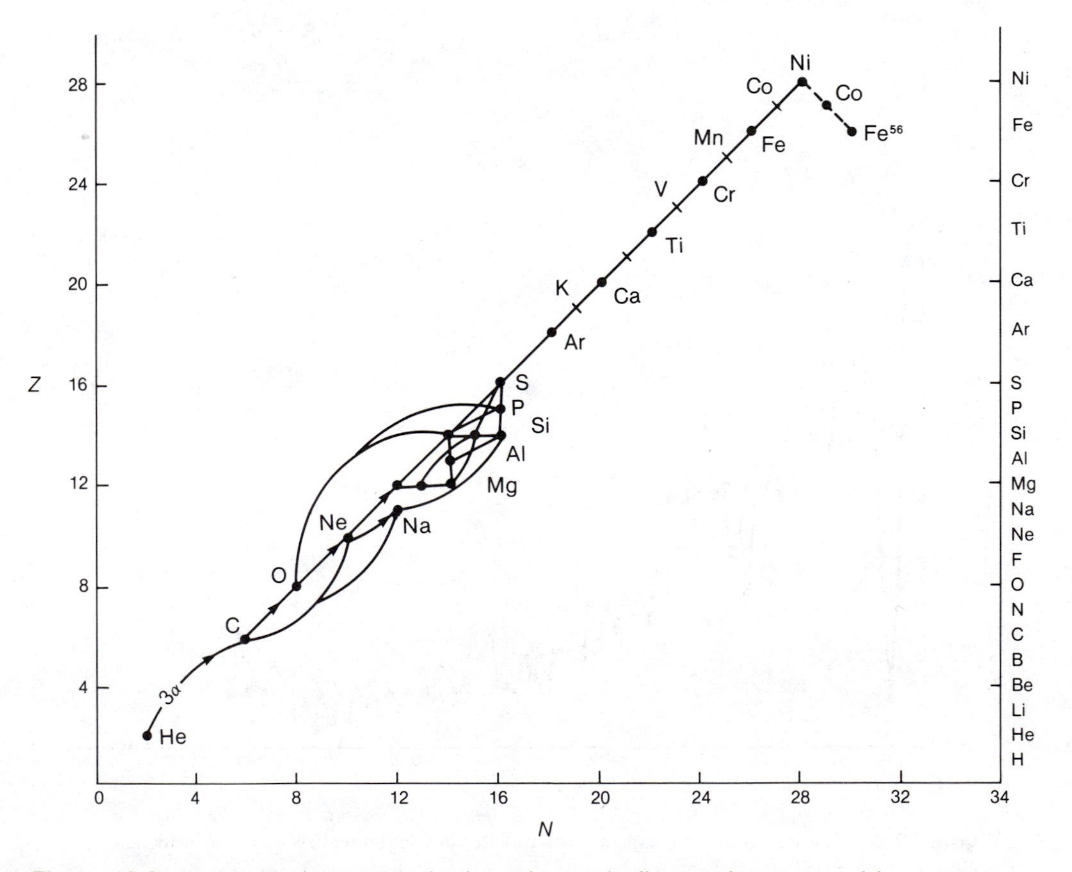

Figure 12.3. Nuclear-burning network, showing major steps leading up to iron-group nuclei.

photodisintegrate S^{32} and ultimately Si^{28}. As a consequence, the process for these heavier elements differs from that for lower-mass fuels. What results is a complex, quasiequilibrium state, set up by competing photodisintegration and particle-capture reactions that lead ultimately to the synthesis of iron-group nuclei (the isotopes of Cr, Mn, Fe, Co, and Ni).

The nuclear processes at these high temperatures are initiated when gamma rays strip protons, neutrons, and α particles from the nuclei present. These liberated particles may recombine with the nucleus they left, or with a nucleus of some other species. In this way, the nucleons can rearrange themselves among the various nuclei, leading to a buildup of those nuclei that have the greatest binding energy. Since the binding energy per nucleon reaches a maximum near the atomic weight corresponding to iron and its neighboring elements in mass number, we may expect to see a fairly high relative abundance of these elements, as indeed we do. Let us now consider these processes in greater detail.

When core temperatures approach 2×10^9 K, absorption of energetic photons by nuclei produces excited states that can eject the light particles n, p, and α (this last being, of course, an ejected He^4 nucleus). The reaction rates for light particle emission are proportional to

$$r_i \sim \exp(-Q_i/kT)\Gamma_{\text{eff}}, \tag{12.1}$$

where Q_i is the binding energy of the n, p, or α particle (as denoted by the index i), and Γ_{eff}, the effective particle width, depends on the particle-emission mechanism in such a way that the rates are greatest for energies near the Gamow peak.

The dominant species in the core after O^{16} burning are Si^{28} and S^{32}. The other products of C^{12} and O^{16} burning play a lesser role in subsequent energy generation, but do contribute to the relative abundance of intermediate-A elements. An analysis of α, p, and n binding energies show that Si^{28} is more tightly bound than is S^{32}. Therefore, when $T \simeq 2.5 \times 10^9$ K, S^{32} rapidly photodisintegrates to form Si^{28}:

$$S^{32} + \gamma \rightarrow Si^{28} + \alpha. \tag{12.2}$$

For purposes of discussion, the core therefore contains predominantly Si^{28} when the temperature reaches 3×10^9 K, at which temperature further photodisintegration begins. A complex scenario follows, in which the state of matter in the core depends on two competing processes. The first process is the ejection by Si^{28} of p, α, and finally n to form nuclei with lower atomic weight (Al, Mg, Ne, and O), since according to (12.1) the light particles that are least tightly bound are the first to be ejected. This process would lead to a significant concentration of light particles in the core. However, not all of the Si^{28} photodisintegrates at one time, and an appreciable concentration remains. These nuclei may capture α, p, or n to form more tightly bound nuclei, with $A > 28$. Some of the newly formed nuclei will rapidly photodisintegrate, but a few may again capture light particles (because of the relatively low Coulomb barrier for the intermediate Z of the target nucleus, and $Z = 0, 1,$ or 2 for the light particle) and further increase their atomic weight. This represents the second competing process, and clearly leads to a gradual buildup in atomic mass in the core. The persistence of nuclei with increasing A is a consequence of the fact that the average binding energy per nucleon in intermediate nuclei increases up to a maximum at Fe^{56}.

The formation of iron-group elements actually involves α, p, and n capture processes. At low temperatures, the α capture process dominates, and may be expressed by the sequence of reactions

$$\begin{aligned} Si^{28} + \alpha &\rightleftarrows S^{32} + \gamma, \\ S^{32} + \alpha &\rightleftarrows Ar^{36} + \gamma, \\ &\vdots \\ Cr^{52} + \alpha &\rightleftarrows Ni^{56} + \gamma. \end{aligned} \tag{12.3}$$

The least tightly bound particles (n, p, and α) are the most readily ejected, as is suggested by the dependence of (12.1) on binding energy, and the concentration of elements with $A > 28$ builds up. The ejected α (and to a lesser extent the nucleons) may be recaptured by any one of the nuclei present. Initially, this will be by Si^{28}, since it is most abundant at the start:

$$Si^{28} + \alpha \rightarrow S^{32} + \gamma. \tag{12.4}$$

Notice that (12.4) is the inverse of the original photodisintegration reaction (12.2); the two processes actually occur simultaneously, and a quasiequilibrium state is established. Once S^{32} has formed, it may capture α to form Ar^{36}, and the quasiequilibrium process

$$S^{32} + \alpha \rightleftarrows Ar^{36} + \gamma \tag{12.5}$$

is established and gradually shifts the composition toward higher A. In the process the supply of light particles is depleted. The photodisintegration of Si^{28} replenishes the supply of light particles as they are locked up in iron-group nuclei. In fact, since Si^{28} is the most resistant of intermediate mass nuclei to photodisintegration, the duration of the quasiequilibrium process $2Si^{28} \rightarrow Ni^{56}$ is set by the abundance and rate of disintegration of the Si^{28} in the core. For this reason the formation of iron-group nuclei from the products of O^{16} burning is called *silicon burning*. When the abundance of Si^{28} becomes small, there are too few free particles to produce capture, and silicon burning terminates.

The binding energy per nucleus reaches a maximum for Fe^{56}; so that energy must be absorbed from the gas whenever additional particles are added to a nucleus with $A > 56$. For this reason, elements above the iron group are not formed by quasiequilibrium silicon burning. If silicon burning is complete, the core will contain iron-group nuclei ($50 < A < 60$), but if burning terminates before Si^{28} is depleted, the core will also contain an admixture of intermediate ($30 < A < 50$) mass elements.

Captures of n and p, as well as α, may build up the iron-group nuclei. Many of the products of these reactions are themselves unstable against β decay. The importance of nucleon captures will generally increase with temperature. When silicon burning persists for more than about 10^4 sec, β decay will occur, and the released electrons shift the proton-neutron equilibrium ratio, which is governed by the weak interaction process $p + e^- \rightarrow n + \nu_e$, toward a greater neutron concentration. These complications are primarily important in establishing the final abundances of iron-group nuclei. Figure 12.4 shows the growth in the abundance of several elements that result from the photodissociation of Si^{28}, and subsequent nuclear rearrangement for a given density and temperature. Notice the shortness of the time-scale, with the abundances becoming nearly constant after 10^{-3} sec.

Once the composition is dominated by iron-group nuclei, the relative isotope abundances are determined by the statistical equilibrium of the various possible reactions, both forward and backward. The establishment of this equilibrium is known as the equilibrium or e-process. After its establishment, the dominant nucleus will probably be either Ni^{56} or Fe^{54}, depending on the precise conditions of temperature and pressure. An important question is how the large natural abundance of Fe^{56} arises, since Fe^{54} is the most favored iron isotope. The nucleus Ni^{56} is, however, unstable; when it is removed from the vicinity of the star, and hence of the region where statistical equilibrium is established, it can decay to Fe^{56}. Alternatively, if the temperature is quite low, Fe^{56} may be made directly. Figure 12.5 summarizes the situation for the most abundant nucleus formed from the photodisintegration rearrangement and the e-process starting with silicon. Notice that at very high temperatures, the nuclei are broken down again into He^4 nuclei, a fact that has some important consequences for the subsequent evolution.

Energy-generation rates accompanying silicon burning depend strongly on the temperature, and are affected by competing neutrino-loss processes. Energy is released by silicon burning as long as the subsequent particle rearrangement leads to n, p, and α-particle binding in iron-group nuclei. Thus silicon burning is like any other fusion reaction, where energy generation results from binding energy that is released by the rearrangement of the nucleons. The efficiency of the process is sensitive to the coexistent radiation field. For $T \lesssim 3 \times 10^9$ K, silicon burning is exothermic, and yields the greatest energy release. An approximate energy-generation rate is given by

$$\epsilon_{Si} \simeq \epsilon_0 X_{Si}{}^{1.143} (1 - X_{Si})^{-0.143} T_9{}^{6.31} e^{-143/T_9}, \quad (12.6)$$

where $\epsilon_0 \simeq 2.9 \times 10^{27}$ in cgs units, and X_{Si} is the Si^{28} mass fraction. For $T_9 \simeq 3$, this yields an energy-generation rate of about $\epsilon_{Si} \simeq 3 \times 10^9$ ergs g^{-1} sec^{-1} if $X_{Si} = 0.5$.

Problem 12.2. Assume that $\epsilon_{Si} = 3 \times 10^9$ ergs g^{-1} sec^{-1} throughout a 5 $M_\odot$ stellar core of uniform density $\rho \simeq 10^7$ g/cm^3. Find the core luminosity due to nuclear energy generation. Estimate the duration of Si burning if it is to eject a 3.5 $M_\odot$ envelope.

For 3×10^9 K $\lesssim T \lesssim 5 \times 10^9$ K, the effective energy-generation rate is reduced by competing neutron processes. When $T > 5 \times 10^9$ K, the high radiation flux leads to a substantial concentration of free nucleons in the core; energy that would otherwise be released when nucleons bind to form iron-group nuclei must now be tied up as thermal energy of the nucleon gas, and so the energy-generation rate is

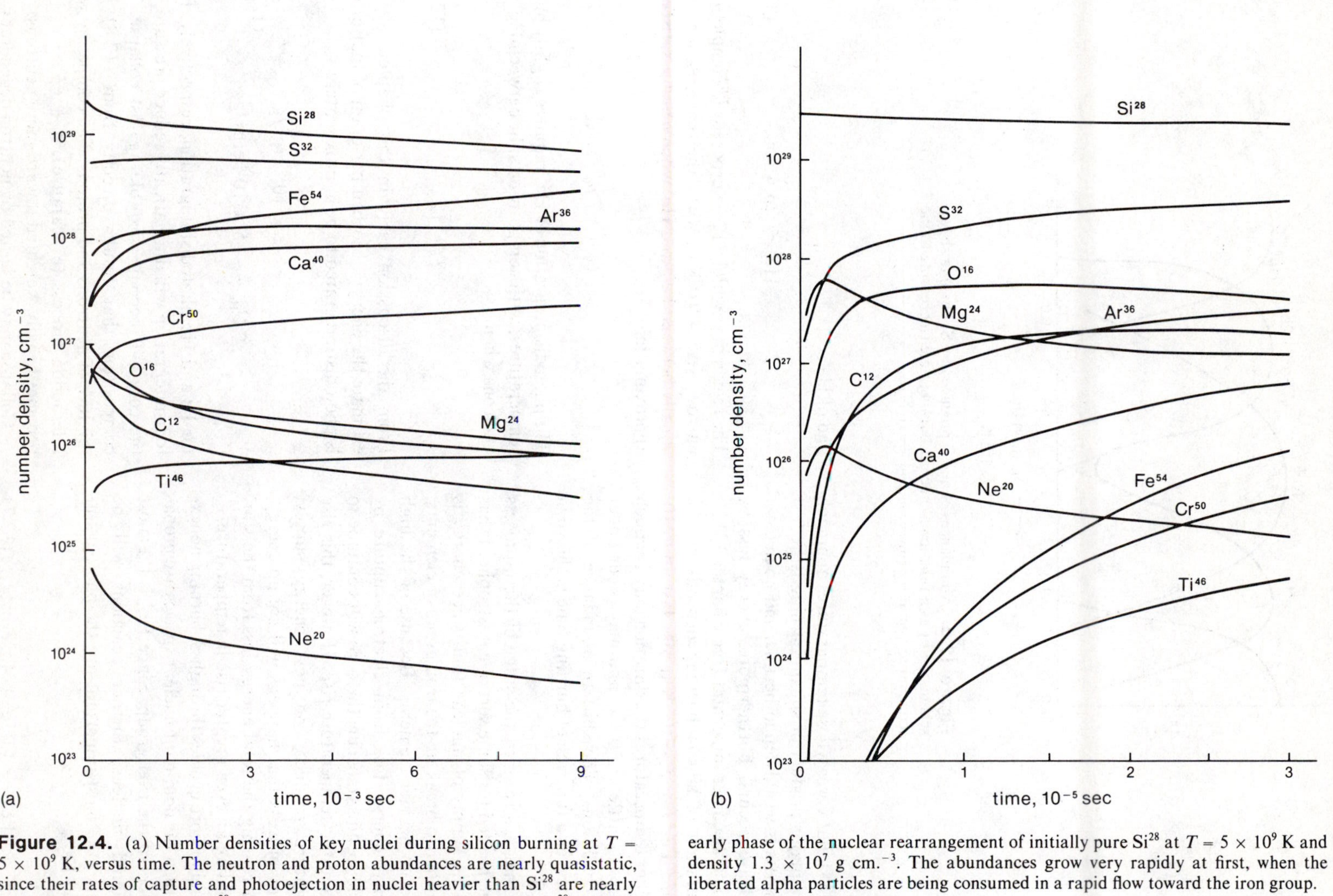

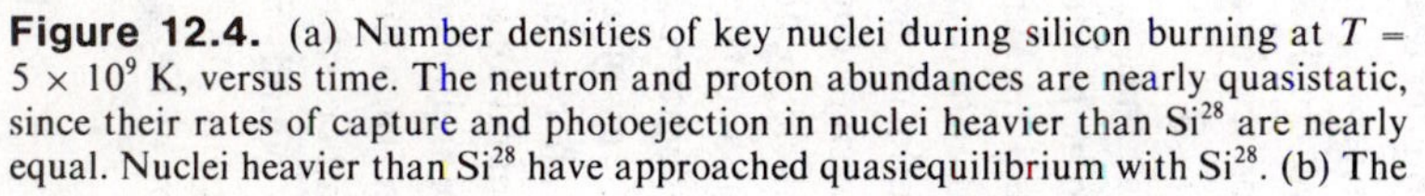

Figure 12.4. (a) Number densities of key nuclei during silicon burning at $T = 5 \times 10^9$ K, versus time. The neutron and proton abundances are nearly quasistatic, since their rates of capture and photoejection in nuclei heavier than Si^{28} are nearly equal. Nuclei heavier than Si^{28} have approached quasiequilibrium with Si^{28}. (b) The early phase of the nuclear rearrangement of initially pure Si^{28} at $T = 5 \times 10^9$ K and density 1.3×10^7 g cm.$^{-3}$. The abundances grow very rapidly at first, when the liberated alpha particles are being consumed in a rapid flow toward the iron group.

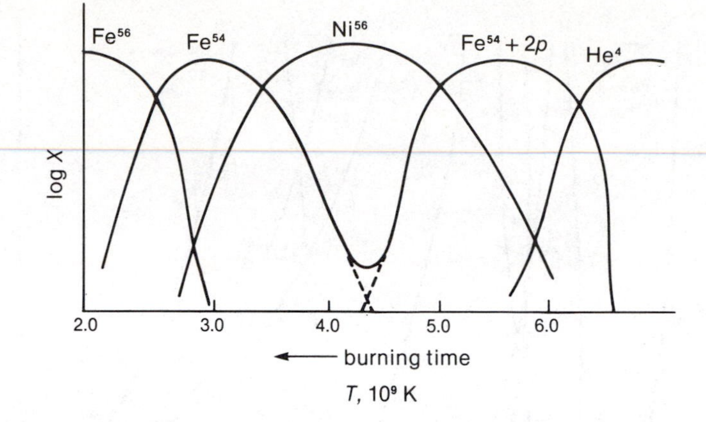

Figure 12.5. Abundances of nuclei produced by Si burning shown schematically as a function of nuclear burning time (increasing to the left) and temperature (increasing to the right).

reduced. In stellar cores whose temperatures approach 10^{10} K, the major product of silicon burning is Fe^{54} + 2p, the process is weakly endothermic, and no nuclear energy release occurs. Furthermore, energy loss by neutrinos now becomes important (see Chapter 13).

The duration of silicon burning depends on the temperature. For $T < 3 \times 10^9$ K it may last up to 10^6 sec. At higher temperatures the duration may be much less, perhaps as little as 10 sec, and may occur under explosive conditions. The shift in abundance of the final products as a function of burning time is shown in Figure 12.5.

A star that has reached the stage of the e-process seems to be headed for some sort of catastrophe. It has exhausted its supply of nuclear fuel, since energy is required either to break up the iron group of elements, or to build up heavier elements. Because of the high core density, energy loss from electron capture by heavy nuclei begins to occur, the core will continue to contract, and the temperature rise. However, this rise of temperature produces no new energy supply; instead, any process that happens will use up energy, thereby increasing the total energy loss from the star, and demanding an even greater rise in temperature.

If the stellar core evolves through Si burning nonexplosively, then the dominant nuclear species are probably members of the iron group. Since Fe^{56} is the most tightly bound nucleus, further reactions will be endothermic; the most important are the photodistintegration of Fe^{56},

$$Fe^{56} \rightarrow 13He^4 + 4n, \qquad (12.7)$$

and of He^4,

$$He^4 \rightarrow 2p + 2n. \qquad (12.8)$$

The region over which (12.7) proceeds to completion is narrow ($\Delta T \approx 10^9$K), with the composition split equally between Fe^{56} and He^4 when ρ and T satisfy the approximation

$$T_9 \sim 5.7\, \rho_6^{0.081}. \qquad (12.9)$$

The photodisintegration of He^4 requires slightly higher temperatures, with composition split between He^4 and nucleons when

$$T_9 \sim 12\, \rho_8^{0.13}. \qquad (12.10)$$

From the discussion early in this chapter, we can estimate the energy absorbed per gram of matter that is photodistintegrated to be approximately

$$\begin{aligned} Q_{Fe} &= -2 \times 10^{18} \text{ ergs/g}, \\ Q_{He} &= -5 \times 10^{18} \text{ ergs/g}, \end{aligned} \qquad (12.11)$$

for Fe and He photodisintegration, respectively. Photodisintegration will dramatically reduce pressure, and the core will become unstable against collapse. The composition that results directly from (12.8) is a mixture of neutrons and protons. The competing weak interaction process (e^- capture), $p + e^- \rightarrow n + \nu_e$, becomes heavily weighted toward more neutrons.

Collapse may be halted by neutron degeneracy, but

this process will probably be accompanied by violent activity in the outer region of the star, possibly explosive, blowing off much material, and perhaps producing a Type II supernova. If the e-processed core is more massive than the mass limit for a neutron star, then presumably the collapse cannot be halted, even by neutron degeneracy, and the remnant will become a black hole. It is significant that a most likely candidate for a neutron star, the pulsar at the center of the Crab nebula, is at the site of a known supernova explosion. Of course, the outer layers of the supernova will have contained material undergoing all the various phases of nuclear burnings, out to possibly unprocessed material near the edge of the object. The ejection of this material into space is obviously a way in which the interstellar medium is enriched by the heavier elements produced by nuclear burning in stars.

Problem 12.3 A star begins its life as primarily H, releases more than 10^{51} ergs of radiant energy (for 1 $M_\odot$) as a result of nuclear fusion, and then ends up as nucleons. Where does the energy for this last step come from?

Problem 12.4. Suppose a stellar core's density was 10^{10} g/cm^3 when photodisintegration of Fe and He occurred, and that the subsequent collapse is dynamic. What energy-generation rate would be needed to offset the absorption rate (12.11)?

There remains the problem of the synthesis of the elements heavier than iron, since, according to the picture just outlined, they cannot arise directly from advanced nuclear burning in a star. Because of the Coulomb barrier, there seems to be no way to make the heavy elements by interaction with charged particles. However, there is no such barrier for neutrons, and heavy nuclei capture neutrons easily, even at relatively low temperatures. Suppose a nucleus of charge Z and atomic weight A captures a neutron. It becomes a nucleus of charge Z and atomic weight $A + 1$. This new nucleus is either stable or unstable. If it is stable, it may eventually capture another neutron and become a nucleus Z, $A + 2$, and so on. If it is unstable, however, two things may happen. Either it decays before there is time for another neutron to be captured, or it captures another neutron before it has time to decay. The rate of neutron capture depends, of course, on the capture cross sections and on the neutron velocities, and hence on the temperatures. If the neutron capture rate is slow compared to the decay rate (usually β decay), then nucleosynthesis will proceed along the chain of stable isotopes of each element until an unstable one is encountered, at which point decay will occur until a stable isotope of a new element is produced. Then neutron capture can take place again, and the whole sequence is repeated. This process, in which neutron capture is slow relative to β-decay times, is called the s-process.

On the other hand, the decay rate of the nuclei may be much slower than the neutron capture rate. An element will then build up heavy isotopes, which are very neutron-rich, until the nucleus cannot capture any more neutrons because of low binding energy or (what is virtually the same thing) until the decay rate becomes comparable with the capture rate. These nuclei then undergo radioactive decay until a stable isotope is reached. This process, in which neutron capture is rapid relative to β-decay times, is known as the r-process.

It is clear that the s-process tends to produce the most stable isotopes at each atomic weight, whereas the r-process tends to produce those stable isotopes at each given atomic weight that are on the neutron-rich side of the most stable. There remain, however, some stable isotopes on the proton-rich side of the most stable isotopes. Such nuclei are not produced by either the r-process or the s-process; but the observed abundance of proton-rich isotopes is always very low.

If the r-process is very rapid, it could build up some very heavy nuclei, the limit being set, of course, by the stability of the nucleus. Isotopes up to those of uranium, and perhaps transuranic elements, could be built up in this way. Of some interest here is the isotope Cf^{254}. This is unstable by fission, with a half-life of about 55 days, which is close to the time-scale for the decay of Type I supernovae light curves. It has been suggested that the r-process, followed by the release of nuclear energy by the spontaneous fission of Cf^{254}, may be responsible for these objects. The suggestion has not been verified (by detection of Cf^{254}, for instance), but remains an interesting hypothesis.

Finally, a major problem in the synthesis of elements by either the s-process or the r-process is the source of neutrons. Neutrons do not play a major role in the principal reactions of stellar energy generation, and any that are present must come from subsidiary reactions of no importance for energy generation. This is obviously a worse problem for the r-process than for the s-process. The seriousness of this problem may be shown by noting that, for the r-process, the time rate of change of the nuclear abundance N_A is described by

$$\frac{dN_A}{dt} = n_N\langle v \rangle \, (\sigma_{A-1} N_{A-1} - \sigma_A N_A). \qquad (12.12)$$

Here σ_A is the neutron-capture cross section, n_N is the neutron number density, and $\langle v \rangle$ the relative speed of the neutrons. To an order of magnitude this is equivalent to

$$\frac{1}{\tau} \simeq n_N \sigma \langle v \rangle, \qquad (12.13)$$

where σ is an average capture cross section, which we take to be of order 10^{-25} cm^2. For temperatures $T \approx 10^9$ K, the thermal neutron speed $\langle v \rangle \simeq 3 \times 10^8$ cm/s, and β-decay times are about 10^{-2} sec. Requiring that $\tau \ll 10^{-2}$, we find that the r-process will be effective for neutron fluxes $n_N\langle v \rangle \gg 10^{28}$ cm^{-2} sec^{-1}.

Problem 12.5. Show that dN_A/dt is given by (12.12) when the capture rates are much greater than the β-decay rates.

Chapter 13

WEAK INTERACTIONS IN STELLAR EVOLUTION

Stellar neutrinos have little direct influence on the static properties of stars, but they are often critical for dynamic stages of evolution. In H burning by the pp chain, for example, the escaping neutrinos carry off only about 2 percent of the total energy produced; the remaining 98 percent of the energy is in photons, which are available to support the overlying matter. However, if produced in sufficient numbers and, as sometimes happens, at the expense of pressure-producing particles, neutrino emission can trigger dynamic collapse in a star. Stages of stellar evolution in which neutrino emission plays a crucial role include:

(1) red supergiants,
(2) planetary nebulae,
(3) cooling rates of hot, white dwarfs and neutron stars,
(4) supernova explosions,
(5) vibrational damping of neutron stars,
(6) collapse of neutron stars and the formation of black holes.

To illustrate the way that neutrinos induce dynamic instabilities, let us consider a hot gas, in hydrostatic and thermal equilibrium that radiates photons and neutrinos. If the neutrinos carry off only energy (and do not substantially alter the pressure), then photons and neutrinos will induce a slight deviation from thermal equilibrium. Subsequent readjustment by the star to this imbalance will be on the thermal time-scale, and the evolution is quasistatic. If, however, the escaping neutrinos reduce (or in some cases increase locally) the pressure, then the star deviates from hydrostatic equilibrium, and dynamic changes in its structure occur on a free-fall time-scale. Unlike most thermonuclear reactions, the weak interaction processes are often highly density-dependent, so that the rate of energy release by neutrinos can also change on a dynamic time-scale. The development of the instability continues unless the pressure can stop it (as in the formation of a neutron star) or a new thermonuclear fuel ignites, possibly under explosive conditions.

Problem 13.1. Use the virial theorem to justify the scenario for neutrino losses outlined in the preceding.

The weak interactions that arise in stellar evolution* are associated with neutrinos, which have zero rest mass (like a photon), travel at the speed of light in vacuum, and have zero electric charge and an intrinsic spin with angular momentum $\hbar/2$. Because of its spin, a neutrino behaves like a fermion. Historically the existence of the electron neutrino was inferred from nuclear decay (β decay)

$$N(Z, A) \to e^+ + \nu_e + N(Z-1, A), \quad (13.1)$$

where e^+ is a positron, and the electron neutrino is denoted by ν_e. The process

$$N(Z, A) \to e^- + \bar{\nu}_e + N(Z+1, A) \quad (13.2)$$

also occurs, where $\bar{\nu}_e$ is the electron antineutrino. Special cases of these reactions are

$$p \to n + e^+ + \nu_e, \qquad n \to p + e^- + \bar{\nu}_e. \quad (13.3)$$

The second of these reactions is the primary mode of decay of the neutron, whose half-life under normal conditions is about 15 minutes. Notice that (13.1) reduces the nuclear change, but (13.2) increases it. Other weak interaction processes of astrophysical importance are the emission and absorption processes

$$n + \nu_e \to p + e^-, \qquad p + e^- \to n + \nu_e, \quad (13.4)$$

$$\mu^- + p \to n + \nu_\mu, \qquad \mu^- \to e^- + \bar{\nu}_e + \nu_\mu, \quad (13.5)$$

and pion decay

$$\pi^- \to \begin{cases} \mu^- + \bar{\nu}_\mu, \\ e^- + \bar{\nu}_e. \end{cases} \quad (13.6)$$

The last three processes introduce the muon neutrino denoted by ν_μ and its antiparticle $\bar{\nu}_\mu$. Recent developments have established the existence of a third neutrino type, the tau neutrino, ν_τ, which is associated with a lepton analogous to the muon. The astrophysically important reactions involving the ν_τ appear to be analogous to those involving the μ and ν_μ.

Theory, supported by recent experimental data, indicates that neutrinos couple directly to the leptons $e^\pm$ and $\mu^\pm$:

$$e^+ + e^- \to \nu_e + \bar{\nu}_e, \qquad \mu^- + \mu^+ \to \nu_\mu + \bar{\nu}_\mu, \quad (13.7)$$

$$e^- + \nu_e \to e^- + \nu_e, \qquad \mu^- + \nu_\mu \to \mu^- + \nu_\mu. \quad (13.8)$$

In addition, neutrinos scatter off nucleons:

$$\nu_e + n \to \nu_e + n, \qquad \nu_e + p \to \nu_e + p. \quad (13.9)$$

The processes in (13.8) are inelastic, and result in degradation of the neutrino energy [note the analogy of (13.8) to Compton scattering]. Although the cross sections for these processes are often much smaller than elastic coherent scattering (Section 13.3), their inelastic nature is important as a mechanism for neutrino thermalization in the core of supernovae.

Problem 13.2. Assume that ν_e produced by electron capture with typical energies $E_\nu \simeq 15$ MeV are scattered by nucleons in a stellar core where $T \simeq 10^{10}$ K. What is the energy change ΔE_ν of the neutrino per collision? Estimate the number of ν_e, n scatterings needed to reduce the neutrino energy to $m_n c^2$.

The process (13.3) has already been encountered as the central step in the formation of deuterium by the fusion of He from H via the pp chain. The processes (13.7) to (13.8) are extremely important in some stages of advanced stellar evolution. The first is analogous to the annihilation $e^+ + e^- \to \gamma + \gamma$, the second to electron scattering of photons.

The importance of these processes lies in the extreme weakness of the interactions. Even though a neutrino has no rest mass, it will carry off energy just as a photon does. Unlike the photon, which couples to matter electromagnetically, with scattering cross sections typically of order 10^{-16} to 10^{-24} cm^2, neutrino processes involve cross sections of order

$$\sigma_{\text{weak}} \simeq 2 \times 10^{-44} (E_\nu / m_e c^2)^2 \text{ cm}^2, \quad (13.10)$$

where E_ν is the neutrino energy. The smallness of σ_{weak} means that matter under most conditions is transparent to neutrinos. To appreciate this, note that the neutrino mean free path in matter of baryon number

*There are some weak interaction processes that do not involve neutrinos but they are of primary interest in cosmology.

density n is

$$l_\nu \sim (n\sigma_{\text{weak}})^{-1} \sim \frac{m_H}{\rho\sigma_{\text{weak}}} \simeq \frac{8 \times 10^{19}}{(E_\nu/m_e c^2)^2 \rho} \text{ cm.} \quad (13.11)$$

The mean free path in the solar core exceeds 6 pc. For white dwarfs with $\rho_c \lesssim 10^9$ g/cm^3 and for neutron stars with $\rho_c \lesssim 10^{15}$g/cm^3, the neutrino mean free path is of order 24 $R_\odot$ and 17 km, respectively.

Problem 13.3. The average density of matter in the Sun is about 1 g/cm^3. Neutrinos are produced in the solar core during hydrogen burning by reactions such as

$$p + p \rightarrow D + e^- + \nu_e \quad 0.42(0.26),$$

$$B^8 \rightarrow Be^8 + e^+ + \nu_e \quad 14\ (7.2).$$

The numbers on the right give the maximum (average) neutrino energy per reaction in MeV. If the cross section for neutrino scattering is of order 10^{-45} cm^2, estimate the fractional number of solar neutrinos that scatter before they can escape from the sun.

Essentially all observational data on stellar structure are derived from electromagnetic radiation that is emitted from the photosphere of stars. Although such data can help to pin down the surface boundary conditions (effective temperature, luminosity, composition, and some information on pressures, temperatures, and densities near the surface), these tell us nothing directly about the stellar interior. Generally speaking, we must rely on indirect arguments (the stellar structure equations!) in order to conclude anything about the internal conditions. Since neutrinos produced in most stellar cores have mean free paths much greater than stellar radii, they may in principle be used as probes of the conditions in the core. This is possible because neutrino production rates are quite sensitive to the specific reactions that occur, and these in turn reflect the internal temperatures, densities, and composition. In this way a detailed understanding of stellar neutrino production complements the data obtained from traditional sources. However, the extreme weakness of neutrino interactions, which allows neutrinos to escape directly from the core, also makes them difficult to detect. In practice, observations of neutrinos from ordinary stars is probably limited to the Sun.

13.1. Solar Neutrinos

We will now consider the use of neutrinos as a probe of the internal structure of the Sun. The ideas are in principle applicable to other stars, but current technology limits their use to the Sun. Solar neutrinos can be detected primarily by the weak capture process

$$_{17}\text{Cl}^{37} + \nu_e \rightarrow {}_{18}\text{Ar}^{37} + e^- \quad (13.12)$$

which has a capture threshold $E_T = 0.81$ MeV. A nuclear energy-level diagram is shown in Figure 13.1 for this process. The quantum numbers of the excited state of $_{18}Ar^{37}$ that has energy $E = 5.1$ MeV are the same as those of the ground state of $_{17}Cl^{37}$; so the transition (arrow) is superallowed. Currently detectable neutrinos are those with energy in excess of 5.1 MeV. Recent efforts to detect solar neutrinos use CCl_4 (about 600 tons!), which contains about 25 percent Cl^{37}, as a detector. The unstable Ar^{37}, which is produced by each absorbed neutrino, is separated by flushing the system with He^4. The Ar^{37} decays, emitting an electron (half-life ~ 35 days), which is then detected. Estimates of the neutrino production rate of the Sun, and of the capture cross section entering into (13.12), suggest that about one Ar^{37} will be produced in every 48 hours in the target (the background rate is less than 0.2 per day).

The primary sources of energy in the Sun are the pp chain and the CNO cycle. The reactions in each of these networks that produce neutrinos are given in Table 13.1, along with the average and maximum neutrino energy, and the reaction's relative importance as a source of ν_e (as deduced from standard models of the Sun). The relative importance of each reaction as a detectable source of neutrinos is conventionally measured in units of 10^{-36} captures per Cl^{37} atom per second (one solar neutrino unit = SNU). We will discuss here only the pp chain.

As we have already noted, reaction $p + p \rightarrow D + e^+ + \nu_e$ sets the time-scale for the release of energy via thermonuclear reactions. It contributes only 10 percent of the total energy per conversion $4H \rightarrow He^4$. It contributes most of the neutrino flux, but the energies are below the threshold for capture. The reaction

$$B^{8*} \rightarrow Be^8 + e^+ + \nu_e \quad (13.13)$$

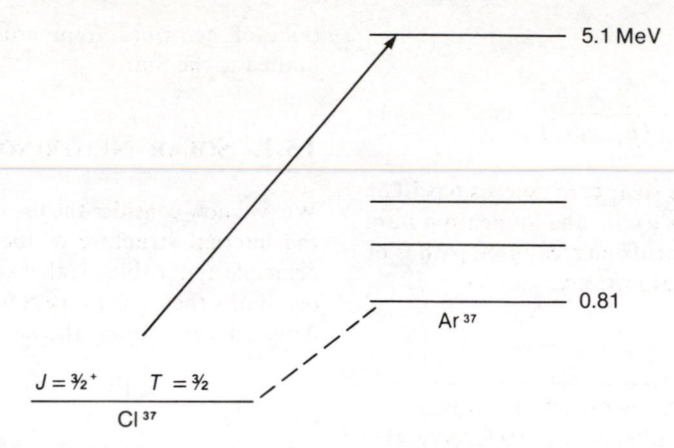

Figure 13.1. Energy-level diagram for ν_e capture by Cl^{37} used in the Davis experiment to detect solar neutrinos.

contributes far less to the total neutrino flux, but the average neutrino energy is in excess of the superallowed transitions ($E = 5.1$ MeV); so these are the neutrinos that are most readily detected.

Under stellar conditions the nuclear energy-generation rates may be approximated by expressions of the form (7.6). Examination of the important neutrino-producing reactions in the Sun shows that the neutrino flux depends only on the temperature profile in the interior, at least to a very good approximation. The photon production rate (the luminosity) is sensitive both to temperature and to density profiles throughout the star. For an assumed model of the Sun (or any star), we can find the temperature-dependence of the change in neutrino flux with luminosity, $d\Phi^{(k)}/dL$, for each nuclear reaction k that produces detectable neutrinos (see Figure 13.2). It is found that large changes in T cause almost no change in the contributions that the process $p + p \rightarrow D + e^+ + \nu_e$ makes to the neutrino flux. The reaction (13.13), which produces essentially all the detected neutrino flux, however, is highly sensitive to the temperature. The weak decay of Be^7, and the O^{15} and N^{13} decays in the CNO cycle, are also sensitive to the interior temperature, but not as strongly, and the emitted ν_e is at or not far above threshold.

We can now discuss how solar neutrinos reflect the detailed internal structure of the Sun. The Sun's observed mass and luminosity place stringent constraints on the internal parameters (the neutrino luminosity may be ignored, because it is small compared to the photonic part). A model of the Sun is therefore assumed that is compatible with $M_\odot$ and $L_\odot$. This gives $L(T)$ from the center to the surface of the core (beyond which L is a constant). A different model (say, based on changes in assumed chemical composition, or with more radical modifications) would yield a different form for $L(T)$, e.g., the one shown by the dashed curve in Figure 13.3. Such a model would obviously have to be consistent with the age of the

Table 13.1
Neutrino-producing nuclear reactions in the solar core.

Reaction	Neutrino energy (MeV) Average	Maximum	Relative importance (solar neutrino units)
$p + p \rightarrow D + e^+ + \nu_e$	0.26	0.42	6.0
$p + p + e^- \rightarrow D + \nu_e$	1.44	1.44	1.55×10^{-2}
$Be^7 + e^- \rightarrow Li^7 + \nu_e$	0.86	0.86	0.4
$B^8 \rightarrow Be^{8*} + e^+ + \nu_e$	7.2	14.0	4.5×10^{-4}
$N^{13} \rightarrow C^{13} + e^+ + \nu_e$	0.71	1.19	3.7×10^{-2}
$O^{15} \rightarrow N^{15} + e^+ + \nu_e$	1.00	1.70	2.6×10^{-2}

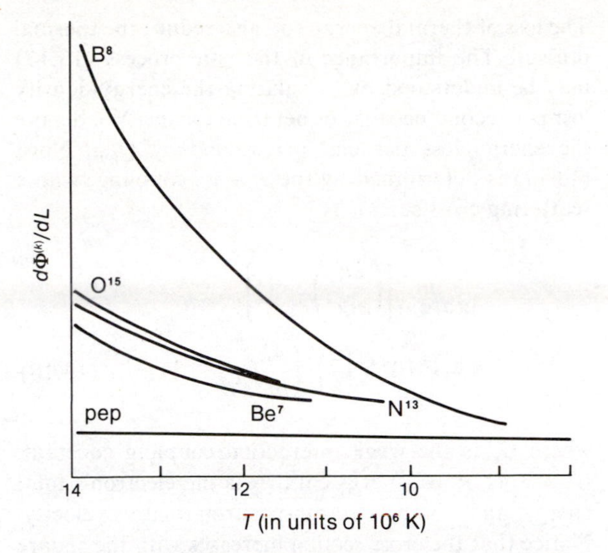

Figure 13.2. Contribution to ν_e flux by specific nuclear processes in the Sun versus temperature. Standard models predict $T_0 \approx 1.2 \times 10^7$ K in the Sun.

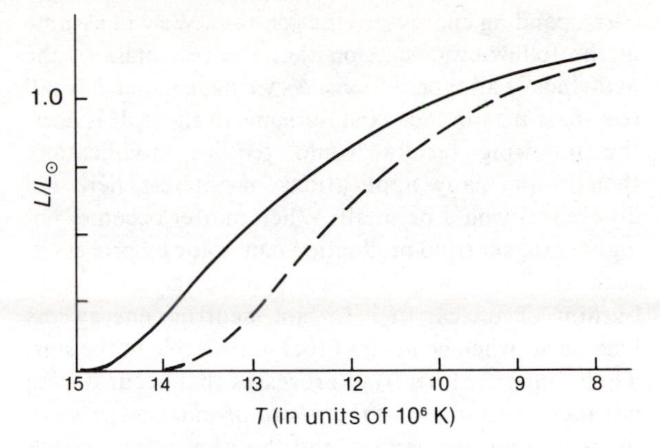

Figure 13.3. Sensitivity of $L(r)$ to changes in solar model.

Sun and its surface composition, as well as with $M_\odot$ and $L_\odot$.

Using the model-based function $L(T)$ and the theoretically derived $d\Phi^{(k)}/dL$, we obtain the predicted neutrino flux Φ_ν from the integral

$$\Phi_\nu = \sum_k \xi_k \int_{T_c}^{T_s} \frac{d\Phi^{(k)}}{dL(T)} dL(T). \qquad (13.14)$$

The sum is over all neutrino-producing reactions, and ξ_k is a weighting factor that reflects the relative importance of the kth neutrino-producing process in the core, T_c is the central temperature, and T_s the temperature of the core-envelope interface, beyond which L = constant. Theoretically predicted neutrino flux rates at the top of the Earth's atmosphere are $\sim 10^{10}$ cm^{-2} sec^{-1}. When this is converted into an observed counting rate for the decay of Ar^{37} produced from reaction (13.12), it is found that standard solar models predict rates of about 7.8 SNU. The observed count rates are 2.1 ± 0.3 SNU; and the predictions from the standard solar models fall outside the observational uncertainties.

There are many sources of potential uncertainty in the standard models of the Sun, including opacities, nuclear cross sections, and energy-generation rates, which in turn predict capture rates on Cl^{37} between 3 and 10 SNU. Attempts to reduce the capture rates below 3 SNU require major revisions of the theory of solar (and presumably stellar) structure. These include incorporating composition inhomogenieties and periodic mixing of nuclear isotopes. It has also been suggested that the neutrino may decay, or change from one type to another periodically ($\nu_e \rightarrow \nu_\mu \rightarrow \nu_e$, for example), which would result in fewer counts for the same core production rate. Recent experiments that indicate the neutrino may have a nonzero rest mass (perhaps as large as a few tens of eV) make this an intriguing possibility. At present, the discrepancy between theory and experiment persists.

The development of Ga^{71} detectors, sensitive to neutrinos produced by the pp chain, should help to establish whether the solar neutrino output is really as low as is indicated by B^8 decay. Detection results here because of the capture process

$$Ga^{71} + \nu_e \rightarrow Ge^{71} + e^-; \qquad (13.15)$$

the Ge^{71} has an electron-capture half-life of about 12 days, and the capture process has a threshold of 0.23 MeV. Standard solar models predict a Ga^{71} capture rate equivalent to 92 SNU, of which 65 SNU is due to the pp reaction, and very little comes from B^8 decay. Because the rate of ν_e production by the pp reaction is relatively insensitive to the central conditions in the Sun (Figure 13.2), the reaction (13.15) is a model-independent measure of the solar neutrino output.

13.2. Neutrino Energy-Loss Rates

When neutrinos are an important source (or sink) of energy, the stellar structure equations must include a corresponding energy-production rate. We will assume in the following discussion that the rest mass of the neutrinos of all types is zero. As we have noted, a small rest mass may in fact exist for some of them. If it does, the following results would require modification, though for many applications of interest here the differences would be small. When matter becomes hot and dense, neutrino production can occur by processes, such as (13.7), which do not involve the concurrent burning of nuclear fuel. In fact, neutrino energy loss may occur when no nuclear fuel is available to the star. Three important neutrino processes that occur during advanced evolution are the *photo production process, neutrino pair production,* and the *plasma process.* A neutrino energy-generation rate ϵ_ν (erg g^{-1} sec^{-1}) may be found for each neutrino process expected to occur in stellar matter. In general, ϵ_ν will differ from nuclear energy-generation rates in two ways: (1) it is strongly sensitive to density as well as temperature; and (2) it is often negative. When $\epsilon_\nu < 0$, its inclusion in (8.7) reduces the photon luminosity, which in turn reduces the temperature gradient. A final difference that becomes important for highly degenerate stars is that some neutrino processes, such as (13.1) and (13.2) occurring alternately, can convert mechanical energy (for example, vibrational energy) into energy radiated by neutrinos with no net change in the star's composition (URCA process).

Neutrino Pair Production

When the temperature in a stellar core reaches $T \simeq 2m_ec^2/k \simeq 6 \times 10^9$ K, electron-positron pairs are produced in large numbers by the equilibrium reaction

$$e^+ + e^- \longrightarrow \gamma + \gamma. \qquad (13.16)$$

A competing process results from the direct electron-neutrino coupling, and permits the e^+e^- pair to decay into neutrinos:

$$e^+ + e^- \longrightarrow \nu_e + \bar{\nu}_e. \qquad (13.17)$$

In hot, dense systems, the reaction (13.16) occurs about once for each 10^{22} times that (13.13) occurs. The photons diffuse outward, contributing to the thermal pressure that helps support the overlying portions of the star; but the neutrinos usually escape directly, robbing the star of thermal energy and thus cooling it. The loss of thermal energy will also reduce the thermal pressure.The importance of the pair process (13.17) may be understood by calculating the energy density lost per second because of neutrino escape. We denote the energy loss per cm^3 per second by Q_{pair}. Now, (13.17) is determined by the $e - \nu_e$, coupling, whose scattering cross section is

$$\sigma = \frac{G_w^2}{3\pi v} m_e^2 \left[\left(\frac{2E_p}{m_ec^2}\right)^2 - 1\right]$$

$$= 1.4 \times 10^{-45} \left(\frac{c}{v}\right)^2 \left(\frac{4E_p^2}{m_e^2c^4} - 1\right) \text{cm}^2, \qquad (13.18)$$

where G_w is the weak interaction coupling constant, $G_w = 1.41 \times 10^{-49}$ ergs cm^3, E_p is the electron's total energy, and v is the electron-positron relative velocity. Notice that the cross section increases with the square of the electron energy. The energy that goes into ν_e and $\bar{\nu}_e$ is the sum of the electron and positron total energies. Therefore Q_{pair} must be given by

$$Q_{pair} = -\int (E^+ + E^-)\, \sigma_l v\, dn_-\, dn_+, \qquad (13.19)$$

where $E^\pm$ is the positron (electron) energy, $n_\pm$ is the positron (electron) number density, and σ_l is (13.18) expressed in a frame of reference fixed with the matter. The number densities $n_\pm$ are Fermi-Dirac distributions given by (2.53). For nonrelativistic electrons, $\sigma_l \approx \sigma \sim 4 \times 10^{-45}/v$ cm^2, $E^\pm \simeq m_ec^2$, and the pair process is given approximately by

$$Q_{pair} \simeq -4.3 \times 10^{-45} \times (2m_ec^2)n_+n_- \text{ erg cm}^{-3}\text{ sec}^{-1}. \qquad (13.20)$$

The electron number density is related to the chemical potential μ by (2.56), which for high temperatures becomes

$$n_- = \frac{e^{\mu/kT}}{\pi^2}\int_0^\infty p^2 e^{-p^2/2m_ekT}\, dp$$

$$= \frac{e^{\mu/kT}}{\sqrt{\pi}}\left(\frac{m_ekT}{\pi}\right)^{3/2}. \qquad (13.21)$$

The positron number density is given by n_- with μ replaced by $-\mu$ (the electron chemical potential is -1 times the positron chemical potential) times a factor $\exp(-2m_ec^2/kT)$. It follows, upon substitution into

Q_{pair}, that

$$Q_{pair} \simeq -2.15 \times 10^{-45}\, m_e(m_e kT/\pi)^3\, e^{-2m_ec^2/kT}$$
$$= -4.9 \times 10^{18}\, T_9^{\,3}$$
$$\times \exp(-11.89/T_9) \text{ ergs cm}^{-3}\text{ sec}^{-1}. \quad (13.22)$$

The equivalent energy-generation rate is given by Q_{pair}/ρ.

When the electrons are degenerate and relativistic (as they usually are in evolved stellar cores), the pair process is approximately

$$Q_{pair} \simeq -1.43 \times 10^{15}$$
$$\times (1 + 5kT/E_F)T_{10}^{\,4} \exp(-E_F/kT) \quad (13.23)$$

where

$$E_F = c\sqrt{(3\pi^2 n_e)^{2/3} + m_e^2c^2}$$

is the relativistic electron Fermi energy. The energy-release rate increases with temperature more strongly than in the nonrelativistic regime, but decreases strongly with increasing density. This may be understood by noting that increasing T aids the first step of the process

$$\gamma + \gamma \rightleftharpoons e^+ + e^- \longrightarrow \nu_e + \bar{\nu}_e, \quad (13.24)$$

but that increasing the density implies fewer available states for the e^+e^- pairs by which the neutrinos are ultimately produced. In the nonrelativistic, high-T regime, (13.22) is independent of density, but increases with increasing T. Exact results for Q_{pair} are shown in Figure 13.4 for several temperatures.

Problem 13.4. Suppose neutrinos are produced by the pair process within the isothermal core of a white dwarf, whose central temperature is 10^9 K. Take the core radius $R_c = 10^{-3}R_\odot$ and the total thermal energy to be stored in radiation: that is,

$$U = aT^4\, V_c,$$

where $V_c = 4\pi R_c^{\,3}/3$ is the core volume. Find the neutrino luminosity, and estimate the cooling time due to neutrino losses.

The pair process is most important in stellar cores at intermediate densities $10^7 \lesssim 10^9$ g/cm^3, and high temperatures $T \gtrsim 10^9$ K, which arise in the core of red supergiants, and in the central stars of planetary nebulae (hot white dwarfs).

Photoneutrino Process

Photon-induced neutrino pair production may also occur when the temperature is too low for significant e^+e^- pair production. The photoneutrino process depends on the possibility that an electron can absorb a photon and then re-emit a neutrino-antineutrino pair,

$$e^- + \gamma \longrightarrow e^- + \nu_e + \bar{\nu}_e. \quad (13.25)$$

The energy available as neutrinos is proportional to:

(1) the initial energies of the e^- and γ, less the final e^- energy;
(2) the cross section; and
(3) the number densities of e^- and γ in the initial state.

It also depends on the number density and degree of degeneracy of the electrons in the final state. The results are summarized in Figure 13.4. Photoneutrino loss rates increase with increasing T at all densities because of the increased γ production, which initiates the reaction (13.25). For a given T, an increase in density leads to an increase in Q, although the rate of increase becomes smaller at higher density. This occurs since increasing ρ at fixed T increases the degree of degeneracy of the electrons, which reduces the number of available final states for the e^-. Finally, note that the Q becomes density-independent in the degenerate limit (high density) corresponding to the dashed lines in Figure 13.4.

At high densities this approach must be modified to include the effects of matter on the photons. In particular, the energy-momentum relation $\omega = kc$* for photons, which was used in obtaining Q, is strictly valid only in vacuum. When light propagates through a medium, the correct energy-momentum relation must include the dielectric constant of the medium $\epsilon(\omega, k)$. When this is included, the correct photo production rates are found to decrease rapidly with

*Here we use k to represent the photon's wave vector, which equals $2\pi/\lambda$.

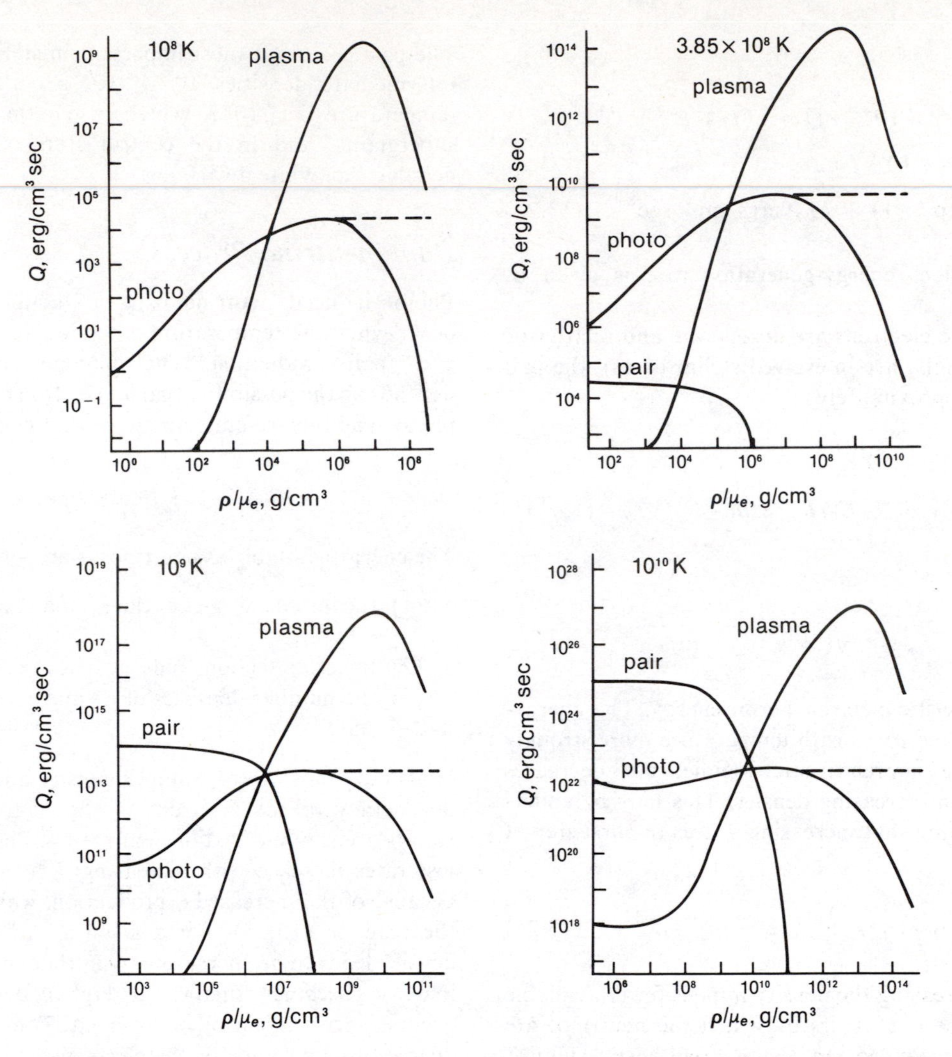

Figure 13.4. Neutrino energy-emission rates at four temperatures as a function of ρ/μ_e. The dashed extension of the photoemission curve represents what would occur if plasma corrections were not included.

increasing density at fixed T, as shown by the solid curves in Figure 13.4.

Finally, we remark that the presence of positrons will lead to an additional contribution to the photo production of neutrinos via the process

$$\gamma + e^+ \rightarrow e^+ + \nu_e + \bar{\nu}_e. \tag{13.26}$$

This is important at sufficiently high T when the equilibrium reaction

$$\gamma + \gamma \rightleftharpoons e^+ + e^- \tag{13.27}$$

occurs. The total photoneutrino-production rate is the sum of contributions from the processes (13.25) and (13.26). The curves in Figure 13.4 include both effects.

Plasma Process

When the stellar core density becomes large, electromagnetic radiation is strongly affected by the dielectric properties of the electron gas. The mutual (collective) interaction of a photon with the plasma is called a *plasmon.* The distinction between a plasmon and a photon becomes important when the dielectric constant of the plasma

$$\epsilon(\omega, k) = 1 - \omega_p^2/\omega^2 \tag{13.28}$$

becomes less than one. The plasma frequency

$$\omega_p^2 \simeq \frac{4\pi n_e e^2 c^2}{E_F}, \tag{13.29}$$

where E_F is the relativistic electron Fermi energy, and n_e the free-electron number density. Equation (13.29) reduces to the plasma frequency for nonrelativistic electrons ($E_F \approx m_e c^2$). The plasmon, $\gamma_{\rm pl}$, can decay directly to form a neutrino-antineutrino pair,

$$\gamma_{\rm pl} \longrightarrow \nu_e + \bar{\nu}_e. \tag{13.30}$$

Problem 13.5. Show that $\epsilon(k, \omega) \approx 1$ for the solar core. How dense would a C^{12}-burning core have to be for $\omega_p/\omega \approx 1$, assuming $T \simeq 6 \times 10^8$ K for C^{12} burning? Assume $E_F = m_e c^2$.

This process can not occur in vacuum, that is, for a photon, since it would not conserve energy and momentum. The plasma process actually involves the intermediate production of a e^+e^- pair, which subsequently decays, forming neutrinos; so (13.30) is a combination of weak and electromagnetic interactions. It is possible to show that the plasmon energy is given by the relation $k^2c^2 = \omega^2\epsilon(k, \omega)$, which, using (13.29), gives

$$\begin{aligned}\omega &= (k^2c^2 + 4\pi n_e e^2 c^2/E_F)^{1/2} \\ &= (k^2c^2 + \omega_p^2)^{1/2}.\end{aligned} \tag{13.31}$$

We see that ω_p acts as an effective mass for the plasmon. Plasmons exist in sufficient numbers to initiate the decay (13.30) whenever the density and temperature satisfy $\hbar\omega_p \gtrsim kT$. The neutrino energy-loss rate via the plasma process depends on:

(1) the energy available for neutrinos ($h\nu$);
(2) the photon number density, i.e., the Planck distribution function with energy (13.31); and
(3) the plasmon decay rate $\tau_{\rm pl}$.

For low density and high temperature, the plasmon energy-loss rate increases like T^3, which is proportional to the photon number density. Since the plasmon decay rate increases with density, so does the loss rate. At high density and fixed T, the loss rate drops rapidly with density, because as ρ increases, so does ω_p. Consequently only the more energetic plasmons have a real wave vector k (13.31), and their numbers are limited by the value of T. Plasma neutrino loss rates are shown in Figure 13.4.

The net neutrino loss rate due to pair, photo, and plasma processes is given by

$$Q_\nu = Q_{\rm pl} + Q_{\rm pair} + Q_{\rm photo}. \tag{13.32}$$

Not all three will be important at the same time, as is obvious from Figure 13.4. Figure 13.5 shows the regions of the ρ, T-plane in which each process dominates. The lines are values of ρ and T for which the processes on either side become equal. Finally, we remark that an energy-generation rate ϵ_ν(erg g^{-1} sec^{-1}) due to these neutrino processes may be obtained immediately from Q_ν by the relation

$$\epsilon_\nu = Q_\nu/\rho. \tag{13.33}$$

A plot of ϵ_ν based on exact expressions is given in Figure 13.6, assuming that the material is characterized by a mean molecular weight per electron $\mu_e = 2$. The numbers above each curve represent the temperature in units of 10^9 K, and T is constant along each curve. The segments of the curve with negative slope at lower densities are dominated by pair processes; the horizontal sections indicate that photo production dominates pair production. The hump at high densities is due to the plasma process. The rapid decrease in ϵ_ν in the lower-density regime results because $Q_{\rm pair}$ is independent of ρ, and thus $\epsilon_\nu \approx Q_{\rm pair}/\rho$ goes as $1/\rho$ until the contribution from the photoproduction process becomes significant.

We noted in the discussion of silicon burning that neutrino energy-loss rates could become competitive with nuclear energy release. Noting that densities of 10^5 to 10^7 g/cm^3 and temperatures of 3 to 5 $\times$ 10^9 K are common during silicon burning, we see from Figure 13.5 that pair annihilation is the dominant neutrino-producing process in such regimes, and find from Figure 13.6 that typical energy-loss rates because of neutrinos are in excess of 10^{12} ergs g^{-1} sec^{-1}. This may be comparable to, or exceed, silicon energy-generation rates.

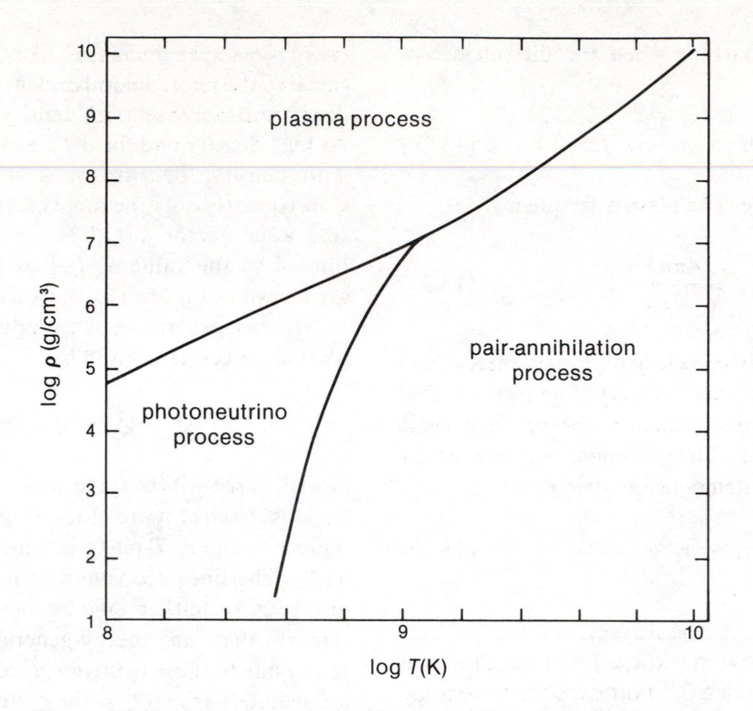

Figure 13.5. Dependence of neutrino energy-release processes on ρ and T.

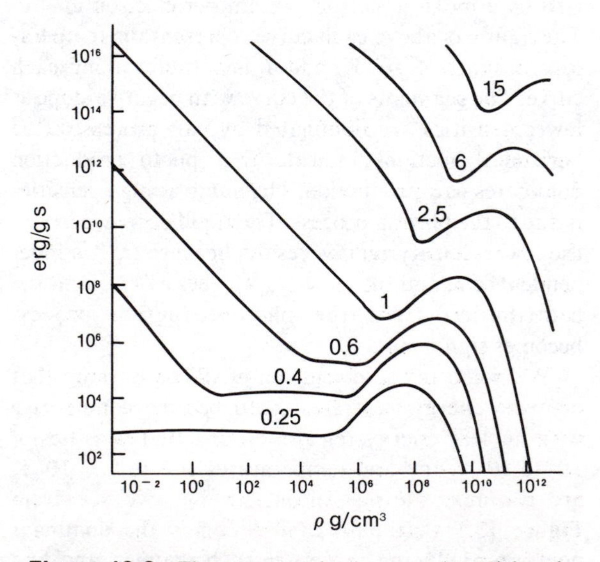

Figure 13.6. The energy-production rates (13.32) based on the three processes in Figure 13.5. The curves are labeled by T in units of 10^9 K.

Neutrino emission may affect other nuclear burning stages as well. For example, C^{12} burning begins when $T \approx 5 \times 10^8$ K, and releases about 14 MeV per reaction. In stars of intermediate mass, the core density may exceed 10^4 g/cm^3 when carbon ignition occurs. Under these conditions the photo neutrino process is the most important (see Figure 13.5), and removes energy from the stellar core at a rate 4×10^5 ergs g^{-1} sec^{-1}. This results in an acceleration in the rate of carbon burning, as follows.

Assuming that the core remains in or near hydrostatic equilibrium, the virial theorem states that gravitational potential energy released during contraction $\Delta\Omega$ is split about equally between increased internal energy ΔU and radiative losses ΔE_L:

$$\Delta\Omega = \Delta U + \Delta E_L. \tag{13.34}$$

The energy radiated from the star, ΔE_L, contains photon and neutrino contributions. Therefore ΔE_L including neutrinos will exceed ΔE_L due to photons alone, and thus ΔU must be greater when neutrinos are included. As the internal energy increases, so does the temperature and the rate of nuclear burning.

We may also estimate the extent to which neutrino

losses shorten carbon burning. The energy released per nucleon by C^{12} burning is $E_C \simeq 14/24 = 0.583$ MeV/nucleon, and the energy per gram of core matter by $E_c N_0$, where N_0 is the number of nucleons per gram. Now, suppose all the nuclear energy released locally is radiated away by neutrinos at rate ϵ_ν. Then the lifetime for the process is roughly

$$t_{C^{12},\nu} \simeq \frac{E_C N_0}{\epsilon_\nu} \simeq 10^{12} \text{ sec} \simeq 3 \times 10^4 \text{ yrs},$$

using the values for E_C, and ϵ_ν quoted above. Carbon burning under similar ρ, T conditions, but without neutrino losses, lasts about 10^6 yrs. Therefore neutrino radiation can reduce nuclear burning time by a factor of 1/25.

Problem 13.6. At $T \simeq 3 \times 10^9$ K, Si burns to Ni^{56} with the release of about 0.18 MeV/nucleon of matter in the core. Suppose that the core density is $\rho_c \simeq 10^7$ g/cm^3. What is the dominant neutrino loss process, and the estimated burning time, assuming that hydrostatic equilibrium is maintained? Is the assumption of hydrostatic equilibrium valid?

13.3. Coherent Scattering Off Nuclei

A neutrino-scattering process that is important during core evolution in massive stars, and that may be associated with supernova explosions, is the coherent scattering of neutrinos by massive nuclei. The theory of neutral-current weak interactions predicts that when heavy nuclei are abundant in stellar cores, then a neutrino will elastically scatter, not off an individual nucleon in the nucleus $N(A, Z)$, for which $\sigma \simeq \sigma_0 \alpha^2 E_\nu^2$, but off all A nucleons in a collective manner. In fact, the scattering process

$$\nu_e + N(A, Z) \rightarrow \nu_e + N(A, Z) \qquad (13.35)$$

is characterized by a cross section

$$\sigma_{\text{coh}} \simeq \sigma_0 \alpha^2 A^2 E_\nu^2 \text{ cm}^2, \qquad (13.36)$$

where E_ν is the neutrino energy in MeV, $\sigma_0 = 1.7 \times 10^{-44}$, and the weak interaction parameter $\alpha = 0.2$. The dependence on atomic weight A^2 reflects the coherent nature of the scattering. Coherent scattering is essentially an elastic process in stellar cores because of the large nuclear mass, but it does result in substantial momentum transfer between the neutrino and the nucleus. In fact, the average scattering angle for (13.35) is about 70°.

The mean free path for a neutrino of energy E_ν is given by

$$l_{\nu,\text{coh}} \simeq (\sigma_{\text{coh}} n_A)^{-1} \simeq \frac{4.9 \times 10^{20}}{\rho Z_A A E_\nu^2} \text{ cm}, \qquad (13.37)$$

where Z_A is the mass fraction of iron-group nuclei in the core with average atomic weight A, ρ is in g/cm^3, and E_ν in MeV, and we have used n_A (the number density of heavy nuclei) $\simeq Z\, \rho/m_H A$. For iron-group nuclei, $A \simeq 56$, and the coherent scattering cross section is about 3,000 times larger than the cross section for scattering off individual nucleons.

When the core of a star that has evolved beyond Si burning begins to collapse hydrodynamically, its average density may exceed 10^{10} g/cm^3. As the core collapses, electron capture on nuclei,

$$e^- + N(A, Z) \rightarrow N(A, Z - 1) + \nu_e \qquad (13.38)$$

produces a strong neutrino flux with average energies $\bar{E}_\nu \simeq 15$ MeV. Since $A \simeq 56$ for Fe-group nuclei, $l_{\nu,\text{coh}} \simeq 4 \times 10^6$ cm or less, and the ν_e produced within this region (whose radius is typically $r_c \simeq 10^8$ cm) will undergo on the order of $(r_c/l_{\nu,\text{coh}})^2 \simeq 7 \times 10^2$ collisions before escaping from the star. Since the scattering is elastic, the neutrino energy remains unchanged. However, because of the relatively small mean free path, a typical neutrino remains within the core long enough to undergo inelastic collisions such as (13.8), which reduce the neutrino energy (Problem 13.7).

Problem 13.7. In a typical presupernova core, $T \simeq 10^{10}$ K and $\rho \simeq 10^{10}$ g/cm^3. Use the results of Problem 13.2 to find out whether an arbitrary ν_e distribution will thermalize.

In this way the neutrinos may reach an equilibrium state at the local temperature of the gas, with a distribution in momentum space given approximately by the Fermi-Dirac distribution (2.6) with energy $\epsilon_\nu =$

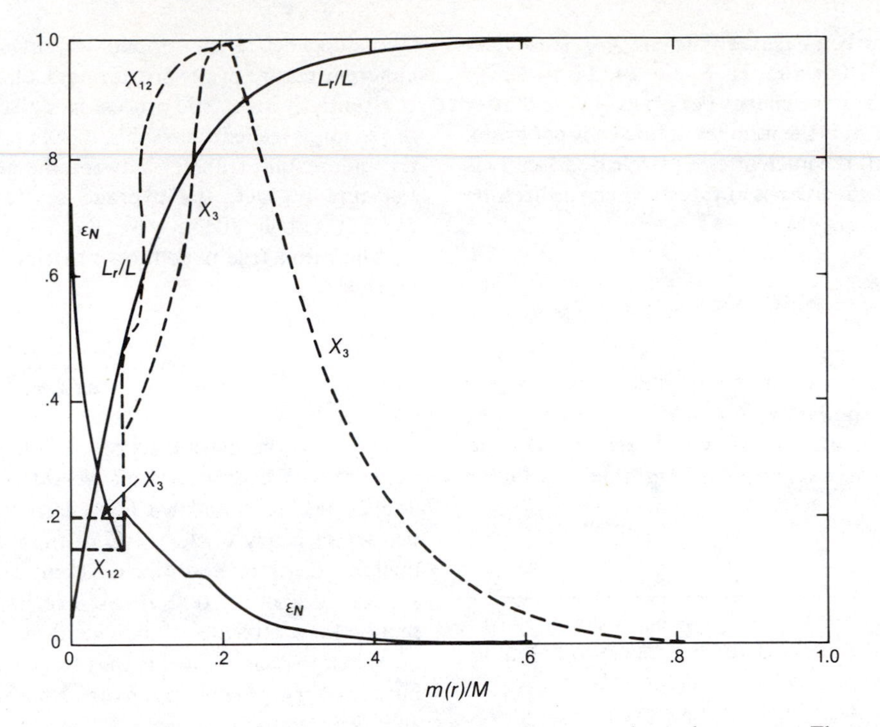

Figure 13.7. Solar model for Problem 13.9. ϵ_N represents the nuclear energy. The model corresponds to the stage for which the luminosity of the Sun is a minimum (the zero age main sequence).

$h\nu$ and a neutrino chemical potential μ_ν. Similar arguments apply to the electron antineutrino also. The energy density and pressure of a degenerate neutrino gas in thermal equilibrium are the same as for an ultrarelativistic electron gas with statistical weight $g = 1$; for nondegenerate conditions ($\mu_\nu \ll kT$), they are given by

$$u_\nu = \frac{7}{16} aT^4,$$
$$P_\nu = \frac{1}{3} u_\nu. \tag{13.39}$$

Problem 13.8. Use (2.38)–(2.41) with $(\epsilon_\nu - \mu_\nu)/kT \approx \epsilon_\nu/kT$, and the identity

$$\int_0^\infty x^3(1 + e^x)^{-1}\, dx = 7\pi^4/120,$$

to establish u_ν and P_ν given above. Show that the entropy per unit volume $S/V = (7a/12)T^3$.

The neutrino number density in thermal equilibrium at temperature T for a nondegenerate system is given by (2.39), and is

$$n_\nu \simeq 7.6 T^3 \text{ cm}^{-3}. \tag{13.40}$$

Note the resemblance of these expressions to P and u for photons.

We may use the results above to estimate the neutrino luminosity L_ν and evolutionary time-scale t_ν of a presupernova stellar core, assuming that the neutrinos are in thermal equilibrium. The time required for a typical neutrino of energy E to escape from the core is $\tau \approx l_{\nu,\text{coh}}/c$ and the neutrino energy loss rate per gram is $dE/dt \approx E_\nu n_\nu/\rho\tau$. The luminosity of the core of mass M_c is therefore $L_\nu \approx E_\nu n_\nu M_c/\rho\tau$. The time-scale for thermal neutrinos to support the core with this luminosity is, to within an order of magnitude, the ratio of the total energy emitted as neutrinos ϵ to L_ν, $t_\nu \approx \epsilon/L_\nu$. We estimate ϵ by noting that if the core is to become a neutron star, then $M_c/2m_H$ protons must be converted to neutrons by inverse β decay (13.3). Thus

$$t_\nu \approx (M_c E_\nu / 2m_H)/(E_\nu n_\nu M_c / \rho\tau) = (\rho\tau / 2m_H n_\nu),$$

which is independent of core mass. Assuming typical presupernova values to be $T \simeq 10^{10}$ K, $E_\nu \simeq 15$ MeV, $\rho \simeq 10^{10}$ g/cm^3, and $A \simeq 56$, the neutrino luminosity $L_\nu \simeq 3 \times 10^{53}$ ergs/sec and the evolutionary time-scale $t_\nu \simeq 0.1$ sec, which is comparable to the free-fall time-scale for the core. Detailed numerical studies show that these estimates are quite reasonable.

Problem 13.9. Periodic mixing of material in the solar core has been proposed as a means of resolving the problem of the low observed Cl^{37} capture rates. Figure 13.7 shows the helium abundance X_{He} in a standard solar model. Suppose an instability gradually builds up in the solar core, to be relieved rapidly by a complete mixing of the region $r \lesssim r_1$. Referring to (7.7)–(7.9), explain how this could account for the low B^8 decay rates implied by the solar neutrino experiments.

Chapter 14

DEGENERATE STARS

The matter in stars near the end point of their evolution may reach densities so great that the ideal-gas equation of state in the form (2.1) is no longer applicable. Typical examples include red-giant cores, some models of presupernovae, white dwarfs, and neutron stars.

Although we will concentrate on stellar objects, the physics of degenerate matter is also crucial for understanding the structure of planetary interiors. Here degeneracy may arise at relatively low density simply because the temperatures are so low.

Several important properties of degenerate configuration result from the Pauli exclusion principle, and may be discussed in general. The basic properties of ideal gases under degenerate conditions are reviewed in Chapter 2. We will develop here the properties of white dwarf stars, discussing their internal structure first, and then the structure of their nondegenerate envelopes, which control their rate of cooling.

In order to describe the more advanced states of stellar evolution, we must digress into the problem of matter at high densities. A coin in the reader's pocket is a typical example of a system in which the electrons are degenerate, and to which some of the following discussion would also apply.

14.1. Degenerate Matter in Stars

The successive contraction of the core of a star, caused by its running out of fuel, will lead to higher and higher core densities. Eventually the matter may become so dense that it ceases to be described by the equation of state (2.1). This happens when electrons in ordinary stars or nucleons in more compact stars are so compressed that their mutual separation is comparable to their Compton wavelength. The star's properties are then determined by the uncertainty principle, which requires that the fermion's momentum, p, and the linear extent of the region to which it is confined, λ, must satisfy $p\lambda \gtrsim \hbar$. Since the volume per fermion is then $V_0 \sim \lambda^3$, the momenta satisfy

$$p \gtrsim \hbar / V_0^{1/3}, \tag{14.1}$$

and the energy (assuming that the fermions are nonrelativistic) must be at least

$$E_0 \sim \frac{p^2}{2m_e} \gtrsim \frac{\hbar^2}{2m_e} V_0^{-2/3} = \frac{\hbar^2}{2m_e}\left(\frac{N}{V}\right)^{2/3} \equiv E_F, \tag{14.2}$$

where the total volume occupied by N fermions is $V = NV_0$, E_0 is the quantum-mechanical zero-point energy of the fermions, and E_F is the Fermi energy. If the temperature is nonzero, then the fermion's thermal energy will be roughtly $E_T \sim kT$. Notice that even when the temperature and thermal energy vanish, E_0 is nonzero. The condition for degeneracy is that E_0 exceed E_T. An equivalent condition is that the chemical potential of the fermion $\mu > kT$; the degree of degeneracy is measured by the ratio μ/kT. Using (14.2) and noting that for a system of ions and electrons the density $\rho \sim m_I(N/V)$, where m_I is the mass of the ions in the core, we find that the condition $E_0 \gtrsim E_T$ implies

$$T \leq \frac{\hbar^2 \rho^{2/3}}{2m_e m_I^{2/3} k}. \tag{14.3}$$

Problem 14.1. Verify equation (14.3). Typically $T < 10^8$ K and $10^4 \lesssim \rho < 10^9$ g/cm^3 in the core of white dwarfs. Would the electrons be nondegenerate under these conditions?

Problem 14.2. Modify the arguments leading to (14.3) for a system of relativistic electrons whose energies are $E_0 \simeq pc$. Show that the corresponding degeneracy condition is

$$T \leq \frac{\hbar c}{k m_I^{1/3}} \rho^{1/3}. \tag{14.4}$$

The two conditions (14.3) and (14.4) may be applied to degenerate nucleons in neutron stars by replacing m_e and m_I by the nucleon mass $m_N \simeq m_p$.

Because the temperature inside a star that is close to hydrostatic equilibrium grows with increasing density roughly as

$$T \sim \rho^{1/3}, \tag{14.5}$$

it follows from (14.3) that the degeneracy limit must eventually be reached by a contracting star. As long as (14.3) or (14.4) is satisfied, the pressure is determined primarily by degeneracy. Therefore, in order to simplify matters, we will consider just the extreme case of degeneracy pressure in very dense matter. Problem 14.3 shows that when (14.3) is satisfied, the ratio of thermal pressure P_T to degeneracy pressure P_D is small.

Problem 14.3. Show that the momentum flux across an arbitrary surface in degenerate electron matter is given by

$$\text{flux} \sim pvn, \tag{14.6}$$

where the electron's average momentum is p, its average velocity v, and its number density n. Noting that pressure is momentum transfer per unit area, show that $P_D \simeq p^2 n/m_e$, and that

$$\frac{P_T}{P_D} \sim \frac{m_e m_I^{5/3} kT}{\hbar^2 \rho^{5/3}}; \tag{14.7}$$

show that this is small when (14.3) is satisfied.

Our treatment of degenerate objects does not imply that they can not radiate, or that their surfaces have zero effective temperature. The star has two sources of internal energy: degeneracy energy, E_D, and the thermal energy, E_T, associated with random motion of the ions, and perhaps a small contribution due to a slight excess in electron energy over E_0. E_D can change only if the volume (or radius) of the star changes [see (14.2)]. Therefore, once a state of hydrostatic equilibrium is reached and the radius is fixed, only the thermal energy can be reduced by radiation from the system. Nevertheless, the amount of energy stored thermally in the motion of the ions can be significant. For example, at an average density and temperature of 10^6 g/cm^3 and 10^6 K, respectively, and a radius of 8×10^8 cm, as is typical for white dwarfs, the total thermal energy content is about 2×10^{47} ergs. This energy supply is sufficient to maintain a typical white dwarf luminosity $L \sim 10^{-3} L_\odot$ for 10 billion years.

We should also note that under conditions of high density and electron or neutron degeneracy, the thermal conductivity is very high. Thermal conduction is impeded when particles scatter off one another. For this to occur, there must be physical states available into which the particles can be scattered. If the particles are fermions (electrons in white dwarfs or red-giant cores; nucleons in neutron stars), then the final states must be empty. In nondegenerate matter, most states are unpopulated, and scattering can occur. In degenerate matter, however, nearly all energetically accessible final states are already occupied; so the fermions, despite their small interparticle separations, have a low probability of scattering. They are extremely efficient conductors of heat energy. Therefore energy is easily transmitted through degenerate material by conduction; in fact, whenever we have degeneracy in stellar cores, we may as a first approximation consider the core to be isothermal.

Because degenerate systems may be treated as nearly isothermal, and because large temperature changes leave the pressure unchanged, the mechanical and thermal properties may usually be decoupled, and their internal structure will be described by the equation of hydrostatic equilibrium. For these reasons, models of white dwarfs and neutron stars are rather easily calculated, once the pressure-density equation of state has been specified. Finally, since the temperature has little effect on the mechanical equilibrium of these stars, their structure may, for a given equation of state and composition, be specified by a single parameter (strictly speaking, this holds only for nonrotating and nonmagnetic stars) and composition. The parameter usually chosen is either the total mass or the central density ρ_c.

14.2. Degenerate Matter in Hydrostatic Equilibrium

The equilibrium properties of a degenerate system of fermions (electrons; nucleons or other baryons) may be considered in general, and the conclusions applied to either white dwarfs or neutron stars with only slight modifications. Each will be considered individually later. The basic structural properties of degenerate stars may be obtained from a simple combination of the uncertainty principle and gravitation, as we show in the following.

The pressure at $T = 0$ is due in part to the uncertainty principle, which states physically that no two fermions in the same quantum state may occupy the same point in space. This is equivalent to requiring that the volume per fermion be proportional to $\lambda_c^3 \sim (\hbar/mc)^3$, where m is the fermion's mass and λ_c is its Compton wavelength. The average number density of the fermions is therefore $n_f \sim \lambda_c^{-3}$. In white dwarfs the density is n_f times the total mass per electron, and in neutron stars it is the nucleon mass times n_f. Denoting the average mass per fermion by μ_f [for electrons in white dwarfs we take $\mu_f \simeq (A/Z)m_n$, and for neutron stars $\mu_f \simeq m_n$, where A and Z are atomic weight and atomic number, and m_n is the nucleon mass], the average density

$$\rho \sim \mu_f \lambda_c^{-3} \sim \mu_f (mc/\hbar)^3. \qquad (14.8)$$

When interparticle interactions become important, this estimate must be modified (as we will see).

When the mass is not too large, degeneracy pressure can support a star against gravity. However, with increasing mass, the fermions become relativistic, and the star becomes unstable. To show this we examine the energetics of the system. The average kinetic energy per fermion is

$$\begin{pmatrix}\text{kinetic energy}\\ \text{per fermion}\end{pmatrix} \simeq \begin{cases} \dfrac{\hbar^2 p_F^{\,2}}{2\,m} & \text{(NR)} \\ \hbar\, c\, p_F & \text{(ER)} \end{cases} \qquad (14.9)$$

in the nonrelativistic and extreme relativistic limits, denoted by NR and ER, respectively. The Fermi wave vector $p_F \simeq N_f^{1/3}/R$, where N_f is the number of fermions in the star. For white dwarfs, the ion kinetic energy is proportional to $kT \lesssim 10^{-2}$ MeV for $T \lesssim 10^8$ K; for densities greater than 10^4 g/cm^3, the electron's kinetic energy is $\gtrsim 10^{-2}$ MeV. Therefore in white dwarfs the electrons contribute the major portion of the energy. In neutron stars the electron Fermi energy is well below the nucleon Fermi energy, because the electron number density has been greatly reduced by electron capture. Consequently the nucleons supply most of the kinetic energy. In either system the total mass M of the star results primarily from the ions or nucleons; so the average gravitational potential energy is proportional to $-M^2\,G/R$. Consider a white dwarf whose density is less than 10^6 g/cm^3. The electrons are then nonrelativistic and the total energy (kinetic plus gravitational potential) is therefore, to order of magnitude,

$$E_T \simeq N_I \frac{\hbar^2 N_f^{2/3}}{2m_f R^2} - \frac{M^2 G}{R}. \qquad \text{(NR)} \quad (14.10)$$

Here N_I is the ion number, $ZN_f = N_I$, and $M \simeq N_I A m_p$. The fermion mass m_f is taken as the electron mass m_e. For neutron stars N_I equals the fermion or baryon number N, and $Z = A = 1$. Therefore (14.10) is applicable to neutron stars with fermion mass $m_f = m_p$.

For a given mass M, the system will be in that state for which the total energy E_T is lowest. Minimizing (14.9) with respect to R yields a relation between mass and radius or mass and average density for the equilibrium configuration. Thus, if M or $\bar{\rho}$ is specified, the structure of the degenerate system is determined.

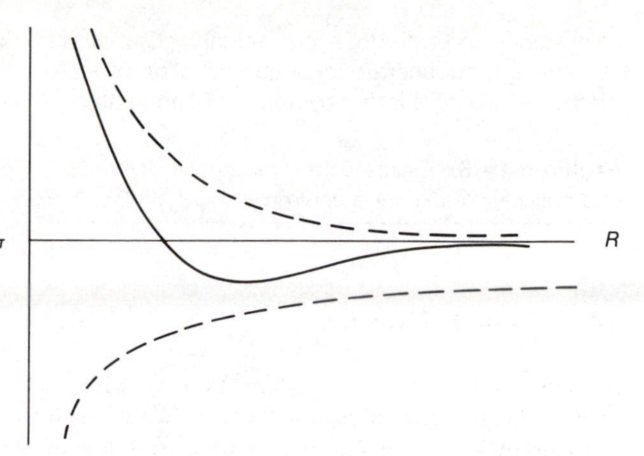

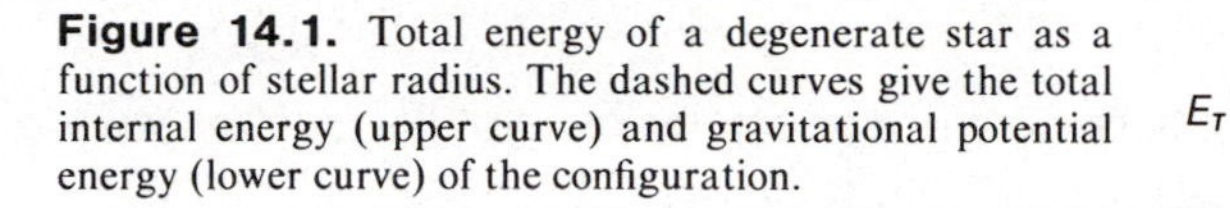

Figure 14.1. Total energy of a degenerate star as a function of stellar radius. The dashed curves give the total internal energy (upper curve) and gravitational potential energy (lower curve) of the configuration.

Problem 14.4. Find the radius and total energy of the equilibrium system whose fermions have average energy E_T.

The total energy E_T is shown schematically in Figure 14.1. Notice that an equilibrium state always exists for the nonrelativistic system, because, as the system expands, its pressure drops more rapidly than do the gravitational forces; so, if expansion was initiated by overpressure, a new state of balance is rapidly regained. Furthermore, the addition of mass to the system leads to a new equilibrium state, as is evident from Problem 14.4.

As more mass is added, the average momentum of the fermions increases, eventually becoming relativistic. In white dwarfs this occurs for densities above 10^6 g/cm^3. The total energy of the star is then, to order of magnitude,

$$E_T \sim \frac{\hbar c\, Z^{4/3}\, N^{4/3}}{R} - \frac{M^2 G}{R}, \qquad \text{(ER)} \qquad (14.11)$$

and the system cannot readjust once it has deviated from equilibrium, since both pressure and gravitational terms are proportional to the same power of the radius. In fact, we can use (14.11) to find an upper limit to the system's mass if it is to be in equilibrium, by finding the largest mass for which $E_T < 0$. The critical mass, for which $E_T = 0$, is

$$M_{\text{crit}} = (Z/A)^2\, m_p\, (\hbar c / m_p^{\,2}\, G)^{3/2}, \qquad (14.12)$$

as follows from (14.11).

When account is taken of the gradual transition from a nonrelativistic to a fully relativistic fermion gas, the relation between stellar total mass and central density appears as shown schematically in Figure 14.2. For white dwarfs, the mass limit is called the Chandrasekhar mass; for neutron stars it is called the Tolman–Oppenheimer–Volkoff mass. Its exact value will depend to some extent on the chemical composition and on the interactions between particles. In white dwarfs and neutron stars, these effects turn out to be of critical importance for stability, and in setting the mass limit.

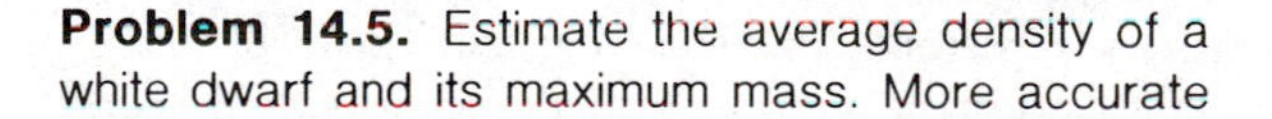

Problem 14.5. Estimate the average density of a white dwarf and its maximum mass. More accurate

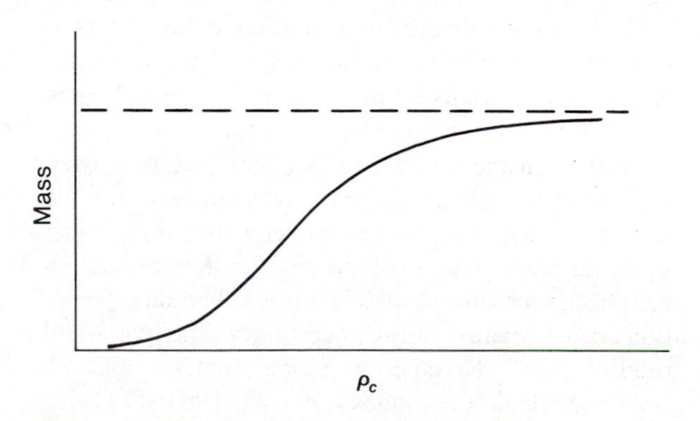

Figure 14.2. Mass versus central density for a star, assuming total pressure due to ideal fermion gas and Newtonian gravitation.

calculations indicate that the actual critical mass (assuming a degenerate ideal-gas equation of state for the electrons) is larger by a factor of about three.

Problem 14.6. Estimate the average density, radius, and maximum mass for a neutron star.

14.3. WHITE DWARFS

Sections 14.1 and 14.2 emphasized the order-of-magnitude properties of degenerate stars. We will now consider in greater detail the interior structure of white dwarf stars.

Stars less massive than about 4 $M_\odot$ probably end their evolution as white dwarfs. Since mass loss (Section 10.6) probably reduces M below 1.4 $M_\odot$ before C^{12} burning is initiated, He is the last nuclear fuel burned before degeneracy halts collapse. In this state, electron degeneracy represents the final source of pressure capable of supporting the star. Since stars that reach this stage have evolved beyond helium burning, the end product will consist primarily of a C^{12}, or C^{12} and O^{16}, core, surrounded by a He^4 envelope and possibly an outer layer of hydrogen. Most of the mass will therefore have evolved beyond H; so we will assume that $A = 2Z$.

Two components of matter may be distinguished in white dwarfs. First, the degenerate electrons supply nearly all the pressure, and their high thermal conductivity represents the most efficient mode of energy transport in the interior. Second, the ions supply nearly all the mass, and contain nearly all the stored thermal energy that is ultimately radiated away as the star cools. Consequently, the hydrostatic and thermal equilibrium equations may be decoupled. The hydrostatic equations are discussed in this section; thermal transport will be considered in Section 14.4.

Matter inside white dwarfs is completely ionized, even when the star has cooled to zero temperature. The source of ionization is the extreme pressure, which squeezes atoms into a volume smaller than the atomic volume under terrestrial conditions. Pressure ionization occurs because of the uncertainty principle, which requires that the electron momentum, p, and the region to which it is confined, $r_e \sim V_0^{1/3}$, satisfy (14.1). The volume per electron V_0 is related to the volume per ion V_i by $ZV_0 = V_i$, so that (14.1) becomes $p \gtrsim \hbar\, Z^{1/3}/V_i^{1/3}$.

Problem 14.7. The energy of a single electron of momentum p bound to an ion of charge Ze is given by

$$E = \frac{p^2}{2m_e} - \frac{Ze^2}{r_i}, \tag{14.13}$$

where r_i is the average distance of the electron from the ion, and may be taken as the radius of the system. Use the uncertainty principle as discussed above to show that E is a minimum if

$$r_i \sim Z^{-1/3} a_0, \tag{14.14}$$

where a_0 is the radius of the first Bohr orbit for hydrogen.

As shown in Problem 14.7, the effective ion radius is of the order $r_i \sim a_0 Z^{-1/3}$. Pressure ionization results when the interior separation

$$d_i \simeq (3/4\,\pi n_e)^{1/3}\, Z^{1/3} < r_i.$$

The condition $d_i < r_i$ leads directly to the density constraint

$$\rho > \frac{3}{4\pi}\frac{AZm_p}{a_0^3} = 3.2\, AZ \text{ g/cm}^3. \tag{14.15}$$

When this inequality holds, the electrons will be completely ionized. Since most white-dwarf interiors are at densities in excess of 10^4 g/cm^3, pressure ionization is usually complete.

Problem 14.8. Derive the density condition for pressure ionization (14.15). Evaluate it for white dwarfs made of He^4, C^{12}, and Fe^{56}. Justify the claim that when this condition holds, all Z electrons of an atom of atomic number A will have been ionized.

The structure of a white dwarf is determined by hydrostatic equilibrium, by the pressure-density equation of state, and by the composition. In general, the total pressure is given by a sum of the ion pressure P_i and the electron pressure P_e, but since P_i/P_e is small (typically a few percent), we will assume that the total pressure is just P_e.

In low-mass white dwarfs, with central densities below 10^6 g/cm^3, the motion of the degenerate electrons is nonrelativistic, and electron pressure may be approximated by the polytrope of index $n = 1.5$:

$$P_e = \frac{(3\pi^2)^{2/3}\hbar^2}{10\pi^2 m_e}\left(\frac{\rho}{m_p\mu_e}\right)^{5/3}$$
$$= 1.0 \times 10^{13}(\rho/\mu_e)^{5/3}. \qquad (14.16)$$

The electron number density n_e and the ion mass density ρ are related by $n_e = \rho/m_p\mu_e$, which defines the mean molecular weight, μ_e. Note that μ_e is not defined in the same way as the mean molecular weight μ, which appears in models of nondegenerate stars. Assuming that ionization is complete and that the composition is limited to a single atomic species throughout most of the mass, we may set $\mu_e = A/Z$.

In more massive white dwarfs, whose central densities exceed 10^6 g/cm^3, the average electron energies are relativistic, and the electron pressure is given approximately by the polytrope of index $n = 3$:

$$P_e = \frac{\hbar c(3\pi^2)^{1/3}}{4}\left(\frac{\rho}{m_p\mu_e}\right)^{4/3}$$
$$= 1.24 \times 10^{15}(\rho/\mu_e)^{4/3}. \qquad (14.17)$$

Notice that the electron pressure is independent of the electron mass in the extreme relativistic limit. Although the electrons supply some mass density to the system, the contribution is small, and the mass density may be expressed as

$$\rho = Am_p n_I = (A/Z)m_p n_e. \qquad (14.18)$$

The structure of low-mass white dwarfs based on the nonrelativistic pressure-density equation of state for electrons follows from the polytropic relations discussed in Section 4.3. In particular, equations (4.50) and (4.51) show that the mass and radius are related by

$$M \sim R^{-3}. \qquad (14.19)$$

An increase in mass will result in a smaller radius, where P_e is well approximated by (14.16).

When the average electron energy is relativistic, (14.17) is a better approximation to P_e, and the results of Section 4.3 show that the white dwarf's mass is then given by

$$M_{\text{Ch}} = \frac{5.80}{\mu_e^2} M_\odot. \qquad (14.20)$$

Problem 14.9. Refer to the section on polytropic stellar models, and show that the mass for an $n = 3$ polytrope is given by (14.20).

This is the Chandrasekhar mass. Notice that it is independent of density, but does depend on composition. For matter that has evolved beyond hydrogen burning, $\mu_e \simeq 2$, and (14.17) reduces to

$$M_{\text{Ch}} = 1.44\ M_\odot. \qquad (14.21)$$

For intermediate masses, the complete expression for the pressure and density of an ideal degenerate electron gas must be used in the equations of hydrostatic equilibrium.

When we take into account the electromagnetic interactions between electrons and ions in high-density white dwarfs the equation of state $P(\rho)$ gives a slightly lower pressure than would an ideal-gas equation of state at the same density. The interactions in this case soften the equation of state, and reduce the maximum mass. We may show that the equation of state is softened by noting that, if each ion has α nearest neighbors an average distance r_i away, then the Coulomb energy associated with ion-ion interactions is given roughly by

$$E_I \sim \frac{1}{2}\alpha^2 N_i \frac{(Ze)^2}{r_i}, \qquad (14.22)$$

where N_i is the number of ions in the star. Each electron has roughly Z nearest neighbor electrons at an average distance r_e. Since the volume per ion is Z times the volume per electron,

$$r_e \simeq Z^{-1/3} r_i,$$

the electron-electron Coulomb interaction is roughly

$$E_e \sim \frac{1}{2}(Z\alpha)^2 \frac{e^2}{r_e} N_i$$
$$\simeq \frac{1}{2}\alpha^2 N_i \frac{(Ze)^2}{r_i} Z^{1/3} = Z^{1/3} E_I. \qquad (14.23)$$

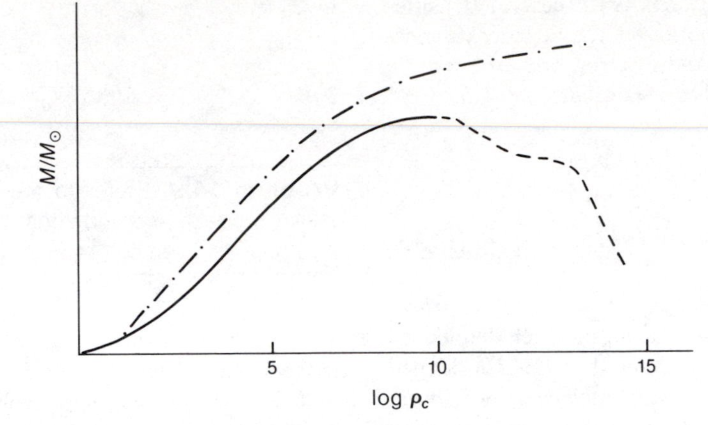

Figure 14.3. Mass versus central density when effects of particle interactions are included through the equation of state.

Finally the electron-ion Coulomb energy is proportional to

$$E_{el} \sim -\alpha\,(Z\alpha)\,N_i \frac{(Ze)e}{r_{ei}}$$

$$\simeq -\alpha^2 N_i \frac{(Ze)^2}{r_i} Z^{1/3} = -2\,Z^{1/3}\,E_I, \qquad (14.24)$$

assuming the average ion-electron separation to be $r_{ie} \simeq r_e$. The total Coulomb energy is the sum of (14.22), (14.23), and (14.24):

$$E_{\text{Coul}} \simeq \frac{1}{2}\alpha^2 N_i \frac{(Ze)^2}{r_i}(1 - Z^{1/3}). \qquad (14.25)$$

For all $Z > 1$, $E_{\text{Coul}} < 0$. That it vanishes for $Z = 1$ is a consequence of the oversimplified treatment here. E_{Coul} may be written in a more convenient form by noting that the volume per ion $V_i = V/N_i$, where V is the volume of the star, and thus $r_i \sim V_i^{1/3} \sim (VN_i)^{1/3}$. Substituting this into (14.25), and taking the derivative with respect to volume at constant N_i, gives the change in the system pressure due to Coulomb interactions:

$$\Delta P_{\text{Coul}} = \frac{1}{6}\alpha^2\,(Ze)^2\,(1 - Z^{1/3})\left(\frac{N_i}{V}\right)^{4/3}$$

$$= \alpha^2 \frac{(Ze)^2\,(1 - Z^{1/3})}{6A^{4/3}\,m_p^{\;4/3}}\,\rho^{4/3}$$

$$= 1.94 \times 10^{12} \frac{Z^2\,(1 - Z^{1/3})}{A^{4/3}}\,\rho^{4/3}. \qquad (14.26)$$

Problem 14.10. Derive (14.26). Evaluate the ratio of ΔP_{Coul} to the ideal gas pressure for $A = 2Z = 12$, $T = 10^8$ K, and $\rho = 10^6$ g/cm³.

The relative importance of the Coulomb interactions increases with density, and theory indicates that the ions will eventually form a crystalline lattice embedded in an electron gas. We can estimate the temperature at which this occurs in white dwarfs by assuming that the lattice melts when the thermal energy per ion E_T exceeds some fraction η of the Coulomb energy of the ions E_{Coul}. Since the interion spacing is roughly

$$r \sim n_I^{\;-1/3} = (Am_p/\rho)^{1/3},$$

where n_I is the ion number density, the Coulomb energy is roughly

$$E_{\text{Coul}} \simeq \alpha^2 \frac{(Ze)^2}{r} \simeq \frac{Z^{5/3}\,e^2}{m_p^{\;1/3}}\left(\frac{\rho}{\mu_e}\right)^{1/3}. \qquad (14.27)$$

Here α is the number of nearest neighbor ions, but is of order unity, and $\mu_e = A/Z$. The preceding argument suggests that we take as a condition for solidification $E_T \lesssim \eta\,E_{\text{Coul}}$, which is equivalent to the following condition on the temperature:

$$T \lesssim \alpha^2\eta \frac{Z^{5/3}\,e^2}{k\,m_p^{\;1/3}}\left(\frac{\rho}{\mu_e}\right)^{1/3}$$

$$= 1.4 \times 10^3\,Z^{5/3}\left(\frac{\rho}{\mu_e}\right)^{1/3} \qquad (14.28)$$

A typical estimate for η is 0.01. For white-dwarf cores that have evolved beyond He burning ($\mu_e = 2$ and $Z \gtrsim 6$), average densities are $\rho \gtrsim 10^6$ g/cm^3, and core temperatures in excess of about 10^7 K would be required to melt a lattice for which $\alpha = 6$. In these cases the maximum mass is reduced. Calculations suggest that an upper limit of about 1.2 $M_\odot$ is more appropriate than 1.44 $M_\odot$, though it may be reduced to 1.1 $M_\odot$ for Fe^{56} or 1.0 $M_\odot$ for nuclei in statistical equilibrium.

An additional complication arises because of electron capture by nuclei, which becomes important at densities above about 10^9 g/cm^3, as was discussed early in Chapter 13. The principal effect of electron capture is to reduce the degeneracy pressure without changing the total mass; so the star must contract. As it does, the degeneracy pressure tends to rise, but the capture rates grow more rapidly. The result is loss of stability, and dynamic collapse will ensue. That the resulting configurations are dynamically unstable can be shown by investigating the adiabatic exponent Γ_1, which is reduced below 4/3 where electron capture occurs.

The relation of mass to central density for cold, degenerate matter is shown schematically in Figure 14.3. The solid curve represents stable hydrostatic equilibrium; the dashed portions unstable equilibrium states. The dash-dotted line is the mass curve obtained for ideal degenerate electrons in Figure 14.2.

Suppose the star can be adequately described by the polytrope $P \sim \rho^\gamma$. If it is in hydrostatic equilibrium, it follows that

$$P_c \sim \frac{MG}{R}\rho_c \sim M^{2/3}\, G\, \rho_c{}^{4/3}, \qquad (14.29)$$

and, since $P_c \sim \rho_c{}^\gamma$, we can solve for $M\ (\rho_c)$ and show that

$$(dM/d\rho_c) \sim (\gamma - 4/3)\, \rho_c{}^{-3(2-\gamma)/2}. \qquad (14.30)$$

Problem 14.11. Prove (14.30).

When $\gamma = 4/3$, the derivative $dM/d\rho_c$ is zero. This result is generally valid as long as $T = 0$, and shows that equilibrium configurations go through a maximum in $M\ (\rho_c)$ when they become dynamically unstable. Some minima may represent a point beyond which the system again becomes stable. These results are expressed in a theorem about degenerate matter that states that $dM/d\rho_c < 0$ is a sufficient condition for instability, but that $dM/d\rho_c > 0$ is necessary but not sufficient for stability. The distinction is important primarily for neutron stars.

The masses of white dwarfs may be calculable when they are members of binary systems. Three have been estimated in this way. They are 40 Eri B ($M = 0.45\ M_\odot$), Sirius B ($M = 1.05\ M_\odot$), and Procyon B ($M = 0.65\ M_\odot$), all well below the limit of 1.2 $M_\odot$ mentioned earlier. The observed mass-radius relation for several white dwarfs is shown in Figure 14.4.

The radius of a typical white dwarf has been estimated in Problem 14.5. Typical values are $\sim 10^{-2}\ R_\odot$. Observations of spectra indicate surface temperatures on the order of 10^4 K; so their luminosities are only about $10^{-3}\ L_\odot$ (Table 14.1).

The properties of a white dwarf composed of Fe^{56}, and having no thermonuclear energy sources, are shown in Figure 14.5. The star's mass is 1 $M_\odot$, and the model has a radius and luminosity $R = 6.1 \times 10^{-3}\ R_\odot$ and $L = 5.67 \times 10^{-4}\ L_\odot$. The ratio of thermal electron pressure $P = n_e kT$ to that given by the equation of state (including degeneracy) is less than 0.1 for $r \lesssim 0.9\ R$. The mass-luminosity ratio $m(r)/L(r)$ is very nearly constant throughout, and the temperature is constant out to $m(r) \simeq 0.7\ M$, dropping significantly only in the outer 3 percent of the mass. This may be understood if it is remembered that the ions, which obey the perfect gas law ($c_V \simeq 3/2\ Nk$), are the source of thermal energy, and that the high thermal conductivity of the electrons results in a nearly isothermal interior.

Problem 14.12. Establish the result $L(r)/m(r) \sim$ constant for degenerate matter whose thermal energy is stored in a nondegenerate component such as ions. To do this, establish the thermal energy generation rate ϵ_T (ergs g^{-1} sec^{-1}):

$$\epsilon_T = (3/2)\, Nk\, (dT/dt). \qquad (14.31)$$

Justify the assumptions that dT/dt is essentially independent of radius. Show that

$$L(r)/m(r) \simeq \epsilon_T(t). \qquad (14.32)$$

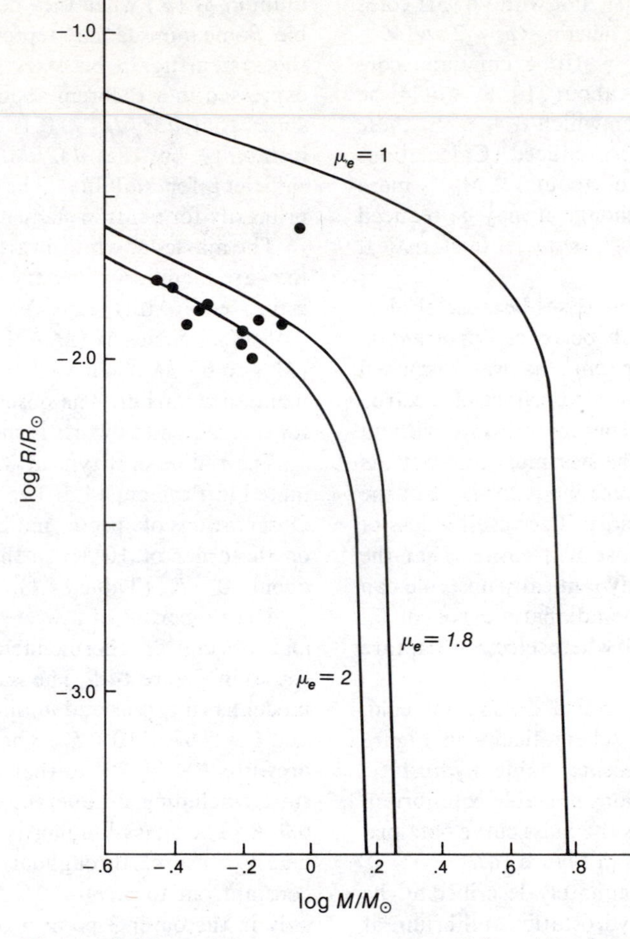

Figure 14.4. Radius versus mass observed for white dwarfs (points). The curves represent theoretical models with mean molecular weight per electron μ_e as shown next to each curve.

Table 14.1
White dwarf parameters (observed).

Star	S	M_V	$B - V$	log $M/M_\odot$	log $R/R_\odot$
V. Maanen 2	DG	14.24	0.56	−0.2	−1.91
L870-2	DAs	11.89	0.33	−0.16	−1.89
40 Eri B	DA	11.01	0.03	−0.44	−1.77
Sirius B	DA	11.3	0.4	−0.01	−1.6
He 3 (Ci_{20} 398)	DA	10.95	−0.07	−0.3	−1.83
Procyon B	DF	13.1	0.5	−0.37	−1.9
L 532-81	DAs	11.9	0.05	−0.2	−1.94
R627	DF	13.8	0.30	−0.18	−2.0
L770-3	DA	10	−0.2	−0.32	−1.84
W 1346	DA	10.8	−0.07	−0.4	−1.79
L 1512-34B	DA	11.3	0.17	−0.09	−1.9

SOURCE: C. W. Allen, *Astrophysical Quantities* (London: The Athlone Press, 1973).

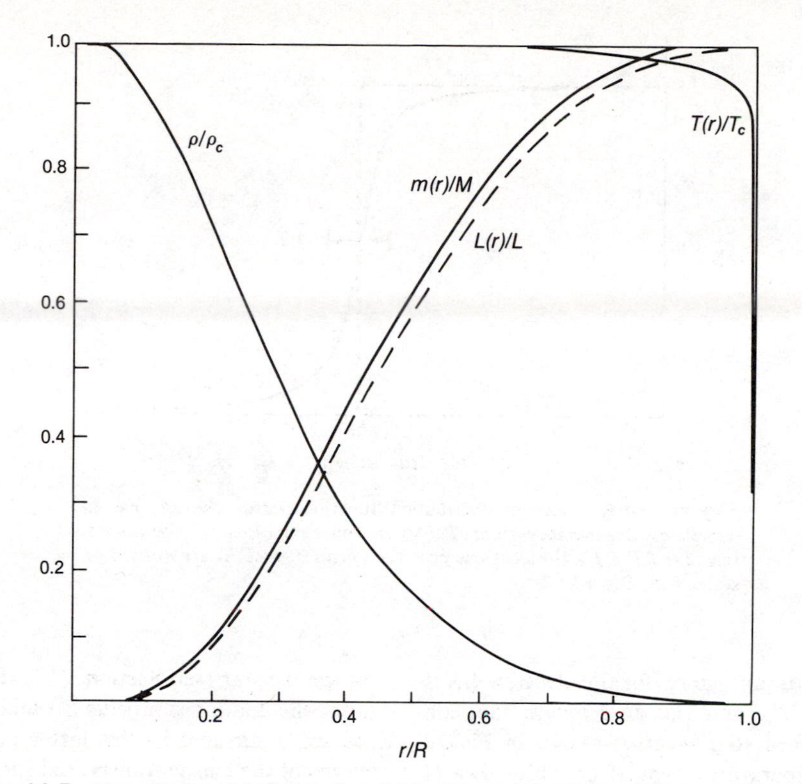

Figure 14.5. Numerical model of white dwarf, $M = M_\odot$, and $R = 6.1 \times 10^{-2}\, R_\odot$, made up of Fe^{56}.

14.4. Envelope Structure

Having discussed the properties of white dwarfs that follow from the equation of hydrostatic equilibrium, we consider those that result from the requirement of radiative equilibrium. This has negligible effect on the star's mass or radius, as we will show. However, since surface temperatures are of order 10^4 K, and typical envelope densities are less than 10^2 g/cm^3, the outermost layers are nondegenerate. An understanding of this region is crucial to the theory of cooling rates, which in turn determine the evolutionary track in the HR diagram. In effect, the nondegenerate envelope controls the rate at which the thermal energy, most of which is contained in the core, escapes from the star.

The thermodynamic internal-energy density of a system of low-temperature electrons and ions in the envelope of a white dwarf may be written as

$$u = \frac{3}{5} n_e E_F + \frac{\pi^2}{4} n_e \frac{(kT)^2}{E_F} + \frac{3}{2} \frac{\rho k T}{\mu m_p}. \qquad (14.33)$$

The last term is the energy density associated with a classical ideal gas of ions at temperature T. To this are added two terms. The first is the energy density of a degenerate electron gas, given by the number of electrons per unit volume n_e times the average energy per electron,

$$(3/2)\, E_F = (3/2)(\hbar^2/2m_e)\, n_e^{2/3},$$

obtained using (14.2). Note that the energy density associated with the first term of (14.33) cannot be removed from the system unless it expands. The middle term in (14.33) represents the removable thermal energy associated with the electrons at low but nonzero temperature, and may be understood as follows.

Problem 14.13. Recalling the relation between E_F and number density, show that the degeneracy energy $\sim n_e E_F$ must remain constant in a static system, and that it must increase (decrease) as the system contracts (expands). Where does the increased internal energy come from in a star?

When $T = 0$, all the electrons lie in the lowest energy states compatible with the Pauli exclusion

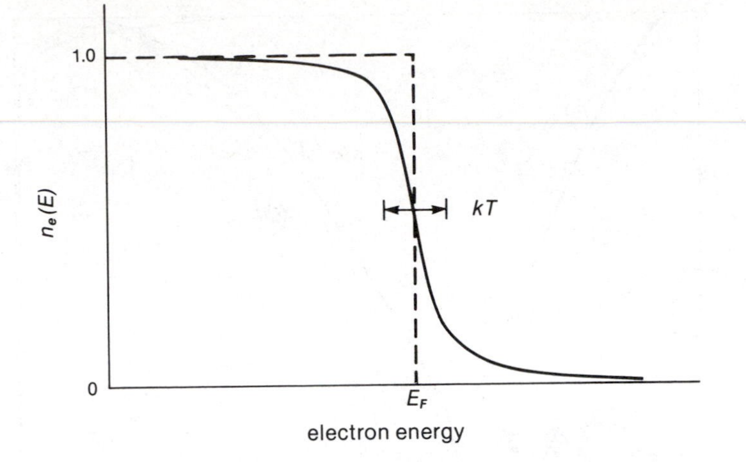

Figure 14.6. Electron distribution function versus energy. For a completely degenerate system ($T = 0$), n_e would correspond to the dashed line. For $kT \ll E_F$, the electrons near the Fermi surface E_F are excited to states with $E \gtrsim E_F$.

principle. The maximum energy of the electron gas is the Fermi energy E_F, and the distribution function resembles the dashed step function shown in Figure 14.6. For the temperatures typical of white dwarfs ($kT \ll E_F$), it resembles the solid curve, differing significantly from the $T = 0$ distribution only in a narrow region of width kT about E_F. Of the n_e electrons per unit volume, only that fraction (kT/E_F) will have an average thermal energy kT. Therefore the average thermal energy density of the electrons is proportional to $(kT/E_F)n_e$ times kT.

Energy equipartition is not achieved for a system composed of a classical gas and a semidegenerate one. In fact, the ion component of (14.33) is normally much larger than the electron component; so the second term in (14.33) may often be ignored in dealing with the star's luminosity. The middle term does play a critical role in maintaining, by high electron conductivity, the approximately isothermal character of the interior as the star gradually cools.

Problem 14.14. Show that the ratio of electron-to-ion thermal energy is proportional to (kT/E_F).

The final process of white dwarf cooling involves two competing processes. First, at sufficiently high temperatures ($T \gtrsim 10^{10}$ K), the weak interactions convert thermal energy into neutrinos, which readily escape the star (see Section 13.2). For lower temperatures, the dominant process is radiative diffusion from the star's surface. In this latter process the thermal energy of the ions is transported (primarily by conduction) from the core to the outer nondegenerate envelope, where it is radiated away. To explain the observed surface properties of white dwarfs, we must take into account the relatively thin outer envelope separating the core from the atmosphere. The scale height in the envelope is roughly

$$l \sim P/\rho g \simeq kT/\mu g, \tag{14.34}$$

where $g = MG/R^2$. For typical dwarfs $g \sim 5 \times 10^8$ cm sec^{-2}, and the temperature in the envelope is $T \lesssim 10^6$ K. Therefore $l \lesssim 10^{-3} R \simeq 1$ to 10 km. The density at the core-envelope interface can be found from the temperature there and the fact that the electron gas is becoming nondegenerate. Equating the electron degeneracy pressure to the ideal gas pressure and solving for density, we find

$$\begin{aligned}\rho_1 &= \left[\frac{5km_e m_p{}^{2/3}}{\hbar^2(3\pi^2)^{2/3}}\right]^{3/2}\left(\frac{\mu_e{}^{5/2}}{\mu^{3/2}}\right)T_1{}^{3/2}\\ &= 2.4 \times 10^{-8}\,\mu_e{}^{5/2}\,T_1{}^{3/2}/\mu^{3/2}\end{aligned} \tag{14.35}$$

for the density and temperature at the core boundary. Taking $T_1 \simeq 10^6$ K, we find $\rho_1 \simeq 10^2$ g/cm^3, and certainly expect it to be below 10^3 g/cm^3. Using the latter as an upper bound, we find that the total mass of the envelope is less than

$$M_s \le 4\pi R^2 l\rho_1 \simeq 2 \times 10^{-4} M_\odot. \qquad (14.36)$$

This clearly justifies the assumption that $m(r) \simeq M$, the total mass, in the envelope region. Equation (14.34) justifies treating the ratio $r_1/R \simeq 1$. Then the gravitational acceleration $g(r) = m(r)G/r^2 \simeq MG/R^2 = g_s$, the surface gravity, throughout the envelope.

Problem 14.15. Derive (14.32), which gives the boundary in the ρ, T-plane between nonrelativistic degeneracy and complete nondegeneracy for an electron gas. The ions are of mean molecular weight per electron μ_e.

The envelope structure can be constructed easily if we assume $m(r) = M$, and $r_1/R \simeq 1$; then the equation of hydrostatic equilibrium becomes

$$\frac{dP}{dr} = -\frac{MG}{r^2}\rho = -g_s\,\rho, \qquad (14.37)$$

and the radiative transfer equation is

$$\frac{dT}{dr} = -\frac{3\kappa\rho}{16\pi ac}\frac{L}{r^2\,T^3}. \qquad (14.38)$$

Since $m(r) \simeq M$, (14.32) implies that $L(r) \simeq L$. The ratio of (14.37) to (14.38) gives an equation for dP/dT that depends only on P, T, κ, and constants. Assuming κ to be of the form (6.86) and integrating, we find

$$P = \left(\frac{a+1}{4+a-b}\right)^{1/(a+1)} \times \left[\frac{64\pi\sigma}{3}\frac{MG}{\kappa_0\,L}\left(\frac{k}{\mu}\right)^a\right]^{1/(a+1)} T^{(4+a-b)/(a+1)}. \qquad (14.39)$$

The density follows from the ideal gas law $P = \rho kT/\mu m$. In deriving (4.39), we assumed that P and T vanish together. Using (14.39) and the ideal gas law in (14.37) and integrating gives the temperature in the envelope:

$$T = \frac{a+1}{4+a-b}\frac{\mu}{k}MG\left(\frac{1}{r} - \frac{1}{R}\right). \qquad (14.40)$$

The envelope structure is given by (14.39), (14.40), and the perfect gas law. For Kramers' opacity, $a \simeq 1$ and $b = -3.5$.

In order to complete the model, we relate the core temperature (which is known from evolutionary calculations) to the effective surface temperature as follows. Substituting (14.40) into (14.39) and evaluating the result at the core-envelope interface yields the luminosity (which is constant throughout the envelope) as a function of T_1 and ρ_1. The dependence on ρ_1 may be eliminated by using (14.35). Finally, because of the isothermal nature of the core, we may set the central temperature $T_0 \simeq T_1$. The result is

$$L = \frac{64\pi\sigma}{3}\frac{a+1}{4+a-b}\left(\frac{\mu}{k}\right)^{1.5a+2.5} \times K_1^{\,2.5(a+1)}\frac{MG}{\kappa_0}T_c^{\,1.5(1-a)-b}, \qquad (14.41)$$

where $K_1 = 10^{13}$ in cgs units. Assuming a Kramers' opacity, with the simple approximation (6.86) and

$$\kappa_0 = 4.34 \times 10^{25}\,(\bar{g}/t)Z,$$

the luminosity (14.41) becomes

$$L = 5.7 \times 10^{25}\frac{(t/\bar{g})}{Z}\frac{\mu}{\mu_e^{\,2}}\left(\frac{M}{M_\odot}\right)T_c^{\,3.5}, \qquad (14.42)$$

with the temperature in units of 10^6 K. For numerical purposes we adopt the typical value $t/\bar{g} \sim 10$. Since T_c and R are determined by the mass and composition, (14.42) gives the luminosity. The effective surface temperature T_{eff} is obtained from $L = 4\pi\sigma R^2\,T_{\text{eff}}^{\,4}$. Combining (14.42) with this last expression gives $L\,(T_{\text{eff}})$, which locates the star in the HR diagram.

Problem 14.16. Treat the core of a white dwarf as a polytrope of index $n = 1.5$, and relate its radius to mass. Use the result to solve for T_{eff} as a function of L.

Problem 14.17. Find the conditions for the opacity $\kappa = \kappa_0\rho^a T^b$ that would make the white-dwarf core and envelope discussed in the preceding stable against convection. What happens if the outer layers are dominated by H^- opacity for which a and b are positive, and b may be large?

Problem 14.18. Estimate the surface temperature T_{eff} of a white dwarf whose core is cool enough to be crystalline. Assume that the core is a polytrope of index $n = 1.5$.

The envelope structure (14.39) to (14.42) is a reasonable approximation when the matter is an ideal gas that is not partially ionized. In main-sequence stars like the sun, H ionization is complete for $T \gtrsim 10^4$ K, and the final stage of He ionization is complete above 4×10^4 K. Stellar evolution theory, and observations of the spectra of some white dwarfs, suggest that the envelope composition should be rich in He and heavier elements, particularly C^{12} and possibly O^{16}. These latter constituents have high ionization potentials, and are likely to occur in varying states of ionization. Furthermore, since envelope pressures are high until the photosphere is reached, the degree of ionization of H and He will be reduced. This follows from the Saha equation, which shows that the degree of ionization x of H, for example, at temperature T and gas pressure P_g, is given by (6.18), or

$$x^2 = (1 - x^2)f(T)/P_g. \qquad (14.43)$$

Using model values for $P(r)$ and $T(r)$ in a typical envelope, we find $x \simeq 0.5$ for temperatures of about 10^6 K. Consequently, partial ionization will occur, perhaps in more than one zone, throughout much of the envelope.

The complications associated with varying composition and partial ionization do not greatly affect the white dwarf's mass or radius. They are, however, crucial for the mode of energy transport in the envelope. The condition for convective instability is that the magnitude of the radiative temperature gradient exceed the magnitude of the adiabatic gradient (5.146):

$$\left|\frac{dT}{dr}\right|_{\text{Rad}} > (1 - 1/\Gamma_2)\frac{T}{P}\left|\frac{dP}{dr}\right|. \qquad (14.44)$$

Since partial ionization reduces Γ_2 from 5/3 to values near unity in a monatomic ideal gas, the envelope will become unstable against convection.

Models of white dwarfs with convective envelopes may be constructed in analogy to models with radiative envelopes. It is still reasonable to treat the core as a degenerate system and to use $T = 0$ models for M and R. In fact, the analysis up through (14.37) remains unchanged. Instead of (14.38), the temperature gradient is given by

$$dT/dr = (1 - 1/\gamma)\,(T/P)\,dP/dr \qquad (14.45)$$

with γ dependent, possibly strongly, upon position, but generally $1 \lesssim \gamma \lesssim 5/3$. As an additional complication, the zero boundary conditions $P(R) = T(R) = 0$ must be replaced by the photospheric boundary conditions (6.120):

$$T_{\text{eff}} = (L/4\pi\sigma R^2)^{1/4}, \qquad (14.46)$$

$$P_{\text{ph}} = \frac{2}{3}\frac{MG}{R^2}\kappa(P_{\text{ph}}, T_{\text{eff}})^{-1}. \qquad (14.47)$$

The energy-transport and hydrostatic-equilibrium equations may be integrated through the envelope to relate T_{eff} to T_c. The core-envelope boundary may be chosen approximately as in (14.35) and T_c related to T_{eff} as before. When this is done for white-dwarf compositions suggested by stellar evolution theory, it is found that core temperatures are lower than would be expected for a radiative envelope having the same M, L, and T_{eff}. Physically this should be evident from (14.45). Incomplete ionization reduces γ to values near unity, which reduces the temperature gradient in the envelope. Since the core, which determines M, L, and R, is insensitive to the envelope structure, the reduced temperature gradient implies a lower T_c for a given T_{eff}. The extent of the convective region is greatest in lower-mass white dwarfs. For those with masses $M \approx 1\ M_\odot$, numerical studies show that convection is limited to the envelope. For low-mass white dwarfs, convection may extend well into the degenerate region, and T_c may be reduced by as much as a factor of four relative to a radiative model with the same L and T_{eff}. A complete treatment of convective envelopes must include modifications of the Saha equation (which describes the degree of ionization) for electron degeneracy, pressure ionization, and Coulomb interactions. It is also likely that model atmospheres that relate T_{eff} to observed spectra must include convection at their bases.

Problem 14.19. The fundamental parameters of a stellar atmosphere include T_{eff} and the surface gravity $g = MG/R^2$. Discuss specific effects that convection in a white-dwarf envelope could have on the assignment of stars to spectral classes.

From the preceding envelope-model calculations, we conclude that degenerate white dwarfs probably have core temperatures as high as 10^7 K. If H were present in these regions, it should burn rapidly. In fact, the hydrogen concentration X must be small, or the luminosity would exceed the range of observed values, as shown in Problem 14.20.

Problem 14.20. Energy generation by the proton-proton chain at high density may be approximated by

$$\epsilon = \epsilon_0 X^2 \rho T_{(7)}{}^n,$$

where $\epsilon_0 = 8.4 \times 10^{-2}$ erg g^{-1} sec^{-1} and $n \sim 4.5$ for T near 10^7 K. Assume that T is constant in the core, and use the observed range of white-dwarf luminosities to bound X.

Problem 14.21. More exact calculations show that $X \lesssim 4 \times 10^{-7}$ for the CN cycle and $X \lesssim 8 \times 10^{-5}$ for the pp chain. The resulting low hydrogen abundance in white-dwarf cores is consistent with predictions from stellar evolution theory.

14.5. Evolving White Dwarfs

The evolution of a white dwarf is described by its track in the HR diagram, and by the time required for it to radiate away its store of thermal energy. The final state is one of astrophysically dead matter, an object devoid of available energy sources and gravitationally self-bound. This section describes the approach to this final state.

The luminosity of a white dwarf whose envelope is in radiative equilibrium has already been obtained (14.41), assuming Kramers' opacity and complete ionization. It may be expressed as

$$\frac{L}{L_\odot} = 1.5 \times 10^{-8}\, \frac{t/\bar{g}}{Z}\, \frac{\mu}{\mu_e^2} \left(\frac{M}{M_\odot}\right) T_{c(6)}{}^{3.5}. \qquad (14.48)$$

Unfortunately, this does not depend explicitly on the time. Since the evolution is quasistatic, we may also express the luminosity starting with (5.54). Before proceeding, we rewrite the term $T\,ds/dt$, assuming that the specific entropy $s = s(T, \rho)$. Then

$$T\frac{ds}{dt} = T\left(\frac{\partial s}{\partial T}\right)_\rho \frac{dT}{dt} + \left(\frac{\partial s}{\partial \rho}\right)_T \frac{d\rho}{dt}$$
$$= c_V \frac{dT}{dt} + \left(\frac{\partial s}{\partial \rho}\right)_T \frac{d\rho}{dt}. \qquad (14.49)$$

The last term may be rewritten using the thermodynamic Maxwell relation

$$\rho^2 (\partial s/\partial \rho)_T = -\,(\partial P/\partial T)_\rho.$$

If this is substituted into (14.49), and the result integrated over $dm = 4\pi\rho r^2\, dr$, we obtain an alternative expression for the luminosity,

$$L = \int_0^M dm \left[\epsilon - c_V \frac{dT}{dt} + \frac{T}{\rho^2}\left(\frac{\partial P}{\partial T}\right)_\rho \frac{d\rho}{dt}\right]. \qquad (14.50)$$

The terms in the integrand correspond to: (1) nuclear plus neutrino energy release; (2) thermal heat energy release; and (3) contractional energy release. For typical core densities and temperatures, neutrino energy losses are negligible. Furthermore, the evolutionary discussions of low-mass stars ($M_c < 1.4\, M_\odot$) indicate that nuclear burning will have converted all core He into C^{12} and O^{16} before degeneracy sets in and that these fuels are not hot enough to ignite. Finally, since the core is supported by degeneracy pressure, no further concentration is possible and $d\rho/dt \simeq 0$. Therefore (14.50) reduces to

$$L = -\int_0^M dm\, c_V \frac{dT}{dt}. \qquad (14.51)$$

The star radiates entirely at the expense of the thermal energy stored in the electrons and ions. This energy is ultimately radiated away, leaving a black dwarf.

Now, we have already shown that the heat energy of the ions greatly exceeds that of the electrons in a degenerate white dwarf. Therefore, assuming that the ions remain an ideal gas, we may express the specific heat c_V as

$$c_V = c_V{}^i = \frac{3}{2}\frac{k}{Am_p} = \frac{3}{2}\frac{n_i k}{\rho}, \qquad (14.52)$$

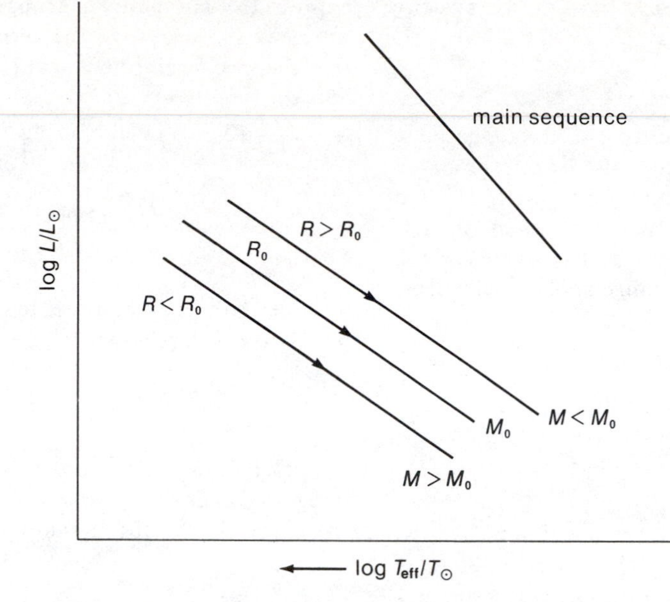

Figure 14.7. Evolutionary path of cooling white dwarfs in an HR diagram.

where $\rho = Am_p n_i$, and n_i is the ion density. Finally, since the core is isothermal, we may set $dT/dt = dT_c/dt$, remove it from the integral, and obtain for the luminosity

$$L \simeq -\frac{3}{2}\frac{k}{Am_p}\frac{dT_c}{dt}\int_0^M dm = -\frac{3}{2}\frac{kM}{Am_p}\frac{dT_c}{dt}. \quad (14.53)$$

This expression, which relates L to dT_c/dt, must be equal to (14.48), which relates L to T_c. Eliminating L from these two equations yields a differential equation for dT_c/dt, which is easily integrated. Assuming that the initial core temperature $T_c(0) \gg T_c(t)$, and using (14.48) to replace T_c by L in the result, yields

$$t = 8.9 \times 10^7 \frac{Z^{2/7}}{A}\left(\frac{\bar{g}}{t}\right)^{2/7}\left(\frac{\mu_e^2}{\mu}\right)^{2/7} \times \left(\frac{M}{M_\odot}\right)^{5/7}\left(\frac{L_\odot}{L}\right)^{5/7} \quad (14.54)$$

in years. For typical white dwarfs, $(t/\bar{g}) \simeq 10$ and $\mu_e \simeq \mu \simeq 2$, and the time required for the luminosity to be reduced by a factor of 10 is nearly 10^{10} years. The final cooling times are therefore comparable to the age of the universe.

The evolutionary track of a white dwarf is given by the mass-radius relation (14.19) and the definition of the effective temperature $L \sim R^2 T_{\rm eff}^4$. Eliminating the radius between these yields

$$\log\left(\frac{L}{L_\odot}\right) = 4\log\left(\frac{T_{\rm eff}}{T_\odot}\right) - \frac{2}{3}\log\left(\frac{M}{M_\odot}\right) + C. \quad (14.55)$$

Note that the more massive stars have evolutionary tracks lying below those of lower mass, as shown schematically in Figure 14.7. For fixed mass and composition, the luminosity decreases in time as $T_{\rm eff}$ decreases, so that white dwarfs evolve downward along tracks of fixed M and R.

Problem 14.22. Carry out the analysis leading to (14.54) for the cooling time, and verify the schematic behavior of evolutionary tracks shown in Figure 14.7.

Problem 14.23. Use the nonrelativistic and extreme relativistic limits of c_V^e for degenerate electrons obtained from (2.50) to evaluate the ratio c_V^e/c_V^i. Discuss circumstances under which electronic heat energy could affect the cooling rate.

Chapter 15

SUPERNOVAE

The most dramatic phenomena in stellar astrophysics involve the catastrophic explosion of a supernova. We have already discussed several other explosive processes (He-flash; H and He shell flash, for example) that do not lead to a radical change in the star's structure, and that may (shell flashes) even recur during the star's lifetime. A supernova explosion, however, leads to a total reorganization of the matter distribution. The fate of a star that becomes a supernova may depend on the star's mass prior to the explosion. Some supernovae may be violent enough to disrupt the star totally, spewing the enriched elements back into the interstellar medium. In others, the stellar core implodes, and only the outer envelope (or a part of it) is ejected into interstellar space. The imploded core will become a hot neutron star, or possibly a black hole. Although the ejected mass eventually melds into the interstellar medium, an expanding nebula may persist for 10^6 or more years after the explosion.

Unfortunately, supernovae are relatively rare (six known in our galaxy in 10^3 years; about 80 well-studied extragalactic events), and observational data are scarce. Furthermore, the physical processes that occur during the supernova stage appear to be subtly related, and involve matter at extremes of density and temperature that have not been studied in the laboratory. Although theories of the supernova process are in a rapid state of flux, a few results have emerged that appear to be generally valid, and these will be emphasized below.

15.1. Observational Features

Observational data come from two primary sources. The first is historical events: supernovae that have occurred in our galaxy and have been recorded during the past thousand years. In all, six are known. Their date of occurrence, distance from the Sun, type, and height above the Galactic plane are shown in Table 15.1. The date for Cas A is not clear, since the event was evidently not recorded (if seen), but the extent of the remnant indicates that it certainly occurred within the last thousand years, as we will see. Unfortunately, no supernova event has been observed within our galaxy during recent times. The primary information obtained from historical events includes the spatial distribution (as indicated by the nebulae) of ejected mass, and the presence of a compact remnant (the pulsar in the Crab nebula, for example).

The six historical supernovae lie near the plane of

Table 15.1
Historical supernovae, giving date of occurrence, distance from the Sun, type, and height above the Galactic plane.

Designation	Date	Distance (kpc)	Type	Height (kpc)
	1006	2.4	I	0.6
Crab	1054	2	I	0.2
	1181	8		0.43
Tycho	1572	6	I	0.15
Kepler	1604	10	I	1.2
Cas A	1667	3	II	0.11

our Galaxy's disk, within a sector $\Delta\theta = 60°$. Assuming that the same density of events has occurred throughout the disk in the previous $\Delta t = 10^3$ years, we obtain an event rate for supernovae in our Galaxy $R_G \simeq (6/\Delta t)\,(2\pi/\Delta\theta) = 1/28$ per year. The second source of data (and the most reliable relating to the explosion itself) is extragalactic events. These are supernovae that have occurred recently in nearby galaxies. These events indicate that at maximum light, the absolute visual magnitude can reach $M_V = -19$, corresponding to $L \approx 10^{10}\ L_\odot$, equivalent to an intermediate-size galaxy. The total time-integrated light output in the visible spectrum may reach 10^{49} ergs. An important characteristic of supernovae is the light curve, which gives the apparent visual magnitude $m_V(t)$ as a function of time. The light curves fall into two easily identifiable classes, called Type I and Type II, examples of which are shown in Figure 15.1. The time taken by supernovae to reach maximum brightness (rise time) is typically several days, and they remain near maximum for several days before declining. After maximum, Type I light curves decline rapidly for about 30 days, after which time their decay becomes almost exactly exponential. Thus in their late stages, the luminosity of a Type I supernova is given by

$$L = L_0 e^{-t/\tau_I}. \tag{15.1}$$

Apart from an overall scale factor set by the maximum brightness, there appears to be little variation in the light curve from one Type I supernova to the next. Empirically the decay rate $\tau_I \approx 70$ days. Type II supernovae are observed only in the arms of spiral galaxies, and not in elliptical galaxies. This suggests that the progenitors of Type II supernovae are relatively massive Population I stars. Assuming that they formed in the spiral arms, their lifetime would be less than about 10^7 years if the event occurred within the same arm; so their main-sequence mass should exceed

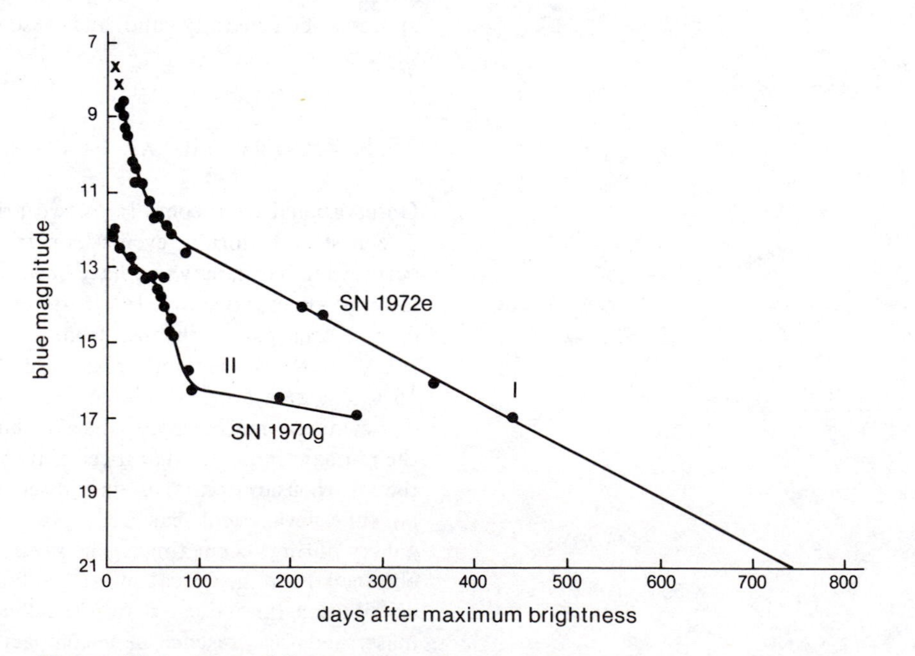

Figure 15.1. Observed light curve for Type I and Type II supernovae. Note the late time-exponential decay for Type I.

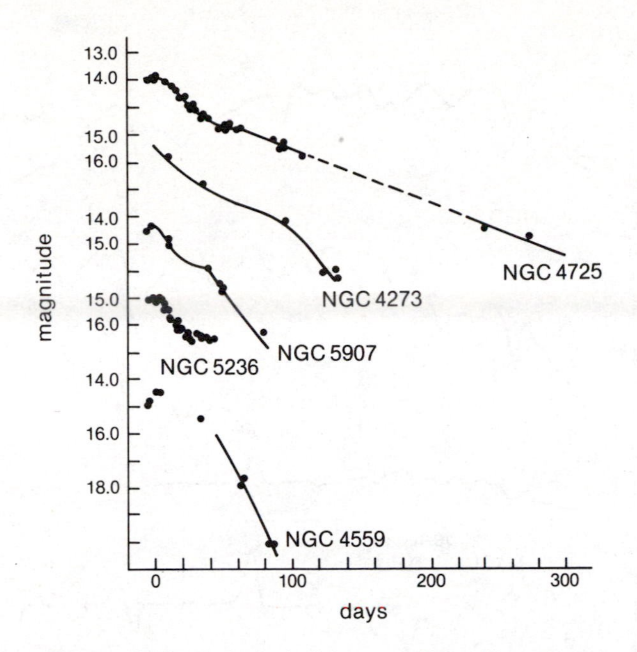

Figure 15.2. Variation in light curve of Type II supernovae.

10 $M_{\odot}$. The decline in brightness following maximum for Type II is qualitatively different, and varies from one supernova to another (Figure 15.2). The initial decline covers about 25 days, after which the brightness levels off for 50 to 100 days. This plateau is followed by a further rapid decline. Finally, the rate of decay may become extremely gradual. Type I supernovae occur in all galaxies, with the rate in spirals comparable to the rate in ellipticals. Since they occur in ellipticals, where star formation has probably ceased, the progenitors of Type I supernovae are assumed to be old, relatively low-mass, Population II stars.

The formation rate of Type II supernovae is currently estimated to be $R_{II} \approx 1/100$ per year per spiral galaxy, and for Type I, $R_I \approx 1/300$ per year per elliptical or spiral. These should be compared with the white-dwarf formation rate $R_{\mathrm{WD}} \approx 3$ per year in our Galaxy. We also note that $R_G \approx 10\ R_I$. The maximum magnitude of Type II supernovae is usually about a magnitude greater than that of Type I, and their integrated visible light an order of magnitude less. Spectral studies (to be discussed later) also indicate that they eject less mass than do Type I supernovae.

The spectra of supernovae contain much more detailed information than do their light curves, but spectroscopic studies are more time-consuming. However, for a few extragalactic supernovae the evolution of the spectrum at wavelengths in the range 3100 Å to 11,000 Å has been obtained, starting near maximum light and continuing for up to two years. These observations yield important data on the velocities and composition of the ejecta, the temperature of the expanding photosphere, and the amount of mass ejected. A knowledge of the photospheric temperature can be used to measure the supernova's distance (independently of local—that is, galactic—distance scales) and hence the distance of the galaxy in which the event occurred, and is an important independent check on methods used to establish intermediate distance scales in cosmology.

During their early stages supernovae exhibit approximately black-body spectra on which are superimposed absorption or emission features, or both. The lines actually present, and their profiles, may vary with time, or they may persist throughout most of the expansion. Representative spectra of Types I and II are shown in Figure 15.3, which gives the log of the relative flux vs. wavelength. The two spectra for each type occurred about a month apart, the first near maximum light. The light curves are given in Figure 15.1. Both types of spectra show broad absorption troughs and emission peaks. The position of prominent lines as measured in the laboratory are also shown. For

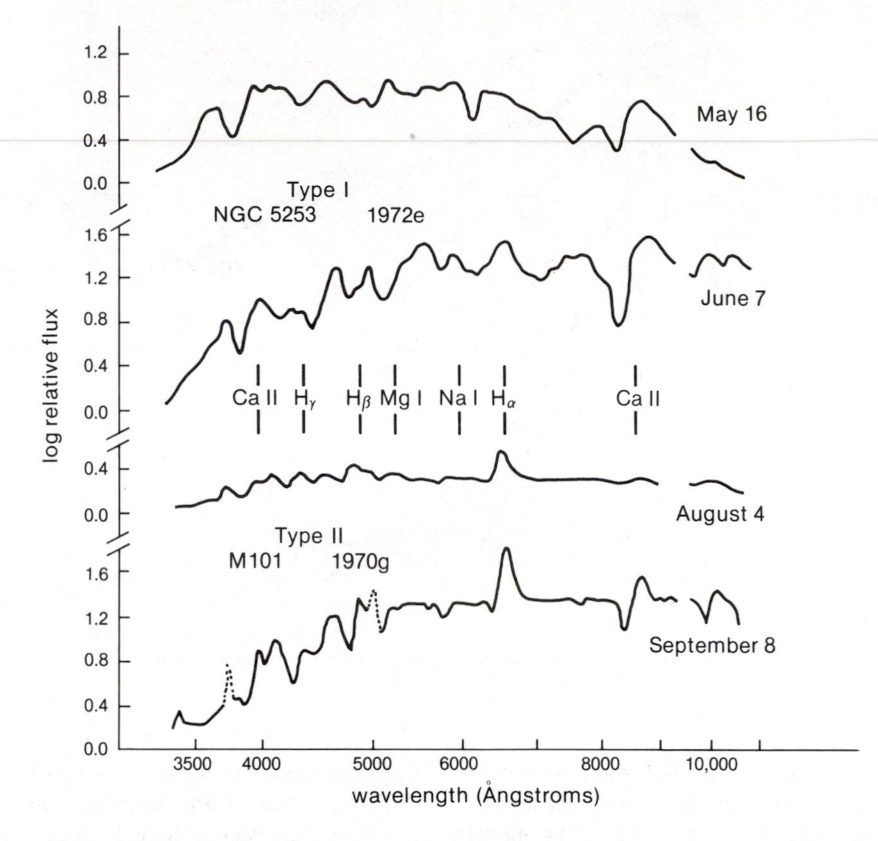

Figure 15.3. Spectra of supernovae. The upper two curves show spectra at two times for Type I; the lower two curves are for Type II, both at comparable stages in their development. Wavelength of prominent lines shown between upper two and lower two sets.

each line, a broad absorption trough is observed that has been blueshifted as a result of the expansion velocity of the ejecta; in these examples, $\Delta\lambda/\lambda \approx 4 \times 10^{-2}$. Assuming the Doppler effect to be the primary source of this blueshift, we can obtain the expansion velocity from

$$v_{\text{exp}} = \frac{\Delta\lambda}{\lambda} c. \tag{15.2}$$

Typically values as high as $v = 10^4$ km/sec are observed.

Problem 15.1. Photospheric temperatures in supernovae are typically less than 10^4 K. Estimate the amount of thermal broadening of a typical absorption line.

Type I spectra show little variation from one supernova event to another. Just before maximum light, the spectra show broad features that may be due to emission, or to a combination of emission and absorption. A characteristic feature of Type I supernovae near maximum light is a strong Si absorption line ($\lambda = 6150$ Å), which is not seen in Type II spectra. The structure at $\lambda \lesssim 5000$ Å is highly stable as the supernova evolves, as is evident in the upper two curves of Figure 15.3. The long-wavelength structure is irregular, with lines appearing and vanishing as in ordinary novae. Prior to the exponential decline in brightness, the spectra are dominated by the continuum, and the line features appear as secondary modifications. Figure 15.4 shows the spectra of SN 1972e over a two-year period, starting 15 days after maximum. The spectra during exponential decline (about 100 days after maximum) is qualitatively different from the earlier distributions, and is dominated by a series of strong emission peaks in the range $4000 < \lambda < 6000$ Å.

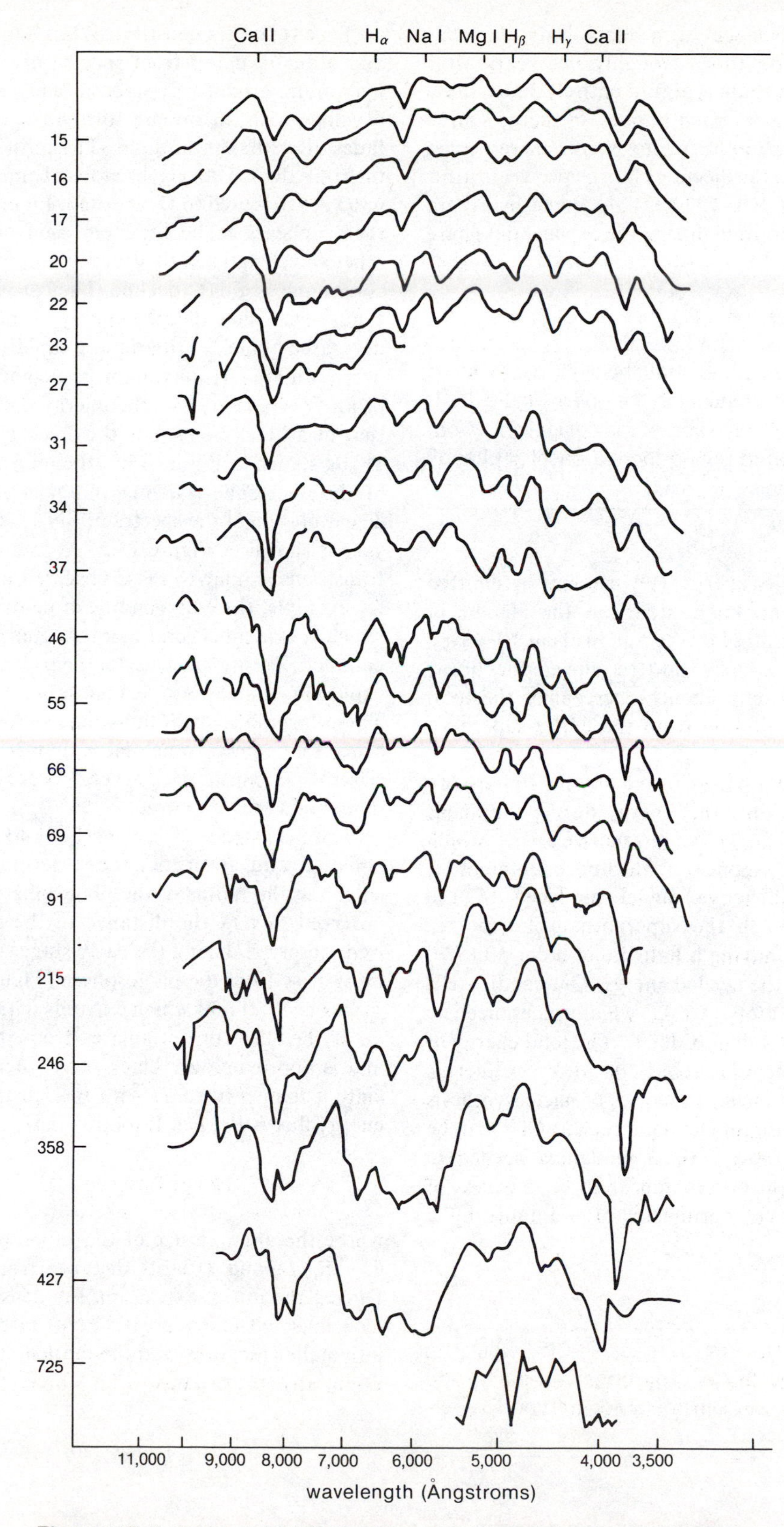

Figure 15.4. Spectra of SN 1972e (Type I) over two-year period. Principal spectral lines are identified at bottom. The numbers at left give the time in days after maximum.

There is little evidence of a black-body-like continuum, and at late times (roughly two years after maximum) the spectrum is almost entirely due to these features. Since line emission implies an energy source interior to the expanding photosphere, energy must still be flowing into the ejecta as late as two years after the explosion. For SN 1972e (light curve in Figure 15.1), the energy emitted during the exponential phase is about 10^{48} ergs.

Problem 15.2. Assume that the light curve for a Type I supernova is identical to the one in Figure 15.1, and that $L_{max} = 10^{43}$ erg/sec at maximum light. What is the energy radiated during the exponential phase? Assume $\tau_I = 70$ days.

The bulk of this energy (99 percent) is emitted during the first year. For comparison, this amount of energy would be emitted by the Sun in about 10^7 years. Nevertheless, the energy emitted during the linear phase amounts to only about 1 percent of the total electromagnetic radiation from the entire supernova explosion.

The presence of a pulsar formed by the stellar-core implosion could supply this energy during the linear phase (see Section 16.3). An alternative model, which also explains the exponential decline in luminosity, assumes that radioactive nuclei produced by the extreme conditions in the supernova explosion (see Section 12.2) and having a half-life of order 50 to 70 days could supply the needed energy. One candidate is the radioactive isotope Cf^{254}, which spontaneously fissions with a half-life of order τ_I. The total energy of these fission fragments is about 200 MeV per nucleus of Cf^{254}. Although current theories of nucleosynthesis show that transuranium elements such as Cf^{254} can be produced in supernovae, the abundances needed to explain Type I light curves appear to be in excess of observed abundances (primarily of uranium) by a factor of 10^2.

Problem 15.3. How much mass of Cf^{254} would be required to supply the energy observed during the exponential decline in luminosity for a Type I supernova?

Type II spectra constitute a less homogeneous class, and differ in detail from spectra of Type I. Prior to maximum, a relatively smooth black-body continuum develops with an intense ultraviolet component that fades after maximum light. The most prominent feature initially is a strong, broadened emission line at $\lambda = 6563$ Å attributed to H_α (bottom, Figure 15.3). In fact, there appears to be little emission or absorption at other wavelengths until after maximum, when the blue continuum begins to decline. The bottom two curves of Figure 15.3 show that the energy in the ultraviolet has decreased about 2.5 times more rapidly than in the red over a month. At maximum, the color resembles that of an O-type star, and the energy distribution is like that of a black body. For the Type II supernova SN 1970g shown in Figure 15.3, the inferred temperature of the photosphere at maximum is about 12,000 K. The evolution of the spectra of SN 1970g over nearly a year is shown in Figure 15.5. At late times forbidden transitions similar to those observed in novae appear, for example, the emission line of neutral oxygen [OI], which is evident several months after maximum. This emission feature persists for nearly a year after the explosion, indicating a continuing energy source. These forbidden transitions require low gas densities. Since they appear much later than in novae, Type II supernovae must eject several orders of magnitude more mass than do novae.

The persistence of a black-body-like continuum as emission and absorption lines develop enable us to calculate the radius of the photosphere and, from the observed flux f_ν, the distance to the supernova. The light observed during the early stages of the explosion must arise from the photosphere (Figure 15.6), whose radius is $R_p(t)$ and which expands with radial velocity v_p. If the spectrum originates from the photosphere, and is approximately black-body, then we may associate a temperature T_p with it such that the emitted energy flux is $B_\nu(T_p)$. It follows that

$$\theta = [f_\nu/\pi B_\nu(T_p)]^{1/2}, \tag{15.3}$$

where the angular size of the supernova photosphere $\theta = R_p/D$, and D is its distance from the observer. During the initial expansion, the density in the envelope must be much greater than in the surrounding interstellar medium, and the motion of the envelope is essentially free expansion, for which

$$R_p(t) \simeq \int_{t_0}^{t} v_p(t)\, dt + R_p(t_0). \tag{15.4}$$

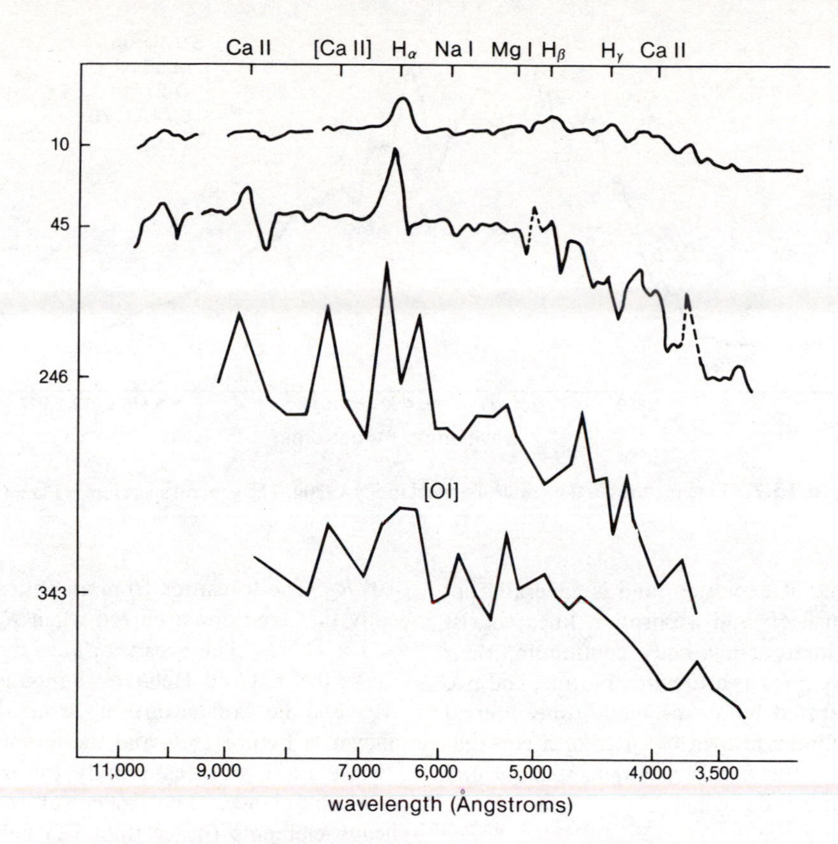

Figure 15.5. Spectra of SN 1970g (Type II) over one-year period.

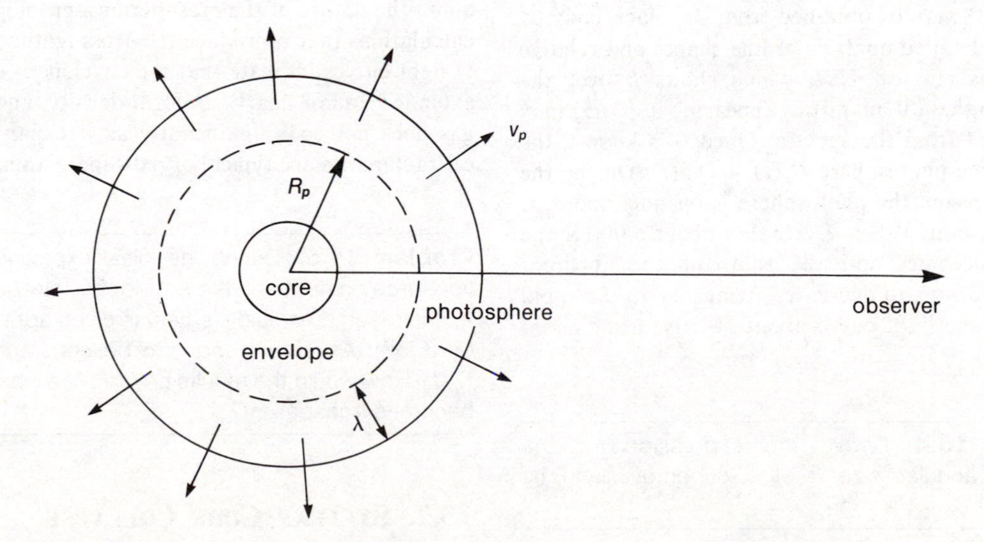

Figure 15.6. Expanding photosphere of a supernova (schematic). Leading edge is a mean free path λ away from R_p, the radius of the photosphere, shown as a dashed line.

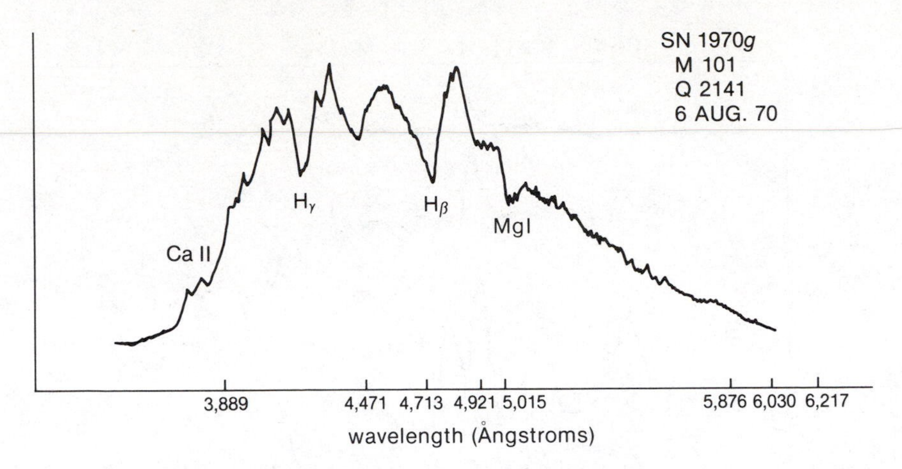

Figure 15.7. Line features in the visible observed in SN 1970g. The general spectrum is Planckian.

Here t_0 is the time of explosion, and is generally not known. Since emission and absorption lines coexist with the approximate black-body continuum, their Doppler shifts give $v_p(t)$ as a function of time, and two observations separated by a reasonable time interval may be used to eliminate t_0 in (15.4). For a constant expansion velocity, the observed angular size θ and (15.4) may be combined to yield

$$D = \frac{v_p(t_2 - t_1)}{\theta_2 - \theta_1}, \tag{15.5}$$

where the subscripts denote the time. The temperature T_p in (15.3) may be obtained from the black-body fit and from detailed analysis of line shapes and relative line widths (Figure 15.7, which shows f_λ over the visible), so that all quantities appearing in (15.5) may be obtained from the spectra. Once D is known, the radius of the photosphere $R_p(t) = D\,\theta(t)$. During the early expansion, the photosphere is opaque, and v_p is nearly constant. When T_p reaches about 6,000 K the envelope becomes optically thin and the apparent radius R_p begins to decrease, giving rise to the rapid decline in the light curve about 30 days after maximum.

Problem 15.4. Derive the expression for the observed angular size θ of a supernova and its distance D.

For Sn 1970g, $L_{max} \simeq 10^{42}$ erg/sec, and the radius of the photosphere at maximum was 3×10^{14} cm $\simeq 4 \times 10^3\ R_\odot$. The transition from an optically thick to optically thin envelope occurred when $R_p \simeq 2 \times 10^{15}$ cm $= 3 \times 10^4\ R_\odot$. The expansion velocity during this time $v_p \simeq 5{,}000$ km/sec. Relative abundances of H, Ca, Na, Mg, and Fe can be inferred from the line structure shown in Figure 15.5, and the temperature T_p. In Sn 1970g, these were essentially the same as observed solar abundances. The absence of lines identified with heavy elements (other than Fe) believed to be produced by the supernova explosion suggests that the early spectra probably contain little information relevant to the details of the core explosion other than the total energy released. The observations do yield clues about the nature of the presupernova envelope. Model calculations that reproduce the gross features of Type II light curves indicate that the envelope is extremely extended and of nearly uniform density, and that the gas does not cool significantly as it expands. These characteristics are typical of red supergiants.

Problem 15.5. During the initial expansion, radiation force on matter in the envelope is strong, and the acceleration on a fluid element is given approximately by (5.57). Assume κ and L to be constant and find $R_p(t)$. How would the results be modified when the gas becomes transparent?

15.2. Stellar Core Collapse

The observational details discussed in Section 15.1 are reasonably well accounted for if the presupernova

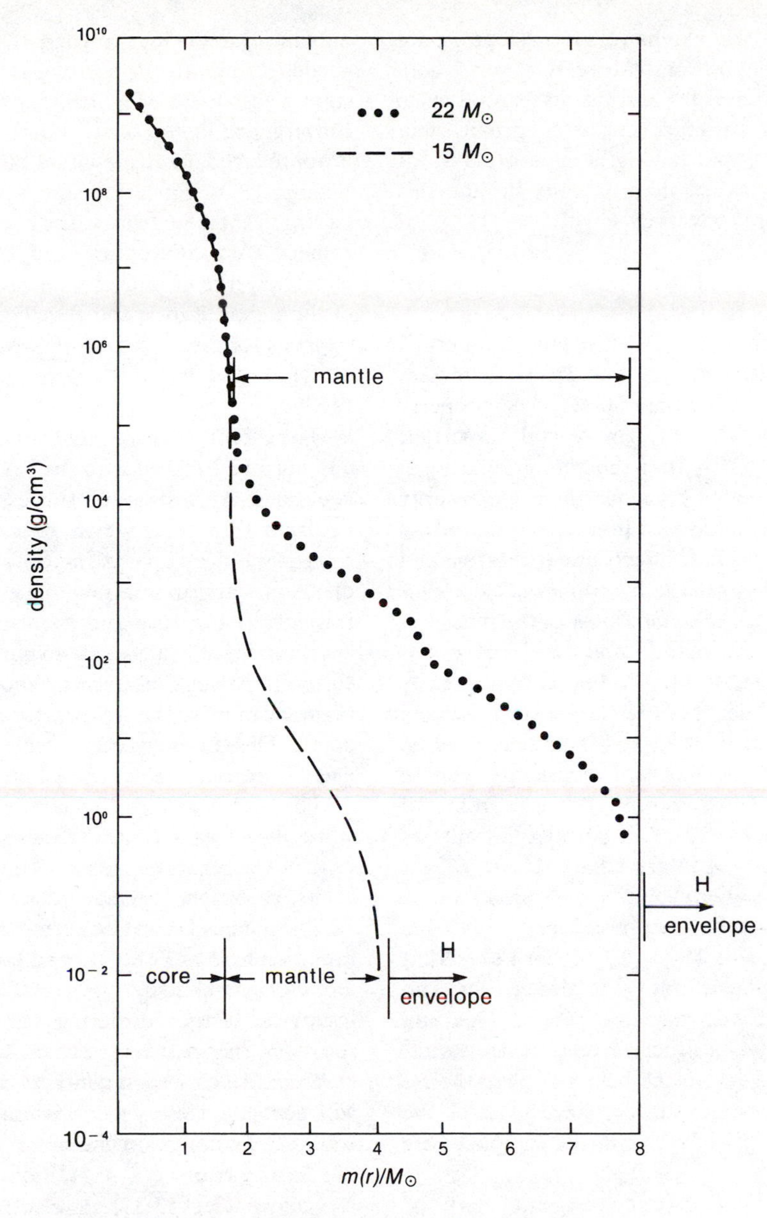

Figure 15.8. Core structure of a 15 $M_\odot$ and a 22 $M_\odot$ Population I star at the onset of core collapse. Only the inner $m(r) \lesssim 8\ M_\odot$ is shown.

envelope is that of a red supergiant, and an explosive release of energy E_{rel} is introduced interior to it. Unfortunately, the core parameters on which the energy release depends appear not to influence the light curve and spectrum, at least during the early years after the explosion. Much later, when the ejected mass has escaped from the system, detailed studies may disclose abundance anomalies resulting from explosive nucleosynthesis accompanying the supernovae, or attributable to the presupernova structure of the core. Many of the conclusions are indirect and model-dependent, but they do impose some constraints on the theory.

The core structures for two highly evolved stars (15 $M_\odot$ and 22 $M_\odot$) are shown in Figure 15.8, just after Si burning, when the core has begun to collapse hydrodynamically. The extensive H envelope is not shown. The first important feature is that in both models the core converges during its evolution to essentially the same structure, consisting primarily of iron-group elements, and having a mass $M_{core} \simeq 1.4\, M_\odot$. This is usually called the iron core. In fact, this tendency toward core convergence with

$$M_{core} \simeq 1.4 - 2\, M_\odot \tag{15.6}$$

appears to be a common feature of detailed numerical models whose initial (i.e., main sequence) masses range from 10 $M_\odot$ to more than 60 $M_\odot$. This property appears to be general, and has several important consequences. The first is that the core gravitational binding energy is greater than the amount of energy that could probably be released in it during the subsequent evolution (we exclude effects due to rotation and magnetic fields, although it is by no means clear that this is always justified). Second, most of the mass lies outside the core in the mantle and envelope, and is spread out over a substantial volume. This greatly reduces the binding energy of the mantle and envelope, so that 10^{50} to 10^{51} ergs of energy is all that need be released by core collapse in order to accelerate them to escape velocities. Third, since the ash from all nuclear-burning stages except Si burning lies outside the core, the elements of intermediate atomic weight synthesized during earlier evolutionary stages may be ejected into the interstellar medium. Finally, since the core mass (15.6) exceeds $M_{Ch} \simeq 1.2\, M_\odot$ for Fe or Ni, it must ultimately collapse until the degeneracy and repulsion associated with nucleons sets in. One end point of core collapse is, therefore, a neutron star. Although $M_{core} \lesssim 2\, M_\odot$, a black hole may nevertheless form after a supernova explosion, since some of the ejecta may eventually accrete onto the compact core remnant.

The phenomenon of core convergence may be understood by considering what happens to a nuclear fuel surrounded by a shell-burning source. Assume that the core is inert; then if the electrons are nondegenerate and M_{core} is less than the maximum stable mass M_{SC} for an isothermal core (Section 9.5), it will evolve quasistatically as mass is gradually added to it by the shell source. When sufficient mass has been added for the core temperature to exceed that necessary for ignition of the next nuclear fuel, the core contraction is halted until the fuel is exhausted. The contracting core will eventually become degenerate, possibly passing through a series of burning stages; its maximum stable mass is then M_{Ch}. However, the mass needed to initiate Si burning is about 0.98 M_{Ch}. For stars whose mass $M > 10\, M_\odot$, $M_{SC} > M_{Ch}$, and core convergence should occur. Furthermore, as Si burning is approached, neutrino losses become substantial (see Section 13.3). Since the star is still effectively transparent to these neutrinos, the energy loss increases the rate of core contraction and core heating, so that effectively the core evolution proceeds independently of the stellar mantle and envelope. It is not clear what happens for intermediate-mass stars (2–4 $M_\odot \lesssim M < 10\, M_\odot$), and we will return to this category separately.

The evolution subsequent to Si burning is dynamic, and must be handled with the full set of hydrodynamic equations. This stage of the evolution must include neutrino transport, which becomes more important than photon transport. The difficulty here is that the diffusion approximation may not be applicable throughout the core and mantle, or for all neutrino energies. Finally, although no nuclear fuels exist in the core, the passage of a strong shock wave through the mantle can raise the temperature there to the ignition point. This in turn may result in further nuclear-energy release, and additional nucleosynthesis.

The core in Figure 15.8 is essentially an iron-nickel white dwarf at a central density $\rho \simeq 10^9$ g/cm^3 and central temperature greater than 10^9 K. At this stage in its evolution, nuclear photodissociation (Section 12.2) becomes important, removing photons from the radiation field as Fe → He and finally He → nucleons; and electron-neutrino pair production (Section 13.3) increases, further depleting the radiation field, and removing energy from the core. According to the virial theorem, the core must contract, and as it does the ions will heat up; these processes mark the onset of core collapse. The electron chemical potential increases as the density rises ($\mu_e \sim \rho^{1/3}$), inducing electron capture by heavy nuclei (13.38). As electrons are captured, the electron pressure drops. All three of these processes keep the adiabatic index below 4/3. The final result of hydrodynamic core collapse is the ultimate conversion of heavy nuclei and He into a gas of nucleons, and the liberation of a substantial amount of energy as neutrinos.

Figure 15.9 shows schematically the radius of selected mass elements versus time once iron-core collapse begins. Curve d represents the core-envelope interface, which evolves very slowly relative to the inner-core mass elements a, b. The mass contained within d is generally greater than 2 $M_\odot$. The primary

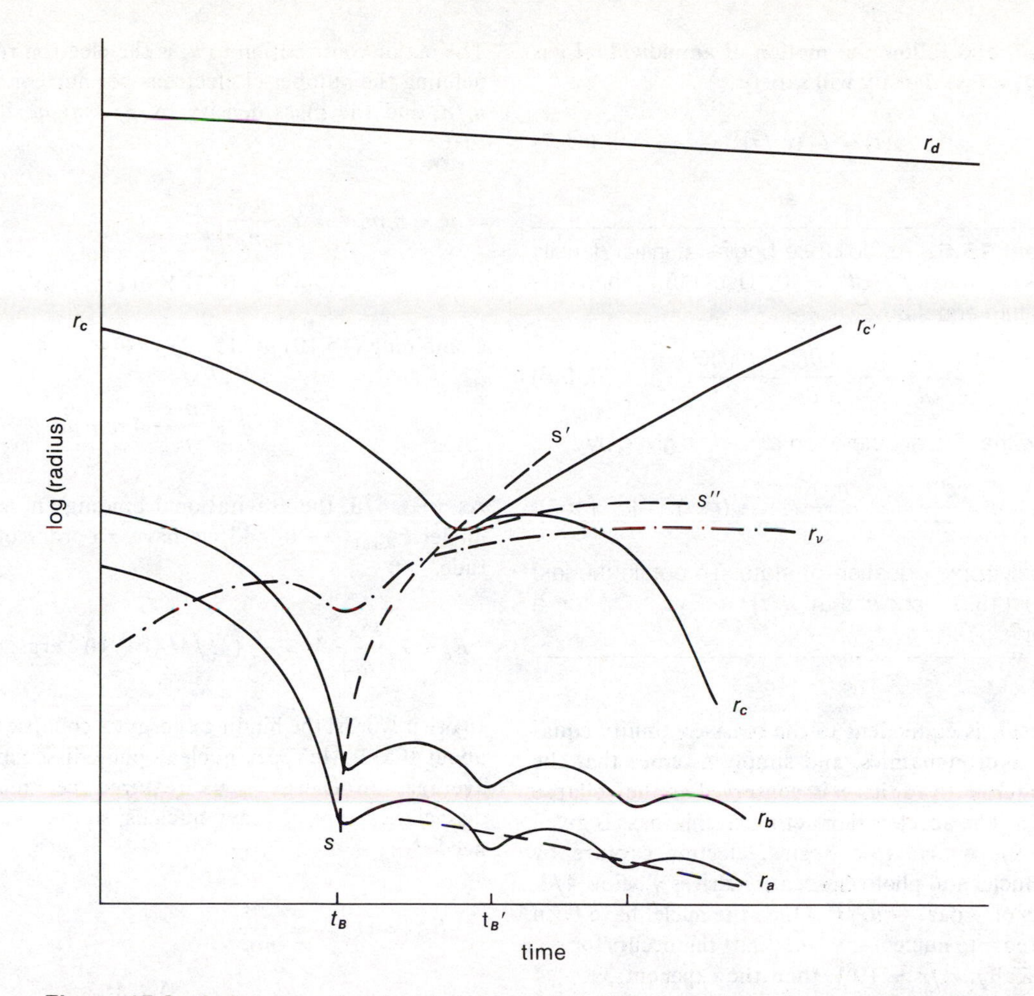

Figure 15.9. Radius versus time for selected mass shells in the core collapse, bounce, and possible explosion of a supernova.

source of neutrino opacity during the early stages of core collapse is coherent scattering off nuclei (Section 13.4) and electron scattering [equation (13.7)]. When the central density reaches a few times 10^{11} g/cm^3, the neutrino opacity is great enough to begin to trap neutrinos within the inner core (radius $r_\nu \approx 30$ km), and a narrow region develops across which the material changes from opaque to transparent. The transition zone has properties analogous to a stellar photosphere, and is called the neutrinosphere. Within the neutrinosphere, the neutrinos are thermalized, and their transport may be approximated by diffusion theory.

By time t_B, the inner core ($r < r_b$) consists of electrons, nucleons, and neutrinos, and $\gamma > 4/3$. Nucleon pressure may now support the mass, and a hydrodynamic bounce may occur. As the core rebounds ($t > t_B$), the overlying layers, which are still infalling, are compressed, setting up a shock wave that travels outward into the mantle (dashed line). The inner-core mass ($r < r_b$) oscillates after bounce about the quasistatically evolving state shown dashed through curve r_a. If the outer envelope and mantle are ejected, this core will be a hot neutron star. The behavior of intermediate-mass regions (curve r_c) will be considered later. The density at bounce may be as high as 10^{14} g/cm^3; in any event, for $t > t_B$, the average core density is comparable to nuclear density.

The core bounce results when the equation of state stiffens at high densities, so that γ changes from $\gamma \approx 4/3$ to $\gamma \simeq 5/3$ for the nucleon gas. To see this we assume a polytropic equation of state for the matter,

$P = K\rho^{\gamma}$, and follow the motion of an individual gas element, whose density will satisfy

$$\rho(t) = \rho_0 (r_0/r)^3. \qquad (15.7)$$

Problem 15.6. As collapse begins at initial density ρ_0, $r \simeq r_0$ and $d^2r/dt^2 \simeq 0$. Use (15.7) and the momentum equation

$$\frac{d^2r}{dt^2} = -\frac{1}{\rho}\frac{\partial P}{\partial r} - \frac{m(r)G}{r^2} \qquad (15.8)$$

to show that the acceleration d^2r/dt^2 is given by

$$a = \frac{d^2r}{dt^2} = -\frac{m(r)G}{r^2}[1 - (r_0/r)^{3\gamma-4}] \qquad (15.9)$$

for a polytropic equation of state. To obtain the last term in (15.9), show that $m(r) = 3\gamma p_0/\rho_0 r_0$ for a polytrope.

This result is equivalent to the mass-continuity equation in hydrodynamics, and simply assumes that the mass interior to radius r is conserved as the collapse develops. The acceleration acting on this mass is given by (15.9). As collapse begins, electron capture by heavy nuclei and photo dissocation drives γ below 4/3, and a is of order $-mG/r^2$. Once the nuclei have been converted into nucleons, $\gamma \rightarrow 5/3$. If this occurs for $r \ll r_0$ (typically, $r_0/r \simeq 10^2$), then the exponent $3\gamma - 4$ changes sign; the second term in brackets becomes large, and the acceleration becomes positive. Therefore, a strong bounce requires that the equation of state be relatively soft ($\gamma \lesssim 4/3$) during the initial collapse stages, and then become stiff ($\gamma \gtrsim 5/3$) rather suddenly at high density. A strong bounce tends to establish a strong outgoing shock wave, which may be essential for an explosion.

We may estimate several useful parameters from the energetics of collapse. For example, the gravitational binding energy is the negative of the total energy,

$$E_B = -\int (3P - u)\, dV, \qquad (15.10)$$

where the total energy density u consists of contributions from nuclei and electrons before collapse begins:

$$u = u_{\text{nuclei}} + u_e. \qquad (15.11)$$

The major contribution to u_e is the electron rest mass; defining the number of electrons per nucleon by $Y_e \equiv n_e/n_b$ and the mass density by $\rho = m_H n_b$, it follows that

$$u_e \simeq n_e m_e c^2 = Y_e \frac{m_e c^2}{m_H}\rho$$
$$= 4.9 \times 10^{17} (Y_e\rho) \text{ ergs/cm}^3. \qquad (15.12)$$

Combining (15.10) to (15.12) yields

$$E_B = E_{B,\text{nuclei}} + \int Y_e \frac{m_e c^2}{m_H} 4\pi r^2 \rho\, dr. \qquad (15.13)$$

As $\gamma \rightarrow 4/3$, the gravitational binding energy of the nuclei $E_{B,\text{nuclei}} \rightarrow 0$, and we have, to order of magnitude,

$$E_B \simeq Y_e \frac{m_e c^2}{m_H} M \approx Y_e (M/M_\odot) \times 10^{51} \text{ ergs}. \qquad (15.14)$$

In other words, the binding energy as collapse begins is about 0.511 MeV per nuclear photodissociation. On average, an energy ϵ_N is required per nucleon to dissociate a typical heavy nucleus; so the total energy needed is

$$E_{\text{nucl}} \simeq \epsilon_N N_N \simeq \frac{\epsilon_N M}{m_H}$$
$$= 1.5 \times 10^{52} \left(\frac{\epsilon_N}{8 \text{ MeV}}\right)\left(\frac{M}{M_\odot}\right) \text{ ergs}, \qquad (15.15)$$

where N_N is the number of nucleons in the core, and ϵ_N is in MeV. Typically, $\epsilon_N = 8$ MeV. Finally, the neutrino energy loss E_ν will be given roughly by the product of the average energy per neutrino ϵ_ν, the number of neutrinos emitted per nucleon due to electron capture, ΔY_e, and N_N:

$$E_\nu \simeq \Delta Y_e \epsilon_\nu N_N$$
$$\simeq \Delta Y_e \left(\frac{\epsilon_\nu}{16 \text{ MeV}}\right)\left(\frac{M}{M_\odot}\right) 3.1 \times 10^{52} \text{ ergs}. \qquad (15.16)$$

Typically, $\epsilon_\nu \simeq 16$ MeV, and $\Delta Y_e \simeq ½$ if all neutrinos escape. The total energy released during collapse is approximately

$$E_{\text{rel}} \simeq E_\nu + E_{\text{nucl}}. \qquad (15.17)$$

E_{rel} is of order 10^{52} or more, and is nearly 10^2 times the binding energy of the original core before collapse. Therefore, the binding energy of the neutron star, E_{NS}, is

$$E_{NS} = E_{B,el} + E_{rel} \simeq E_{rel}.$$

As we will see, the nucleon gas in the core at bounce is nondegenerate; so we approximate it by a polytrope with $\gamma = 5/3$. Therefore the results of Section 4.3 apply, and

$$E_{NS} \simeq \frac{3}{7}\frac{M^2G}{R}. \qquad (15.18)$$

Furthermore, the central pressure for $\gamma = 5/3$ is given by

$$P_c = \frac{\alpha MG\rho_c}{R}, \qquad (15.19)$$

where $\alpha = 0.54$. Using the ideal-gas equation of state (2.1) to eliminate P_c, and (15.18) to eliminate R, we find

$$E_{NS} = \frac{3}{7\alpha}\frac{M_c}{\mu m_H}kT_c$$

$$= 1.3 \times 10^{52}\left(\frac{M_c}{M_\odot}\right)\frac{T_{c(11)}}{\mu}\ \text{ergs}, \qquad (15.20)$$

where μ is the mean molecular weight, and $T_{c(11)}$ is in units of 10^{11} K.

Problem 15.7. Verify equations (15.18) to (15.20).

Equating E_{NS} and E_{rel}, and solving for T_c, we find

$$T_c \simeq \frac{7\alpha}{3k}\mu\,(\epsilon_N + \Delta Y_e\,\epsilon_\nu)$$

$$= 2.3 \times 10^{11}\,\mu\left(\frac{\epsilon_N + \Delta Y_e\epsilon_\nu}{16\ \text{MeV}}\right)\text{K}. \qquad (15.21)$$

For $\epsilon_\nu \simeq 16$ MeV, $\epsilon_N = 8$ MeV, and $\mu \simeq \frac{1}{2}$ the core temperature at bounce, $T_c \simeq 1.2 \times 10^{11}$ K. To obtain the density at bounce, rewrite (15.19) in the form $P_c = \alpha' M^{2/3} G\rho_c^{4/3}$, where $\alpha' = 0.43$, equate it to the pressure (2.1) of an ideal gas, and rewrite

$$\rho_c = \left(\frac{k}{\mu m_H G\alpha'}\right)^3 \frac{T_c^3}{M_c^2}$$

$$\simeq 8 \times 10^{13}\left(\frac{T_c}{1.2 \times 10^{11}}\right)\left(\frac{M_\odot}{M_c}\right)^2 \text{g/cm}^3. \qquad (15.22)$$

Problem 15.8. Estimate the thermal energy in the nucleons after bounce. Show that the nucleons are nondegenerate at bounce.

The analysis above indicates that the core ($r < r_b$ in Figure 15.9) increases its binding energy by about 10^{51} ergs$/M_\odot$ during collapse, reaching a central temperature and density of order 10^{11} K and 10^{14} g/cm^3, respectively. In the process about 10^{52} ergs of energy are liberated as neutrinos. After bounce the core undergoes nearly adiabatic oscillations of period $t_{pul} \simeq 2\pi/\omega$, where t_{pul} is given by (11.24). Thus

$$t_{pul} \simeq \frac{2\pi}{\omega} = \frac{2\pi}{(3\gamma - 4)^{1/2}}\left(\frac{R_c^3}{M_cG}\right)^{1/2}$$

$$= 1.7 \times 10^{-2} R_{(7)}{}^3\left(\frac{M_\odot}{M_c}\right)^{1/2} \text{sec}. \qquad (15.23)$$

As in the expressions above, M_c is the core mass and R_c is the core radius. In most numerical models, immediately after bounce $M_c \simeq 0.5 - 1.0\ M_\odot$ and $R_c \lesssim 10^7$ cm. Finally, since the core consists primarily of nucleons, the neutrino scattering cross sections are small, and the mean free path through the core is of order R_c or less.

Now consider the evolution of matter initially at r_c (Figure 15.9) when collapse begins. The central question for the subsequent evolution is whether or not this layer of mass (and that above it) will be ejected. For a nonmagnetic, nonrotating core of the type considered here, there are four principal mass-ejection mechanisms available. These result from neutrino momentum deposition, neutrino energy deposition, the bounce shock, and thermonuclear burn. The final result may depend on several or all of these.

The shock wave (shown as the dashed line ss' in Figure 15.9) produced by the bounce advances into the mantle with a speed of a few times 10^9 cm/sec, compressing and heating the matter to temperatures of order 10^{11} K. The material through which the shock travels consists of heavy nuclei, which will dissociate when the temperature exceeds about 10^{11} K ($kT \sim 8$ MeV is the binding energy per nucleon in a typical nucleus). This removes thermal energy from behind the shock and lowers the adiabatic index. The net effect of nuclear dissociation is to weaken the shock. Additional shock weakening occurs in the neutrinosphere, where $\nu_e \bar{\nu}_e$ pair production removes energy at a rate proportional to T^9. When the shock reaches the neutrinosphere, the neutrino emissivity removes energy from behind the shock, reduces the pressure difference across it, and reduces the amount of work that it can do on the overlying material. This process is known as neutrino damping of the shock. If the shock is relatively weak (dashed curve ss'' in Figure 15.9), it will only temporarily halt the infalling shells of matter (such as r_c) when it intercepts them, and the matter will ultimately fall onto the core. A strong shock, however, may survive its passage through the neutrinosphere with enough energy to accelerate the overlying mass to escape velocities (curve r_c'). Clearly, since the escape velocity goes as $\sqrt{2\, m(r)G/r^2}$, those layers that are still beginning their collapse are most easily ejected. Furthermore, the shock energy that can be transmitted to regions of low gravitational potential will be maximized if most of the core bounces at essentially the same time. Additional shock waves may be set up if more mass falls onto the core, and these may give outer layers an additional outward thrust.

Envelope ejection may also result from the deposition of neutrino momentum in matter outside the core. This depends on the difference in neutrino opacity κ (proportional to the neutrino scattering cross sections) in the core and mantle. To illustrate this, we assume that neutrino transport may be described as diffusion, in which case the gradient in neutrino pressure P_ν depends on the neutrino energy flux F,

$$\frac{\partial P_\nu}{\partial r} = -\frac{\kappa \rho F}{c}, \qquad (15.24)$$

where κ is the mean neutrino opacity.

Problem 15.9. Demonstrate (15.24) for the case of spherical symmetry. Note that the neutrino momentum flux is F/c.

Writing the total pressure as the sum of gas and neutrino pressure $P = P_g + P_\nu$, substituting into (15.8) and using (15.24), we find

$$\ddot{r} = -\frac{1}{\rho}\frac{\partial P_g}{\partial r} - \frac{mG}{r^2} + \frac{F\kappa}{c}. \qquad (15.25)$$

The gas-pressure gradient is nearly equal to the gravitational acceleration as collapse begins; so the first two terms on the right side of (15.25) tend to cancel. Thus the maximum acceleration due to neutrino momentum deposition is of order

$$\ddot{r} \simeq \left(\frac{\kappa F}{c}\right) \simeq \left(\frac{\kappa L_\nu}{4\pi r^2 c}\right)_e; \qquad (15.26)$$

the subscript specifies that all quantities are evaluated in the envelope, and we have assumed that the neutrino luminosity $L_\nu = 4\pi r^2 F$. If this is to result in escape velocity, then $\ddot{r} > m(r)G/r^2$, or

$$L_\nu > \frac{4\pi M_c G c}{\kappa_e}, \qquad (15.27)$$

where the core mass is M_c, and κ_e is evaluated in the envelope. This should be recognized as the Eddington limit applied to neutrinos. The neutrino opacity in the envelope arises primarily from electron scattering, and scattering off nuclei of atomic weight A. Denoting the cross sections for these processes by σ_e and σ_A, respectively, and noting that the electron number density $n_e = Y_e\, n_b$, and the number density of nuclei $n_A = X_A \rho / A m_H$, we have

$$\kappa = \sum_A \frac{n_A \sigma_A}{\rho} + \frac{\sigma_e n_e}{\rho}$$

$$= \sum_A \frac{X_A}{A}\frac{\sigma_A}{m_H} + \frac{Y_e \sigma_e}{m_H}. \qquad (15.28)$$

In the hot neutron-rich core behind the shock wave, $X_A \simeq 0$ and the opacity is due primarily to electron scattering. In the mantle, where nuclei of large A may be formed as a result of electron capture, X_A may be large. Since σ_A is much larger than σ_e, the opacity will be larger there. It is this last effect that contributes most to the Eddington limit (15.27).

In principle, neutrino momentum deposition alone could eject the mantle. It can be shown that for this to happen the time t_D required for the neutrino radiation force

$$F_{\text{Rad},\nu} = \frac{\kappa L}{4\pi r^2 c} \tag{15.29}$$

to accelerate the mantle to escape velocity is of the order of the core pulsational period (15.23). The maximum neutrino luminosity occurs shortly after bounce. Referring to Figure 15.9, we see that the mass to be ejected (curve r_c) must reach escape velocity within a time t_D after bounce. Momentum deposition alone will probably not occur rapidly enough to eject the mantle, but it may contribute enough so that, in conjunction with an intermediate-strength bounce shock, ejection can occur.

When neutrinos are absorbed by nuclei or scatter off electrons, energy is deposited in the matter (coherent scattering is highly elastic, and does not contribute to energy deposition). Energy deposition in the mantle increases the gas pressure and its gradient, which tends to promote mass ejection. Because of coherent elastic scattering, the high-energy neutrinos are downscattered before they can diffuse far. Therefore energy deposition appears to be far less effective in mass ejection than does momentum deposition.

Finally, thermonuclear burn outside the core could release enough energy to cause, or contribute significantly to, the explosion; primarily because, as the shock wave passes through the mantle, where active shell sources separated by regions of unburned fuel lie, it raises the temperature; this increase in temperature can greatly increase the burning rates, or even ignite the gas in regions that were inert before the passage of the shock. We noted earlier that Fe^{56} is produced directly by Si burning at temperatures of order 3×10^9 K. If the passage of the shock wave through the Si shell raises the temperature to 5×10^9 K, the energy-generation rate (12.6) increases by a factor of about 10^9. Furthermore, at these high temperatures the end products are mostly Ni^{56}. Although this decays weakly to Fe^{56} with emission of an electron and antineutrino, the presence of electrons in the dense mantle will impede the decay. However, once the ejecta have expanded and reached lower densities, the decay will occur. This may represent one source of energy input to the ejected gas following the explosion, and may help account for the presence of emission lines late in the spectra of Type II supernovae.

Numerical models of supernovae can produce maximum neutrino luminosities of 10^{51}–10^{52} ergs per sec just after bounce. The kinetic energy of the envelope can be several times 10^{51} ergs, and the remnant (neutron star) mass tends to fall in the range 1.1–1.5 $M_\odot$.

Whether or not these mass-ejection mechanisms actually produce supernovae explosions is unclear, since theoretical models require a wide variety of physical input, some of which is not well established. The stellar-collapse scenario described above requires that the presupernova mass exceed 10 $M_\odot$ or so, and indicates that the explosion ejects a substantial amount of mass; so its identification with Type II supernovae is reasonable. This leaves open the fate of stars in the mass range $4\ M_\odot \lesssim M \lesssim 10\ M_\odot$. Since these stars evolve to C^{12} and O^{16} burning, which occurs under degenerate conditions, and involve energy-generation rates that are highly temperature-sensitive, a thermal runaway leading to detonation could result. Theoretical studies indicate that C^{12} burns at densities of 2 to 5×10^9 g/cm^3, which, in stars less massive than about 9 $M_\odot$, can release a total of 2×10^{51} ergs if detonation occurs. This exceeds the gravitational binding energy of the star, and could result in total stellar disruption, leaving no compact remnant. Whether or not such a detonation occurs depends on the efficiency of the neutrino emission that is taking place simultaneously via the URCA process, (13.1) and (13.2). If a balance is achieved between these competing processes, stellar disruption may be avoided, and a neutron-star or black-hole remnant might remain. We don't currently know whether enough energy is released to actually produce a detonation wave (which travels sonically and can set up a shock front), or whether the energy travels subsonically as a combustion front into the unburned fuel.

There remains the problem of explaining Type I supernovae. Since these appear to arise from low-mass stars ($M \gtrsim 1\ M_\odot$), which will not burn Si, and eject only a small amount of mass ($\simeq 0.2\ M_\odot$), a different scenario must be involved. Furthermore, there is an apparent conflict with the evidence that stars less massive than a few $M_\odot$ end up as white dwarfs. It is unlikely that Type I supernovae represent a stage in the evolution to the white-dwarf state for all stars, since the formation rates ($R_{\text{WD}}/R_I \simeq 10^3$) are so different. One model that may circumvent this difficulty assumes that the Type I presupernova is a maximum-mass, hot white dwarf. Although the mass limit for cold white dwarfs $M_{\text{Ch}} \simeq 1.2\ M_\odot$, in the hot system the pressure of the nondegenerate gas of nuclei supplies additional support, and the maximum mass is increased slightly. To order of magnitude, this increase

ΔM should be proportional to the ratio of the ideal gas pressure for the nuclei to the pressure of the degenerate electrons. Since the electrons will be relativistic,

$$\frac{\Delta M}{M_{\mathrm{Ch}}} \simeq \frac{P_{\mathrm{nuclei}}}{P_e} \simeq \frac{m_H{}^{1/3} kT}{\hbar c \rho^{1/3}}. \tag{15.30}$$

Assuming that $\Delta M / M_{\mathrm{Ch}}$ is small, we make set ρ equal to the maximum value for a cold white dwarf. A more detailed analysis shows that ΔM is given by

$$\Delta M \simeq 2 \times 10^{-11} \, ATM_{\mathrm{Ch}}, \tag{15.31}$$

where A is the mean atomic weight of the nuclei. Therefore the critical mass for a white dwarf will be approximately

$$M_{\mathrm{Ch}}(T) \simeq M_{\mathrm{Ch}}(0)\,(1 + 2 \times 10^{-11} \, AT), \tag{15.32}$$

and will decrease as the star cools. Once $M_{\mathrm{Ch}}(T)$ drops below the star's actual mass, dynamic collapse will be initiated by electron capture. The subsequent evolution is unclear. Possibly neutrinos emitted by core bounce can heat the infalling envelope of low-Z elements to their ignition point, ejecting the outer 0.1–0.2 $M_\odot$, and leaving a neutron star remnant. Or, if the original white dwarf were magnetic (white dwarfs with fields as large as 10^7 gauss have been observed), the magnetic fields can transfer much of the gravitational energy released during collapse to the envelope, heating the gas and possibly detonating it.

The preceding scenario is consistent with the distribution of Type I supernovae in elliptical and spiral galaxies. In ellipticals, star formation has ended; 1.5 $M_\odot$ stars would have evolved off the main sequence to form hot white dwarfs ($M \approx 1.2 \, M_\odot$) surrounded by planetary nebulae. The mass of a few hot dwarfs will exceed M_{Ch} at $T = 0$. It is estimated that if $\Delta M \approx 3 \times 10^{-5} \, M_{\mathrm{Ch}}$, these white dwarfs will have cooled enough so that core collapse would occur during present times, producing Type I supernovae. The disparity between formation rates of white dwarfs and Type I supernovae may then be due in part to the relative scarcity of hot white dwarfs with $M \gtrsim M_{\mathrm{Ch}}$. The same arguments apply to 1.5 $M_\odot$ stars formed during the early stages in spiral galaxies. In addition, recently formed hot white dwarfs with a mass excess of $\Delta M \simeq 3 \times 10^{-3} M_{\mathrm{Ch}}$ and initial mass of about 1.5 $M_\odot$ would now be undergoing collapse to form Type I supernovae as well. The larger mass excess implies that these white dwarfs will evolve to collapse in as little as 10^4 years. Consequently, the Type I formation rate should be greater per star in spiral than in elliptical galaxies, as is observed.

Chapter 16

COMPACT STELLAR AND RELATIVISTIC OBJECTS

In this chapter we will deal with the following classes of objects: neutron stars, pulsars, and black holes. We will also consider their relation to compact supernova remnants. The properties of neutron stars are important for theories of pulsars and compact x-ray sources (Chapter 17). Neutron stars also represent the most compact objects known that can resist gravitational collapse to a black hole. In order to discuss black holes, we will introduce the concept of strong (or relativistic) gravitational fields, and consider several unusual properties associated with them. Finally, we will consider the production of gravitational radiation by compact objects, particularly by binary pulsars; such production lends strong support to Einstein's theory of gravitation, which predicts the existence of black holes.

16.1. Compact Supernova Remnants

A supernova explosion releases $E_{rel} = 10^{52}$ ergs or more energy as photons, neutrinos, and gravitational radiation. Since the binding energy of the presupernova core is typically $B_W \simeq -10^{48}$ ergs, the remnant (if one is left) must have a binding energy $B_{rem} = B_W - E_{rel} \simeq -E_{rel}$. For remnant masses M_{rem} of order $M_\odot$, the imploded core must have average densities well in excess of those typical of the most massive white dwarfs. The only known objects that can be stable under such conditions are neutron stars.

Many stars show signs of rotation, and some are also magnetic. Although these effects have seldom been incorporated in models of stellar structure and evolution, they can become important in compact supernova remnants. Figure 16.1 shows the average linear velocity at the equator of rotating main-sequence and giant stars as a function of spectral type. Supernova progenitors ($M \gtrsim 5\ M_\odot$) correspond roughly to spectral type B5 and earlier. If these stars rotate more or less rigidly, then their angular velocity $\Omega = v_e/R \approx 10^{-4}$ sec^{-1}. Consider the collapse of a stellar core with this initial angular velocity. Its angular momentum is roughly $J \sim M_c R_c^2 \Omega$; assuming that J is constant during core collapse, the angular velocity of the remnant will be

$$\Omega_{rem} \simeq \Omega\,(R_c/R_{rem})^2 \sim 2 \times 10^4\ \text{sec}^{-1}, \quad (16.1)$$

where we used $R_{rem} \sim 10^6$ cm, as is typical for neutron stars. There is a limit to how rapidly a star can rotate, obtained by equating the gravitational and centripetal forces acting on an element of matter at the star's

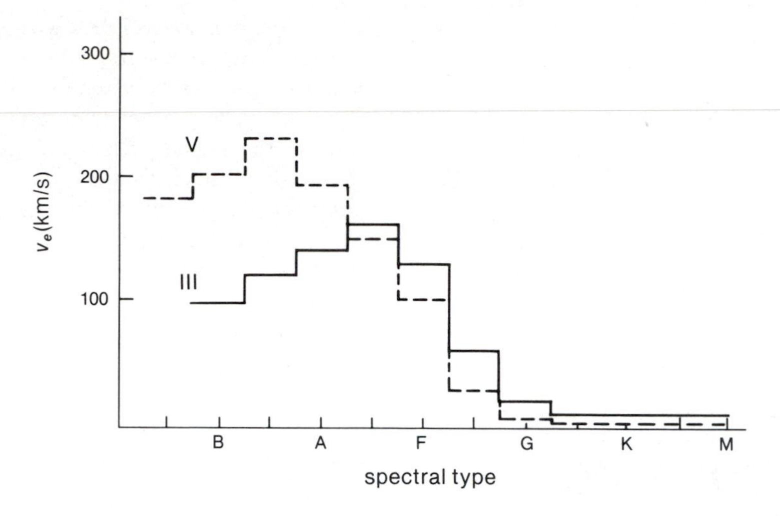

Figure 16.1. Equatorial velocity of rotating stars as a function of stellar spectral type.

equator:

$$\Omega_{\text{crit}} \simeq (MG/R^3)^{1/2}. \tag{16.2}$$

For a neutron star $\Omega_{\text{crit}} \approx 2 \times 10^4$ sec^{-1}, which is the same as Ω_{rem} in (16.1). This suggests that compact supernova remnants may be formed in as rapid a state of rotation as possible. Rotation would then be dynamically important during collapse, since large angular momenta generate centrifugal forces, which may balance gravitation near the equator. The subsequent evolution of the remnant mass might then be quite different than for nonrotating configurations.

However, the core of a star need not rotate with the same Ω as do the surface layers; and the core angular momentum need not remain constant during the evolution. We note from Figure 16.1 that early-type giants rotate more slowly than main-sequence stars of the same spectral type. Taking the core angular velocity of a giant to be equal to its equatorial value, and assuming conservation of angular momentum during core collapse, we find $\Omega_{\text{rem}} \approx$ several sec^{-1}, and corresponding periods $P_{\text{rem}} \approx 1$ sec. Since theory and observations give no clear indication which of these extremes is correct, we can only say that supernova remnants are probably formed with initial periods in the range 10^{-4} sec to a few seconds.

Zeeman splitting has been observed in spectral lines from Am and Ap stars, which is consistent with large-scale magnetic fields of order 10^3 gauss, and in a few cases of more than 10^4 gauss. Magnetic fields have also been detected in several white dwarfs by measuring circular polarization, yielding values as high as 10^7 gauss. The cause of these fields, and why they appear in only a few stars, is not clear. We do not know whether they penetrate the entire stellar mass, or whether they arise in the mantle or atmosphere. However, if we suppose that these fields are maintained during stellar evolution, and that they move with the matter as the star contracts (Chapter 22), the magnetic flux will be conserved, and the average field during two different stages of evolution will satisfy

$$B_2 \sim B_1 \, (R_1/R_2)^2. \tag{16.3}$$

Assuming that this relationship applies to Am and Ap stellar types with $B_1 \simeq 10^3$ gauss and $R_1 \simeq R_\odot$, then for $R_2 \simeq 10^{-2} \, R_\odot$ as is typical of white dwarfs, $B_2 \simeq 10^7$ gauss.

If we further suppose that the magnetic field in the collapsing core of a presupernova satisfies (16.3), then the neutron-star remnant will contain fields as large as 10^{12} gauss. The magnetic energy associated with these fields is small compared to neutron-star binding energies ($E_B/E_G \approx 10^{-13}$) and probably remains so during core collapse if the magnetic fields are uniform throughout the matter. Nevertheless, fields of this magnitude will have a dramatic influence on the less-dense plasma surrounding the newly formed neutron star.

Problem 16.1. Show that the ratio of magnetic to gravitational energy

$$E_B/E_G \approx B^2R^4/3M^2G$$

for uniform density and magnetic field. Then show that if $E_B \ll E_G$ when core collapse begins, it will hold during collapse as well.

If the phenomenon of core convergence discussed in Section 15.2 holds in general, then compact supernova remnants will have masses of $1.5 - 2\ M_\odot$ and, as we will see, may end their evolution as neutron stars. However, if the remnant mass is too great, then a black hole will be formed. It can be shown that a black hole cannot maintain an external magnetic field, in which case the field will be expelled, carried off presumably by the ejected matter.

The arguments above indicate that compact supernova remnants may be rapidly rotating, magnetic neutron stars. The formation of a black hole is also possible, either directly or as a result of mass accretion from matter originally ejected by the supernova. Mass transfer from a companion star may also result in black-hole formation (Chapter 17).

16.2. Neutron Stars

A complete treatment of static neutron-star structure is complicated by two effects that are of only secondary importance for white dwarfs. First, an equation of state is needed for matter at and above nuclear density, where the strong interactions (nuclear forces) can not be neglected. Second, because of the high matter densities, the effects of general relativity become important. Nevertheless, we may obtain order-of-magnitude estimates of the bulk properties of cold neutron stars from the following simple observations: (1) primary pressure support results from a combination of nuclear forces between baryons and the Pauli exclusion principle; (2) most of the material inside neutron stars is degenerate. The second observation permits us to treat the problem of mechanical equilibrium separately from such matters as neutron-star cooling.

The strong interactions between nucleons or other baryons are more complicated than electromagnetic coupling between changed particles. Basically, the strong interactions are characterized by an attractive force at distances of the order of the pion's Compton wavelength $\hbar/m_\pi c$, and a stronger repulsive component at shorter distances characterized by the Compton wavelength of vector mesons whose masses are comparable to the nuclear mass. A measure of the range of the repulsive part is thus given by $\hbar/m_p c$. An interparticle potential that shows these effects is given schematically in Figure 16.2. The equilibrium interparticle separation will be somewhere between $\hbar/m_\pi c$ and $\hbar/m_p c$. For definiteness we will adopt $r_b \simeq \hbar/2m_p c \simeq 0.7 \times 10^{-13}$cm. Then the average density of matter will be given roughly by

$$\rho \sim 3m_p/4\pi r_b^3 \simeq 1.2 \times 10^{15}\ \text{g/cm}^3. \qquad (16.4)$$

No simple equation of state exists for strongly interacting nucleons at these densities, but we can estimate the pressure by assuming that the nucleons are a nonrelativistic degenerate gas. The pressure is then

$$P \sim \frac{\hbar^2 (3\pi^2)^{2/3}}{5\ m_p^{8/3}} n^{5/3} \simeq 5.4 \times 10^9\ \rho^{5/3}, \qquad (16.5)$$

which, when evaluated at the density (16.4), gives

$$P \sim 7.4 \times 10^{34}\ \text{dynes/cm}^2. \qquad (16.6)$$

Problem 16.2. The average energy per nucleon in degenerate nuclear matter is (noninteracting model)

$$\frac{\hbar^2 (3\pi^2)^{3/2}}{2\ m_p} n^{2/3}. \qquad (16.7)$$

Show that this leads to the pressure in (16.5).

Actually at densities $\sim 10^{15}$ g/cm^3, the strong interactions are repulsive and (16.5) is therefore an underestimate.

The arguments leading to (14.4) may be easily modified to estimate the temperature below which an ideal nuclear gas will be degenerate. Assuming nonrelativistic energies, the analogue of (14.4) is

$$T < \frac{\hbar^2}{2\ m_p^{5/3}\ k} \rho^{2/3} \approx 17.1\ \rho^{2/3}\ \text{K}, \qquad (16.8)$$

with ρ in g/cm^3. Assuming $\rho \sim 10^{15}$ g/cm^3, we can suppose that nucleons in neutron stars may be treated

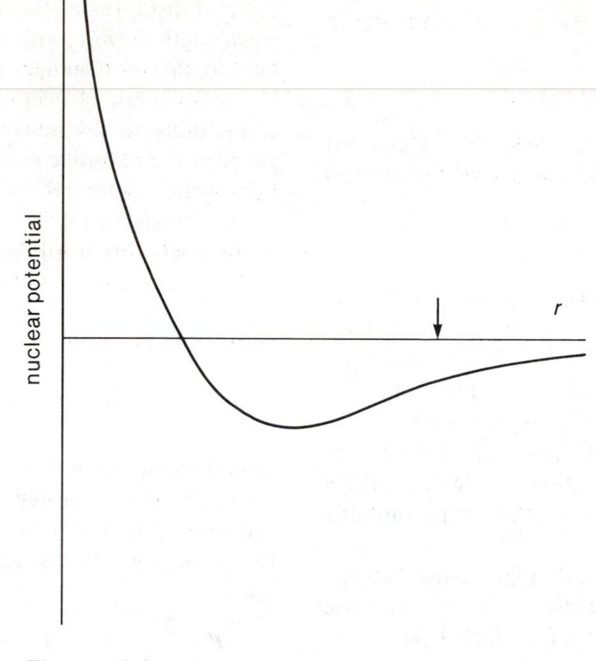

Figure 16.2. Schematic of nuclear potential versus relative separation of two nucleons. The arrow shows the relative separation of a neutron and proton in the ground state of a deuteron.

as degenerate for temperatures below about 10^{11} K. Since neutron stars are a special case of the degenerate stars discussed in Section 14.2, critical-mass estimates may be obtained directly from (14.12) with $A = Z = 1$:

$$M_{\rm crit} \simeq m_p(\hbar c/m_p{}^2 G)^{2/3} \simeq 1.86\ M_\odot, \quad (16.9)$$

which is slightly higher than the Chandrasekhar mass limit for white dwarfs. There is more spread in theoretically obtained values for the critical mass of neutron stars than in those for white dwarfs because of the uncertainties in the underlying equation of state above nuclear density. Values of $M_{\rm crit}$ range from about 0.7 $M_\odot$ for noninteracting neutrons (Tolman–Oppenheimer–Volkoff limit) up to about 2.5 $M_\odot$.

Using the maximum mass (16.9) and average density (16.4) of a neutron star, we find that its radius should be

$$R \sim (3M/4\pi\rho)^{1/3} \simeq 9.2\ \text{km} \quad (16.10)$$

or about 10^{-5} $R_\odot$. Furthermore, we observe that the ratio $2MG/Rc^2 \simeq 0.6$, indicating that effects associated with general relativity will be important, especially for massive neutron stars.

Problem 16.3. Assuming that Newtonian gravitation is sufficient for order-of-magnitude estimates, find the gravitational binding energy of a massive neutron star. Compare this to the rest-mass energy and total radiated energy (on the main sequence) of the Sun.

Another useful parameter is a neutron star's moment of inertia, since this may be measured indirectly for the Crab pulsar. The conditions of extreme degeneracy in the interior imply that matter has been compressed to the limit set by the uncertainty principle. Consequently, the density is nearly uniform, and the moment of inertia may be approximated by that of a sphere of uniform density:

$$I \simeq \frac{2}{5}MR^2 = 1.3 \times 10^{45}\ \text{g cm}^2. \quad (16.11)$$

Table 16.1
Parameters for neutron stars.

Quantity	Order-of-magnitude estimate of value	Range of best current estimates of value
Density (g/cm^3)	1.2×10^{15}	10^{14} to 7×10^{15}
Pressure (dyne/cm^2)	3.4×10^{35}	10^{34} to 3×10^{36}
Mass ($M_\odot$)	1.86	0.5 to 2.7
Radius (km)	9.2	7 to 20
Moment of inertia (g cm^2)	1.5×10^{45}	10^{44} to 4×10^{45}

These estimates agree well with estimates from observational data, and with predictions obtained from current descriptions of superdense nuclear matter and numerical studies of neutron star structure. The results are summarized in Table 16.1, which also gives ranges of the parameters as currently obtained from observations and theory.

The internal composition of a degenerate neutron star is quite different from that of a white dwarf. Since the pressures throughout all but a thin outer shell of the star exceed the ionization pressure, there are no bound electrons left. The schematic cross-section of a typical neutron star interior is shown in Figure 16.3. At densities below about 10^4 g/cm^3, the composition probably depends primarily on external conditions, the presence of strong magnetic fields (10^{12} gauss), and the star's previous history. If neutron stars are the imploded cores of more massive stars that went supernova, then most of the matter will have been processed beyond Si and possibly as far as Fe-group nuclei.

The interior consists of at least two regions; the outermost is probably a solid crust of ions permeated by an extremely relativistic degenerate electron gas, and the inner fluid core composed dominantly of electrons, nucleons, and more massive baryons. The inner core may contain superfluid neutron and superconducting proton phases, and in lighter neutron stars may be solid rather than fluid. The composition may be explained by considering what happens to matter as its density is increased (maintaining zero temperature) from about 10^4 g/cm^3 to 10^{15} g/cm^3. Initially there will be a mixture of heavy nuclei [denoted by their atomic numbers and weights as $N(Z, A)$] and a degenerate electron gas. For matter in its lowest energy state, the nuclear configuration (A, Z) will be the one that minimizes the total energy per gram,

$$E = E_n(A, Z) + E_e + E_c, \qquad (16.12)$$

where E_e is the electron energy per gram and E_c

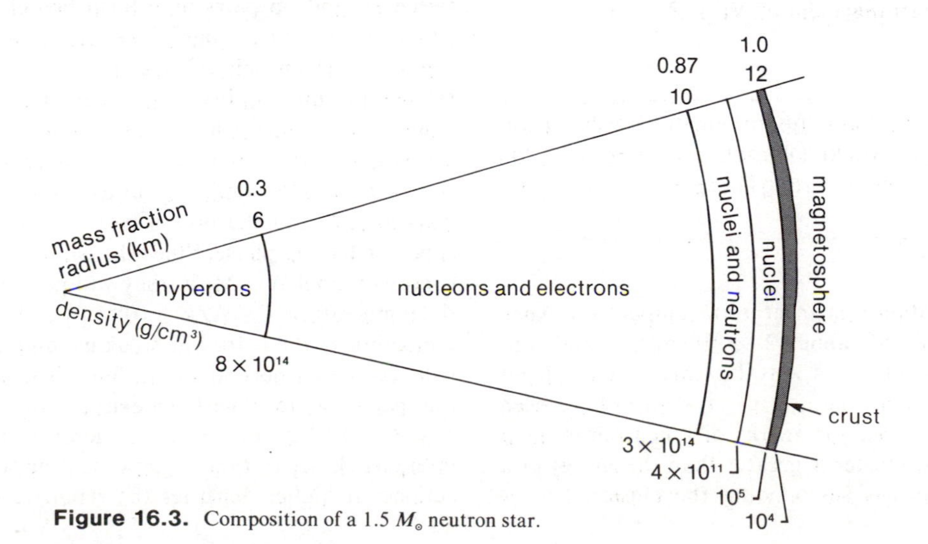

Figure 16.3. Composition of a 1.5 $M_\odot$ neutron star.

contains the Coulomb energy of the ion lattice and electrostatic screening effects. In vacuum ($E_e = E_c = 0$), E_n is a minimum for $_{26}Fe^{56}$. The additional terms in (16.12) shift the minimum to higher values of A, and the result is a shift to increasingly neutron-rich nuclei as ρ increases. Although these neutron-rich nuclei would be unstable against β decay in the laboratory,

$$N'(A, Z) \rightarrow N(A, Z + 1) + e^- + \bar{\nu}_e, \quad (16.13)$$

the presence of degenerate electrons in the star effectively blocks the decay process, and stabilizes the nucleus.

Problem 16.4. A special case of (16.13) for $A = 1$ and $Z = 0$ is the weak decay of the neutron

$$n \rightarrow p + e^- + \bar{\nu}_e. \quad (16.14)$$

In vacuum this reaction takes about 12 min. on average. Find a lower limit to the mass density above which neutron β decay will not occur.

As shown in Problem 16.4, degenerate matter can stabilize particles that would normally decay under laboratory conditions. The process of neutron enrichment, known as neutronization, can be thought of as the reverse of (16.13),

$$N(A, Z) + e^- \rightarrow N(A, Z - 1) + \nu_e, \quad (16.15)$$

followed by rearrangement of $N(A, Z - 1)$.

Problem 16.5. Find the minimum density above which neutrons would interact, assuming that the maximum range of the strong interactions is $\hbar / m_\pi c \sim 2 \times 10^{-13}$ cm.

Neutronization sets in at zero temperature when the density exceeds about 2×10^7 g/cm^3, and continues up to about $\rho_D = 4 \times 10^{11}$ g/cm^3, by which point nuclear clusters having a large A/Z ratio have been formed. For $\rho > \rho_D$, the energy of a neutron inside a typical nuclear cluster is greater than the energy of a neutron in the gas surrounding the cluster. Consequently, a new phase of matter appears, consisting of clusters embedded in a noninteracting gas of neutrons. With increasing ρ, the cluster size grows until they begin to overlap, at which point the system undergoes a phase transition to a fluid mixture of neutrons, protons, and electrons. This transition occurs slightly below nuclear density, and the number densities are roughly $n_p = n_e \approx 0.1 n_N$. The neutrons, which would decay by the weak process (16.14), are stabilized by the presence of the degenerate electrons. When the density reaches about half nuclear density, the average kinetic energy of the electrons exceeds the rest-mass energy of the μ^- lepton, and the reaction

$$e^- \rightarrow \mu^- + \nu_e + \bar{\nu}_\mu \quad (16.16)$$

can occur. At about twice nuclear density, the neutrons will capture an electron to form the hyperon Σ^-:

$$n + e^- \rightarrow \Sigma^- + \nu_e, \quad (16.17)$$

and at higher densities the additional hyperons Λ^0, Σ^0, Σ^-, and Δ^- will be produced. Each of these particles is a fermion, and each species will supply a term to the pressure. The core of a neutron star therefore consists of nucleons (n, p) and more massive hyperons, as well as the electron and muon gas.

The nature of the strong interactions between nucleons is such that nn and pp bound states do not occur under laboratory conditions. In the dense core of neutron stars, however, the strong interactions are modified by the presence of other baryons, and the resulting density-dependent effective interaction between nn and pp pairs may form bound states. The phenomenon is analogous to the formation of a bound electron-electron pair (Cooper pair) in metals, and is related to superfluidity and superconductivity. The bound nucleon states have integer spin, and therefore are bosons rather than fermions. Roughly speaking, the neutron pairs condense out as a superfluid analogous to He^4, and the proton pairs condense out as a superconducting phase. Since the pairs have a binding energy of order $\Delta \sim$ MeV, they will persist even when the temperature $T \lesssim \Delta/k \sim 10^{10}$ to 10^{11} K. For higher temperatures the pairs will break up, and the material will revert to a normal Fermi fluid. Theory indicates that pairs may form within a density range $8 \times 10^{13} \lesssim \rho \lesssim 8 \times 10^{14}$ g/cm^3. At lower densities the average interparticle separation exceeds the range of the interactions; at higher densities the repulsive part of the strong interactions dominates, and the pairs will not form.

Superfluid and superconducting nucleons probably have a greater effect on the energy-transport (cooling) and magnetic properties of neutron stars than on their macroscopic structural properties. Theory suggests that when super phases occur, they involve only a small fraction (~10 percent) of the nucleons. It is therefore likely that the pressure energy-density equation of state that ultimately describes the star's equilibrium mass and radius will be insensitive to these phenomena.

Problem 16.6. The composition described in the preceding will result only if the matter in the neutron star has had sufficient time for the processes of electron capture, neutrino emission, and nuclear rearrangement to occur. They probably will not occur during the dynamic core collapse associated with a supernova, since the neutrinos form a degenerate Fermi gas, and the rate of the reaction (16.15) will be governed to a large degree by the rate of escape of neutrinos from the core. What effect is this likely to have on the final (observable) structure of a neutron star?

Figure 16.4 shows typical neutron-star equations of state. The attractive strong interactions dominate for densities $10^{13} < \rho \lesssim 3 \times 10^{14}$ g/cm^3, and above $\rho \simeq 10^{15}$ g/cm^3 the repulsive parts dominate. Representative mass and density profiles are shown in Figure 16.5 for a typical intermediate-mass ($M \simeq 1.2\ M_\odot$) and high-mass ($M \simeq 2.4\ M_\odot$) neutron star as obtained from the theory of relativistic stellar structure. The density profile represents the total energy density divided by c^2, and includes rest, kinetic, and interaction energy. Nevertheless, it is instructive to compare the result with Figure 16.5 for a typical theoretical model of a white dwarf. We see that the density in neutron stars is more nearly uniform, and that white

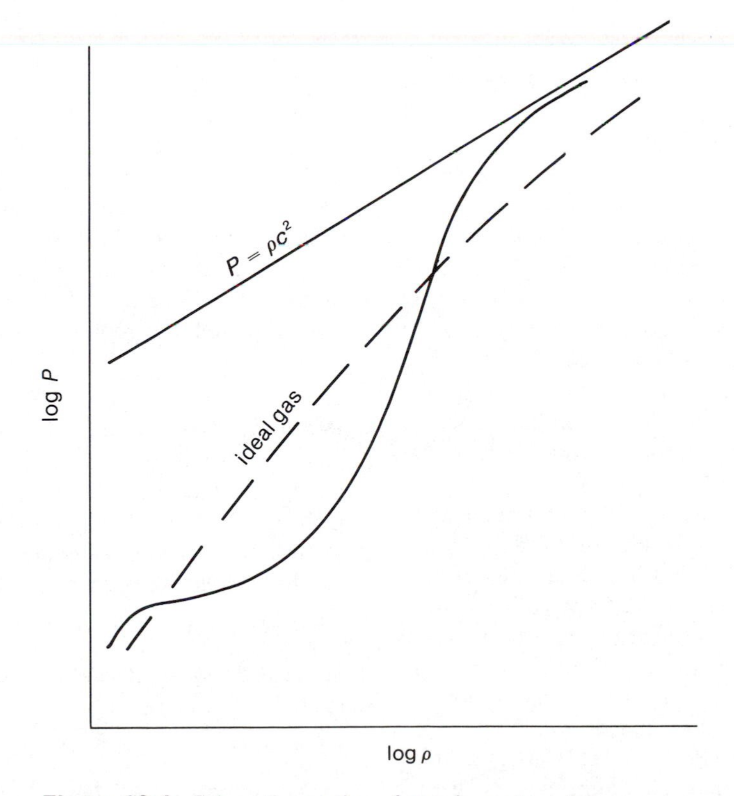

Figure 16.4. Schematic equations of state for neutron stars.

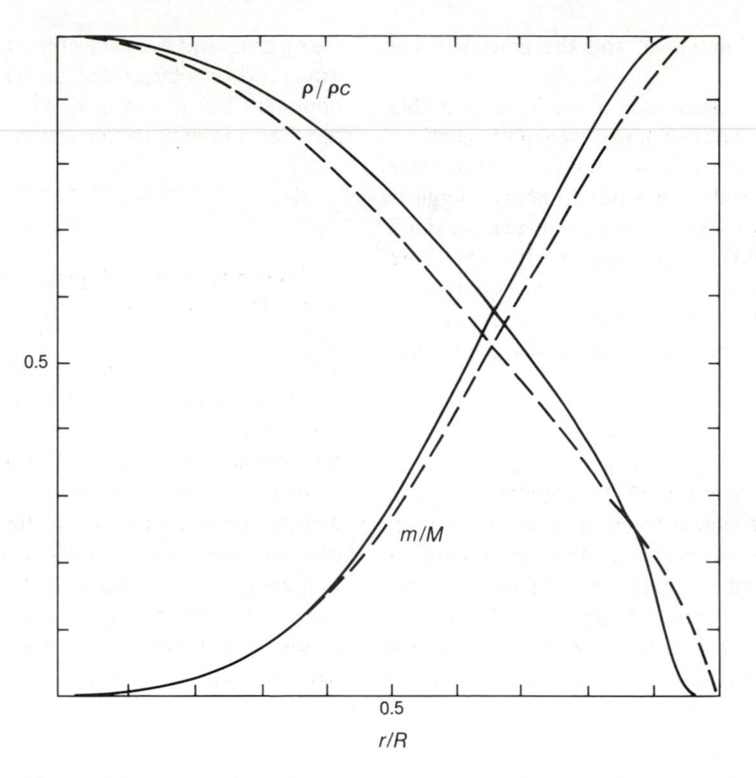

Figure 16.5. Structure of neutron stars of total mass 1.15 $M_\odot$ [R = 10.3 km and $\rho_c = 1.26 \times 10^{15}$ g cm^{-3} (solid curve)] and 2.38 $M_\odot$ [R = 12.7 km and $\rho_c =$ 1.41×10^{15} g cm^{-3} (dashed curve)].

dwarfs show a higher degree of central mass concentration.

In principle, a neutron star could have a mass as low as 0.05 $M_\odot$. However, the binding energy for masses less than about 0.2 $M_\odot$ are less than those for typical white dwarfs. Therefore, if a neutron star were to form with such a low mass, some mechanism would be required to add energy to the system without increasing its mass. Since this appears unlikely, we are probably safe in assuming that neutron stars of mass $M < 0.2\ M_\odot$ do not exist.

Pulsars are believed to be rapidly rotating magnetic neutron stars whose surface fields exceed 10^{12} gauss. Fields of this magnitude will have a significant effect on the atomic properties of the crust ($\rho \simeq 10^4 - 10^5$ g/cm^3), since the motion of atomic electrons will be restricted perpendicular to the field lines. In fact, for $B > 10^9$ gauss, the quantum-mechanical motion normal to B will be less than the Bohr radius a_0 of a normal hydrogen atom (Problem 16.7). As a result, the electron distribution in such atoms will be greatly distorted, forming elongated prolate spheroids. The electrons now spend a greater fraction of their time near the nucleus, and are thus more tightly bound than they would be under laboratory conditions. In fact, for $B \approx 10^{12}$ gauss the electronic binding energy is of the order of 160 eV, and surface temperatures approaching 10^9 K would be required to ionize the electrons.

Problem 16.7. An atom is immersed in a magnetic field of strength B. Assume that the magnetic force acting on the electron is much greater than the Coulomb force, and use semiclassical arguments to show that the radius of the electron's orbit perpendicular to the field is of order

$$a \sim (\hbar c/Be)^{1/2} = 2.6 \times 10^{-10}\ B_{(12)}{}^{-1/2}\ \text{cm.} \quad (16.18)$$

Show that neglect of the Coulomb force is reasonable for $B \gg 10^9$ gauss.

An additional feature of the reduced atomic radius normal to B is the formation of a tightly bound crystal lattice, whose Coulomb binding energy will be of order 15 keV per atom. We will see that this places important restrictions on the regions immediately surrounding pulsars. Finally, strong magnetic fields imply that the crust behaves like a conductor perpendicular to B, that is, outwardly, but is an extremely good insulator normal to B.

16.3. GRAVITATIONAL COLLAPSE AND BLACK HOLES

We now consider the fate of matter that has become too compact for material forces to establish a stable equilibrium configuration of finite density. We will also discuss several phenomena associated with strong gravitational fields, and the production of gravitational radiation.

Two important astrophysical consequences of Einstein's theory of general relativity, considered as a theory of gravitation, are the existence of gravitationally collapsed objects (black holes) and of gravitational radiation from time-varying mass distributions. Both results have no analogy in Newtonian gravitation. A consistent description of general relativity requires the machinery of tensor analysis, which can not be developed here. We will instead rely on heuristic arguments, and summarize the primary results.

We note that general relativity includes as a special case the theory of special relativity. Einstein's theory includes the fundamental constants G and c, but Newtonian theory includes only G. Therefore, the characteristic length and time scales over which general relativistic corrections to Newtonian theory are expected to be important, for a given mass distribution M, are of order

$$l \sim MG/c^2, \qquad \tau \sim MG/c^3. \qquad (16.19)$$

General relativity is usually important only when the linear dimension of the mass distribution is comparable to l, and the difference between Newtonian and Einstein gravitation is small at distances $r \gg l$. For most astronomical systems, these scales are too small to have major effects (for the sun $l_\odot \sim 3$ km and $l_\odot/R_\odot \sim 10^{-6}$).

Although we can not develop Einstein's theory of general relativity here, we may compare the form of the theory with the Newtonian theory of gravitation. Consider an arbitrary mass distribution in which the density ρ ($\mathbf{r}$) is related to the pressure $P(\mathbf{r})$ by an equation of state. Referring to (4.1), we can write

$$\frac{d\mathbf{v}}{dt} = -\nabla P = -\nabla\Phi, \qquad (16.20)$$

where the Newtonian gravitational potential Φ satisfies

$$\nabla^2\Phi = 4\pi G\rho. \qquad (16.21)$$

In Newtonian theory, the gravitational field is determined by one potential, Φ ($\mathbf{r}$), according to an equation linear in Φ. There is nothing in this formulation that requires a mass distribution to have a maximum mass limit, nor is there any reason why a distribution of matter can not be compressed to arbitrarily high densities and remain stable against further collapse. This follows from (16.20) by noting that for any $\rho(\mathbf{r})$, the potential Φ is fixed, and in principle P can be made large enough that $d\mathbf{v}/dt = 0$ [this, of course, implies an equation of state $P(\rho)$]. That maximum mass limits do exist in nature (see, for example, Chapter 14) is a consequence of the properties of matter (that is, a restriction on the equation of state), not of the Newtonian theory of gravitation.

The general theory of relativity replaces the single Newtonian scalar potential Φ by the second-rank tensor g_{ij}, and replaces the left-hand side of (16.21) by $R_{ij} - \frac{1}{2}g_{ij}R$, where R_{ij} is a second-order nonlinear set of differential equations (which behave like components of a tensor under coordinate transformations) in g_{ij}, and R is a scalar constructed from the components of R_{ij}. The source term in the new field equations, which replaces the right-hand side of (16.21), is the energy-momentum density tensor $(8\pi G/c^2)T_{ij}$ whose diagonal components ($i = j$) contain the total energy density and pressure, and whose off-diagonal components ($i \neq j$) contain the momentum density of the matter distribution. The general-relativistic analogue of (16.21) is

$$R_{ij} - \tfrac{1}{2}g_{ij}R = \frac{8\pi G}{c^2} T_{ij}. \qquad (16.22)$$

The tensors above are symmetric ($R_{ij} = R_{ji}$, for example); so general relativity contains up to ten coupled, nonlinear partial differential equations for the gravitational field produced by the matter distribution T_{ij}. The important point here is that T_{ij} for a continuous distribution of matter contains the pressure as well as the energy density. Therefore, if a configuration is contracting gravitationally, we can not simply increase the pressure to stop it as in Newtonian theory, since P acts through T_{ij} as a source for the gravitational field. This point will be illustrated in what follows for matter at uniform density.

Although the analogy between g_{ij} and potentials can be pursued, it is often more convenient (and traditional) to interpret the effects of general relativity geometrically. The justification for this interpretation is that in relativity (special or general), the complete specification of any event or action consists of identifying when and where it occurs, that is, giving its spacetime coordinates $x^i = (x^0, x^1, x^2, x^3)$. One of the coordinates x^i is a time axis, and the rest lie in an orthogonal spatial plane. It is a consequence of relativity that the spacetime separation ds between two events, one at x^i, the second at $x^i + dx^i$ an infinitesimal distance away, is given by

$$ds^2 = \sum_{ij} g_{ij}\, dx^i\, dx^j, \qquad (16.23)$$

where g_{ij} is the metric tensor that describes the local geometric properties of spacetime. In Einstein's theory of relativity, g_{ij} is identified with the g_{ij} in (16.22), and gravitation acts to alter the four-dimensional geometry of spacetime. In the absence of matter, g_{ij} reduces to the Minkowski metric of special relativity and is, in Cartesian coordinates,

$$g_{ij} \longrightarrow \eta_{ij} = \begin{pmatrix} 1 & 0 & 0 & 0 \\ 0 & -1 & 0 & 0 \\ 0 & 0 & -1 & 0 \\ 0 & 0 & 0 & -1 \end{pmatrix};$$

$$dx^i = (ct, x, y, z). \qquad (16.24)$$

The path followed by a photon according to special relativity is described by

$$ds^2 = \sum_{ij} \eta_{ij}\, dx^i\, dx^j = 0. \qquad (16.25)$$

It follows at once from (16.24) and (16.25) that dx/dt for photons is c. The experimentally observed fact that all massive particles travel with speeds less than c is incorporated into relativity by requiring that their spacetime paths satisfy [for the choice of η_{ij} in (16.24)]

$$ds^2 > 0. \qquad (16.26)$$

Geometrically, the surface formed by all points near x^i satisfying (16.25) defines the light cone about x^i. This is shown for the x, t-plane in Figure 16.6; the observer is located at $x = 0$ (the y and z coordinates are supressed) and his time axis is the ordinate ct. The light cone consists of the two lines $x = \pm ct$ of slope ± 1. Also shown are the trajectories (spacetime paths or world lines) of three other particles. Two of these, A and B, are at rest relative to the observer for his time $t \leq 0$, and are a distance $x = a$ and $x = b$ away, respectively (all times and distances are relative to the observer at $x = 0$). At $t = 0$, particle B undergoes a constant acceleration for all $t > 0$. Although his relative speed continues to increase from zero, it never quite reaches c (according to the observer at $x = 0$, B would never quite catch up to a photon emitted from $x = 0$ at $t = 0$). Particle A remains at rest at $x = a$ until $t = t_b$, when it decays, emitting a photon that travels away from the observer at $x = 0$; the recoil particle travels with $v = c$ toward $x = 0$. The dashed line would be its path if its velocity were c (note that the dashed line and wavy line form a light cone relative to the decaying particle at $t = t_b$). Finally, particle c (which was at rest with the observer at $x = 0$ for $t < -t_a$) accelerates to a speed $-c < -v$, travels at uniform speed, and decelerates to rest at $x = -C$ when $t = 0$. It then accelerates toward $x = 0$, reaching a maximum speed $v < c$, which is maintained until the final deceleration brings it to rest back with the observer at $x = 0$ $(t = t_a)$.

Figure 16.6 is a spacetime diagram in the absence of gravitational fields. The path of particles on which no forces act (such as A) are straight lines in this case, and particles that accelerate in response to external forces follow curved paths. The straight lines that free particles follow are called *geodesics*. Massive particles in Newtonian theory follow curved paths through space in response to gravitational forces. Gravitational forces are replaced in general relativity by modifications in the geometry of spacetime, and free particles (those on which no inertial forces act) follow geodesics. Figure 16.7 shows the spacetime path for three particles in a spherically symmetric gravitational field. Here particle S is at rest on the surface of a neutron star of mass M centered at $r = 0$; T is a test particle in

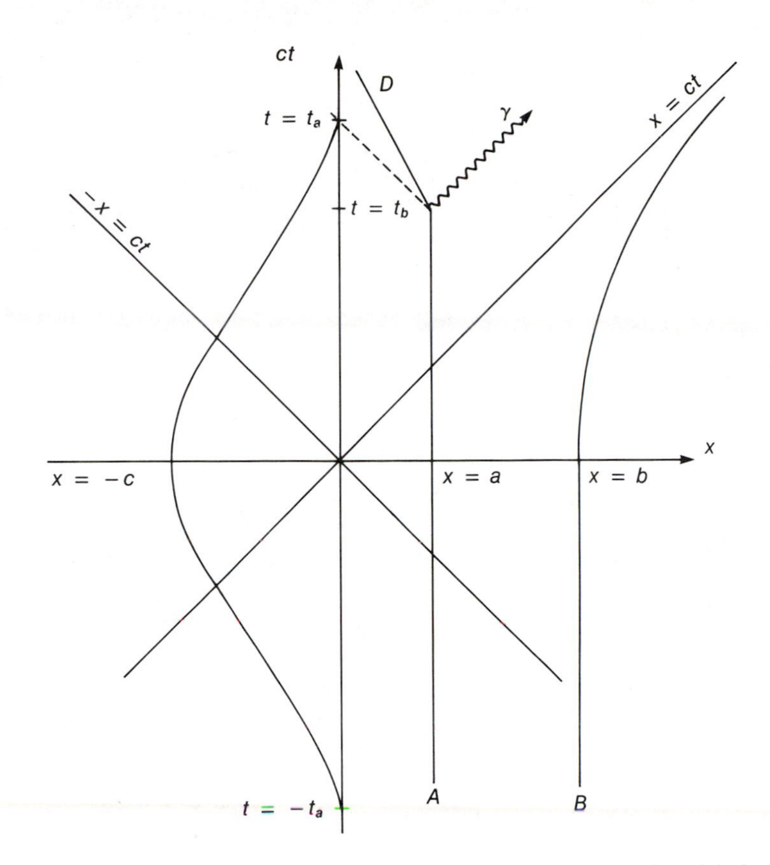

Figure 16.6. Spacetime trajectories as seen by an observer at the origin in the absence of gravitation (special relativity).

free fall in the gravitational field of M; and O is a distant observer, who is in Keplerian orbit around M. The paths O and T are geodesics; but the path S is not, since forces associated with the pressure in the neutron star act on the particles to maintain the mass distribution against collapse. Notice that when gravity curves spacetime, the light cones (shown along T in Figure 16.7) no longer make the same angle with the t axis. In general, however, the path followed by photons is given by (16.23) with $ds^2 = 0$, and the path of a massive particle must satisfy (16.26).

The geometric character of gravitation may be illustrated by a simple example. For weak fields, (16.23) can be written as

$$ds^2 = c^2 (1 + 2\Phi/c^2)\, dt^2 - dx^2 - dy^2 - dz^2. \quad (16.27)$$

In the absence of noninertial forces, the spacetime path of a mass m will be a geodesic. Of all possible paths between two points $x_A{}^i$ and $x_B{}^i$, the one that corresponds to a maximum or minimum length (an extremum) is a geodesic, and is given as the solution of the variational

$$\delta \int_{x_A{}^i}^{x_B{}^i} ds = 0.$$

Problem 16.8. Explain why the spacetime path O in Figure 16.7 is a geodesic.

The path satisfying this constraint in the geometry corresponding to (16.27) is obtained from Euler's equations

$$\frac{d}{ds}\frac{\partial L}{\partial \dot{t}} - \frac{\partial L}{\partial t} = 0, \qquad \frac{d}{ds}\frac{\partial L}{\partial \dot{x}^k} - \frac{\partial L}{\partial x^k} = 0, \qquad (16.28)$$

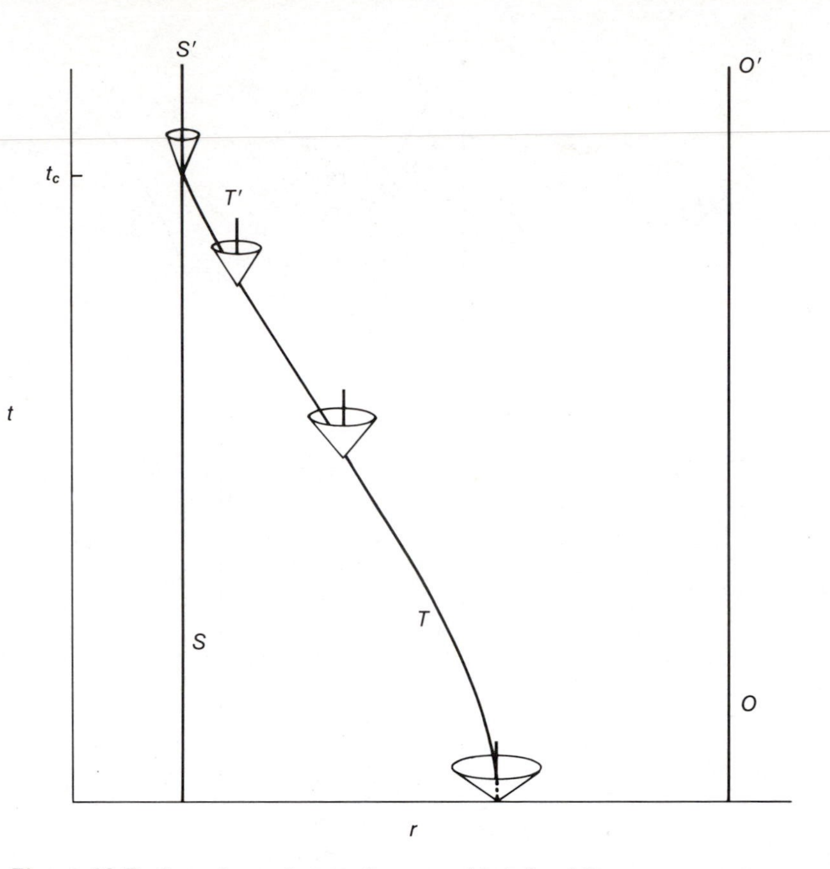

Figure 16.7. Spacetime trajectory of a test particle in free fall near a neutron star.

where

$$L = c^2(1 + 2\Phi/c^2)\,\dot{t}^2 - \dot{x}^k\dot{x}^k;$$

here $k = 1, 2$, or 3; $\dot{t} = dt/ds$; and $\dot{x}^k = dx^k/ds$. The solution of (16.28) is easily shown to be

$$\frac{d^2x^k}{dt^2} = -\frac{\partial\Phi}{\partial x^k}. \tag{16.29}$$

This is Newton's second law, and Φ is the gravitational potential. The preceding heuristic arguments illustrate the way in which the geometry of spacetime responds to the presence of matter. Finally, we recall that the path of photons is given by the solution of $ds^2 = 0$; from (16.27), for a photon traveling in the x direction,

$$\frac{dx}{dt} = c(1 + 2\Phi/c^2)^{1/2}.$$

This is not the speed of light, since dt is not the change in physically measured time (proper time). In the approximation (16.27), proper time (time measured by any kind of real clock) is measured in intervals

$$d\tau = (1 + 2\Phi/c^2)^{1/2}\,dt. \tag{16.30}$$

Therefore, the speed of light is $dx/d\tau = c$, as expected. For an arbitrary but diagonal metric $g_{ij} = g_{ii}$ ($g_{ij} = 0$ for $i \neq j$), the proper distance (or time) between two near-by points whose coordinate separation is dx^i is, in analogy with (16.30)

$$dl_i = \sqrt{g_{ii}}\,dx^i. \tag{16.31}$$

Black Holes

The simplest type of black hole, and the one on which the following discussion will be based, results from the collapse of a spherically symmetric mass distribution.

According to Einstein's equations, if the initial mass is large enough, the configuration will continue collapsing until it has been compressed to a point mass. We do not know of any physical phenomenon that can alter this outcome. It can be argued that quantum gravitational effects (which are not contained in Einstein's equations) might alter the collapse. Presumably the only physical constants in quantum gravity are $\hbar$, c, and G, from which the characteristic length $l_g \equiv (\hbar G/c^3)^{1/2}$ follows. Thus Einstein's equations undoubtedly break down over distance scales $l \lesssim 10^{-33}$ cm and time intervals $\tau \lesssim 5 \times 10^{-44}$ sec; however, these scales seem far too small to influence astronomical process. For practical purposes, we must assume that once collapse progresses far enough, it must continue until the entire mass distribution has contracted to a point. A heuristic description of a black hole was introduced in Chapter 1: a black hole is a mass M confined within a region whose characteristic linear extent is less than $2MG/c^2$. If the mass distribution is of finite extent and is confined within a volume of order $(2MG/c^2)^3$, then it must collapse to a point in a time of order $\tau = MG/c^3$. If the resulting mass motion during this time leads to time-varying quadrupole or higher moments, then gravitational energy will be radiated away. Eventually (relative to the time τ) all mass motion will cease, and a spherically symetric black hole remains. Its characteristic radius,

$$r_g \equiv 2MG/c^2 = 2.96\,(M/M_\odot)\text{ km}, \quad (16.32)$$

represents the black hole's event horizon. Thus a static, spherically symmetric black hole consists of a point mass M surrounded by a spherical event horizon of area $4\pi r_g^2$. The event horizon is not a material surface. Nonetheless, an external observer (whose position $r_0 > r_g$) can, by physical means, learn nothing about the interior ($r < r_g$) other than the amount of mass and electric charge that it contains. In particular, a black hole can have no magnetic field (the likelihood that sufficient charge separation could occur during collapse to be observable is small).

Let us summarize the primary characteristics of a spherically symmetric black hole. Since the region $r < r_g$ is inaccessible to the external observer, we will exclude it from most of our discussion. We will frequently refer to a distant observer in the following discussions. A distant observer is one whose separation from the source of a gravitational field is large enough ($r \gg r_g$) that the field he observes is essentially Newtonian (radial force $\sim MG/r^2$) and has no effect on local clocks or lengths. For notational convenience, we will denote the position of a distant observer by r_∞ and the time measured on his clock by t_∞. For distances $r_\infty > r \gg r_g$ the gravitational field is essentially Newtonian. However, as $r \to r_g$ (the strong field regime), general relativity has a dramatic effect on objects nearby. If dt_∞ denotes the interval between ticks of a standard clock as measured by a distant observer, then the interval he observes, if an identical clock is placed at rest at the point r, is given by

$$dt = (1 - 2MG/rc^2)^{1/2}\, dt_\infty. \quad (16.33)$$

Similarly, a standard ruler of length dr_∞ at r_∞, when measured by a distant observer, would appear to that observer to be of length

$$dr = dr_\infty/(1 - 2MG/rc^2)^{1/2} \quad (16.34)$$

when placed at r. It is an immediate consequence of (16.33) that clocks near a black hole appear to run slow when observed from a large distance. In fact, (16.33) applies to any clock (such as a cesium atom) in even a weak gravitational field. Thus a clock in the upper atmosphere will appear to run more rapidly than an identical clock on the Earth's surface. Experiments comparing Earth-based and airborne clocks have shown that the gravitational time dilation described by (16.34) occurs. In a series of 15-hour flights at 30,000 ft. the time difference observed was 47.1×10^{-9} sec.

Problem 16.9. How much faster would a clock at an altitude of 30,000 ft. run relative to one on the Earth's surface?

The behavior of a test particle released from rest at r_0 in a field of a black hole is suggested by (16.33) and (16.34). In fact, the instantaneous coordinate r as seen by a distant observer is given approximately by

$$r \simeq r_g + r_0\, e^{-ct_\infty/2r_g}, \quad (16.35)$$

where t_∞ is the time as seen by the distant observer. We find that infalling matter (including photons and neutrons) are never observed to reach r_g.

A consequence of the gravitationally induced slowdown of clocks in strong gravitational fields (16.33) is that spectral lines emitted at rest at $r \gtrsim r_g$ will undergo a gravitational redshift described by

$$\frac{\Delta\lambda}{\lambda} \equiv \frac{\lambda_\infty - \lambda}{\lambda} = (1 - 2MG/rc^2)^{-1/2} - 1. \quad (16.36)$$

Here λ is the characteristic wavelengths of the emitter located at r, and λ_∞ is the apparent wavelength seen at r_∞. Note that $\Delta\lambda > 0$, that is, no blueshift is possible, and that $\Delta\lambda$ can be a significant fraction of λ for a neutron star (for $M \sim 1.5\ M_\odot$ and $R \simeq 10$ km, $\Delta\lambda/\lambda \simeq 0.3$). The redshift of spectral lines emitted from white dwarfs, and gamma-ray lines from e^+e^- pair annihilation near neutron stars, have been used to estimate stellar masses.

The preceding results suggest that unusual physical processes occur at $r = r_g$. This, however, is not the case. The anomalies appear only to the distant observer. If an observer were to ride a test probe toward a black hole, he would notice no unusual local effects as he crossed $r = r_g$, which he would do smoothly. However, once inside r_g he could never return to the region $r \geq r_g$. Furthermore, he could not even remain at a fixed distance from the point mass M, but would be constrained to fall toward $r = 0$. Most of these phenomena are shown by means of a spacetime diagram, such as Figure 16.8, which shows the region near a black hole. In this diagram u and v are functions of radius and time. Lines of constant t are radial, beginning at $u =$

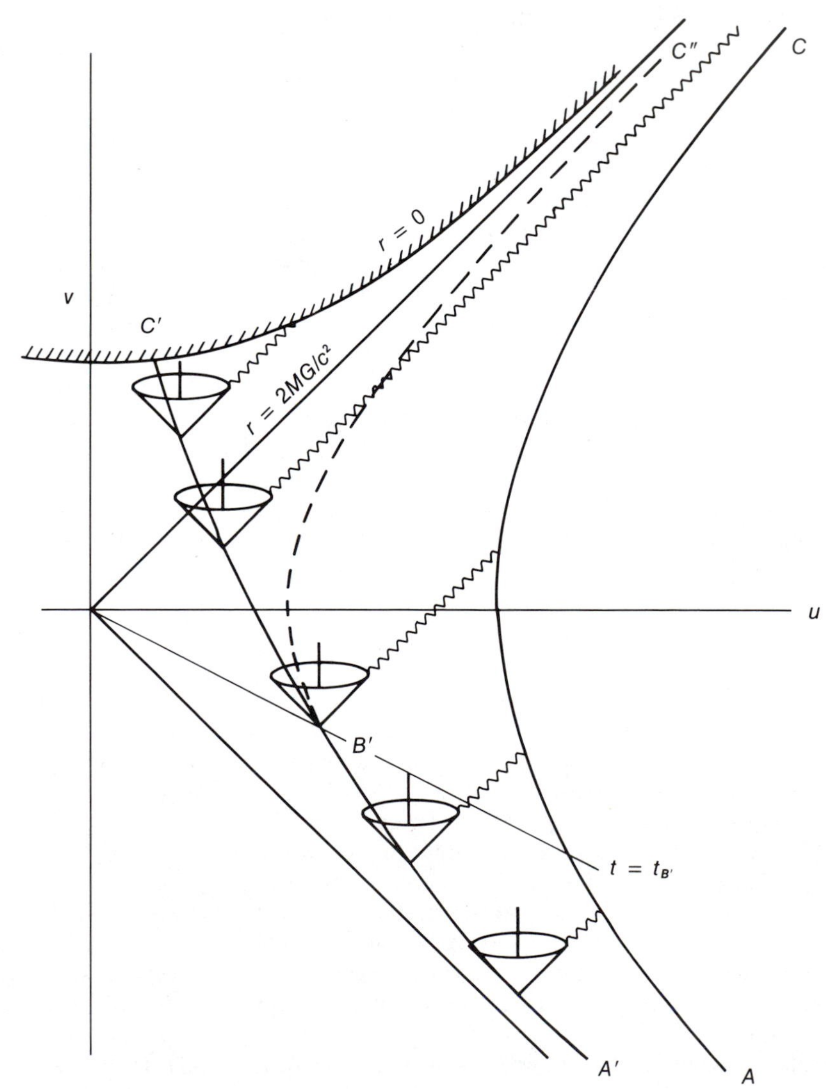

Figure 16.8. Spacetime trajectories near a black hole. Lines of constant t are radial in u, v coordinates. Curves AC and $A'C'$ are curves of constant r.

$v = 0$, with lines of greater slope corresponding to later times. The line $v = 0$ is identified as $t = 0$. Two lines of constant radius, AC (corresponding to r_∞) and $A'C''$ (radius r_1) are also shown. The advantage of the coordinates u, v (called Kruskal coordinates) is that photons travel along paths $u = \pm v$ at all points in the diagram. The line labeled $r = 0$ represents the location of the point mass M, and the line $u = v$ (labeled $r = 2MG/c^2$) is the event horizon of the black hole. The curve AC may be taken as the spacetime path of a distant observer. The path $A'C'$ is the path of a test particle near the black hole ($2MG/c^2 < r_1 \ll r$), moving initially in an orbit of constant radius. At time t'_B, the test particle begins to accelerate toward the central mass M. The last photon emitted before passing through the event horizon approaches the distant observer only as $t \to \infty$. Once inside $r = 2MG/c^2$, all photons or other particles emitted by the test mass will fall back to $r = 0$ in a finite time.

Problem 16.10. Use the spacetime diagram in Figure 16.8 to show that a test mass at $r < 2MG/c^2$ can not remain in an orbit at a fixed distance away from M.

Collapse of a Uniform Density Distribution

The equation of hydrostatic equilibrium for spherically symmetric mass distributions can be obtained from Einstein's theory. The pressure gradient is given by

$$\frac{dP}{dr} = -\frac{m(r)G}{r^2} \times \frac{(1 + P/\rho c^2)[1 + 4\pi r^3 P/m(r)c^2]}{[1 - 2m(r)G/rc^2]}, \quad (16.37)$$

where ρc^2 is now the total energy density [$\rho = n_b m_b (1 + \epsilon/c^2)$, with n_b and m_b the baryon number density and rest mass, respectively, and ϵ the specific energy density]. The total gravitational mass within a sphere of radius r is given by $m(r)$ with

$$\frac{dm}{dr} = 4\pi r^2 \rho. \quad (16.38)$$

The pressure P and the energy density ρc^2 are measured by an observer at rest at the mass element under consideration. As in the Newtonian limit ($r \gg r_g$ and ϵ or $P \ll \rho c^2$), the Newtonian equation of hydrostatic equilibrium is recovered. Although we can not derive (16.37) here, its general form can be suggested by the following arguments.

The energy content of an element of matter of rest-mass density $n_b m_b$ at pressure P is $n_b m_b \epsilon + P = u_m$. But according to the equivalence of mass and energy, this corresponds to an effective mass density u_m/c^2. Therefore the inertia of the element of matter that is effective in its response to the gravitational field includes u_m/c^2. The factor $(1 - r_g/r)$ in the denominator arises in part because gradients involve locally measured distances rather than coordinate distances. Finally, the pressure correction to the source mass $m(r)$, which is represented by the term $4\pi r^3 P/m(r)c^2$, reflects the fact that the mass equivalent of material energy also contributes to the source of the gravitational field. This pressure term has a dramatic effect on the static structure of compact stars. In Newtonian theory an increased mass can be supported by steepening the pressure gradient. In relativity, the increased pressure gradient means a corresponding increase in the pressure term on the right side of (16.37).

In general, (16.37) and (16.38) can be solved only numerically. A special case that can be solved analytically, and that illustrates some essential features of hydrostatic configurations in general relativity, is obtained by assuming a constant energy density $\rho = \rho_0$. Substituting this into (16.38) yields $m(r)$, and (16.37) may then be rewritten as

$$\frac{dP}{dr} = -\frac{4\pi G}{c^4} r \frac{(P + \rho_0 c^2)(P + \rho_0 c^2/3)}{1 - (8\pi G/3c^2)\rho_0 r^2}. \quad (16.39)$$

This may be integrated directly to obtain $P(r)$. Assuming that the radius of the mass distribution is defined by $P(R) = 0$, we find

$$P(r) = \rho_0 c^2 \left\{ \frac{\sqrt{1 - r_g r^2/R^3} - \sqrt{1 - r_g/R}}{3\sqrt{1 - r_g/R} - \sqrt{1 - r_g r^2/R^3}} \right\} \quad (16.40)$$

The central pressure, $P(0)$ is given by

$$P_c = \rho_0 c^2 \left\{ \frac{1 - \sqrt{1 - r_g/R}}{3\sqrt{1 - r_g/R} - 1} \right\}. \quad (16.41)$$

Problem 16.11. Carry out the solution to (16.39). Compare it with the corresponding Newtonian limit.

The equation of state used to solve (16.39) is highly idealized. Nevertheless, the pressure profile is reasonable as long as $R > 9r_g/8$. However, as $R \rightarrow 9r_g/8$ the value of P_c becomes arbitrarily large. Configurations with $R = 9r_g/8$ would require infinite central pressure to be in equilibrium. Since this can not occur physically, the stars must become unstable and collapse. For fixed ρ_0 and M, no stable configuration can be found if $R < 9r_g/8$, and gravitational collapse to a black hole is inevitable.

It was noted in Chapter 1 that general relativistic effects usually become negligible when $r \gg r_g$, but that small terms can be important for systems on the edge of instability (such as supermassive stars). It can be shown that the adiabatic index corresponding to marginal stability is given by

$$\gamma_{\text{crit}} = \frac{4}{3} + \kappa \frac{MG}{Rc^2} \tag{16.42}$$

when $r_g \ll R$; in (16.42) κ is a constant. Therefore an ideal, fully relativistic gas of electrons and positrons (for which $\gamma = 4/3$) will be unstable. Notice that the gravitational correction to γ_{crit} need not be large to be significant, since stability depends on the sign of $\gamma - \gamma_{\text{crit}}$.

Gravitational Radiation

The analogy between electromagnetic and gravitational forces at the Newtonian level can be used to create an expression for the gravitational energy radiated by accelerated masses. Both forces obey Coulomb-like laws in the Newtonian regime; for electrons

$$F_g = -\frac{m_e^2 G}{r^2}, \qquad F_{\text{el}} = \frac{e^2}{r^2}. \tag{16.43}$$

The ratio of the magnitude of the accelerations is

$$\frac{a_g}{a_e} = \frac{m_e^2 G}{e^2} \simeq 10^{-43}.$$

The analogy between F_g and F_e suggests that we consider $m_e\sqrt{G}$ to be the gravitational charge. The smallness of the ratio of gravitational to electric charge might suggest that gravitational radiation (if it is to be associated with accelerated gravitational charges) is so small (relative to electromagnetic radiation, for example) that it would be astrophysically unimportant. The converse, however, is generally true. Since two electric charge units ($\pm e$) exist in nature, matter is nearly always electrically neutral. Consequently very little electromagnetic radiation is expected to arise from accelerated astronomical objects (pulsars represent one exception; see Section 16.4). Since the gravitational charge is of one sign only, screening does not occur, and the net gravitational charge of a massive object is proportional to its mass. Finally, we note that gravitational radiation will be observed far from the emitting region, where the field of the mass distribution is essentially zero, and that the radiation travels to the observer at the speed of light.

A radiation field carries a fixed amount of energy per unit time from the source to infinity. Suppose that the amplitude of a field disturbance $A \sim r^{-n}$. Then the energy associated with it goes $A^2 \sim r^{-2n}$, and the energy crossing an arbitrary distant surface enclosing the source

$$E \sim \int A^2 r^2 d\Omega \sim r^{2-2n}.$$

If this is to be a constant for all r (we assume no absorbing material to be present), then $A \sim 1/r$. The electric and magnetic fields associated with dipole radiation are of the form

$$\mathscr{E}_{\text{EM}} \approx \frac{ea}{rc^2}, \tag{16.44}$$

where $a = \ddot{r}$ is the acceleration of the dipole er. Electromagnetic radiation is associated with accelerated charge distributions, which are most conveniently represented by multipole expansions. The monopole is just the total electric charge. Since this is conserved exactly, its time-rate of change is identically zero, and electromagnetic monopole radiation does not occur. Since electric and magnetic dipole moments can vary with time, these represent the lowest-order components. The energy flux associated with (16.44) is proportional to $\mathscr{E}_{\text{EM}}{}^2 c$, and the power radiated by it is proportional to the integral of the flux over a closed surface surrounding the dipole:

$$\left(\frac{dE}{dt}\right)_{\text{EM}} \sim \int c\,|\mathscr{E}_{\text{EM}}|^2 r^2 d\Omega \sim \frac{e^2 a^2}{c^3} \tag{16.45}$$

[recall (6.43)]. Now consider the effect of an accelerated gravitational charge $m\sqrt{G}$. In analogy with (16.44) we would expect

$$\mathcal{E}_g \sim \frac{m\sqrt{G}a}{rc^2} \tag{16.46}$$

for the radiation component of the gravitational field. An important difference between electromagnetism and gravitation must be noted here. Gravitational radiation, as suggested by (16.46), is associated with accelerated mass distributions. In analogy with electromagnetic theory, these mass distributions can be decomposed into mass multipole moments. The monopole and dipole moments correspond to the total mass and angular momentum of the source. In the weak-field limit, these quantities are conserved, and the lowest-order gravitational multipole that can give rise to radiation is the quadrupole. Therefore (16.46) can not be applied as it stands. Analogy with electromagnetic theory is useful in describing how (16.46) must be modified. A single test charge $\pm e$ will experience a measurable acceleration due to the passage of an electromagnetic wave of amplitude $\mathcal{E}_{\text{EM}}$. According to the principle of equivalence, there is no physical way to distinguish between the inertial and gravitational acceleration of a single test mass; only relative accelerations (tidal accelerations) between two or more masses can be attributed to gravitational fields. Further, because gravitational effects propagate with finite speed c, there is a relative phase delay between the acceleration of two nearby test masses. For example, if the source is a binary system of masses $M_1 = M_2 = M$, period P, and semimajor axis l, then there is a phase delay

$$\Delta\phi \sim \frac{2\pi l}{Pc} \sim \frac{\omega l}{c} \tag{16.47}$$

between the acceleration of one test mass and that of its neighbor. Since only the relative acceleration is measurable, the component of the field that we must identify with energy transport is proportional to the product $\mathcal{E}_g\Delta\phi$. The acceleration appearing in $\mathcal{E}_g$ for a binary system will be of order $a \sim \omega^2 l$; therefore

$$\mathcal{E}_{g,\text{rad}} \sim \frac{G^{1/2}M\omega^3 l^2}{rc^2}. \tag{16.48}$$

In analogy with electromagnetism, the energy flux carried by the gravitational wave at large distances from the source (the wave zone) will be proportional to $|\mathcal{E}_{g,\text{rad}}|^2 c$, and the total power radiated is given by the integral of this flux over a surface enclosing the source:

$$\left(\frac{dE}{dt}\right)_{\text{grav}} \approx \int c\,|\mathcal{E}_{g,\text{rad}}|^2 r^2 d\Omega$$

$$\approx \frac{G}{c^5}(M\omega^3 l^2)^2 \approx \frac{G|\dddot{Q}|^2}{45c^5}. \tag{16.49}$$

In the last step we have introduced the quadrupole moment of the mass distribution $Q \sim Ml^2$, and noted that each time-derivative of Q is proportional to $\omega Q \sim Q/P$ for a periodically varying quadrupole moment. The last form (including the numerical factor) is the correct expression for the total energy radiated by gravitational waves from a mass distribution with a time-varying quadrupole moment. It must be remembered that this expression applies only at large distances from the source, where the source's Coulomb field MG/r^2 is essentially zero. When applying (16.49) to specific mass distributions $|\dddot{Q}|^2$ means $\Sigma_{ij}\,\dddot{Q}_{ij}^{\,2}$ and the (tensor) quadrupole moment has components

$$Q_{ij} = \int \rho(3x_i x_j - r^2\delta_{ij})\,dV. \tag{16.50}$$

The gravitational-energy loss rate (16.49) may be applied to a distant binary stellar system, to rotating or oscillating stars (white dwarfs or neutron stars), or to asymmetrically collapsing mass distributions.

Problem 16.12. A binary system consists of two masses M_1 and M_2 in circular orbits, with mutual separation r. Use (16.50) to find the quadrupole moment, and show that the gravitational energy loss by the system is given by

$$\left(\frac{dE}{dt}\right)_{\text{grav}} = -\frac{32G}{5c^5}\frac{M_1^2 M_2^2}{(M_1 + M_2)^2} r^4\omega^6, \tag{16.51}$$

where the orbital period $P = 2\pi/\omega$.

The radiation-energy loss given by (16.51) has been averaged over angles. The angular distribution averaged over the orbital period is proportional to $(1 + 6\cos^2\theta + \cos^4\theta)$, which is shown in Figure 16.9. The angle θ is measured with respect to the normal to the orbital plane. The peak intensity occurs normal to the plane, exceeding that emitted in the plane by a factor of eight.

The gravitational power radiated by orbital motion is normally very small. Using Kepler's third law to relate ω to r, we find that (16.51) yields an energy-loss rate of 5×10^{10} erg/sec for Jupiter's orbital motion around the Sun. For a stellar system with each mass $\approx M_{\odot}$, $a \approx 200 R_{\odot}$, and period $\sim 8d$, $E_g \approx 4 \times 10^{25}$ erg/sec. It would require about 10^{12} years for 1 percent of the binary system's binding energy to be radiated away. However, for two neutron stars in orbit with $r \approx R_{\odot}$, $\dot{E}_g \approx 10^{34}$ erg/sec, and the time required for the system to radiate its binding energy away as gravitational waves is about 10^7 yrs. The analysis leading to (16.51) can be generalized to include orbits of ellipticity ϵ. The result is to multiply (16.51) by a function $f(\epsilon)$ that is unity for $\epsilon = 0$, but increases rapidly as ϵ approaches unity. For example, $f(.5) = 10$. Thus much more power than is radiated from a circular orbit can be radiated if the orbit is eccentric.

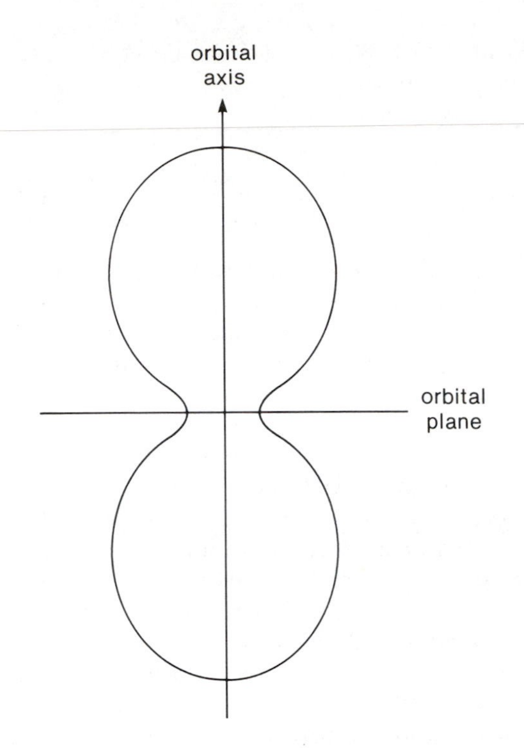

Figure 16.9. Angular distribution of quadrupole gravitational radiation from a binary system.

Problem 16.13. The binary system containing V Pup has a period of 1.45 days, semimajor axis $a = 12.1\ R_{\odot}$, and contains stars of mass $1.66\ M_{\odot}$ and $9.8\ M_{\odot}$. Find the gravitational energy radiated from the system and its flux at the Earth (assume zero eccentricity and a distance of 390 pc).

The Earth is bathed in gravitational radiation from binary and multiple star systems throughout the universe. A first approximation for the flux received at the Earth from systems within a distance R is given by

$$f_{\text{grav}}(R) \simeq \sum_i \int_0^R \frac{L_i(r)}{4\pi r^2} n_i 4\pi r^2\, dr, \qquad (16.52)$$

where $L_i(r)$ is the quadrupole power of the system at r, n_i is the number density of such systems, and i is a general index denoting the type of multiple systems. It is apparent from (16.52) that the flux depends on distant systems, that is, $f_{\text{grav}}(R) \sim nRL$. Thus distant stellar systems in our Galaxy contribute more than nearby ones. Although (16.52) suggests that the flux from other galaxies should be even more important, this is not true. The cosmological redshift resulting from their recessional motion reduces the flux. Thus sources within our Galaxy dominate, and the integral (16.52) is finite. It is estimated that at the Earth, $f_{\text{grav}} \sim 10^{-9}$ erg/cm^2 sec from Galactic binary systems. This corresponds to 10^{-7} times the flux of visible light from stars.

The preceding discussions, (16.51) in particular, are valid only in the weak-field limit. We remarked after (16.38) that according to general relativity a mass can be associated with any energy; so a strong gravitational wave can generate a gravitational field of its own, whose source is associated with the mass equivalent of the field energy. Such effects are excluded in the weak-field limit, but they may be significant in problems involving the asymmetric collapse of massive objects. The formation of massive black holes from stellar collisions and coalesence in globular clusters and in galactic nuclei may represent sources of gravitational waves in the strong-field limit. General arguments indicate that up to about 30 percent of the rest energy of a mass distribution could be radiated as gravitational waves. However, detailed calculations suggest that the effective energy loss may be much less. It has been suggested that the violent activity in galactic nuclei is associated with massive black holes ($M \approx 10^5 - 10^8 M_{\odot}$). The formation of such objects could release as much as 10^{60} ergs of energy in several minutes if 1 percent of the rest energy were radiated away.

Search for Black Holes

A black hole is, by its very nature, a perfect absorber. Only its gravitation field (or angular momentum, if it is rotating) is externally observable. The most practical way to observe one is to search for its effects on nearby stars or gas. There may be many black holes (on the order of 10^9) produced at the endpoint of the evolution of massive stars in our Galaxy. Of these some are likely to be members of binary systems (Cyg X $-$ 1 is a possible example), and may be identified by their interaction with a normal companion (see Chapter 17). Black holes may also form within the center of globular clusters as a result of stellar collision and coalecence. A black-hole mass in excess of $10^3 M_\odot$ would have an observable influence on the cluster's luminosity distribution. Similar processes involving stars and gas may form massive black holes in galactic nuclei with $M \approx 10^7$ to $10^8 M_\odot$. Dust and gas that concentrates in the vicinity may accrete onto them. The energy released by these processes may be an important energy source for active galactic nuclei. In the early universe, some massive gas condensates ($M \approx 10^4 - 10^5 M_\odot$) may contract without fragmentation to form supermassive stars. These stars, powered primarily by hydrogen burning, would have evolved rapidly because of their high mass, and may have ended their evolution as massive black holes. It has also been suggested that extreme conditions during the instants following the Big Bang could produce black holes with a range of masses extending many orders of magnitude below $M_\odot$ (mini black holes). Although black-hole masses below about $10^{18}g$ would have evaporated by now because of quantum processes, those with $M > 10^{18}g$ ($r_g \approx 10^{-10}$ cm $= 10^{-2}$ Å) might still persist. Finally, we note that if the average mass density of the universe exceeds $3H^2/8\pi G \simeq 10^{-29}$ g/cm^3, where H is Hubble's constant, then the universe itself may be a black hole.

16.4. PULSARS

The theory developed in Section 16.2 may now be used to develop a consistent model of radio pulsars based on rotating magnetic neutron stars.

The discovery of radio pulsars in 1967 ranks among the most important and influential observations in modern astrophysics. These unique objects established the existence of neutron stars and their relation to supernovae. They were also a newly discovered source of cosmic rays, and offer a new method of estimating distances within our Galaxy. Observations of a binary system containing the pulsar PSR 1913+16 are the strongest current support for the existence of gravitational radiation as described by Einstein's theory of general relativity.

Apart from PSR 1913+16, and several recently discovered pulsars, the currently observed radio pulsars are not members of binary systems, and only two (the Crab and Vela pulsars) are known to emit outside radio frequencies ($\nu < 10^{12}$ Hz). The x-ray pulsars (which do not exhibit significant radio emission) appear to involve a different emission mechanism, and will be discussed in Chapter 17. Two recently discovered radio pulsars will be considered at the end of this section. First we consider pulsars that are not members of binary systems.

The signal observed from a pulsar is broad-band radio noise whose maximum intensity occurs at exactly periodic intervals of about 1 sec, typically. Figure 16.10(a) shows data for the pulsar PSR 0329+54 (the conventional nomenclature for a radio pulsar is PSR hhmm$\pm$*xx*, where the first four integers give its right ascension in hours and minutes, and *xx* denotes degrees of declination). The pulse intensity exhibits random variations, with some pulses lying below the noise level. Nevertheless, the time interval between pulses is 0.7145s, or multiples thereof. Measurements with time resolutions $\lesssim$1ms show subpulse structure with characteristic widths of several percent of the period; for resolutions $\lesssim 10\ \mu$s the subpulses of some pulsars exhibit microstructure with widths of order $10^{-3}P$. The subpulse structure is generally variable. However, if many individual pulses are added in phase, the resulting integral profile has a stable shape characteristic of that pulsar. Figure 16.10(b) shows several representative integral profiles. Essentially all the radio energy is contained in the main pulse (or pulses, if they are multiple), which for most pulsars corresponds to a small fraction of the period.

Pulsar periods have been observed to range from 0.033 sec (for the Crab, PSR 0531+21) to 3.745 sec (for PSR 0525+21). Figure 16.11 shows the period distribution for 149 pulsars. For many the period has been measured to within a few parts in 10^9 (comparable to the precision of atomic clocks). The periods of all known pulsars show a gradual rate of increase,

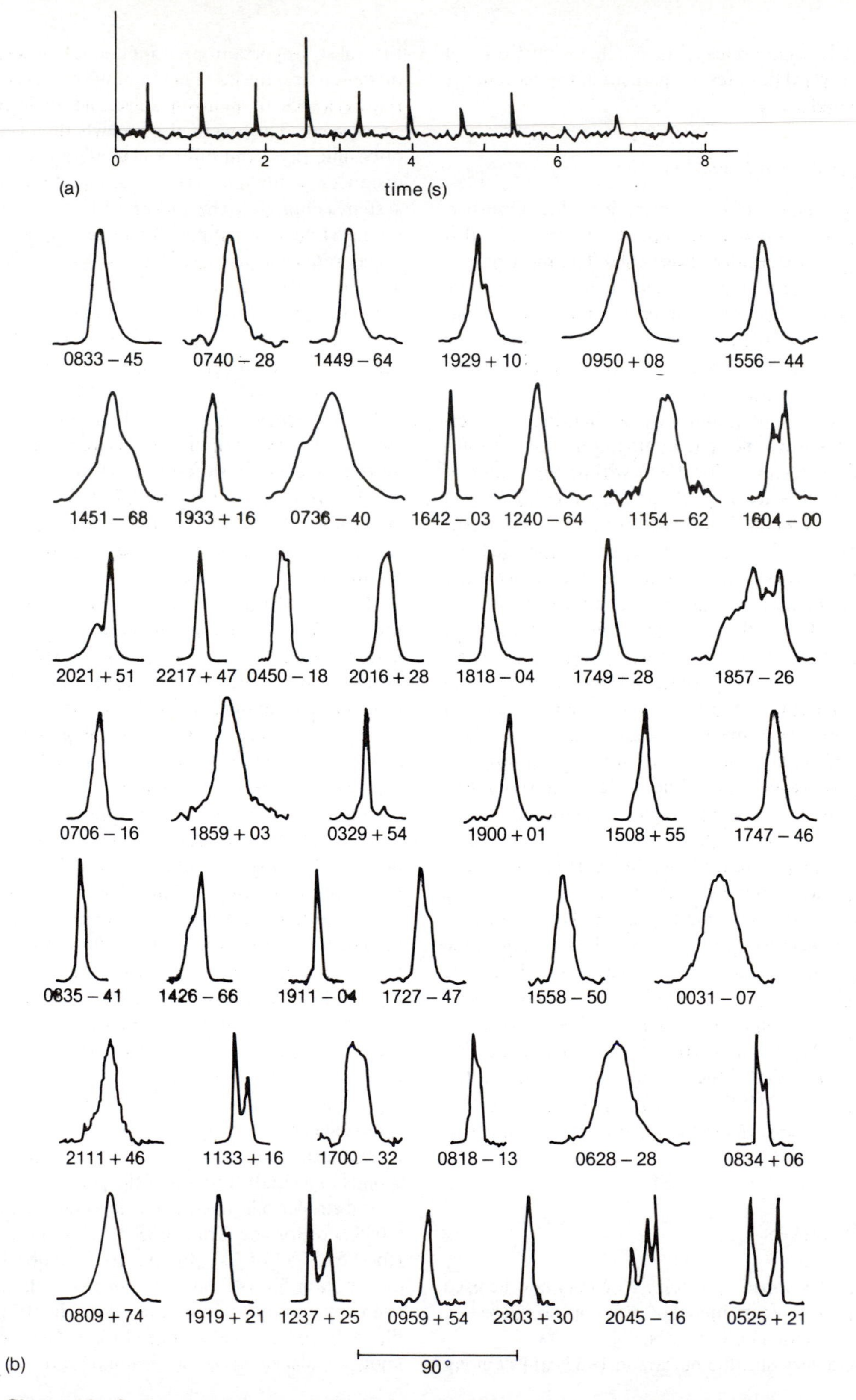

Figure 16.10. Pulsar signals. (a) Individual pulses from PSR 0329+54 at 410 MHz. The pulse period is 0.714 sec. The instrumental time constant is 20 ms. (b) Integrated pulse profiles for several pulsars plotted on the same longitude scale. A 90° bar is shown at bottom. The profiles were recorded at frequencies between 400 and 650 MHz, and are arranged in order of increasing pulse period.

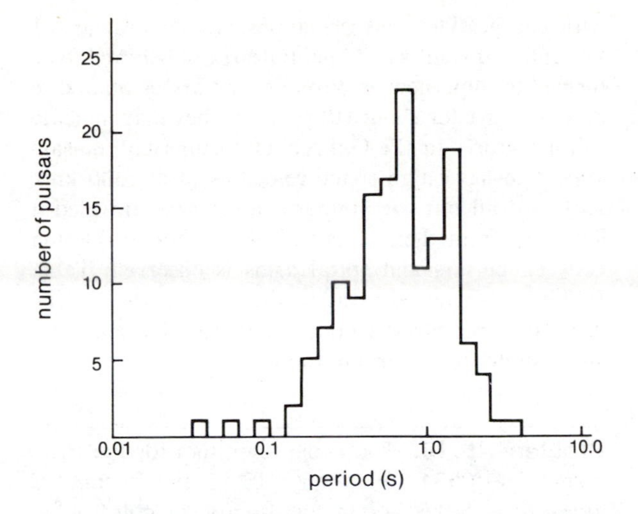

Figure 16.11. Pulsar period distribution for 149 pulsars.

amounting typically to $10^{-13} - 10^{-16}$ sec per sec. (about $10^{-6} - 10^{-9}$ sec per year). The extreme regularity and gradual rate of slowdown of the pulse are key features that theory must explain. Pulsars have power-law spectra $F \sim \nu^{\alpha}$ at radio frequencies. The spectral index α lies between about -1 and -3, with a low-energy cutoff (Figure 16.12). The flux density F is normally of order 10^{-26} $\mathrm{Wm^{-2}\,Hz^{-1}}$ (a unit called the Jansky, Jy). The negative number next to each curve is the spectral index α. Most pulsar observations are made at 400 MHz (arrow on the ordinate of Figure 16.12).

The radio emission from many pulsars shows linear and often circular or elliptical polarization. In some cases (Vela, PSR 0833−45) essentially all the integrated pulse profile is linearly polarized; circular polarization, however, seldom exceeds 20 percent of

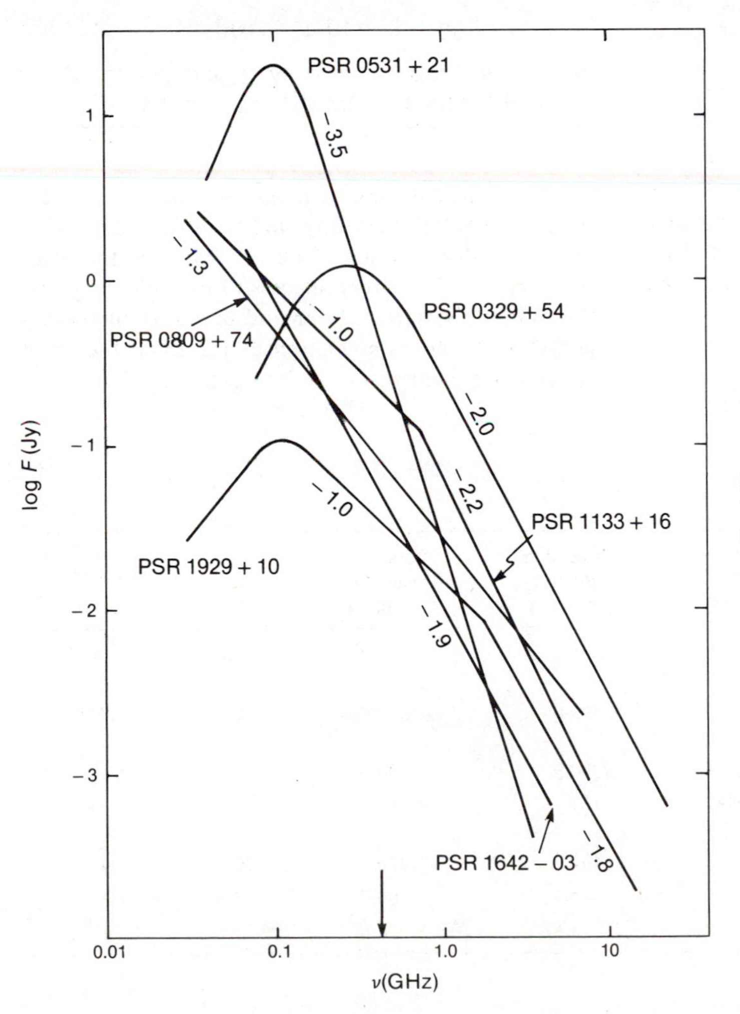

Figure 16.12. Radio spectra of several pulsars. The numbers next to each curve are the approximate spectral index α.

the integrated pulse profile. The source of radio emission is therefore coherent. Coherent emission is also implied by the high intensity of the radio sources. For the Crab pulsar, the brightness temperature estimated for the radio spectrum exceeds 10^{30} K. If the emission mechanism involved incoherent (thermal) processes, then the energy per particle would exceed 10^{26} eV, nearly 10^6 greater than the most energetic cosmic rays.

Problem 16.14. Calculate the brightness temperature for the Crab pulsar (PSR 0531+21; see Table 16.2), assuming $F(\nu) = 2 \times 10^{-25}\mathrm{Wm^{-2}\,Hz^{-1}}$ at $\nu = 10^8$ Hz. Assume an emission region of order 10 km.

Problem 16.15. The observed optical flux density of the pulsar in the Crab nebula is $F(\nu) \simeq 4 \times 10^{-29}$ $\mathrm{Wm^{-2}\,Hz^{-1}}$. What can be said about the relation of optical radiation mechanisms to those involved for radio emissions?

Finally, distance measurements based on electron dispersion (Section 18.4) place pulsars within our Galaxy. Like most young stars, they are concentrated near the Galactic disk, with a scale height of 230 pc. Thus the number of pulsars decreases with height z above or below the Galactic plane roughly as

$$N_{\text{pulsar}} = \exp(-|z|/230),$$

with z in pc. They have been observed as near as 0.1 kpc and as distant as 20 kpc. Pulsars are believed to be formed by supernova explosions, and are estimated to remain active for about 10^7 years; so they may indicate spiral features in the Galaxy. Unfortunately, pulsars appear to have high space velocities (100–500 km/sec), and all but the youngest must have traveled a long way from their places of origin. No correlation between pulsars and spiral arms is observed. Table 16.2 gives the period P and its time-rate of change; the dispersion measure (Section 18.4); rotation measure, and distance for several pulsars.

Problem 16.16. The radio spectrum for the Crab pulsar PSR 0531+21 (Figure 16.12) has a spectral index $\alpha \simeq -3.5$ and a low-frequency cutoff ν_0 at about 10^8 Hz. The observed flux density can be reproduced roughly by

$$F(\nu) = 2 \times 10^3\,\nu^{-3.5}\,\mathrm{Wm^{-2}Hz^{-1}} \qquad (16.53)$$

What is the radio luminosity of the pulsar? (Note assumptions made in arriving at the answer.)

The principal observational features of typical radio pulsars and their range are summarized in Table 16.3. The identification of pulsars with rotating neutron stars has been very successful in explaining why the radio pulses are regular and of slowly increasing period, but there is no consensus yet about the exact emission mechanism.

Table 16.2
Observed parameters for seven pulsars.

PSR	Period (sec)	dP/dt ($\times 10^{-15}$ sec per sec)	Dispersion measure ($\mathrm{cm^{-3}\,pc}$)	Rotation measure ($\mathrm{rad\,m^{-2}}$)	d (kpc)	$\frac{1}{2}P/\dot{P}$ (10^6 y)
0531 + 21 (Crab)	0.0331	422.69	56.7	−42.3	2.0	.0012
0833 − 45 (Vela)	0.0892	125.03	69.0	+33.6	0.5	.011
0525 + 21	3.7454	40.06	50.9	−39.6	1.9	1.5
1913 + 16 (binary pulsar)	0.0590	0.0088	167	——	6.2	——
1859 + 03	0.6554	7.50	402.9	−238.	20.	1.4
0329 + 54	0.7145	2.05	26.7	−63.7	2.6	5.5
1929 + 10	0.2265	1.16	3.17	−8.6	0.1	3.1

Table 16.3
Radio pulsar characteristics.

Feature	Observed range
Periodicity	$1.6 \times 10^{-3} \lesssim P \lesssim 4$ sec
Regularity	$10^{-13} \lesssim \dot{P} \lesssim 10^{-16}$ sec per sec
Compact nature of source emission region	$10^5 \lesssim r \lesssim 10^9$ cm
Spectrum	power law ($\lesssim$ 100 percent linear polarization) radio luminosity $\sim 10^{30}$ erg/s or more
Space velocities	$100 \lesssim v_s \lesssim 500$ km/s
Location	Galactic; distribution similar to that of disk-type stars

Rotating Magnetic Neutron Stars

Cosmic radio sources have been known to exist for about forty years. The startling aspect of pulsars that theory first addressed was the clocklike precision of their signal, that is, periods on the order of tenths of a second that remained constant to within a few nanoseconds per year. This observation, coupled with the substantial amounts of energy emitted, implied that the source must be the rotation or pulsation of a massive object, or a binary system. Radio emission is observed from interstellar plasma, but the extreme regularity of pulsars ruled this out. Sharp features in the radio pulses of duration $\Delta t \lesssim 10$ μsec are observed. For coherent radiation, this implies an emitting region smaller than $c\Delta t$, or about 10 km in extent.

Stellar pulsations (vibrations) are one mechanism capable of producing a regular train of signals. The period of the fundamental mode is comparable to the free-fall time. Therefore

$$P \sim t_{\rm ff} \sim (\bar{\rho}G/3\pi)^{1/2},$$

where $\bar{\rho}$ is the average density of the compact source. For typical white dwarfs, $\bar{\rho} \sim 10^6$ g/cm^3 or less, which implies a fundamental mode period of 10 sec or more. Higher modes could reduce this, but it is difficult to argue convincingly that these should dominate the fundamental. Even for $\bar{\rho} \sim 10^8$ g/cm^3, we find $P \gtrsim 1$ sec, which is far too long for most pulsars (see Figure 16.9). For neutron stars $\bar{\rho} \sim 3 \times 10^{14}$ g/cm^3, and the period associated with the fundamental mode is of order 10^{-4} sec, more than a factor of 10^2 too fast. Although white-dwarf pulsations could provide the longest observed periods, they can not explain the majority of radio pulsars.

The period of rotation of a compact mass must be such that the equatorial angular acceleration does not exceed the gravitational acceleration at the surface. This requires that

$$P_{\rm rot} \gtrsim (3\pi/\bar{\rho}G)^{1/2} = 1.2 \times 10^4/\sqrt{\bar{\rho}}\ \text{sec}, \qquad (16.54)$$

which begins at the period of the fundamental rotational mode. We see that rotating white dwarfs are no better than vibrating ones. However, rotating neutron stars easily cover the entire observed range of periods.

A binary system might also sustain a regular pulsed signal. By means of Kepler's third law, the period P, semimajor axis a, and stellar masses may be related. Periods of less than a second require

$$a = 1.5 \times 10^8 (P/1\ \text{sec})^{2/3} \times [(M_1 + M_2)/M_\odot]^{1/2}\ \text{cm}. \qquad (16.55)$$

Only for a neutron star and a black hole, or for two black holes, would a exceed the radius of the stars. In either case, the binary system would be a strong emitter of gravitational radiation. However, the resulting energy loss would lead to a decrease in orbital radius and a gradual decrease in period, contrary to observations.

Other mechanisms, such as planets or planetoids orbiting a neutron star or a black hole, have been considered for pulsars, but all fail to produce periods in the observed range with $\dot{P}$ positive. As long as pulse-emission models require magnetic fields, rotating black holes (which can not possess magnetic fields) must be excluded. Therefore the only viable candidate for radio pulsars is a rotating neutron star.

The pulse-emission mechanism must be able (a) to produce the observed power-law spectrum, which may be 100 percent linearly polarized, and (b) to produce broad-band radio luminosities of 10^{30} ergs/sec or more. Because of the high implied brightness temperatures, the mechanism is also expected to be coherent. Electromagnetic radiation from charged particles in a magnetic field or from magnetic multipole moments satisfies some of these requirements, as we will see.

A rotating magnetic dipole field will radiate if its dipole moment makes an angle θ with the rotation axis. Denoting the dipole moment by $\mathbf{m}$, we can describe the energy loss by

$$\left(\frac{dE}{dt}\right)_M = -\frac{2}{3c^3}\left(\frac{d^2m_\perp}{dt^2}\right)^2, \qquad (16.56)$$

where

$$M_\perp = |\mathbf{m} \times \hat{e}_z| = \frac{1}{2} Ba^3 \sin\theta \qquad (16.57)$$

is the component of **m** perpendicular to the rotation axis, and a is the radius of the configuration that sets up the dipole field. The time-rate of change

$$d^2m_\perp/dt^2 = m_\perp(P/2\pi)^2,$$

where $P = 2\pi/\omega$ is the rotation period. Combining the preceding results yields

$$\left(\frac{dE}{dt}\right)_M = -\frac{B^2\omega^4a^6}{6c^3}\sin^2\theta$$

$$= -9.6 \times 10^{30}(B_{(12)}a_{(6)}{}^3/P^2)^2$$

$$\times \sin^2\theta \text{ erg sec}^{-1}. \qquad (16.58)$$

Magnetic field strengths $B \approx 10^{12}$ gauss, and typical neutron-star radii $a \sim 10$ km, give an energy-loss rate of 10^{32} ergs/sec, which is comparable to loss rates deduced from radio spectra. As noted in Section 16.1, such field strengths are not unreasonable for neutron stars.

An additional feature of the nonaligned rotator we have described is the sign of dP/dt, which is positive. Assuming that the magnetic-energy loss is supplied by rotational energy, we find the time-rate of change of the period to be

$$\frac{dP}{dt} = \frac{\pi}{3}\frac{B^2\omega a^6}{Ic^3}\sin^2\theta. \qquad (16.59)$$

For $B = 10^{12}$ gauss, $a = 10$ km, a neutron-star moment of inertial $I = 10^{45}$ g cm^2, and a period of 1 sec, (16.59) gives $\dot{P} = 2 \times 10^{-16}$ sec per sec, which is comparable to observed values. A rotating magnetic dipole field therefore accounts for a pulsar's observed radio luminosity, its highly stable period, and its gradual spin-down as rotational energy is transferred into electromagnetic radiation. It will be seen in the next section that the details of this picture are not consistent; that is, a pulsar's emission can not be due to a rotating dipolar magnetic field surrounded by a vacuum. However, a model in which the surrounding vacuum is replaced by a plasma appears plausible.

Problem 16.17. Assume that magnetic dipole radiation is the principal cause of pulsar slowdown, and obtain an expression for the angular velocity as a function of time and the observed parameters $\omega_0 = 2\pi/P$ and $\dot{\omega}_0 = (d\omega/dt)_0$. Show that

$$\frac{1}{\omega^2} - \frac{1}{\omega_0{}^2} = -\frac{2\dot{\omega}_0}{\omega_0{}^3}(t - t_0). \qquad (16.60)$$

Here t_0 is the present. Estimate the age and initial spin rate for the Crab and Vela pulsars, and compare your results with other estimates. Assume that the pulsar's magnetic moment remains constant during slowdown.

Problem 16.18. Pulsar slowdown will also result from gravitational radiation if the neutron star has an energy loss due to a quadrupole moment Q. Assume that the quadrupole distribution remains constant as the neutron star rotates, and estimate the age of the Crab pulsar if this were its sole means of slowdown. What conclusions can be drawn from this estimate?

Problem 16.19. The braking index n for a pulsar is defined by the loss rate

$$\frac{dE_{\text{rot}}}{dt} = -K\omega^{n+1}. \qquad (16.61)$$

Show that the braking index is related to ω and its first two time derivatives:

$$n = \ddot{\omega}\omega/\dot{\omega}^2. \qquad (16.62)$$

It is straightforward to show that an energy-loss rate of the form (16.61) for constant K and n predicts a pulsar age

$$\tau \approx -\frac{(\omega/\dot{\omega})}{n-1} = \frac{(P/\dot{P})}{n-1}.$$

Table 16.2 gives τ for a braking index $n = 3$. Finding a value for the braking index based on observations requires measurement of $\ddot{\omega}$, which is extremely difficult. However, for the Crab pulsar enough observations exist to suggest that

$$n = 2.515 \pm 0.005, \tag{16.63}$$

which is close to the value $n = 3$ favored by magnetic-dipole models. Note that gravitational quadrupole radiation requires a braking index $n = 5$. Although the observed n for the Crab strongly favors magnetic dipole radiation, other factors clearly must contribute to the energy-loss mechanism.

The energy-loss rate (16.58) can be inverted to obtain the field intensity B. For a neutron-star radius $R = 10^6$ cm and moment of inertia $I = 10^{45}$ g cm^2,

$$B \approx 3.2 \times 10^{19} (P\dot{P})^{1/2} \text{ gauss} \tag{16.64}$$

with P in sec and $\dot{P}$ in sec per sec.

Unfortunately, no single model developed for the pulse-emission mechanism explains all the observed radio features. Nevertheless, the simple rotating-dipole model illustrates how pulses of high regularity may arise. Figure 16.13(a) shows schematically a neutron star rotating about the $\boldsymbol{\omega}$ axis with a non-aligned dipolar magnetic field whose associated dipole moment is $\mathbf{M}$. The magnetic field lines near the neutron star corotate with angular velocity ω. At a distance r from the magnetic axis, the field lines will move with tangential speed $v = \omega r$, which must be less than the speed of light. The cylinder—shown dashed in Fig. 16.13(a)—about the magnetic axis of radius

$$r_L = \frac{c}{\omega} = 5 \times 10^9 P \text{ cm}, \tag{16.65}$$

where P is in seconds, is called the *light cylinder,* and represents the furthest point at which the field can corotate with the neutron star. Beyond the light cylinder, the magnetic fields gradually trail the field in the corotating region.

If the neutron star is surrounded by plasma, as is probable, then bunches of charged particles accelerated by the strong magnetic fields will radiate. The most likely locations for coherent emission are along the magnetic poles, where the field intensity is greatest, or just outside the light cylinder, where velocities approach the speed of light. Coherent emission from the poles—Figure 16.13(b)—can produce narrow, pencil-like beams; recall that observed pulses often have half-widths $\Delta t/P \lesssim 0.03$. Emission from the light cylinder tends to produce fan-like beams—Figure 16.13(c). The emission will probably be continuous, although instabilities or irregularities in plasma flow could produce random variability in intensity. The reason for regular pulse timing follows immediately from the model. The beam will sweep past an observer located at an angle θ with the rotation axis once every P seconds. The observed pulse structure—see Figure 16.10(b)—will depend on the exact line of sight and the detailed geometry of the emission region.

Another consequence of the model is that not all pulsars are observable from a single point in space. If the emission region is a pencil beam of angular extent $\Delta\theta_B$, then the fraction of solid angle swept out is $2 \times 2\pi \sin\theta\, \Delta\theta_B$. Assuming that the pulsar rotation axes are randomly oriented in space, the probability of observing a given pulsar is roughly $\overline{\sin\theta}\, \Delta\theta_B$, where $\overline{\sin\theta}$ is the average angle between $\mathbf{M}$ and $\boldsymbol{\omega}$ for all pulsars. If we assume $\Delta\theta_B \approx 10°$, the probability is roughly 0.1 (assuming $\overline{\sin\theta} \approx 2/3$). And if we assume that pulsars are distributed uniformly throughout the Galactic disk, with a density equal to that observed within about 2 kpc of the Sun, 10^4 to 10^5 should be active in the Galaxy today. Theory suggests that a typical pulsar will become inactive after about 10^7 years. Combining this estimated lifetime for pulsars with the preceding observations, and assuming a constant pulsar birthrate during the last 10^{10} years, we find that the total number of active and inactive pulsars in the Galaxy would be about 10^7 to 10^8. To within an order of magnitude, this figure is comparable to the estimated number of supernovae that have occurred (at a rate of about one every 30 to 300 years).

The Magnetosphere

In the absence of magnetic fields, the structure of a neutron-star atmosphere is determined by the pressure scale height $h \sim kTR^2/m_H MG$. A typical scale height for surface temperatures of 10^6 K is about one centimeter. A nonmagnetic neutron star would be surrounded by a few centimeters of plasma. The plasma density would be of order 10^4 g/cm^3 at the surface, but would decrease rapidly away from the neutron star, reaching values comparable to interstellar matter density within a meter. The presence of strong magnetic fields in pulsars dramatically changes this picture, as we can see by considering the electrodynamics of a

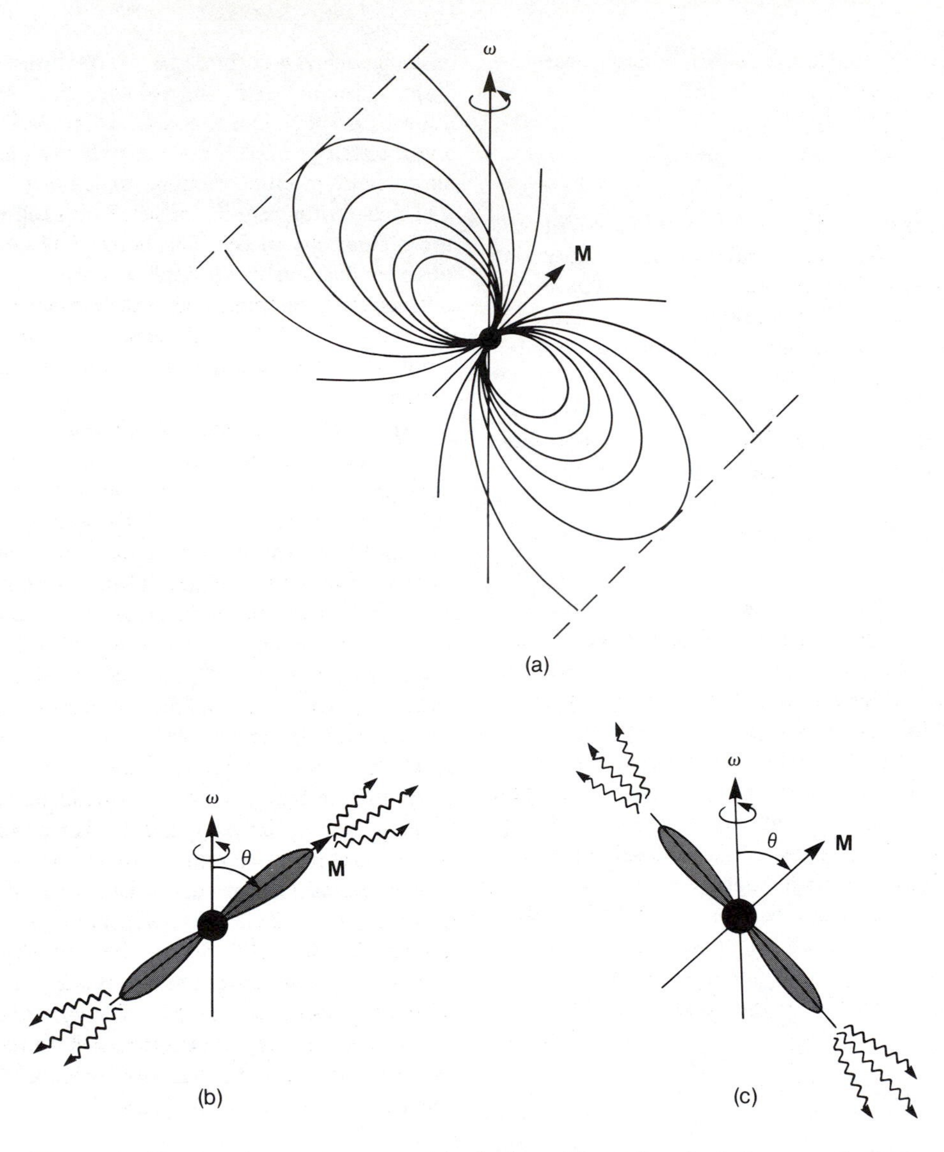

Figure 16.13. Rotating neutron-star models of pulsars: (a) nonaligned rotating magnetic dipole, showing magnetic field lines out to the light cylinder (dashed); (b) pencil-beam emission model; (c) fan-shaped emission region normal to magnetic moment.

rotating magnetized conductor. We recall that the electrical conductivity of matter in neutron stars is extremely large; so this simplified model is not unreasonable.

In a frame at rest relative to the star whose conductivity is σ, the current density j_* and electric field E_* are related by Ohm's law,

$$j_* = \sigma E_*. \qquad (16.66)$$

The reference frame in which E_* is measured moves relative to a frame at rest with respect to the Galaxy with instantaneous velocity

$$\mathbf{v} = \boldsymbol{\omega} \times \mathbf{r}. \qquad (16.67)$$

Here $\boldsymbol{\omega}$ is the neutron star's angular relativity and $\mathbf{r}$ is measured from the rotation axis. Performing a Galilean transformation from the star's rest frame to the

Galaxy's rest frame yields the relation

$$\mathbf{E}_* = \mathbf{E} + (\mathbf{v} \times \mathbf{B})/c, \tag{16.68}$$

where **E** and **B** are measured in the Galaxy's rest frame. We assume that the neutron-star matter is a perfect conductor, in which case σ is infinite. Since j_* must be finite inside a conductor, it follows that $\mathbf{E}_*$ is zero. If we denote the magnetic field set up by the neutron star by **B**, (16.68) indicates that the region surrounding the neutron star must contain an electric field given by

$$\mathbf{E} = -\frac{\mathbf{v} \times \mathbf{B}}{c} = \frac{\mathbf{B} \times (\boldsymbol{\omega} \times \mathbf{r})}{c}. \tag{16.69}$$

Before proceeding further, let us consider the consequences of (16.69) for a pulsar. For typical magnetic fields,

$$E \sim B\omega R/c \approx 2\pi BR/cP \approx 2 \times 10^8 P^{-1} \text{ statvolts/cm}.$$

For the Crab pulsar, this yields $E \approx 6 \times 10^9$ statvolts/cm or 2×10^{12} volts/cm. At the end of Section 16.2, we noted that the ionic binding energy of an atom at the surface of a neutron star is expected to be $W_B \approx 15$ keV. The electric field required to remove an ion from the surface is therefore

$$E \sim \frac{W_B}{Zel} \sim \frac{1.5}{Z} \times 10^{13} \text{ volts/cm} \tag{16.70}$$

for an assumed lattice spacing $l \approx 10^{-9}$ cm, and Z is the ionic charge, which is probably of order 56. It is immediately obvious that ions will be drawn out of the crust into the surrounding region. Once a charged particle has been removed from the crust, the ratio of the electric force to the gravitational force acting on it is of order

$$\frac{F_e}{F_G} \sim \frac{ZeE}{Am_H MG/R^2} \sim \left(\frac{e}{m_H cG}\right) \frac{Z}{A} \frac{B_0 \omega R^2}{M}. \tag{16.71}$$

For electrons the result above applies if $A = Z = 1$, and m_H is replaced by m_e. In general F_e greatly exceeds F_G for any charged particle. Therefore the region just outside a typical pulsar will be permeated by strong electric and magnetic fields. If the electric field has a nonzero component in the radial direction, then charged particles (electrons, protons, and ions) will be pulled out of the crust, and will have sufficient energy to escape from the surface. These charged particles form a relatively dense plasma known as the magnetosphere, extending many radii from the neutron star. It is this plasma that is believed to play a key role in the actual emission mechanism of pulsars.

The construction of pulsar magnetosphere models is extremely complex. Realistic models require that the magnetic and rotation axes be separated by an angle θ, as shown in Figure 16.13. Furthermore, charge flow into the magnetosphere produces currents that induce additional magnetic fields. These in turn modify the structure of the magnetosphere. A self-consistent solution to the problem has not yet been obtained. However, a simplified model can be developed that probably contains the essential features of actual magnetospheres. Consider a neutron star, assumed to be a perfect conductor, with coincident rotation and magnetic axes. Inside the star the magnetic field is $\mathbf{B}_{\text{in}} = B_0 \hat{e}_z$, but outside ($r > R$) it is the dipole field

$$\mathbf{B}_{\text{out}} = \frac{B_0 R^3}{2r^3} (2 \cos\theta\, \hat{e}_r + \sin\theta\, \hat{e}_\theta). \tag{16.72}$$

For simplicity we will assume that magnetic and rotation axes are the same: $\boldsymbol{\omega} = \omega \hat{e}_z$. Therefore we can obtain the electric field inside the star by substituting (16.72) into (16.69):

$$\mathbf{E}_{\text{in}} = -\frac{B_0}{c} \omega r \sin\theta\, (\hat{e}_r \sin\theta + \hat{e}_\theta \cos\theta). \tag{16.73}$$

We note that $\mathbf{E} \cdot \mathbf{B} = 0$ inside the neutron star, as expected for a conductor. The electric field outside the star follows from $\mathbf{E}_{\text{in}}$ and the boundary conditions at the surface. The result is

$$\mathbf{E}_{\text{out}} = -\frac{B_0 \omega R^5}{cr^4} \times [\hat{e}_r P_2(\cos\theta) + \hat{e}_\theta \sin\theta \cos\theta], \tag{16.74}$$

where $P_2(x) = (3x^2 - 1)/2$.

Using (16.72) and (16.74), we readily find the component of **E** parallel to **B** outside the pulsar:

$$(\mathbf{E} \cdot \mathbf{B})_{\text{out}} = -\left(\frac{\omega R}{c}\right)\left(\frac{R}{r}\right)^7 B_0^2 \cos^3\theta. \tag{16.75}$$

The magnitude of $\mathbf{B}_{\text{out}}$ may be obtained from (16.72);

$E_\parallel$ is equal to $B_0\omega R/c$ at the poles, and gradually decreases towards the magnetic equator.

Problem 16.20. The electrostatic potential ϕ outside a neutron star can be expanded in Legendre polynomials $P_n(\cos\theta)$ if the magnetic and rotation axes coincide. Use the boundary condition

$$E_{\theta,\text{in}}(R) = E_{\theta,\text{out}}(R)$$

to show that

$$\phi_{\text{out}} = -\frac{B_0\omega R^5}{3cr^3} P_2(\cos\theta). \tag{16.76}$$

Use Laplace's equation to find the net charge inside and outside the rotating neutron star.

The idealized model developed in the preceding is of limited applicability to pulsars, because the magnetic and rotation axes were taken to be parallel, and because the fields induced by the plasma flow in the magnetosphere have not been taken into account (the solution is not self-consistent). Nevertheless, the qualitative and order-of-magnitude features contained in it are also found in more nearly realistic models. With this in mind, we consider the field structure and behavior of the plasma in the magnetosphere of the aligned magnetic rotator. Figure 16.14 shows schematically the magnetosphere. The shaded region contains closed dipole field lines that lie within the light cylinder of the radius r_L. The tangential velocity of points on these lines is less than the speed of light; so they corotate with the neutron star. The plasma in this region is, to first approximation, bound to the magnetic field and must also corotate with the neutron star. The outermost closed field line leaves the neutron-star surface at an angle θ_p, and is tangent to the light cylinder at the equator. The region (plasma plus fields) contained inside it represents the corotating magnetosphere. Assuming that the field inside the magnetosphere is dipole, we can use the fact that $\sin^2\theta/r$ is a constant along magnetic dipole field lines (Problem 16.21) to find the angle θ_p. Thus, using (16.65), we find

$$\frac{\sin^2\theta_p}{R} = \frac{\sin^2\pi/2}{r_L} = \frac{\omega}{c},$$

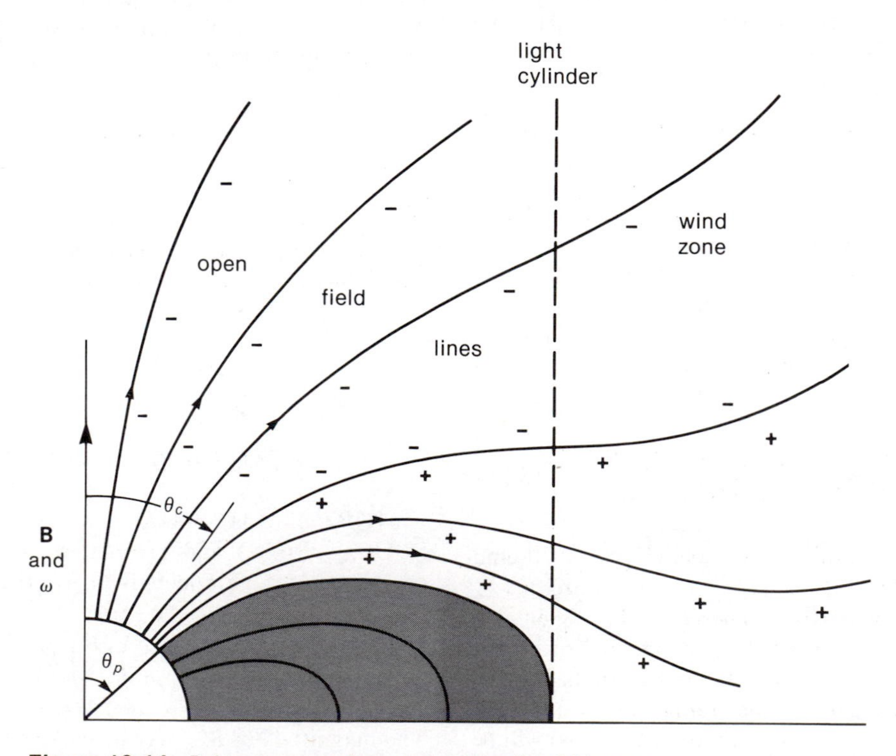

Figure 16.14. Pulsar magnetosphere, and surrounding field structure.

which yields

$$\theta_p = \sin^{-1}(\omega R/c)^{1/2}. \qquad (16.77)$$

Problem 16.21. The magnetostatic potential for a dipole field is proportional to $\phi_B \sim \cos\theta/r$. Show that the surfaces of constant $\psi \sim \sin^2\theta/r$ intersect surfaces of constant ϕ_B at right angles. Note that $\mathbf{B}_{\text{out}}$ is normal to the surface ϕ_B = constant.

For the Crab pulsar $\theta_p \simeq 5°$, but for a pulsar with a period of order 1 second, θ_p is less than a degree. Field lines leaving the neutron star at angles less than θ_p would attain tangential velocities in excess of c if they were to corotate. Instead, they deviate from dipolar near the light cylinder, cross it, develop an azimuthal component, and remain open (in nonaligned rotators, these open lines would join onto the radiation field at large distances from the pulsar). The radius of the polar cap through which the open lines pass is given by

$$r_p \simeq R\sin\theta_p = R(2\pi R/cP)^{1/2}, \qquad (16.78)$$

corresponding to an area of less than 0.1 km^2 for a period of one second.

The magnetic field line leaving the neutron-star surface at a polar angle θ_c (Figure 16.14), called the *critical field line,* intersects the light cylinder at right angles $[B_z(r_L) = 0]$. The critical field line at the neutron-star surface is at the same electrostatic potential as the interstellar medium. For $\theta < \theta_c$, the field lines at the surface are at a lower potential than the surrounding medium. Therefore, electrons stream out along the magnetic field lines, which pass through the polar cap of radius $r_c = R\sin\theta_c$. For $\theta > \theta_c$ the electrostatic potential exceeds the interstellar value, and positive ions stream outward along field lines lying within the annular region defined by $\theta_c \le \theta \le \theta_p$. The value of θ_c is fixed by requiring that the net current flow through the polar cap (radius r_p) vanish.

The drop in electrostatic potential from the pole to the edge of the polar cap ($\theta = \theta_p$) can be obtained from (16.76), and is

$$-\Delta\phi = \frac{1}{2}B_0R(\omega R/c)^2$$
$$= 6\times 10^{12}/P^2 \text{ eV}. \qquad (16.79)$$

For the Crab pulsar, the energy attained by an electron accelerated through this potential is about 6×10^{15} eV.

We arrive at the following picture of charged particle flow in the magnetosphere. A negative current flows out along the poles ($\theta \lesssim \theta_c$), and an equal positive current (protons and ions) flows out in the annular sheath $\theta_c < \theta \lesssim \theta_p$. Each current distribution induces a magnetic field that is toroidal about the magnetic axis. Near the light cylinder the toroidal field caused by the electron current is larger than that caused by the ion current, and is comparable to the magnetic dipole field; so the net magnetic field bends backward as it passes the light cylinder trailing the corotating magnetosphere. This model does not include the modifications to the magnetosphere needed because of these torodial fields. To order of magnitude, the azimuthal component of $\mathbf{B}$ near the light cylinder is

$$B_\phi \approx \frac{v}{c}B_r \simeq \frac{\omega r}{c}\sin\theta\, B_r. \qquad (16.80)$$

The azimuthal component will give rise to electromagnetic radiation in the wind zone (Figure 16.14). For a radiation field propagating in the radial direction, $E_\theta \approx B_\phi$. Outside r_L, the radial component of $\mathbf{B}$ is no longer dipolar but takes the form

$$B_r = \frac{\psi(\theta)}{r^2}. \qquad (16.81)$$

Using the results above, we can easily show the Poynting vector to be

$$\mathbf{S} = \frac{c}{4\pi}(\mathbf{E}\times\mathbf{B}) \simeq \frac{c}{4\pi}\left(\frac{\omega r}{c}\sin\theta\right)^2\frac{\psi^2}{r^4}\hat{e}_r.$$

Integrating over a surface of radius r yields the total electromagnetic energy radiated from the pulsar:

$$\frac{dE}{dt} = \int_0^{2\pi} d\phi \int_0^{\pi} \mathbf{S}\cdot\hat{e}_r \sin\theta\, d\theta$$
$$= 2\int_0^{2\pi} d\phi \int_0^{\theta_0} \frac{\omega^2}{c}\sin^3\theta\, \psi^2(\theta)\, d\theta,$$

where in the last step it is assumed that the open field lines are confined to the region $0 \le \theta \le \theta_0$ near the north and south magnetic poles. For order of magnitude purposes we further assume θ_c is small, and that

$\psi(\theta)$ varies slowly in the range of integration. Thus

$$\frac{dE}{dt} \simeq \frac{\omega^2}{c} \psi^2(\theta_0) \frac{\theta_0^4}{4}. \qquad (16.82)$$

The quantity $\psi(\theta_0)$ may be eliminated by noting that the magnetic flux through the surface is

$$\int_0^{\theta_c} \sin\theta \, d\theta \int_0^{2\pi} d\phi \, r^2 B_r \approx \pi\psi(\theta_0)\theta_0^2$$

for each polar cap. The flux through the stellar surface, which is roughly $B_0 R^2 \theta_0^2$, must equal this value, from which $\psi(\theta_0) \approx B_0 R^2/\pi$ follows. Substituting this into (16.82) finally yields

$$\frac{dE}{dt} \approx \left(\frac{\omega R}{c}\right)^3 \frac{B_0^2 R^3 \omega}{4\pi^2}. \qquad (16.83)$$

The result is, to order of magnitude, equal to the energy-loss rate from a dipole in vacuum (16.58). This is encouraging, since the simple dipole model does yield the correct energy-loss rates for the Crab pulsar if it is to power the nebula.

Since the magnitude of the electric and magnetic fields near the light cylinder are comparable, charged particles that were accelerated initially along **B** near the neutron star's surface will escape into the interstellar medium as cosmic rays.

The preceding simplified model (aligned rotator) must be modified in at least two ways if a reasonable description of a pulsar is to be obtained. First, it must be generalized to nonaligned rotation and magnetic axes; second, the model must be solved self-consistently by including the induced fields. No such model has as yet been developed.

Nonradio Emission

The signal from most pulsars decreases so rapidly with increasing frequency (recall Figure 16.12) that no detectable emission above the radio is observed. For the pulsars in the Crab and Vela nebulae, however, emission at higher than radio frequencies is observed (Figure 16.15). The energy spectrum of the Crab pulsar, shown in Figure 16.16, indicates that more than 7.5×10^{36} erg/sec is radiated at or above the infrared, exceeding the estimated radio emission by a factor of about 10^6. The break in the spectrum between the radio and the infrared appears to be real. The integrated pulse profile for the Crab pulsar in Figure 16.15(a) shows the same qualitative shape throughout the spectrum, containing a main pulse followed by an interpulse 13.4 ms later. The ratio of main pulse to interpulse energy, however, varies from about 1.5 at 410 MHz to about 0.5 in the gamma-ray region. These observations, together with low brightness temperatures at optical frequencies and above, suggest that the radio and nonradio emission are due to different mechanisms. Emission has been detected from the Vela pulsar in the radio, optical, and gamma ray regions but not in the x-ray region. The obvious difference between the radio and nonradio signal is further evidence of different emission mechanisms.

Seismological Effects

Neutron stars formed by supernova explosions are extremely hot ($T \sim 10^{10}$ to 2×10^{11} K). At temperatures below 10^{10} K, the ions and electrons will form a Coulomb lattice. Since neutron stars cool rapidly, the matter at densities $\lesssim 10^{14}$ g/cm^3 that contains ions will crystallize while the newly formed core is rapidly rotating. The fluid at higher densities consists of free baryons and is not thought to crystallize. Thus, for low-mass neutron stars ($\rho_c \lesssim 10^{14}$ g/cm^3), the entire star may solidify shortly after formation. For more massive ones a fluid core will result, surrounded by a crystalline crust several hundred km in thickness. Ignoring magnetic effects, we can suppose that the initial configuration (while fluid) will probably be an oblate spheroid whose moment of inertia about the rotation axis is

$$I = I_0(1 + \xi), \qquad \xi = (5/8\pi)(\omega^2/\rho G), \qquad (16.84)$$

where $I_0 = (2/5)MR^2$ is the moment of inertia of a spherical distribution. The deviation ξ is greatest for the most rapid pulsars; for the Crab pulsar $\xi \approx 10^{-3}$ or less. The shape of the solid crust is such that pressure and centripetal forces balance gravity when solidification occurs. As the pulsar spins down, an imbalance develops and stresses build up within the crust, until its yield strength is exceeded. The crust then fractures, rearranging its mass distribution on a time scale

$$t_{\text{yield}} \sim R/c_s \approx 0.1 \text{ sec},$$

where c_s is the speed of a shear wave in the Coulomb lattice. This type of seismic rearrangement, or star quake, results in a sudden change in the moment of

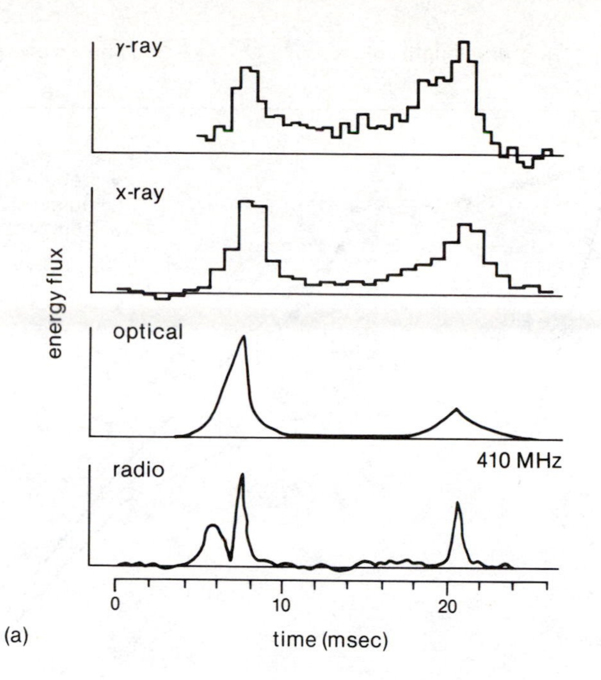

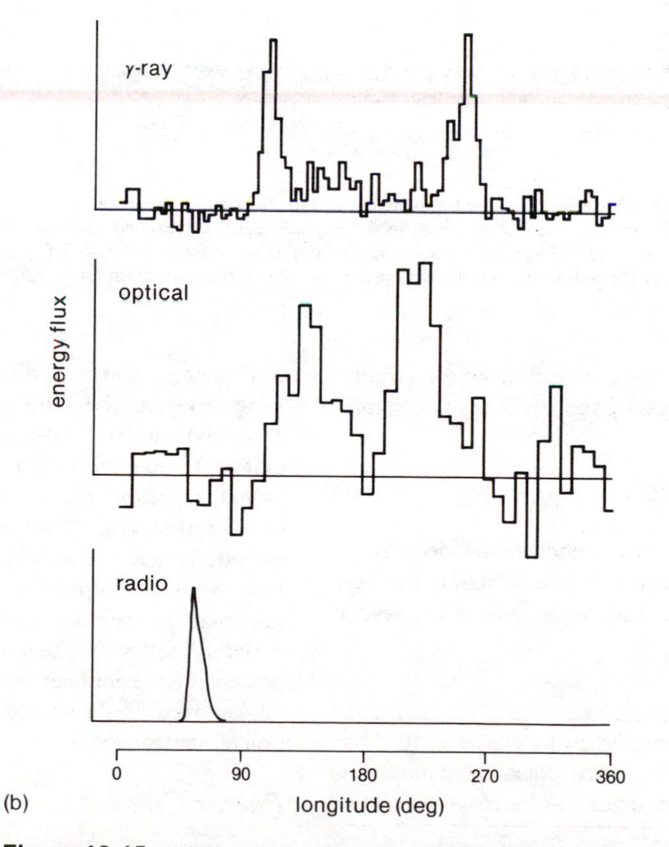

Figure 16.15. (a) Integrated pulse profiles from the Crab pulsar. The radio-frequency pulse components are, from the left, the precursor, main pulse, and interpulse. Only the main pulse and interpulse are present at higher frequencies. (b) Integrated pulse profiles for the Vela pulsar. The pulses are aligned in relative phase as they are emitted from the pulsar.

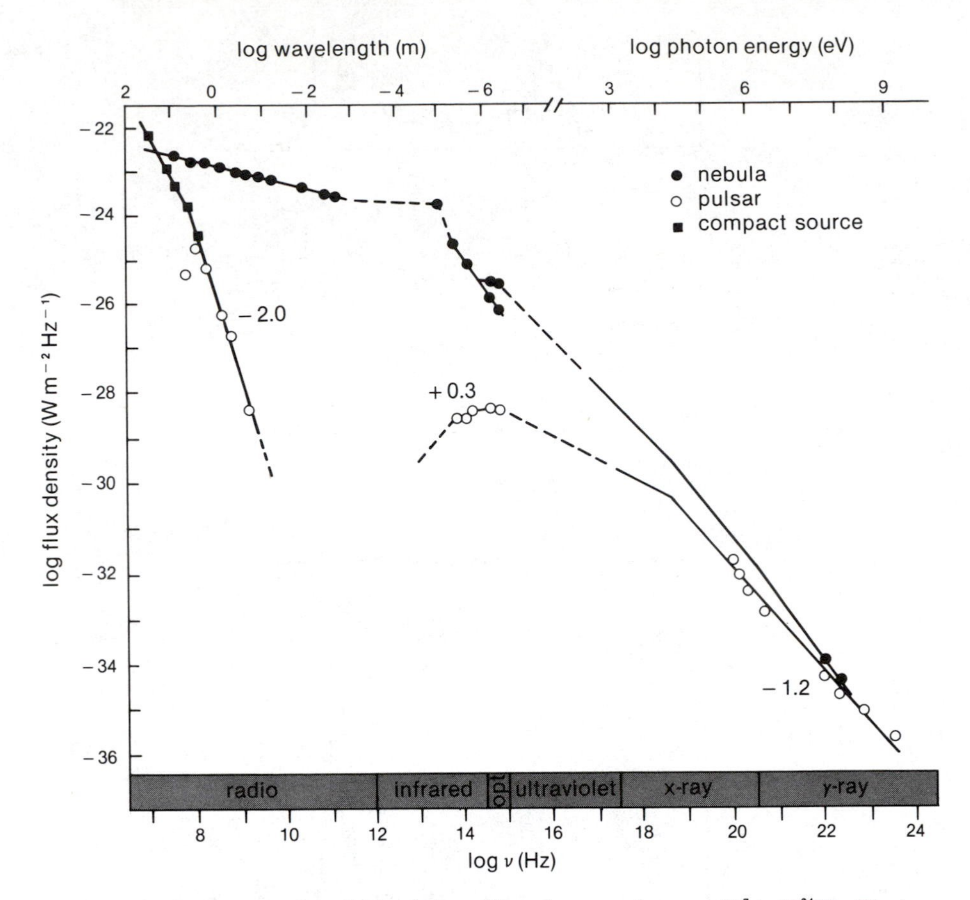

Figure 16.16. Spectra of the Crab nebula and its pulsar over the range 10^7 to 10^{24} Hz. The two lines drawn for the nebular spectrum in the optical region correspond to the observed spectrum (lower curve), and the spectrum corrected for interstellar extinction (A_ν = 1.6 mag). The optical spectrum for the pulsar has also been corrected for interstellar extinction (A_ν = 1.6 mag).

inertia ΔI, and (assuming conservation of angular momentum in the process) a sudden change in angular velocity $\Delta\omega$, given by

$$\Delta\omega/\omega = -\Delta I/I. \tag{16.85}$$

Since ΔI is negative, the period should decrease, a phenomenon known as pulsar spinup. Spinup has been observed in the Crab and Vela pulsars on several occasions.

Problem 16.22. Spinup of order $\Delta\omega/\omega \approx 10^{-6}$ has been observed in the Vela pulsar. Estimate the change in neutron-star radius needed to explain this.

The model explains some of the qualitative features of star quakes. More detailed models consider the coupling between the fluid core (containing superfluid neutrons) and the solid crust. Superfluids do not rotate uniformly like normal fluids. Instead, vortex lines (which represent rotational quanta of the superfluid) form in the core. When the crust yields, it spins up essentially instantaneously. The core, however, may be only weakly coupled to the crust; so a gradual exchange of angular momentum between the two probably follows. The quantitative details of these processes have not been established; once they are, star quakes may offer a way to probe the interior composition of neutron stars.

Pulsar Classes

Two classes of pulsars are well established: the radio pulsars discussed in this section, and x-ray pulsars discussed in Chapter 17. The recent observation of two new, fast pulsars (millisecond pulsars) suggests that a third class may exist. The two new ones are PSR 1937 + 214 ($P = 1.558 \times 10^{-3}$ sec; $dP/dt = 1.26 \times 10^{-19}$ sec/sec) and PSR 1953 + 257 ($P = 6.1337 \times 10^{-3}$ sec; $dP/dt = 5.8 \times 10^{-16}$ sec/sec). The second of these is a member of a binary system whose orbital period is about 120 days. The near proximity of the first one to a planetary nebula suggests that it may once have been a member of a binary system as well. Pulses in the radio and the visible spectra have been observed from PSR 1937 + 214. It is notable that dP/dt for the fastest millisecond pulsar is at least 10^{-4} times smaller than for the isolated pulsars (Table 16.2), suggesting that its energy loss rate may be quite small.

Chapter 17

CLOSE BINARY SYSTEMS

About 50 percent of the stars observed in the solar neighborhood are members of binary or multiple-star systems. In many cases the component stars are well separated (semimajor axis $a \gg$ stellar radius R), and each star has little or no effect on the structure and evolution of the other. However, some astrophysical phenomena are direct consequences of the near proximity of two stars (where $a \sim R$). A few examples are the mechanisms thought to be responsible for many compact x-ray sources, probably all novae, and the selective heating and tidal distortion of some ordinary stars by their companions.

We will concentrate on two aspects of binary systems below. The first is mass exchange, which depends to some extent on the individual structure of the component stars, and is one possible mechanism for the production of novae and of x-ray, gamma, and ultraviolet radiation originating in close binaries. The second involves conditions in which a binary system may be disrupted, or its character significantly changed, if one of its members undergoes a supernova explosion. This represents one scenario for the formation of x-ray sources containing neutron stars, or black holes. A slight modification of the model was originally used to explain the occurrence of OB-type runaway stars. Before proceeding to these cases, we briefly review aspects of the mechanics of binary systems. Newtonian gravity will be assumed, unless otherwise stated.

17.1. Mechanics of Binary Systems

The basic force between members of a binary system is given by Newton's expression for the law of universal gravitation,

$$F = -\frac{M_1 M_2 G}{r^2}, \tag{17.1}$$

where the member masses are M_1 and M_2, and the orbital elements include a (semimajor axis), ϵ (orbital eccentricity), and period P. From this one obtains Kepler's law relating the masses, a, and P, which may be written in several convenient forms applicable to close binary systems:

$$P = \frac{365.25\,(a/a_\odot)^{3/2}}{(M/M_\odot)^{1/2}} = \frac{0.116\,(a/R_\odot)^{3/2}}{(M/M_\odot)^{1/2}}\ \text{days}. \tag{17.2}$$

Here $a_\odot$ is one astronomical unit (1.496×10^{13} cm), and the total mass $M = M_1 + M_2$. This form is useful whenever a is comparable to the stellar radii. For circular orbits ($\epsilon = 0$) the relative velocity of either mass with respect to the other is

$$v = (MG/a)^{1/2} = 940 \frac{(M/M_\odot)^{1/2}}{(a/a_\odot)^{1/2}}$$

$$= 437 \frac{(M/M_\odot)^{1/2}}{(a/R_\odot)^{1/2}} \text{ km/sec.} \qquad (17.3)$$

The angular momentum of the system M_1 and M_2 about the center of mass is given by

$$J = M_1 M_2 (aG/M)^{1/2}$$

$$= 8.85 \times 10^{52} \frac{(M_1/M_\odot)(M_2/M_\odot)(a/a_\odot)^{1/2}}{(M/M_\odot)^{1/2}} \qquad (17.4)$$

erg sec.

Problem 17.1. Verify the expressions for v and J given by (17.3) and (17.4).

In several interesting cases the orbits are elliptical instead of circular (see Figure 17.1). Equation (17.2) is still applicable, but (17.3) must be replaced by a more complicated expression in which the magnitude of the radius vector is given by

$$r = \frac{a(1-\epsilon^2)}{1 + \epsilon \cos \theta}. \qquad (17.5)$$

Equation (17.5) and the properties of an ellipse may be used to obtain the motion of the stars in this more general case.

In order to calculate the mass of each component of a binary system, we must know the system's distance (to obtain a), and the individual orbits of each star must be observed. The last requirement is equivalent to knowing the velocity curve for each star, that is, $v_i(t)$ with $i = 1, 2$ correpsonding to mass M_1 or M_2. Often we deal with single-line spectroscopic binaries, for which observations give only the mass function for the system, which is defined by

$$f(M_1, M_2, i) \equiv \left(\frac{M_2}{M_1 + M_2}\right)^2 M_2 \sin^3 i$$

$$= \frac{4\pi^2}{G} \frac{(a_1 \sin i)^3}{P^2}. \qquad (17.6)$$

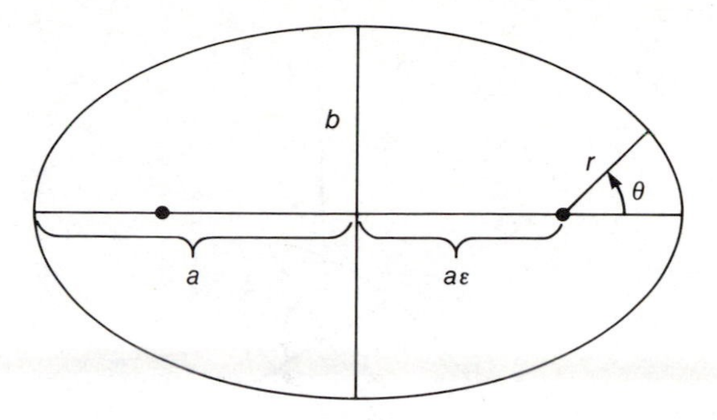

Figure 17.1. Geometry of an elliptical orbit seen face on.

In this case, the motion of M_1 is observed, and the orbital inclination is denoted by i ($i = \pi/2$ corresponds to the orbits seen edge-on). If $a = a_1 + a_2$ is known, then Kepler's third law gives the system's total mass $M_1 + M_2$, and (17.6) gives a relation between M_2 and i. For eclipsing binaries ($i = \pi/2$) this relation fixes M_2; in general, some limits on i may be available, and (17.6) may be used to bound M_2.

A complete specification of the binary system requires, in addition to the elements a, P, and ϵ that characterize its shape, four additional elements specifying its orientation with respect to the observer. The complete set of orbital elements is then:

(1) semimajor axis,	a	
(2) eccentricity,	ϵ	
(3) period,	P	
(4) inclination of orbit,	i	Euler angles of orbit
(5) angle of nodes,	Ω	
(6) longitude of periastron,	ω	
(7) time of periastron passage,	T	

The orbital elements are illustrated in Figure 17.2. The observer is assumed to lie along the z axis, so that the orbit is edge-on if $i = \pi/2$. The cartesian axes pass through the system's center of mass.

A velocity curve represents the time-variation of the star's velocity projected onto the line of sight (usually measured by Doppler-shifted spectra), and depends on the orbital parameters 1 through 6. Figure 17.3 shows the position vector of the star as measured from the center of mass at a given instant (θ measured from periastron). The component of $\mathbf{r}$ along the z axis (line of sight of observer) is given by

$$r_\perp = r \sin(\theta + \omega) \sin i. \qquad (17.7)$$

The radial velocity, denoted by $v_R = dr_\perp/dt$, is easily

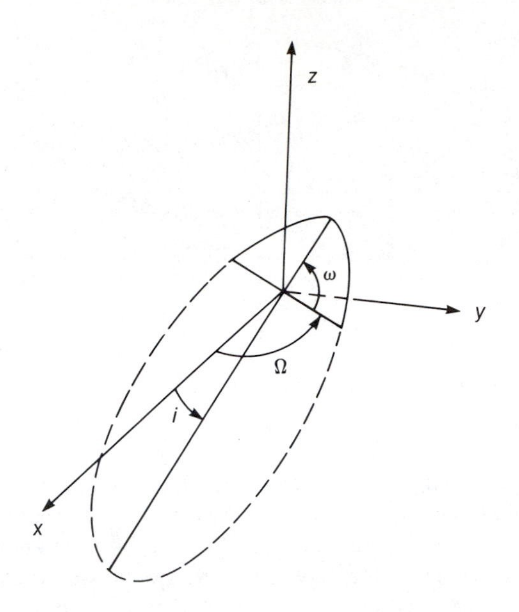

Figure 17.2. Orbital elements for binary system.

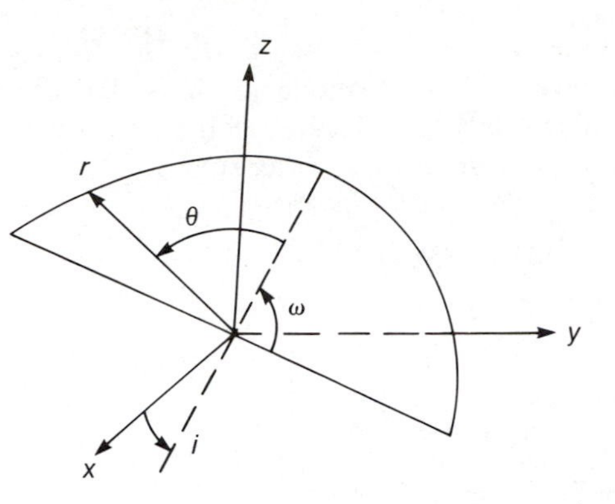

Figure 17.3. Instantaneous position vector of a star moving in absolute orbit as seen by a distant observer.

shown to be

$$v_R = \sin i \left[\frac{dr}{dt} \sin(\theta + \omega) + r \cos(\theta + \omega) \frac{d\theta}{dt}\right]. \quad (17.8)$$

Differentiation of (17.5) gives a relation between dr/dt and $d\theta/dt$:

$$\frac{dr}{dt} = \frac{a\epsilon(1 - \epsilon^2) \sin\theta}{(1 + \epsilon \cos\theta)^2} \frac{d\theta}{dt}. \quad (17.9)$$

Kepler's second law may be used to eliminate $d\theta$. It states that the radius vector sweeps out equal areas in equal times, and is an expression of the conservation of angular momentum. Denoting an element of area by dA and the time taken to sweep it out by dt, Kepler's second law may be written as $dA/A = dt/P$, where P is the period, and the orbital area $A = \pi ab = a^2\pi \cdot (1 - \epsilon^2)^{1/2}$. The element $dA = (1/2)\, r^2 d\theta$, so that

$$\frac{1}{2} r^2 \frac{d\theta}{dt} = \frac{\pi a^2 (1 - \epsilon^2)^{1/2}}{P}. \quad (17.10)$$

Using the last two equations in v_R, and simplifying the trigonometric relations, yields

$$v_R = K\,[\cos(\theta + \omega) + \epsilon \cos\omega], \quad (17.11)$$

where K, the amplitude of the radial velocity, is given by

$$K \equiv \frac{2\pi a \sin i}{P(1 - \epsilon^2)^{1/2}}. \quad (17.12)$$

If the binary system's center of mass is moving with velocity v_c along the observer's line of sight, it must be added to (17.11).

To obtain the velocity curve we need $v_R = v(t)$. Combining (17.9) and (17.10) and noting that P, a, and ϵ are constants, we find

$$\frac{2\pi t}{P(1 - \epsilon^2)^{3/2}} = \int_{\theta_0}^{\theta} \frac{d\theta}{(1 + \epsilon \cos\theta)^2}, \quad (17.13)$$

with $\theta(0) = \theta_0$. Eliminating θ between this equation and (17.11) for v_R gives the desired relation $v_R(t)$. Although this must usually be done numerically, it may often be used to constrain the orbital elements in binary systems. A careful study of $v_R(t)$ often indicates that one star in the binary is tidally distorted, or its surface heated by its companion.

17.2. Structure of Close Binary Systems

The analysis in Section 17.1 is usually adequate when the members of a binary system are well separated. However, when either of the stellar radii is of the same order of magnitude as the orbital radius, the presence of a near companion can dramatically alter the stars' appearance and its evolution. The morphology of close binary systems is illustrated in Figure 17.4, in which two stars of radii R_1 and R_2 are in circular orbits of radius a, assuming that a is of the same order of magnitude as R_1 and R_2. The center of mass is denoted by (x), and we suppose that the system rotates counterclockwise with angular velocity $\mathbf{\Omega}$ as shown. The motion of a test particle of mass m ($m \ll M_1$ or M_2) as seen by an observer at the center of mass rotating with the system appears to be determined by

$$\mathbf{F}_m = \mathbf{F}_{M_1} + \mathbf{F}_{M_2} + \mathbf{F}_{\text{centrifugal}} + \mathbf{F}_{\text{Coriolis}}. \quad (17.14)$$

The last two are pseudoforces that arise from the choice of a rotating reference frame and vanish in an inertial frame. They have an important effect on the motion of test particles near M_1 and M_2.

The four contours surrounding the masses in the figure are surfaces of constant gravitational potential. The innermost ones surrounding the two masses become more nearly spherical as each mass is approached. Those shown differ very little from equipotential surfaces around an isolated point mass. One unique surface intersects itself at the point L_1, called the *inner Lagrange point*. This surface defines two regions, one surrounding each star, which are called *Roche lobes*. Equipotentials surrounding only one star lie within that star's Roche lobe, but surfaces that lie outside a Roche lobe will surround both stars in most cases (see Figure 17.5). Generally speaking, a Roche lobe is not a sphere; however, it may not differ by much from a sphere of suitable radius. It is convenient to define the effective radius of a Roche lobe around the ith star by the relation

$$\begin{pmatrix}\text{volume of} \\ \text{Roche lobe}\end{pmatrix} \equiv \frac{4\pi}{3} r_i^3, \qquad i = 1, 2. \quad (17.15)$$

The radius will depend on the masses M_1 and M_2 and on their separation a. An analytic approximation for r_i accurate to better than 1 percent is

$$r_i = a(0.38 + 0.2 \log M_1/M_2). \qquad (17.16)$$

The expression for r_2 follows by interchanging M_1 and M_2 in (17.16). The effective radius of the Roche lobe grows if the mass is increased.

There are three categories of close binary systems, in which neither, one, or both of the Roche lobes are filled by the star's envelope. Figure 17.4 shows the case

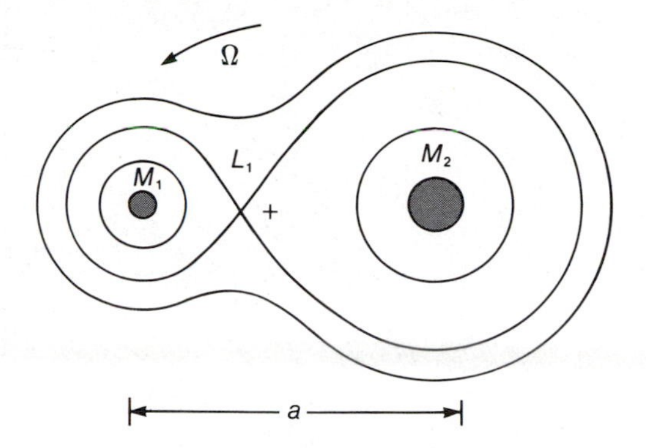

Figure 17.4. Equipotential contours in the equatorial plane of a binary system rotating with angular velocity Ω about the center of mass (+) of the system.

where neither lobe is filled. The three types are called detached, semidetached, and contact binary systems.

In detached binary systems, both stars may be on the main sequence, with the more massive appearing to be the more evolved. Examples include YY Gem, WW Aur, and Y Cyg. Neither lobe is filled; so the stellar radii R_i are less than r_i. When one star fills its Roche lobe, the system is called semidetached. Sometimes the more massive star appears to be near the main sequence, but the less-massive star fills its Roche lobe and resembles a subgiant. Emission lines are common (denoting probable mass exchange, as we will see). Frequently there is evidence of a gaseous ring around the less-massive star. Examples include β Per, V Pup, and some compact x-ray sources, such as Her X-1 and Cyg X-1. In semidetached systems $R_1 \simeq a_1$ and $R_2 < a_2$, where a_i is the distance of M_i from the binary center of mass. In contact binary systems, both stars fill their Roche lobes ($R_1 \simeq a_1$ and $R_2 \simeq a_2$). This category is probably restricted to systems containing lower-mass stars such that $M_1 + M_2 \lesssim 4\ M_\odot$. Both stars may appear close to the main sequence. Examples include W Ursae Majoris stars.

Finally, it is possible that one (or both) of the stars in the course of evolution will overshoot its Roche lobe, creating an envelope of material that completely surrounds both stars. It is not difficult to envision situations in which enough matter resides in a common envelope that the usual definition of a binary system becomes inapplicable. The result might better be described as a single star with two regions of maximum density. Virtually nothing is known about such systems.

The observed properties of primary stars in close

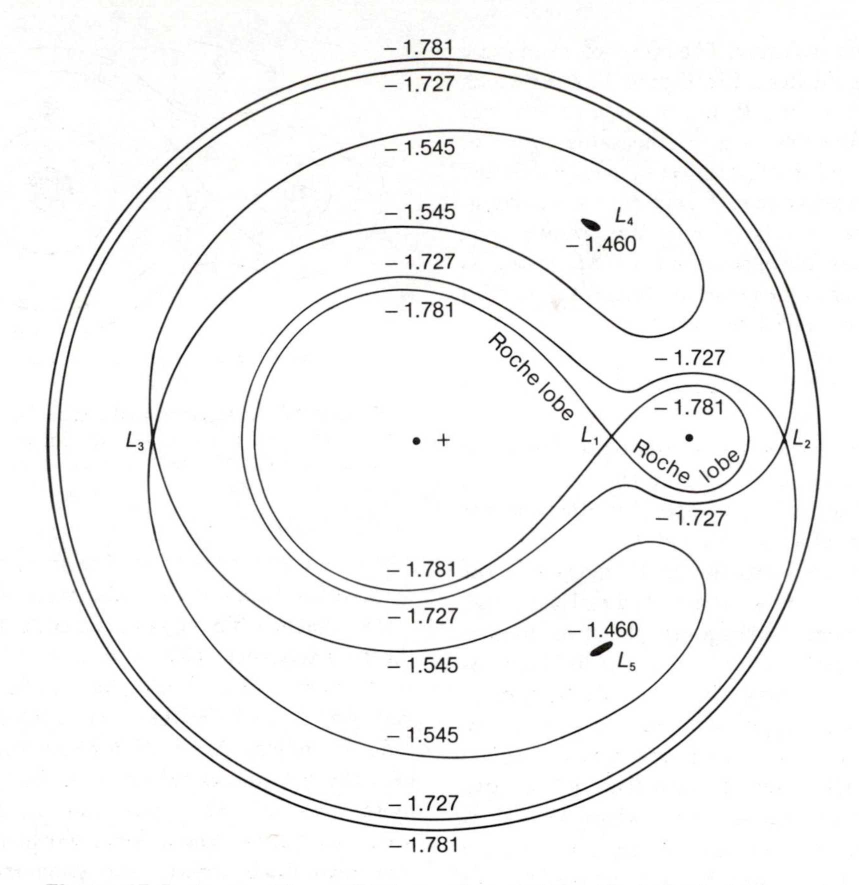

Figure 17.5. Same as Figure 17.4, but equipotentials outside the Roche lobes are also shown. The Lagrange points are denoted by L_i, and the relative values of the contours are shown next to each curve. Note the relative minimum in the gravitational potential formed at L_4 and L_5.

binary systems during or subsequent to mass exchange are summarized in Figures 17.6 and 17.7. As Figure 17.7 shows, the primary is less massive than a main-sequence star of the same luminosity would be. Also, the luminosity excess is greatest for stars with masses less than or about 1 $M_\odot$. The H-R diagram (Figure 17.6) shows that the primary falls in the subgiant region and has a greater luminosity for a given spectral class than would the corresponding main-sequence star. Generally the primary is more luminous and less massive than a normal star in the same stage of evolution. We will see below that these properties are a consequence of close binary structures.

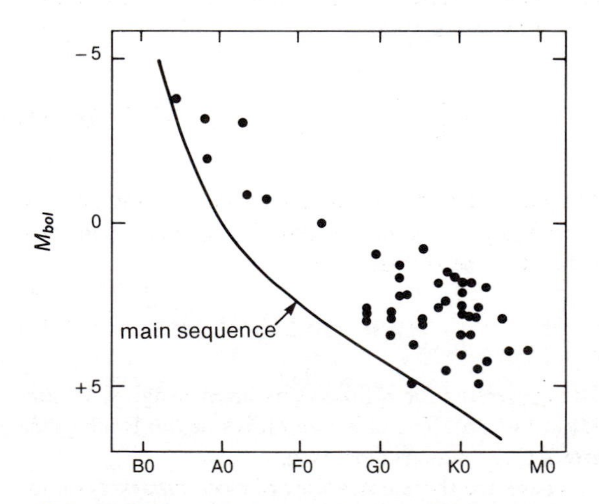

Figure 17.6. H-R diagram for the subgiant components of eclipsing binaries. The solid line is the main sequence.

Mass exchange results when matter from the outer portions of a star expands or is driven into a region where the companion's gravitational field dominates. For example, as seen by an observer moving with the binary system's center of mass, an element of mass (possibly part of the star's envelope) $m \ll M_1$ or M_2 will

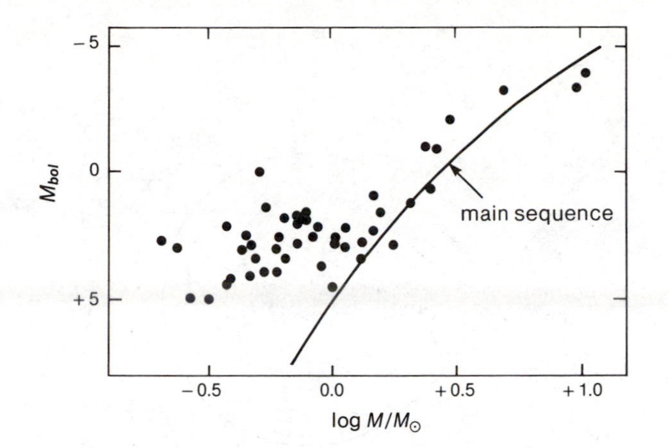

Figure 17.7. Observed mass-luminosity relation for the subgiant components of eclipsing binaries. The solid line is the main sequence.

respond to the net force given by (17.14). Restricting attention to motion in the plane of the orbits, we find that the first three forces may be obtained from the potential

$$\Phi(r) = -\frac{M_1 G}{|\mathbf{r}_1 - \mathbf{r}|} - \frac{M_2 G}{|\mathbf{r}_2 - \mathbf{r}|} - \frac{1}{2}|\mathbf{\Omega} \times \mathbf{r}|^2. \quad (17.17)$$

The last term leads to a centrifugal force; $\mathbf{r}$ is the vector from m to the center of mass; and $\mathbf{\Omega}$ is the angular velocity. The net force is

$$\mathbf{F}_m = -(\nabla\Phi + 2\mathbf{\Omega} \times \mathbf{v})m, \quad (17.18)$$

where $\mathbf{v}$ is the velocity of the test mass, and the last term is the Coriolis force.

Problem 17.2. Use (17.17) and show that the force on a test mass m is given by

$$\mathbf{F}_m = -\frac{mM_1G}{|\mathbf{r}_1 - \mathbf{r}|^3}(\mathbf{r} - \mathbf{r}_1) - \frac{mM_2G}{|\mathbf{r}_2 - \mathbf{r}|^3}(\mathbf{r} - \mathbf{r}_2) - 2m(\mathbf{\Omega} \times \mathbf{v}) - m\mathbf{\Omega} \times (\mathbf{\Omega} \times \mathbf{r}). \quad (17.19)$$

By analyzing the terms appearing in $\mathbf{F}_m$, we can easily show that the gravitational pull of M_1 and M_2 tends to channel the motion of m toward the line joining the two stars, but the centrifugal force tends to move the mass away from the center of mass along $\mathbf{r}$. The effect of the Coriolis force depends on the test-mass velocity $\mathbf{v}$. The motion of m is finally determined by the relative sizes of the four forces in (17.19), which depend on a, $\mathbf{\Omega}$, $\mathbf{v}$, and the masses. One important possibility is shown in Figure 17.8 and accompanies mass exchange.

Consider a star of mass M_2 (Figure 17.4) that expands until it just fills its Roche lobe. If matter in the vicinity of the inner Lagrange point L_1 experiences the slightest perturbation, it will spill over into the lobe surrounding M_1. Once this process begins, it continues, often rapidly, and can result in the transfer of a significant fraction of the mass originally associated with M_2 to the companion star. There is evidence that the mass ratio M_1/M_2 can be nearly inverted as a result of mass exchange. The material moving in the orbital plane will be governed by (17.19). If Ω and a are properly related, the infalling mass may set up a ring surrounding M_1 as shown in Figure 17.8; the ring will rotate in the same sense as the binary system itself.

Some of the matter may move with high enough velocities to escape the system through the outer Lagrange point L_2. When this happens, the mass and angular momentum of the system are reduced. How much mass loss will occur in general is not known with certainty, but will depend on the specific case.

The ring of matter surrounding M_1 will tend to rain gradually down onto M_1. Its rate of infall may be

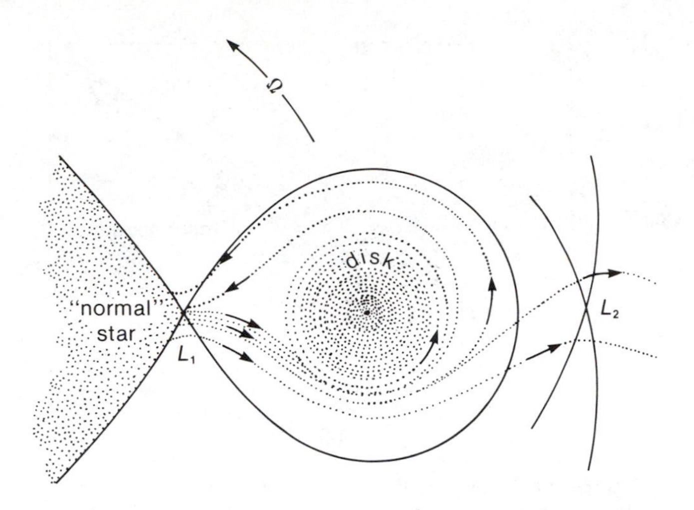

Figure 17.8. Accretion disk formed around compact star. The inner and outer Lagrange points are L_1 and L_2. Notice that the sense of rotation of the binary and of the accretion disk is the same.

significant enough to explain novae and many compact x-ray sources.

Suppose initially that $r_i > R_i$ and that during its evolution away from the main sequence (core contraction and envelope expansion), the star fills its Roche lobe and begins transferring matter to its lower-mass companion (secondary). If there is no mass loss from the binary, and there is conservation of angular momentum, (17.4) may be used to show that the binary's relative orbital radius

$$a \sim \frac{1}{M_1^2 M_2^2} = \frac{1}{M_1^2 (M - M_1)^2} \tag{17.20}$$

and thus that, if $M = M_1 + M_2$ is constant, the mutual separation of the stars changes as a result of mass exchange. In fact, it is easily shown (in Problem 17.3) that a is a minimum when the two stellar masses are equal.

Problem 17.3. Show that a_{min} occurs when $M_1 = M_2$, subject to the constraints discussed in the preceding. Furthermore, show that if M_1 is the primary ($M_1 > M_2$) and loses mass because of mass exchange, the stellar separation a decreases.

17.3. Evolving Binary Stars

As a star evolves off the main sequence, its radius R_1 begins to increase rapidly. First, as long as R_1 satisfies

$$R_1 < r_1 = a(0.38 + 0.2 \log M_1/M_2), \tag{17.21}$$

the star's evolution may be well approximated by that of an isolated system. An isolated star capable of reaching advanced stages of thermonuclear burning will undergo several expansion phases. For example, the variation in radius of a star massive enough to reach carbon burning ($M_1 \sim 5\ M_\odot$) is shown in Figure 17.9. During core hydrogen burning, the radius increases somewhat (less than a factor of two here) but probably not enough to cause mass exchange. The first significant expansion phase is shell hydrogen burning, just prior to He core burning. This occurs on a time-scale short compared to 10^7 yrs. The next expansion phase precedes core carbon ignition, and is again rapid. Each of these two expansion stages results in about a tenfold increase in R over the previous phase.

Problem 17.4. A binary consists of two stars of masses M_1 and M_2 with $M_1 = 5\ M_\odot$ and $M_2 = 2.5\ M_\odot$ and has an observed period of 146 days. Both stars are initially on the main sequence. During which phases of M_1's evolution could mass exchange occur? Assume circular orbits.

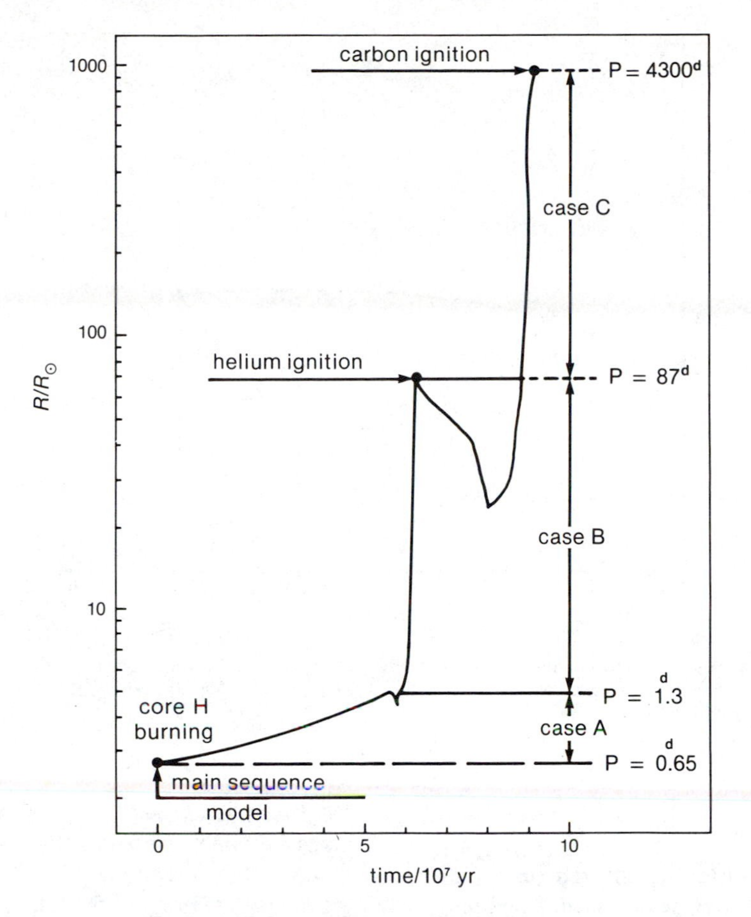

Figure 17.9. Radius versus time for a 5 $M_{\odot}$ star.

In order to get a better idea of the process of mass exchange, consider the change in stellar radius and in the radius of the Roche lobe as the star's mass decreases. The track of an isolated star in the mass-radius plane would be vertically upward (*A–C*) as shown in Figure 17.10. In a close binary, however, R increases until the Roche lobe is reached ($R = r_{in}$), where r_{in} corresponds to the initial mass before mass exchange (point *B*). At point *B* mass exchange begins. During its initial stages the stellar radius remains approximately constant, and then decreases as mass loss continues (*BDE*). Actually, this occurs for stars whose envelopes are in radiative equilibrium, and which start to lose mass on a dynamical time-scale. Once mass loss starts (point *B*), the Roche lobe r_1 decreases until $M_1 = M_2$ (*F* on dashed curve), and then increases again as the secondary becomes the more massive star in the binary. Since $R > r_1$ between points *B* and *D*, it follows that mass loss must reduce the primary's mass from its initial value ($M_{1,A}$) to a value $M_{1,D}$. Calculations show that this probably occurs on a thermal time-scale, at least for radiative envelopes (convective envelopes are more difficult to handle, but evidence suggests that the mass loss in them may occur on a dynamical time-scale). Beyond point *D*, the stellar radius (for the primary, now the less-massive star) is less than the effective radius of the Roche lobe; so two subsequent possibilities exist. First, the radius $R_1 < r_1$, and the binary may become detached. Second, if a new phase of nuclear burning sets in, the star's radius could grow on the nuclear time-scale; the star would gradually expand, filling its Roche lobe while gradual mass loss occurred, with the stellar radius R_1 essentially following r_1 (line *DG*). Thus, two stages of mass exchange may be associated with the evolution of one star: an initially rapid change in which the mass ratio of the two stars is reversed, followed by a gradual continued growth in the mass of the secondary. Studies suggest that the rapid phase could be as short as 10^4–10^5 years (recall thermal and nuclear time-scales

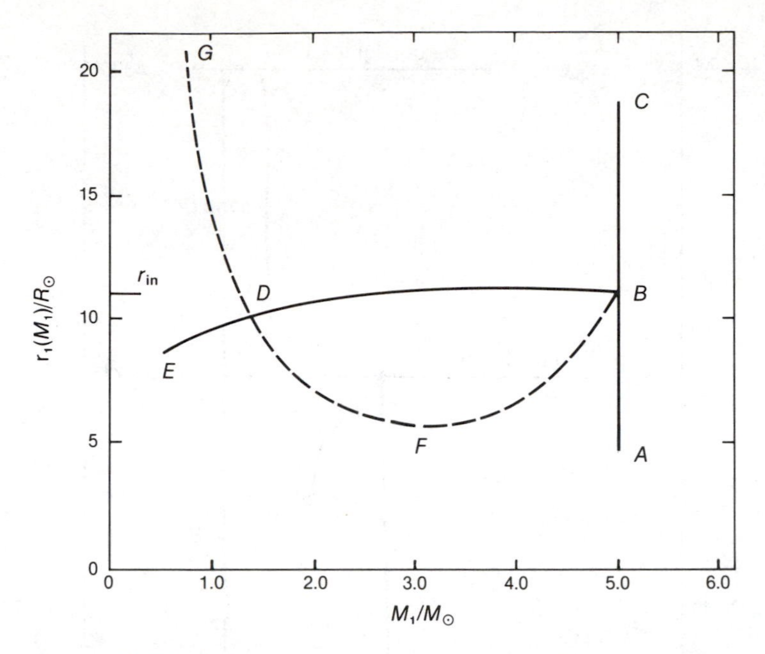

Figure 17.10. Radius versus mass for a star ($M_1 = 5\ M_\odot$) losing mass to its companion in a close binary system ($M_1 + M_2 = 7\ M_\odot$). The initial orbital period is 5 days.

tend to be of the order of 10^7 and 10^{10} years for 1 $M_\odot$ stars).

The amount of matter transferred (assuming no loss from the system) may be obtained if we know the dependence of the stellar radius R_1 on the mass M_1. Since M_1 and M_2 are known prior to mass loss (as are the period and orbital radius), we can find the new values of M_1' and M_2' at which $R_1 = r_1$ (point D). The difference

$$\delta M_1 \equiv M_1 - M_1' \tag{17.22}$$

represents the amount of matter lost (on the dynamical time-scale) by star 1. Problem 17.5 illustrates a general feature of mass exchange, namely, that the mass ratio M_1/M_2, where M_1 is the primary, is easily reversed. As a result, the less-massive member of the system often appears to be the more evolved, a conclusion that does not hold for single stars near the main sequence.

Problem 17.5. A binary system consists of stars $M_1 = 5\ M_\odot$ and $M_2 = 2.5\ M_\odot$ revolving in circular orbits with period 5 days. Assume that both stars are on the main sequence initially, but that M_1 has just exhausted H burning in its core (see Figure 17.9). Assume that the total mass $M = M_1 + M_2$ remains constant during mass exchange, and note that the angular momentum of the system J given by (17.4) remains constant. Find the masses of both stars after completion of the rapid mass-exchange phase (*BD* of Figure 17.10), assuming that the radius of star 1 (the primary) remains fixed during the process. (Hint: it will be easiest to plot r_1 against M_1 during mass exchange to find its approximate intercept with R_1 = constant.)

Certain binary systems have long been known in which the least-massive star is apparently the more evolved, Algol-type systems being the most common. In these the more-massive star appears to be near the main sequence, but the less-massive is an overluminous subgiant (radius and luminosity greater than expected for a star of the same mass on the main sequence). The subgiant is clearly the more-evolved star, and its lower mass is explained by mass exchange. The luminosity excess may be attributed to the star's filling its Roche lobe.

Problem 17.6. Draw a schematic evolutionary track for the star of mass M_1 (in Problem 17.5) in the H-R diagram. Assume that point A (Figure 17.10) corresponds to the zero-age main sequence, and that the evolution subsequent to rapid mass exchange (that is, along *DG*) is the same as that of an isolated star of about 2 $M_\odot$. Also show the evolutionary tracks for an isolated star of $M = 5\ M_\odot$ and $M = 2\ M_\odot$ in the absence of mass exchange for comparison.

Problem 17.7. Find the final period of the binary system after the mass exchange considered in Problem 17.5. Remember that the total angular momentum is assumed to be conserved.

Finding values for time scales for mass transfer is difficult, but basically they fall into the usual categories of slow nuclear, thermal, or rapid dynamical times. Usually the nuclear time-scale

$$t_n \sim (M/M_\odot)(L_\odot/L) \times 10^{10} \text{ yrs}$$

governs the rate of mass transfer if it results from expansion of the stellar radius because of core hydrogen burning. This, as noted earlier, ususally follows the initial mass transfer accompanying mass inversion. The mass and luminosity are to be taken at the onset of mass transfer, so that L will be somewhat in excess of the zero-age main-sequence value for the given mass. Thus, the relevant nuclear time will be less than that usually applied to main-sequence evolution.

If the primary's envelope is in radiative equilibrium during the onset of mass transfer, the characteristic time will be set by the Kelvin-Helmholtz or thermal scale

$$t_K \simeq \frac{M^2G}{RL} \sim 3 \times 10^7 \frac{(M/M_\odot)^2}{(R/R_\odot)(L/L_\odot)} \text{ yrs}, \qquad (17.23)$$

where R and L are again characteristic of the primary at the onset of mass exchange. Usually M and L will be given roughly by the main-sequence mass-luminosity relation (though this underestimates L and thus overestimates t_K), but R will be significantly greater than the corresponding main sequence value (see Figure 17.9 for a 5 $M_\odot$ star). Consequently t_K can be several orders of magnitude less than the usual main-sequence value of 10^7 yrs. Actual numerical calculations of t_K are in excellent agreement with this simple result. Typically it is found that t_K is between 10^3 and 10^7 yrs.

Problem 17.8. Compare t_K for a 5 $M_\odot$ primary that commences mass exchange during core H burning, just before He ignition, and just before C ignition. In all cases, estimate L from the main-sequence mass-luminosity relation. Calculate in each case the mass-loss rate in solar masses per year (see the following for dM_1/dt).

Whenever the primary's envelope is convective rather than radiative, mass loss should occur on the dynamical time scale

$$t_{ff} \sim (R^3/MG)^{1/2} \sim 5 \times 10^{-5} \frac{(R/R_\odot)^{3/2}}{(M/M_\odot)^{1/2}} \text{ yrs}.$$

For stars with relatively small R (5 $M_\odot$ undergoing mass exchange during core H burning), the characteristic time for mass loss can be on the order of days, but for the largest radii and average masses t_{ff} is generally less than several years.

The actual mass-loss rate may be estimated from the relation

$$\frac{\dot{M}}{M_\odot} = \frac{dM/M_\odot}{dt} \approx \frac{M/M_\odot}{t}, \qquad (17.24)$$

where t is the relevant time-scale. For radiative envelopes we use t_K given by (17.22), and find

$$\frac{dM/M_\odot}{dt} \simeq 3 \times 10^{-8} \frac{(R/R_\odot)(L/L_\odot)}{(M/M_\odot)} \frac{M_\odot}{\text{yr}}. \qquad (17.25)$$

All quantities are evaluated at moment mass exchange begins. This simple expression agrees quite well with detailed-models of mass-loss rates.

Advanced Evolution in Close Binary Systems

The details of the system's evolution will depend on whether the primary fills its Roche lobe during H burning or preceding He or C ignition, as well as on the behavior of the star itself during mass exchange. As an example, we will consider a system consisting initially of a 20 $M_\odot$ primary and a 6 $M_\odot$ secondary in circular orbits with period $P = 4.4$ days. Both stars are initially on the main sequence and the most significant stages in the evolution are outlined as follows (see Figure 17.11).

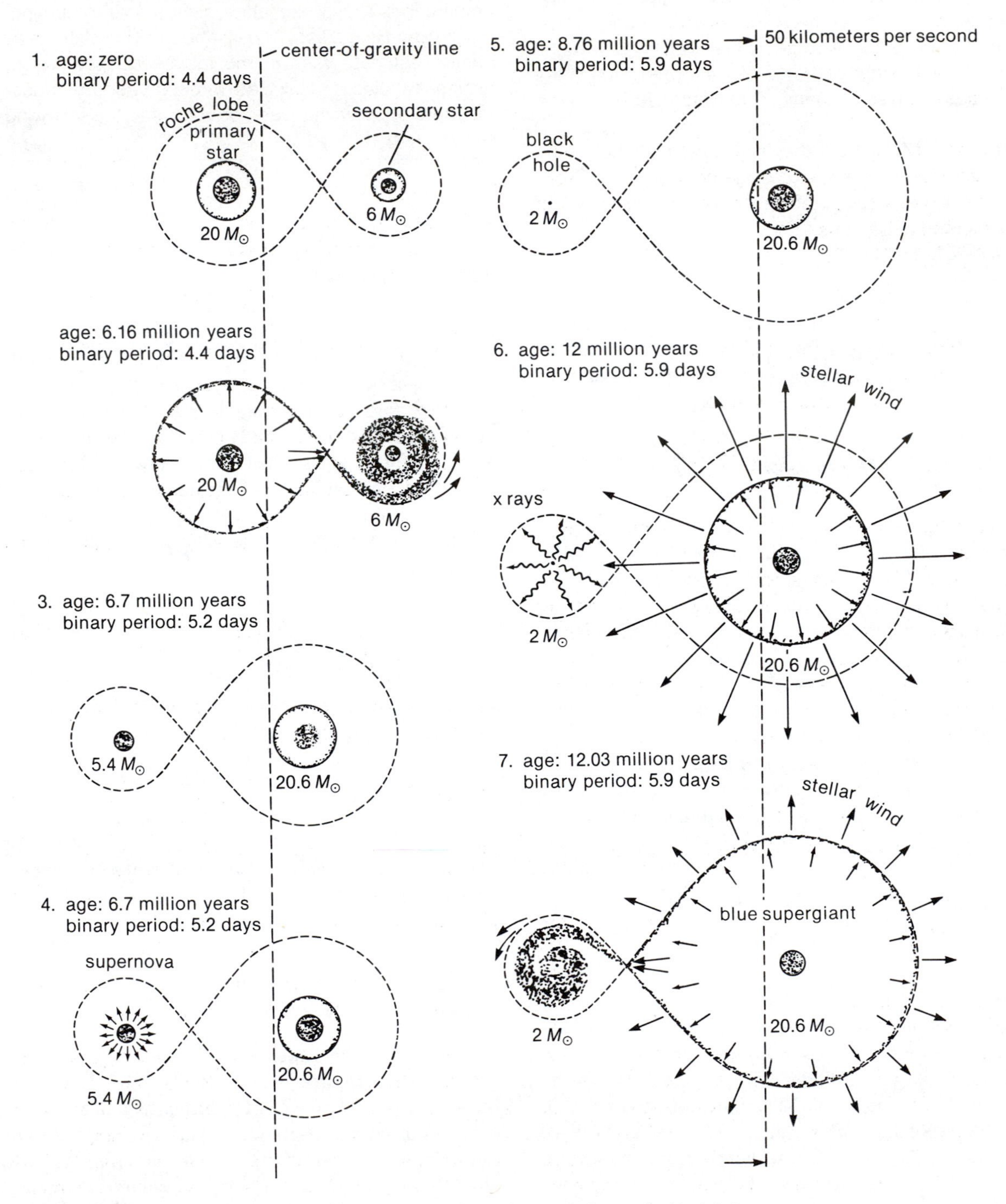

Figure 17.11. Evolution of a binary system containing a blue supergiant and a black hole.

1. Initial configuration (main sequence). The more-massive star evolves fastest, exhausting H core burning after 6.16×10^6 yrs. The main-sequence lifetime of the 6 $M_\odot$ star is about ten times longer.

2. As the envelope expands following core H burning, the star fills its Roche lobe. Matter is transferred on a dynamical time-scale (process completed in about 2×10^4 yrs!) to an accretion disk surrounding the secondary. Time elapsed $\sim 6.18 \times 10^6$ yrs.

3. After about 5×10^5 yrs., the matter surrounding the secondary has been transferred, forming a 20.6 $M_\odot$ hydrogen-rich star. The primary (now the less-massive star, with $M_1 = 5.4$ $M_\odot$) is a He-burning star that evolves through subsequent stages of nuclear burning, never expanding enough to fill its new Roche lobe. The secondary, meanwhile, proceeds to evolve as if it were a young H-burning star of about 20.6 $M_\odot$. The He-burning primary completes core burning in about 6×10^5 yrs., then proceeds on to core C, O, and Si burning, each with a shorter lifetime. The star is now on the verge of becoming a supernova. Meanwhile the secondary continues its relatively slow main-sequence evolution. Notice that we now have a system in which the more-massive star is of main-sequence type, but the less-massive is the more-evolved. Time elapsed $\sim 6.7 \times 10^6$ yrs.

4. The evolved star goes supernova, expelling a shell of matter of mass 3.4 $M_\odot$ and leaving a remnant of mass $M_R = 2$ $M_\odot$. The remnant implodes to densities in excess of 10^{12} g/cm^3 and will rapidly (on a dynamical time-scale) collapse, becoming either a neutron star or a black hole. (There may be no qualitative differences between the observed properties in these two cases, particularly if there are no magnetic fields present). As shown in the next section, the supernova does not disrupt the binary, but the mass loss will increase the size of the orbit. It also gives the entire system a recoil momentum, and may produce significant orbital eccentricity.

5. The new equilibrium orbits of the 20.6 $M_\odot$ main-sequence secondary and collapsed remnant have been reached. The system is a detached binary, possibly with eccentric orbits. Tidal interactions between the two stars will circularize the orbits in several million years. Time elapsed $\sim 6.76 \times 10^6$ yrs.

6. While on the main sequence, the secondary appears to be an O or B type star. Eventually it completes core H burning, its envelope expands, and it becomes a blue supergiant. During the first 3×10^4 yrs or so of the expansion phase, the star loses matter in the form of a stellar wind driven by the high radiation pressure in its outer layers. As both stars rotate around the center of mass, the compact object (primary) accretes matter, much of whose gravitational potential energy is converted into x rays (see Section 17.4). During this stage of its evolution the stellar wind from the blue supergiant produces a matter density around the compact object that is small, and radiation escapes having undergone little or no interaction. It is therefore observed as x rays of luminosity comparable to that of x-ray sources such as Cygnus X-1. Time elapsed $\sim 12 \times 10^6$ yrs.

7. After 3×10^4 yrs the expanding blue supergiant fills its Roche lobe and a new stage of mass exchange begins, qualitatively similar to what occurred in stage 2. An accretion disk forms around the compact star and converts much of the x radiation into ultraviolet or even visible wavelengths. The only x rays likely to escape the system will be those that are emitted by the primary normal to the orbital plane (actually, normal to the plane of the accretion disk). Time elapsed $\sim 12.03 \times 10^6$ yrs.

The intense radiation emerging from the infall of matter onto the compact star's strong gravitational field ($\sim 10^5$ times greater than the potential energy of the surface of the 6 $M_\odot$ main-sequence secondary in stage 2) regulates the rate of accretion as (discussed in Section 24.4) and limits the luminosities to values near the Eddington limit of 10^{38} ergs/sec. It is therefore likely that the compact object will grow in mass gradually.

Disruption of Binary Systems

When a star in a binary system goes supernova, it expells a shell of matter (typical velocities 2 to 7×10^3 km/sec) that ultimately escapes from the system. This decreases the gravitational force between the two remaining stars in a time roughly equal to the binary's radius divided by the velocity of the ejected material $v_{\rm ej}$:

$$t \simeq \frac{a}{v_{\rm ej}} = \frac{R_\odot}{v_{\rm ej}}\left(\frac{a}{R_\odot}\right)$$
$$= 10^{-3}(P\sqrt{M/M_\odot})^{2/3}\ \text{days}, \qquad (17.26)$$

with P in days and $v_{\rm ej} = 10^4$ km/sec. The quantity $a/R_\odot$ has been eliminated using Kepler's law. Typically $P \sim$ days and $M \sim 10$ $M_\odot$ for close binaries; so, to order of magnitude, $(P\sqrt{M/M_\odot})^{2/3} \sim 6$ days and t will be small compared to the binary period. For this reason the mass loss from the supernova will be considered instantaneous. As a result of the decreased gravitational force between the stars, whose orbital velocities are essentially the same, the system may become unbound. We explore this possibility for the special case of spherically symmetric supernova explosions and circular orbits.

The mass of the remnant that survives a supernova explosion is not yet known (by "remnant" we mean the imploded stellar core, not the expanding envelope of material, which presumably becomes a nebula as in the Crab). Current estimates suggest that masses in a narrow range around $M_R = 1.4\ M_\odot$ may result for a wide range of initial masses. When necessary we will assume for the sake of argument that the remnant mass is around 1 or 2 $M_\odot$. This means that massive presupernova stars will shed much of their mass, but the low-mass progenitors will lose less. The lower limit for supernova progenitors is also open to question; so we adopt the value 4 $M_\odot$.

The initial (presupernova) configuration is shown in Figure 17.12. The motion is most readily analyzed in a frame at rest relative to the supernova progenitor M_1. Circular orbits are assumed for simplicity. Before eruption the binary's total energy (kinetic plus potential with respect to M_1) is

$$E_{\text{total}} = \frac{1}{2}\frac{M_1 M_2}{M_1 + M_2} v^2 - \frac{M_1 M_2 G}{a}$$

$$= -\frac{M_1 M_2 G}{2a}; \tag{17.27}$$

the last form is obtained using $v^2 = (M_1 + M_2)G/a$, which is equivalent to equating the centripetal and gravitational forces acting on M_2.

The post-supernova stage is best divided into two parts, the first during which the expanding supernova shell is still within the relative orbit of M_2 (Figure 17.12b), and the second when it has passed the original orbit of M_2 (Figure 17.12c). During the first stage the secondary M_2 continues to move in its orbit of radius a, since the net gravitational potential acting on it is equivalent to a mass $M_1 = M_R + \delta m$ located at the center of the orbit. The amount of ejected mass is defined to be

$$\delta m \equiv M_1 - M_R. \tag{17.28}$$

The system's total energy is still given by (17.27). Notice that the kinetic energy of the expanding shell is

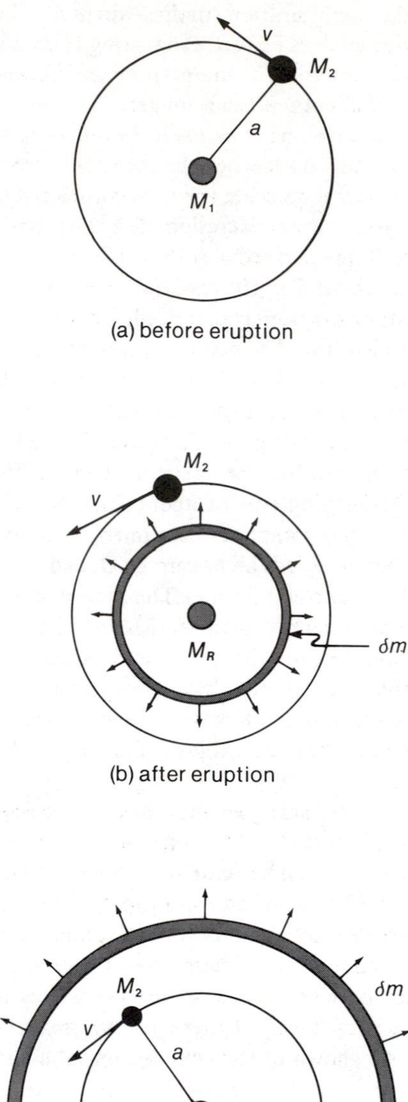

Figure 17.12. Supernova in a binary system (a) Relative orbit of M_2 about supernova progenitor M_1 before explosion. (b) Binary after supernova explosion, but before ejected mass shell reaches relative orbit of M_2. (c) Binary after ejected mass passes relative orbit.

supplied by processes internal to M_1 and may be ignored in our analysis. The orbital velocity of M_1 is still v.

Stage two starts as soon as the shell passes the orbit of M_2, which it does essentially instantaneously, as we saw in discussing (17.26). Therefore the orbital velocity of M_2 remains unchanged, and the total energy of the binary is

$$E_{\text{total}}' = \frac{1}{2}\frac{M_R M_2}{M_R + M_2}v^2 - \frac{M_R M_2 G}{a}, \quad (17.29)$$

corresponding to a binary of total mass $M_R + M_2$, but in which M_2 still moves with a velocity given by

$$v^2 = (M_1 + M_2)G/a \quad (17.30)$$

with respect to M_R. Using (17.30) to eliminate v in (17.29) we find

$$E_{\text{total}}' = \frac{M_R M_2 G}{2a}\left(\frac{M_1 + M_2}{M_R + M_2} - 2\right). \quad (17.31)$$

The new system M_R, M_2, will be bound only if the total energy E_{total}' is negative, and will be unbound (that is, the original system will have been disrupted), if $E_{\text{total}}' > 0$. Therefore (17.31) gives as the condition under which the binary will be disrupted

$$M_R < \tfrac{1}{2}(M_1 - M_2). \quad (17.32)$$

Using the definition of δm given in (17.28), we obtain the equivalent condition for disruption as

$$\delta m > \tfrac{1}{2}(M_1 + M_2). \quad (17.33)$$

We conclude that if the orbits are circular, then a supernova in a binary system must eject at least half the total initial mass of the system if disruption is to occur.

The analysis is more involved if the orbits are not circular ($\epsilon \neq 0$) or if the supernova is asymmetric. For noncircular orbits, the right-hand side of (17.33) is multiplied by the quantity $1 - \epsilon < 1$, so that less mass loss is needed for disruption. However, tidal effects in close binaries are usually sufficient to circularize orbits in a relatively short time. As a rule of thumb, the mass is lost rapidly enough by the expanding envelope if it occurs in something less than a quarter of a period, as it ususally does.

After the passage of the supernova ejecta, M_2 will gradually begin to adjust its orbit to the new gravitational force of M_R. Assuming $\delta m < (M_1 + M_2)/2$, a new equilibrium configuration will be reached in which the centripetal force on M_2 just balances the gravitational force due to M_R. The new orbit's radius b is related to the mass loss δm by

$$\frac{b - a}{a} = \frac{\delta m}{M_1 + M_2 - 2\delta m}. \quad (17.34)$$

The new equilibrium system will have a larger orbital radius [which follows from (17.34) and the fact that $\delta m > 0$] and a new period given by

$$P_b = P_a\left(\frac{b}{a}\right)^{3/2}\left(\frac{2b - a}{b}\right)^{1/2}, \quad (17.35)$$

where P_a is the original period (orbital radius a) before the supernova, and $P_b > P_a$ as expected. Observe that these results agree with the fourth and fifth evolutionary stages discussed earlier.

Problem 17.9. Derive (17.34), noting that the orbits characterized by a and b are equilibrium, circular orbits. Hint: The energy of the system is conserved during the orbital rearrangement. Use Kepler's third law to obtain (17.35).

Problem 17.10. Equations (17.34) and (17.35) are valid for ellipitical orbits as well as circular ones if a and b are interpreted as the respective semimajor axes of the systems. Explain why this is true by carefully noting how the eccentricity of the orbits would enter the derivation in the previous problem.

OB Runaway Stars

Most stars in the disk of the Galaxy have space velocities whose magnitudes, when compared to a local standard of rest, are $\lesssim 20$ km/sec. A few, however, have velocities that may exceed 200 km/sec. Most of the latter are older, Population I stars, moving in orbits that do not lie within the Galactic plane. These stars belong to the halo, and presumably originated from the matter out of which the Galaxy itself formed. There are no young, blue, OB-type stars in this category, since such stars have lifetimes typically $\sim 10^6$–10^7 yrs,

Table 17.1
Properties of x-ray binary systems. The x-ray luminosities L_x are estimates based on most probable distances. The variation in apparent

X-ray source, optical companion	Distance (kpc)	Orbital period (days)	L_x (erg/s)	Optical star: Spectral type	Optical star: $m_V(\Delta m_V)$	Optical star: $M/M_\odot$
SMC X-1 SK 160	60 ± 10	3.8927 ± 10	1.4×10^{38}	B0.5I	13.3 (0.09)	26–30
Vela X-1 HD 77581	1.4 ± 0.3	8.95 ± 2	3.2×10^{36}	B0.5Ib	6.9 (0.07)	18.5–24
Cen X-1 Krzeminski's star	5–10	2.087129 ± 7	6×10^{37}	O6.5II	13.4 (0.08)	16.5–20
Sco X-1 V 818 Sco	0.3–1	0.787313 ± 1	10^{37}	?	13 (0.2)	<2
Her X-1 HZ Her	2–6	1.700165 ± 2	7×10^{36}	A7-B0	~14 (~1)	~2
3U 1700-37 HD 153919	1.5 ± 0.5	3.4120 ± 3	6×10^{37}	O6f	6.6 (0.04)	10
Cyg X-1 HD 226868	2.5 ± 0.5	5.5999 ± 9	1.4×10^{37}	O9.7Iab	8.9 ——	>10

SOURCE: Adapted from R. N. Manchester and J. H. Taylor, *Pulsars* (San Francisco: W. H. Freeman and Co., 1977).
[a]Mass estimate for HeII and H_α emission lines.

whereas the galaxy is $\sim 10^{10}$ yrs old. Nevertheless a statistically significant number of OB-type stars do have velocities in the range 30 to 200 km/sec or more. These "runaway" stars of OB type could be formed with observed velocities by disruption of a binary system.

Suppose that the close companion of an OB-type star goes supernova, ejecting an amount of mass in excess of $\frac{1}{2}(M_1 + M_2)$, so that the binary is disrupted. The orbital velocity of the OB star relative to the system's original center of mass is then given by

$$v_{OB} = \frac{M_1}{M_1 + M_2} v$$

$$= 437 \left(\frac{M_1}{M_1 + M_2}\right)\left(\frac{M/M_\odot}{a/R_\odot}\right)^{1/2} \frac{\text{km}}{\text{sec}}, \quad (17.36)$$

where (17.3) was used in the last step. When the system is disrupted, the OB star will move with linear speed v_{OB} relative to the original center of mass. For typical masses and stellar separations ($M_1 \simeq 5\ M_\odot$, $M_2 \simeq 10\ M_\odot$, and $a \simeq 20\ R_\odot$), v_{OB} will be ~ 125 km/sec. If this mechanism for runaway production is correct, we would expect to find such stars appearing to recede from relatively old supernova remnants. Although the evidence is not conclusive some runaways do appear to be associated with such remnants.

17.4. X-RAY SOURCES

Several close binary systems have recently been observed in which one member (the primary) is an OB or late main-sequence star, and the other (the secondary) is believed to be a compact star. The secondary is not observed optically, but is a bright, often pulsed, x-ray source whose luminosity is between 10^{36} and 10^{38} erg/sec. Table 17.1 summarizes the parameters of seven systems of this type. One of the most important is Cyg X-1, which is believed to be a black hole. The systems containing pulsed x-ray sources enable reasonably good estimates of neutron-star masses to be made.

The compact x-ray sources listed in Table 17.1 vary in several ways, but they share certain characteristic features. The principal optical features include: low-amplitude variations in brightness of the primary; Doppler-shifted stellar absorption lines; primary spectral changes correlated with the binary orbital phase;

magnitude Δm_V is in mag.; the eclipse duration is in fraction of orbital period; the inclination angles are also estimates.

	X-ray source			
$M/M_{\odot}$	Eclipse duration	Short-term variability	Inclination angle	Remarks
2.2–4.2	0.14	periodic 0.7157 sec.	70°	X-ray heating of optical star evident in light curve.
1.35–1.9	0.19	nonperiodic to 1s; but periodic at 283 sec.	90°	Variations in maximum intensity, suggests apsidal motion in an eccentric orbit.
0.6–1.8	0.25	periodic 4.842 sec.	90°	
~1	?	nonperiodic to 1 sec.	—	
~1	0.16	periodic 1.23782 sec.	90°	Strong x-ray heating of optical star.
0.6	0.32	nonperiodic to 0.1 sec.	90°	Optical is under massive O6 star. Peculiar absorption features possibly due to eccentricity, or to high-density gas streams near the secondary.
9–15[a]	none observed	quasiperiodic to 1 msec.	27°	

and Doppler-shifted line emission. The x-ray emission shows variability on time-scales down to several ms, and in some systems is pulsed (pulse period ~ seconds). In all but two cases the x-ray emission is eclipsed by the primary. Finally, the binary period is typically measured in days. Figure 17.13 shows several of these features schematically. The orbital phase (shown at bottom) is the fraction of the period through which the binary has rotated. By convention the phase at mideclipse is taken to be zero. The observer is located at the bottom of the figure. At phase 0.5 the x-ray source partially eclipses the primary.

The Secondary (Neutron Star or Black Hole)

The nature of the secondary can be inferred from the millisecond variability and luminosity of the x-ray source. As with radio pulsars, variability down to milliseconds indicates an emission region less than 10^7 cm in extent. The x-ray luminosity is most likely due to accretion of matter from the primary (either overflow through the inner Lagrange point, or capture from a stellar wind), as shown by the following simplified argument.

Denoting the mass accretion rate onto the secondary by $\dot{M}_a$ and assuming that the gravitational potential energy of this matter is efficiently converted into radiation, we find that the luminosity is roughly

$$L_x \simeq \frac{M_x G \dot{M}_a}{R} \tag{17.37}$$

$$= 8.7 \times 10^{45} \frac{(M_x/M_{\odot})(\dot{M}_a/M_{\odot})}{R_6} \text{ erg sec,}^{-1}$$

where M_x and R are the mass and radius of the secondary; in the last expression $\dot{M}_a$ is in $M_{\odot}$/yr. This must be taken as an upper limit to the luminosity, since only a fraction of the gravitational potential energy released during infall will be converted into radiation (accretion is discussed in greater detail in Section 24.4). An accretion rate

$$\dot{M}_a > 10^{-5}\ M_{\odot}/\text{yr}$$

would be required to produce the observed luminosities $L_x \approx 10^{38}$ erg/sec if the secondary were a white dwarf. Such high mass-loss rates have never been observed in

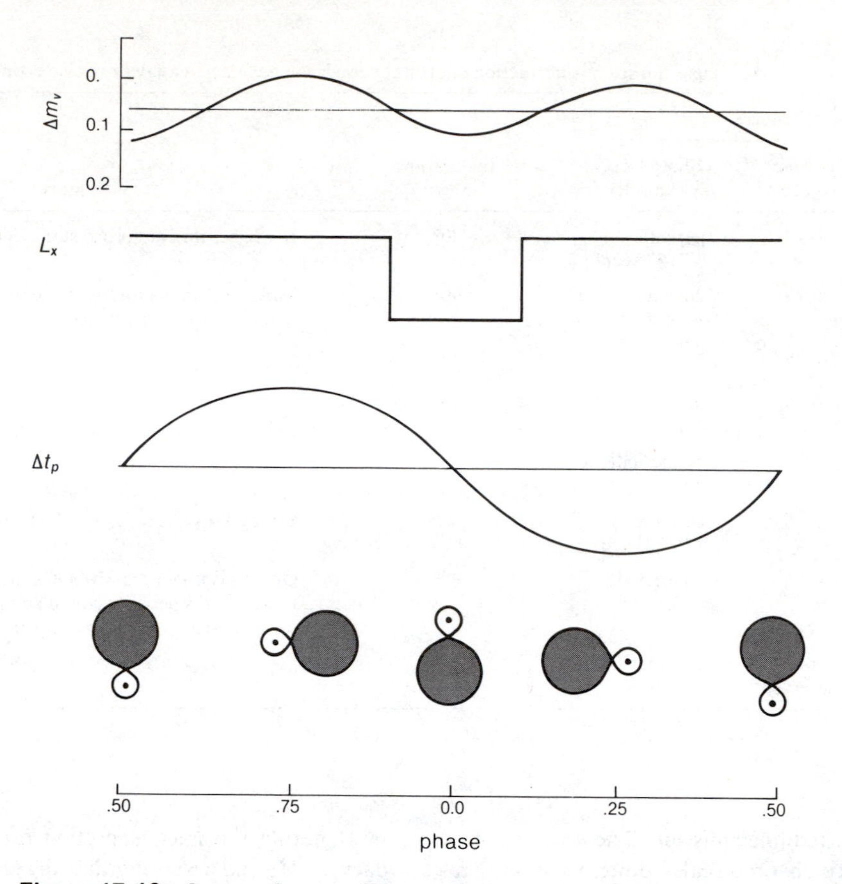

Figure 17.13. Common features of binary systems containing compact x-ray sources.

any stellar system. If the secondary is a neutron star or a black hole, then rates

$$\dot{M}_a \approx 10^{-6} \text{ to } 10^{-8}\, M_\odot/\text{yr}$$

are sufficient. These values are typical of mass-loss rates due to stellar winds or mass exchange. The emitted electromagnetic radiation must flow outward through the accreting gas. Detailed models indicate that a significant portion will be in the 2 to 20 keV range, as observed. We therefore assume that the x-ray source (the secondary) is a neutron star or a black hole.

Mass-Loss Mechanisms

Two of the systems in Table 17.1 contain low-mass ($M_0 \lesssim 2\, M_\odot$), late-spectral-type primaries, but the remainder contain massive early-type (usually giant or supergiant) stars with $M_0 \gtrsim 15\, M_\odot$. No systems containing intermediate-mass primaries are known. It is likely that two different primary mass-loss mechanisms operate. In those systems containing massive primaries, mass overflow from the inner Lagrange point (assuming that the primary fills its Roche lobe) would produce $\dot{M}_a$ in excess of $10^{-6}\, M_\odot/\text{yr}$, which would blanket the x-ray emission from the secondary. Because of their high luminosity, OB-type stars are capable of driving stellar winds. The mass loss from the primary due to a stellar wind $\dot{M}_w$ is radial, except in the immediate vicinity of the inner Lagrange point (Figure 17.14a). It is estimated that in a close binary (typical of those in Table 17.1) the mass-accretion rate onto the secondary due to a stellar wind is $\dot{M}_a \approx 10^{-3}\, \dot{M}_w$. Evolutionary calculations indicate

$$10^{-7} \lesssim \dot{M}_w \lesssim 10^{-5}\, M_\odot/\text{yr}$$

for primary masses in the range 15 to 60 $M_\odot$. If the neutron star mass $M_x = 1.5\, M_\odot$, then stellar winds

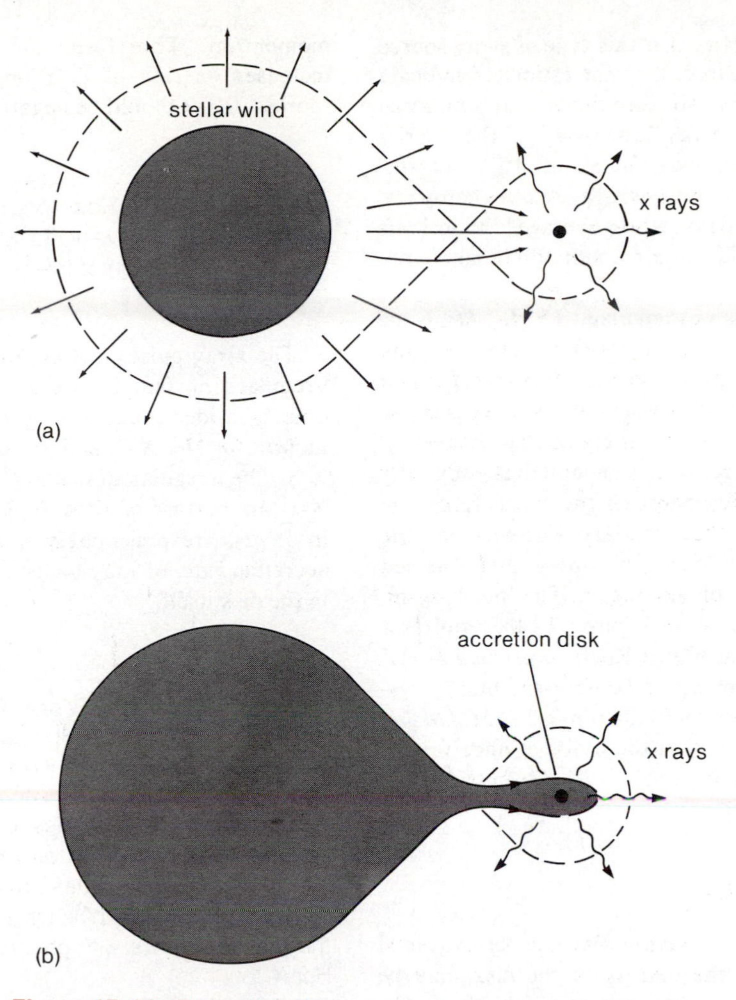

Figure 17.14. Two models of x-ray sources in binary systems: (a) stellar wind from optical star incident on neutron star or black hole generates x rays; (b) lower-mass optical star fills Roche lobe; mass transfer forms accretion disk around neutron star or black hole. Matter rains gradually onto the compact star, producing x rays.

could produce L_x in the range 10^{36} to 10^{38} erg/sec as observed. Direct evidence for stellar winds is observed in the x-ray emission from Cen X-3. The binary is seen nearly edge-on, and the x-ray source (which is pulsed) is eclipsed during a quarter of the binary period. In the absence of a stellar wind, the transition into eclipse (for example, phase 0.75 to 0.0 in Figure 17.13) would be quite rapid. If a stellar wind is present, then, as the neutron star moves away from the observer into eclipse, the emitted x-rays reaching the Earth pass through a progressively thicker column length of matter escaping from the primary; so the observed x-ray luminosity would not drop abruptly, as in Figure 17.13, but would instead decrease gradually over roughly the last hour before eclipse. Just such behavior is observed for Cen X-3. Evolutionary models predict that OB-type stars ($M \gtrsim 15 M_\odot$) can sustain a stellar wind for 10^4 to 10^6 years, after which the star expands, filling its Roche lobe. The subsequent mass overflow is probably rapid enough to extinguish the x-ray source. The effective lifetime of compact x-ray binary systems containing OB-type primaries is accordingly less than

10^6 years, which implies that this type of x-ray source may be relatively scarce. Current estimates indicate that there are probably no more than about a hundred in our Galaxy. The x-ray luminosity of the nearby spiral galaxy Andromeda is about 10^{39} erg/sec. Assuming that it also contains x-ray sources comparable to those in our Galaxy, whose average luminosity is 10^{37} erg/sec, it should contain about 100 binary x-ray sources.

For stars less massive than about 15 $M_\odot$, $\dot{M}_w$ is too small to supply the observed x-ray luminosities, and another mechanism must operate. Mass transfer in a semidetached system whose primary is more massive than about 2 $M_\odot$ will result in blanketing. For $M_0 \lesssim 2\ M_\odot$, the mass transfer is slow enough that essentially all the matter passing through the inner Lagrange point is captured by the secondary. Furthermore, the accretion rate is sufficient to supply the observed luminosities without blanketing it. This mechanism, illustrated schematically in Figure 17.14b, requires a low-mass primary that fills its Roche lobe, such as HZ Her. Evidently, there should be no x-ray binary systems with primary masses in the range $2 \lesssim M_0/M_\odot \lesssim 15$. We note that, to within the limits of uncertainty, the primary masses in Table 17.1 lie outside this range.

The Accretion Disk

The formation of an accretion disk has observational consequences, since the density in the disk may be large enough to absorb or scatter x rays from the secondary. If the rotation axes of the secondary and of the accretion disk are not parallel, then the disk may precess. This could result, for example, if the secondary formed with a slight tilt in its rotation axis. Precession of either the disk or the secondary could lead to a periodic eclipse of the x-ray source. The eclipse frequency and duration will be set by the geometry and the relative motion of the disk and secondary. Her X-1 shows, in addition to its 1.24-sec x-ray pulse period and 1.7-day orbital period, a 35-day on-off cycle in which the x-ray luminosity drops below background for about 24 days (Figure 17.15). It is possible that the process we have been discussing is responsible for this phenomenon.

The presence of an accretion disk has another observable consequence. The disk is stabilized primarily by rotation. The accreting matter, which originates from inner portions of the disk, carries angular momentum. Therefore, the secondary gradually increases its rate of rotation, and in pulsed x-ray sources dP/dt should be negative.

Problem 17.11. How long would a typical x-ray source remain active if its luminosity were derived entirely from rotation of the secondary?

The x-ray pulses in Cen X-3 and Her X-1 exhibit irregularities, but an overall average decrease with time is evident, amounting to about 6 μsec in 14 months for Her X-1, and 3msec in 21 months for Cen X-3. The irregularities, which are substantial for Her X-1 (an increase of about 6 μsec in about two months in 1972), are presumably due to fluctuation in the accretion rate, or may be associated with instabilities in the disk itself.

Problem 17.12. Consider an accretion disk surrounding a neutron star of a typical mass and moment of inertia. Matter is gradually accreted from Keplerian orbits of radius $r_d = 10^8$ cm near the disk's inner edge. Suppose that the process of accretion conserves angular momentum, and estimate dP/dt for the neutron star, assuming a mass-accretion rate $\dot{M}_a = 10^{-8}\ M_\odot$/yr, and an x-ray pulse period of 1 sec. How does the result compare with observations of Cen X-3 and Her X-1?

The accretion disk will also emit x rays (in binaries containing a black hole instead of a neutron star, the disk may be the major source of x-ray emission). The disk material moves essentially in Keplerian orbits around M_x with linear velocity $v \sim r^{-1/2}$ rather than as a rigid body for which $v \sim r$. The presence of even a small amount of viscosity will result in energy transport from the inner, rapidly moving material outward, producing disk temperatures of several times 10^6 K. The opacity of disk matter is generally high enough to produce roughly black-body radiation from the surface. Therefore the disk removes energy from its innermost boundary (the matter there no longer has sufficient energy to remain in Keplerian orbit, and rains down onto the compact secondary), thermalizes some of it throughout the disk, and finally radiates a portion of it away as x rays.

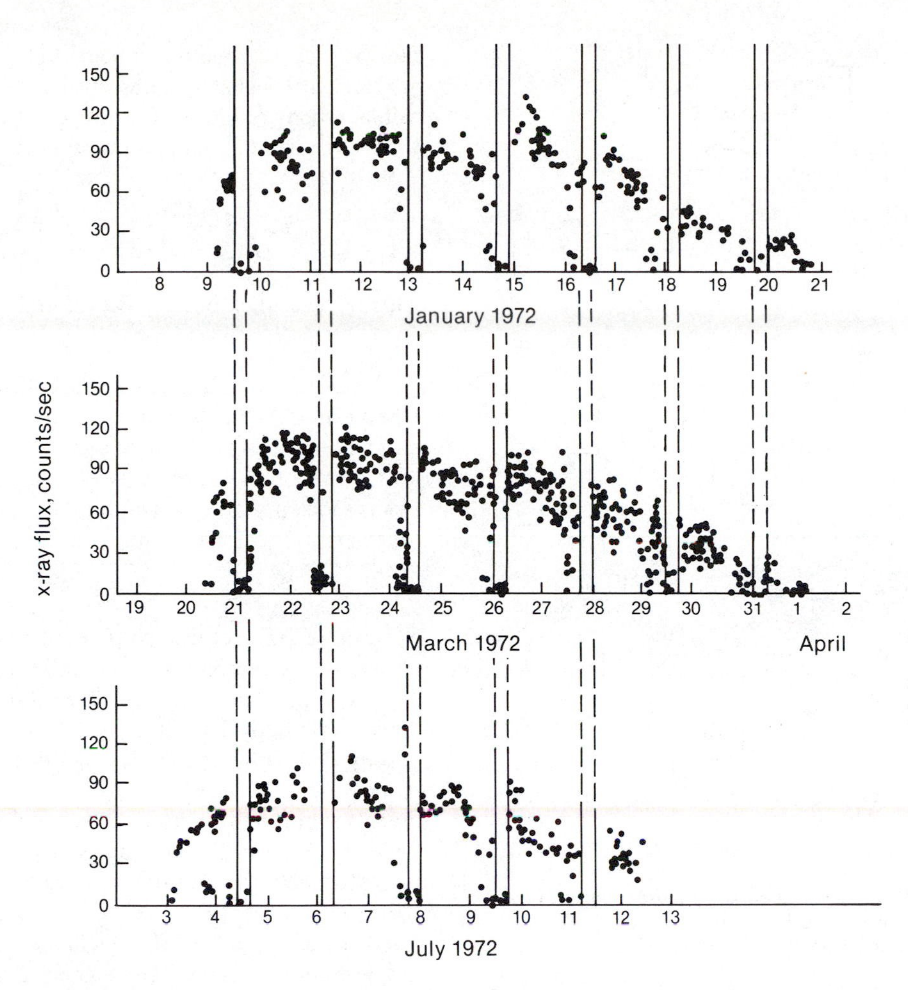

Figure 17.15. X-ray emission from Her X-1 showing 1.7-day orbital eclipse and 35-day on-off cycle.

Problem 17.13. Estimate the temperature of an accretion disk of radius $R_d = 10^8$ cm that would produce a black-body luminosity of 10^{38} erg/sec. What is the average energy of the radiation?

X-Ray Pulsars

Four of the systems in Table 17.1 contain x-ray pulsars. Since the secondary is believed to be a neutron star, it is reasonable to suppose that it may contain a strong magnetic field as do radio pulsars. This complicates the physics of accretion, since the motion of ions in the disk will be influenced by the circumstellar magnetic fields (recall that disk temperatures may exceed 10^6 K; so the matter must be ionized). The neutron star will be surrounded by a magnetosphere, but the details of its structure, and the behavior of matter in its vicinity, will be strongly influenced by the disk.

An integrated pulse profile (Figure 17.16) may be defined for x-ray pulsars just as it was for radio pulsars (Section 16.4). The pulse profiles for Cen X-3, Her X-1, and SMC X-1 share two features in common: they are double-peaked, and their widths are a substantial fraction of the orbital period; these two peaks are separated by 0.5 in phase (contrast Figure 16.10 for radio pulsars, and Figure 16.15 for the x-ray emission from PSR 0531+21).

The presence of the two subpulses differing by 0.5

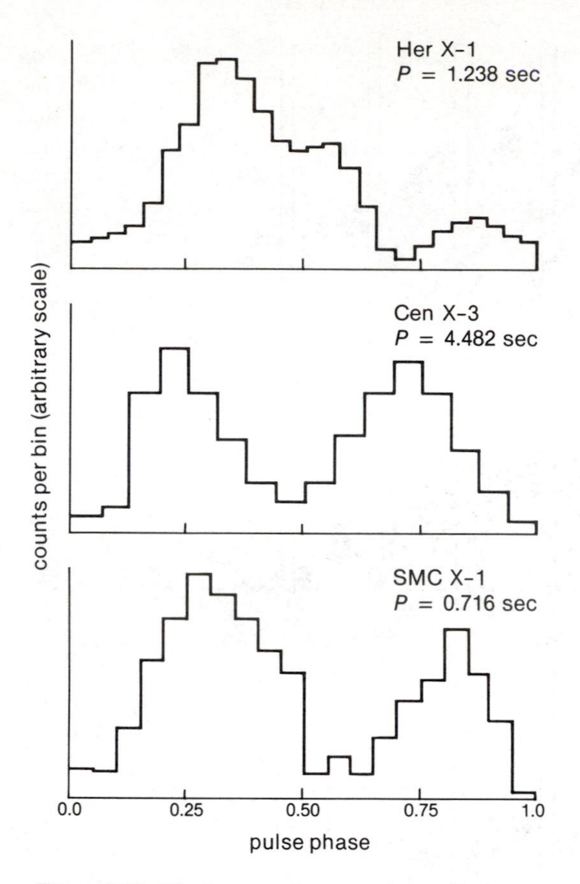

Figure 17.16. Integrated x-ray pulse profile (counts versus pulse phase) for Her X-1 (2 to 6 keV), Cen X-3 (3 to 9 keV), and SMC X-1 (1.6 to 10 keV).

in phase suggests that the two emission regions exist on or near the neutron star's surface, and that they lie in a plane making a large angle with the stellar rotation axis. If the emission regions lie near the orbital plane, the x-rays from the secondary will sweep across the near side of the primary with a period equal to the pulse period (seconds), and substantial heating of the primary may result. For example, the spectral temperature of HZ Her during eclipse (phase 0.0) is estimated to be about 7,000 K, typical of an F-type star. Analysis of the optical-light variations and other factors (as we will see) suggests that the orbital separation between HZ Her and Her X-1 is $a \sim 5.7\ R_\odot$, and that the radius of HZ Her is $R_O \approx 2\ R_\odot$. The x-ray pulse should irradate the near side of HZ Her during a time equal approximately to $(2 \sin^{-1} R_O/a)/2\pi \simeq 0.11$ times the pulse period (assuming Her X-1 is a rotating neutron star). Therefore the energy incident on HZ Her per second should be about $0.11\ L_x \simeq 7.7 \times 10^{35}$ erg/sec. If this energy is absorbed and then reradiated as black-body radiation, the effective surface temperature of the irradiated region would be

$$T_{\text{eff}} \approx \left(\frac{0.11 L_x}{\pi R_O^2 \sigma}\right)^{1/4} = 22{,}000\ K,$$

which is typical of an early B-type main-sequence star.

Problem 17.14. Models of the Cen X-3 system suggest that the orbital radius is $a \sim 8.6\ R_\odot$ and that the radius of the primary is $R_O \simeq 7R_\odot$. Spectra at eclipse indicate an effective temperature of 20,000 K for the primary. Estimate the x-ray heating effect.

Spectra of HZ Her around phase 0.5 indicate surface temperatures comparable to that we have estimated. The heating effect is present even during the 24-day down period for Her X-1, which suggests that the 35-day cycle observed in Her X-1 does not involve an on-off cycle for the x-ray emission, and supports models that attribute it to eclipse of the secondary by the disk. In actuality, the heating is more complex than indicated here. In addition to x rays from Her X-1, there will be heating due to x rays from the disk. Furthermore, not all the incident x rays will be absorbed in the atmosphere; some will scatter to form a diffuse x-ray background.

The binary systems that contain pulsed x-ray sources are extremely important because they yield perhaps the best estimates of neutron-star masses. The regularity of a pulsar can be measured by the time interval between pulses. X-ray pulsars in binary systems move in an orbit about the binary center of mass; so their velocity will undergo periodic variations about some average value (the translational velocity of the center of mass). X-ray pulsars like SMC X-1, Her X-1, or Cen X-1 emit about 10^5 regular pulses per orbital period; so they are equivalent to clocks. As they move away from the observer (from phase 0.5 to 0.0 in Figure 17.13) the interval between pulses Δt is Doppler-shifted to longer intervals; as they move toward the observer (phase 0.0 to 0.5) the interval decreases. The curve Δt against observing time is therefore equivalent to a velocity curve, whose shape and amplitude give the eccentricity of the orbit and the projected radial velocity of the secondary. The natural pulse period is equal to the value observed at phase 0.5, corrected for the linear velocity of the binary system. Given a velocity curve for the primary (obtained, for example, from variations in spectral lines) and its distance (estimated from apparent magnitude and spectral type), the two masses M_O and M_x can be calculated as a function of the orbital inclination. For systems whose primary does not exhibit excessive x-ray heating, we can estimate the mass fraction M_O/M_x as a function of inclination from detailed photometry of the primary. Combining the results of these two approaches yields the individual masses and the inclinations of the orbits (see Table 17.1). The secondary mass is always less than 2 $M_\odot$, with the exception of Cyg X-1 and possibly SMC X-1. Since the latter exhibits pulsed x rays, which are believed to require a magnetic field rotating with the secondary, it is probably not a black hole. Both Cen X-3 and Her X-1 exhibit essentially sinusoidal time-delay curves, indicating circular orbits.

The emission mechanisms operating in x-ray binary systems are different from those associated with the x-ray pulses seen in several radio pulsars. As we noted at the beginning of this section, the x radiation is believed to be released gravitational potential energy of accreted gas from the primary. The details of the gas motion, once an accretion disk has formed, and the specific processes by which radiation is emitted and transported away from the neutron star, are still under investigation; so we will describe only general features.

At large distances from the neutron star, the motion of gas in the disk is determined primarily by the disk itself; in particular the stellar magnetic field has little effect. However, at smaller radii the magnetic field strength increases, and eventually the magnetic pressure $B^2/8\pi$ becomes comparable to the radial kinetic energy of the gas; hereafter its motion is strongly influenced by the magnetic field. This defines the Alfvén surface of the neutron star (analogous to the magnetosphere in radio pulsars), and its characteristic radius r_A in the plane of the disk is typically 3×10^8 cm for a surface magnetic field of strength 10^{12} gauss. For $r > r_A$ there is a gradual tendency for disk gas to drift radially inward. Near r_A the gas motion tends to move along the magnetic field lines, eventually falling to the neutron star's surface at the magnetic poles. During the final infall, several complex processes may occur; these include electromagnetic radiation in strong magnetic fields, the formation of shock waves near the surface; and possibly thermonuclear fusion of H-rich gas into deuterium and finally He. Thermal instabilities associated with these nuclear reactions have been suggested as a source of transient x rays. Besides the inherent complexity of these processes, it is likely that portions of the disk and of the Alfvén surface may become unstable, resulting in nonsteady phenomena. The latter may explain some of the variations observed in the x-ray intensity.

Cyg X-1

The x-ray source Cyg X-1 deserves special comment, since it is widely believed to be a black hole. The x-ray emission from Cyg X-1 is neither pulsed nor eclipsed (see Table 17.1). Consequently the identification of the x-ray source as the secondary of a binary system is somewhat circumstantial. Optical searches near the x-ray position indicate that the supergiant HD 226868 is, to within positional uncertainties, coincident with Cyg X-1. Photometric investigations show periodic variations in absorption lines of HD 226868 with a period of 5.6 days, indicating that it is a member of a binary system. The shape of the optical light curve reveals tidal distortions that are most easily explained if the secondary is massive and nearby.

Evidence for mass transfer (and additional support for binary character) comes from the observed emission lines for HeII and H_α, which vary with a 5.6-day period, but with nearly opposite phase to the absorption lines originating from HD 226868. Phase-shifted emission lines of this type are observed in stellar systems in which mass transfer occurs; the difference in phase between the emission and absorption lines strongly suggests that the former originates from matter surrounding the secondary. If so, the binary is for all practical purposes a double-line spectroscopic binary, and we can find the masses of the two stars as a function of orbital inclination. Analyses that consider only the tidal distortion of the primary light curve and spectral characteristics yield a primary mass M_O in excess of 10 $M_\odot$, and probably $M_O \gtrsim 20\ M_\odot$. The absence of eclipses restricts the inclination (for a late O-type supergiant) to values greater than about 30°. The mass of the secondary must exceed 5.6 $M_\odot$. However, the HeII and H_α emission data require M_x to be in the range 9 to 15 $M_\odot$. The occurrence of x-ray variations on a millisecond time-scale implies a secondary radius (actually, emission region extent) of less than 10^7 cm, which rules out a normal star or a white dwarf. Since it is unlikely that any neutron star can be more massive than about 5 $M_\odot$ (realistic estimates set the limit below 3 $M_\odot$), it is generally concluded that Cyg X-1 is a black hole.

Several alternative interpretations of the observations of HD 226828 and Cyg X-1, which do not require the latter to be a black hole, have been advanced, but each has encountered serious difficulties or appears unduly contrived.

Problem 17.15. Consider a binary system in which no eclipse is observed. Set a lower limit to the mass M_x of the secondary based on the observed mass function $f(M_x, q, i)$ (17.6), and an assumed relationship $R/r_0 = \alpha d$ between the radius of the primary R, and the observed projected semimajor axis of the primary's relative orbit ($r_0 = a_0 \sin i$), where α is a constant, d is the distance to the binary in kpc, and $q = M_0/M_x$. No assumption about the value of M_0 is needed, but circular orbit is assumed. For HD 226868, $\alpha \simeq 2.54$, the mass function $f = 0.22$, and $d \approx$ 2kpc.

17.5. The Binary Pulsar

The radio pulsar PSR 1913 + 16 (mentioned in Section 16.4) is unusual because it is a member of a binary system. Its binary character is deduced from the periodic time-delay (see Section 17.4) observed in the radio pulses, from which is derived the first three parameters in Box 17.1, which also gives the pulsar period P, its rate of change dP/dt, and the mass function f_1. The pulsar's companion has not been observed despite extensive radio, optical, and x-ray searches. These facts, and theoretical modeling of the system's evolution, suggest that the unseen companion is also a compact object, probably a neutron star. Assuming this to be true, the binary may be considered to contain two point masses, since typical neutron-star radii are $R \approx 10^6$ cm $\approx 10^{-5}\, a$.

Problem 17.16. Estimate the tidal distortion experienced by a neutron star in orbit around another neutron star. Assume a circular orbit of radius $a = 7 \times 10^{10}$ cm. Compare this to the rotational distortion of a pulsar of period 0.059 sec.

The general theory of relativity predicts that two massive objects in close orbit will exhibit several nonclassical (i.e., non-Newtonian) features. For such a system, the effective mutual gravitational force between the two masses will deviate from the Newtonian inverse-square law. Consequently the orbits will no longer be strictly Keplerian, but will precess, and the periastron will rotate about the center of the mass with angular velocity $\dot{\omega}_p = d\omega_p/dt$. The effects of general relativity are of order 2 MG/c^2a, and the effect of orbital motion (Doppler shift) on the pulse is of order $v/c \approx 2\pi a/Pc = 5 \times 10^{-4}$. Since these corrections are small, their effect on the orbit can be derived from Einstein's equations in the weak-field limit.

Box 17.1
Parameters for the binary pulsar.

$P_b = 7.75$ hr
$a_1 \sin i = 7.0 \times 10^{10}$ cm
$\epsilon = 0.617$
$d = 5$ kpc
$P = 0.059$ sec
$dP/dt = 8.8 \times 10^{-18}$ sec/sec
$f_1 = 0.13\ M_\odot$
$dP_b/dt = -3.2 \times 10^{-12}$ sec/sec
$\dot{\omega}_p = 4.22$ deg/yr
$\gamma = 0.0047$ sec
$\sin i = 0.81$

Source: Adapted from R. N. Manchester and J. H. Taylor, *Pulsars* (San Francisco: W. H. Freeman and Co., 1977).

The Doppler shift observed in spectral lines of most binary systems is proportional to v/c, where v is the star's instantaneous linear velocity in its orbit. This is known as longitudinal Doppler shift. In addition, relativity predicts that an additional spectral shift [of order $(v/c)^2$] should occur (transverse Doppler effect) as well as a gravitational redshift of order $2MG/ac^2$. The transverse Doppler and gravitational redshift effects also follow from Einstein's equations.

The phase $\phi(t)$ of the radio pulse observed at the Earth will vary with time because of the orbital motion. This variation depends on the parameters listed in Box 17.1. In particular, it can be shown (using Einstein's equations) that the shift in periastron is described by a parameter proportional to

$$\dot{\omega}_p = \frac{6\pi G M_p}{a_1 c^2 P_b (1 - \epsilon^2)}, \tag{17.38}$$

where M_p is the pulsar's mass and a_1 is the semimajor axis of its absolute orbit. The time-rate of change of the transverse Doppler and gravitational redshift will also affect $\phi(t)$ by an amount proportional to

$$\gamma = \frac{2\pi a_1{}^2\epsilon}{c^2 P_b}(2 + M_p/M_c), \qquad (17.39)$$

where M_c is the companion mass. The total of the binary system is $M_T = M_c + M_p$. Observations of $\phi(t)$ over a four-year period have been used to find values for the parameters in Box 17.1. The last four are known to about 20 percent, but continued observations will reduce this error. The other parameters (with the exception of the distance d) are known quite accurately.

We now have sufficient data to calculate the masses M_p and M_c, even though the companion has not been observed directly. Two relations follow immediately from the periodic nature of $\phi(t)$. These are the value of $a_1 \sin i$ and the mass function $f(M_c, M_x, i)$. The orbital eccentricity follows from the velocity curve; consequently $\dot{\omega}_p$ and (17.38) yield the value of $M_p \sin i$, which, when combined with the mass function, gives the total mass

$$\begin{aligned} M_p + M_c &= (\dot{\omega}_p/2.11)^{3/2}\, M_\odot \\ &= 2.83\, M_\odot. \qquad (17.40) \end{aligned}$$

The observational uncertainty in $\dot{\omega}_p$ is less than 1 percent; so the total mass is known to better than 3 percent. Next, we may combine (17.38), (17.40), and the measured values of γ and $a_1 \sin i$ to obtain an expression for either M_p or M_c. Choosing to eliminate M_c, we find

$$M_p/M_\odot = (2.0 + 1.367 \times 10^3\, \gamma)^{1/2} - 1.41. \qquad (17.41)$$

The observed value of γ (Box 17.1) results in a mass $M_p \approx 1.49\, M_\odot$ for the pulsar. The uncertainty in γ is about 15 percent, and the uncertainty in M_p about 10 percent. The companion mass $M_c \approx 1.34\, M_\odot$ with a similar uncertainty. It is noteworthy that these values are just what one would expect for a neutron star formed by a type II supernova.

General relativity predicts that two masses in mutual orbit should radiate gravitational energy at a rate given by (16.52) multiplied by a factor $f(\epsilon)$, which accounts for the orbital eccentricity (Section 16.3). For $\epsilon = 0.617$, $f = 12$. Because the energy-loss rate is small compared with the binary's binding energy, the usual expressions for Keplerian orbits may be used to estimate the effect of $(dE/dt)_{\text{grav}}$ on the period. The total energy of the binary is related to the semimajor axis by

$$E = -\frac{M_p M_c G}{2a}, \qquad (17.42)$$

where M_c is the companion mass. Using Kepler's third law and (17.42), we find that the rate of change of the orbital period is

$$\begin{aligned} \frac{dP_b}{dt} &= -3\pi \frac{64}{5} \frac{G^{5/2}}{c^5 a^{5/2}} \\ &\quad \times M_p M_c (M_p + M_c)^{1/2} f(\epsilon) \\ &= -7.98 \times 10^{-13} \left(\frac{R_\odot}{a}\right)^{5/2} \\ &\quad \times \left(\frac{M_p M_c}{M_\odot{}^2}\right)\left(\frac{M_p + M_c}{M_\odot}\right)^{1/2} f(\epsilon), \qquad (17.43) \end{aligned}$$

where a is the semimajor axis of the relative orbit.

Problem 17.17. Carry out the analysis leading to (17.43) for dP/dt, and evaluate the result for the binary pulsar.

Using the masses already obtained and the data from Box 17.1, we find that (17.43) predicts $\dot{P}_b = -2.4 \times 10^{-12}$ sec/sec, which is within the uncertainty in the observed value. The accumulated shift in the phase delay $\Delta\phi(t)$ must be due primarily to the radiation of gravitational waves if the two stars are indeed compact, since tidal effects and other mutual interactions will be small. Therefore measurement of $\Delta\phi$ over several years should yield the rate of change dP/dt, which may be compared with the prediction of Einstein's theory (17.43). Figure 17.17 shows the phase delay accumulated over a four-year period of observations. The curve represents the accumulated phase shift, assuming that the two masses $M_c \approx M_p = 1.4\, M_\odot$. The result is particularly important for three reasons. First, it represents added support for the mass estimates of the pulsar and its companion, and may be taken as the best current measurement of a neutron-star mass. It is remarkable that we have a better mass estimate for neutron stars than for white dwarfs. Second, the measured decrease in P_b implies a concur-

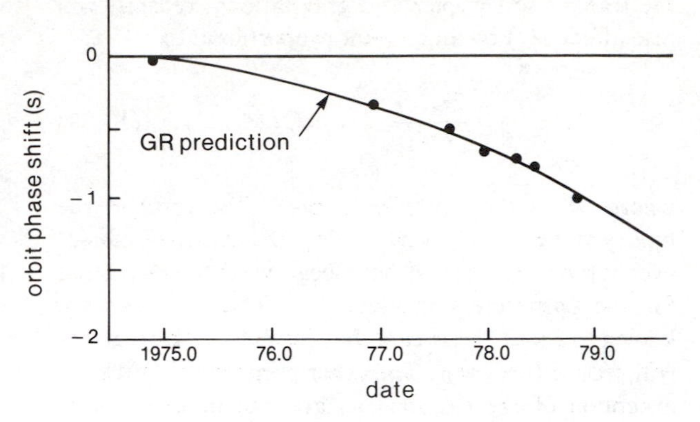

Figure 17.17. Orbital phase shift versus time for binary pulsar (dots). The curve through the data points is prediction of Einstein's general theory of relativity, assuming orbital shift is due to gravitational radiation.

rent decrease in the binary's gravitational binding energy because of gravitational wave emission, and probably represents the first unambiguous demonstration that gravitational radiation occurs in nature. Finally, several alternative theories of gravity have been advanced since Einstein's general theory of relativity, each of which predicts a rate of change dP_b/dt. The data shown in Figure 17.17 are inconsistent with some of these. Improved data, in particular measurements of d^2P_b/dt^2, may conclusively establish Einstein's theory as the correct description of gravitation.

Problem 17.18. Estimate the time required for the binary pulsar to radiate a significant fraction of its initial binding energy away as gravitational waves.

17.6. Novae

Other classes of unusual stars associated with close binary systems in which mass exchange is occurring are novae, and probably some types of flare stars (U Geminorum, for example). A nova is a transient, probably recurring, outburst of a hot, low-luminosity star. The prenova is usually observed to be an ultraviolet dwarf. The absolute visual magnitude M_V of the hot dwarf changes rapidly (~hrs) by as much as eleven magnitudes, reaching a maximum between -6 and -9, which corresponds to a luminosity in the range $2.5 \times 10^4\ L_\odot$ to $4 \times 10^5\ L_\odot$.

Novae are observed primarily near the disk in our Galaxy at a rate of about 2 or 3 per year, which implies that up to 50 may occur throughout the Galaxy each year. A schematic light curve (visual magnitude m_V versus time), which contains the general characteristics found in many novae, is shown in Figure 17.18. The rise to maximum light occurs typically over several hours in the faster novae, but the decline from maximum can take from a fraction of a year to decades (the time axis in Figure 17.18 is not uniform). For example, Nova Puppis 1942 declined from maximum light by about 6 mag in 40 days, but the subsequent decline occurred at a rate of about 3.5 mag/yr. A great deal of variation occurs among novae. The rate of decline is largest for the brightest novae (fast novae) and least for the fainter ones (slow novae). Rapid fluctuations appear during the transition period (Figure 17.18), particularly in fast novae. In others, principally slow novae, a sharp dip in brightness may occur during transition.

Spectral investigations of close binary systems containing novae reveal three principal components in the spectra. The first is the spectrum of a late-type (K or M) red dwarf with a mass of about $1\ M_\odot$. The second is a continuum typical of a hot dwarf of about the same mass. Many novae have masses below $1\ M_\odot$ (DQ Her is probably less massive than $0.5\ M_\odot$), which is typical of white dwarfs. Finally, broad emission bands, which are typical of an accretion disk around a hot star, are observed. These and other data suggest that a nova is a hot white dwarf (ultraviolet dwarf) in a close binary system whose companion is a red dwarf. Mass transfer from the red dwarf (which presumably fills its Roche lobe) produces a massive disk surrounding the white dwarf. Accretion of H-rich matter from the disk leads to unstable H burning in the surface layers of the white dwarf. Theory suggests that the resulting energy release is sufficient to eject 10^{-4} to $10^{-5}\ M_\odot$. If the rate of mass transfer from the red dwarf or the

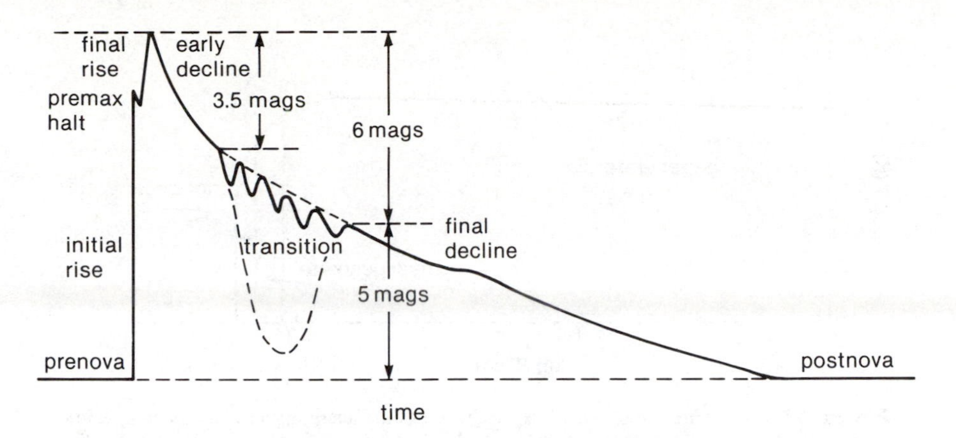

Figure 17.18. Schematic light curve for typical novae. The time axis is not uniform. Magnitudes are visual magnitudes. The transition phase does not occur in all novae.

accretion rate from the disk varies, then recurrent outbursts would be expected. Several novae are known to be recurrent, and it is suspected that all novae, given enough time, will recur. The binary character of novae is reasonably well established. Orbital periods are usually measured in hours (1.4 hr for WZ Sge; 4.6 hr for DQ Her), though a few are fractions of a year.

Evolutionary arguments support this model for novae. Consider an initial system containing stars, each of which is less massive than a few $M_\odot$; the more massive reaches the red-giant stage first, sheds some of its mass (possibly transferring it to the secondary), and becomes a hot white dwarf. The secondary then evolves off the main sequence, toward the red-giant branch. In the process it fills its Roche lobe, and begins to transfer mass to the hot white dwarf. The formation of a disk would follow in much the same manner as for compact x-ray sources (Section 17.4). We note that DQ Her, which rotates about its axis with a period of 142 sec, has been observed to be decreasing in period, as would be expected when accretion occurs.

The relation between accretion, unstable hydrogen burning, and novae outbursts can be illustrated by a simple model. Consider a hot white dwarf (mass M and radius R) whose nondegenerate envelope is of thickness $l \ll R$, and denote the rate at which it accretes mass by $\dot{M}_a$. The thermal energy density in the envelope is roughly

$$u \sim c_v T \sim n_e k T, \tag{17.44}$$

where T is the envelope temperature, and n_e the number of atoms per unit volume in the envelope. The initial gravitational energy of the infalling gas is converted into thermal energy of the envelope at a rate

$$\frac{dU}{dt} \simeq (4\pi R^2 l)\, n_e k \frac{dT}{dt}. \tag{17.45}$$

If the gravitational potential of the accreted mass is converted entirely into thermal energy, then the time-rate of change $\dot{T}$ is roughly

$$\frac{dT}{dt} \simeq \frac{A_e G m_p}{k} \frac{M}{M_e} \dot{M}_a$$

$$= 8.1 \times 10^{-16} \frac{A_e \dot{M}_a M}{M_e} K\ \mathrm{sec}^{-1}, \tag{17.46}$$

where A_e is the atomic weight of the matter in the envelope, and the envelope mass is

$$M_e \sim (4\pi R^2 l)\, A_e\, m_p\, n_e.$$

For hot white dwarfs $M \approx M_\odot$, $R \approx 10^9$ cm, $A_e \simeq 4$, and $M_e \approx 10^{-4}\, M_\odot$. Therefore an accretion rate

$$\dot{M}_a \approx 10^{-5} \text{ to } 10^{-7}\, M_\odot/\mathrm{yr}$$

would heat the envelope at a rate of 10^7 to 10^9 K/yr.

Figure 17.19 shows the temperature profile and composition predicted by theory for an accreting hot dwarf with mass 0.75 $M_\odot$ and effective temperature

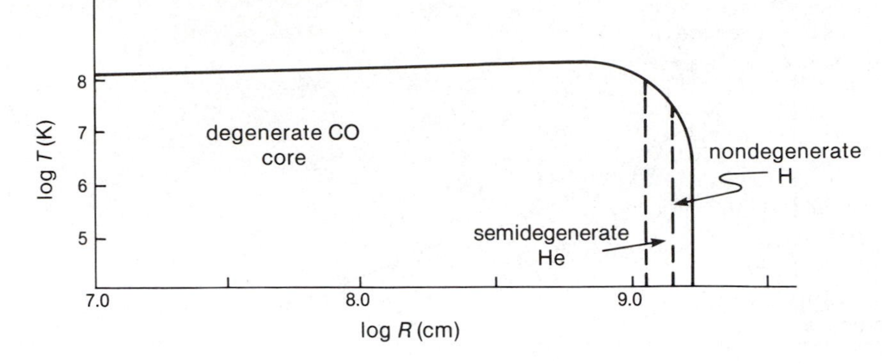

Figure 17.19. Structure of a 0.75 $M_\odot$ hot white dwarf accreting H from its companion, showing a H-burning shell (outer dashed), and a semidegenerate He layer beneath it. The core is degenerate, containing C and O. Note the gradual increase in T with increasing radius in the CO core.

$T_e \simeq 10^5$ K. When the temperature at the base of the H-rich envelope (at a density of 200 g/cm^3) reaches the ignition point, an unstable H shell-burning phase begins and establishes pulsational modes (see Section 11.7). It has been estimated that as much as 6×10^{45} ergs of energy released by nuclear reactions can be stored as pulsational energy. Numerical calculations show that these pulsations build to large-enough amplitude to eject the envelope wth velocities in excess of 10^3 km/sec. The mass of hydrogen contained in the thermally unstable shell is several times $10^{-6}\ M_\odot$, and the shell's luminosity reaches several times $10^5\ L_\odot$ just before ejection, although the total luminosity of the star may only increase to $10^4\ L_\odot$. The amount of material ejected per outburst is in the range 10^{-4} to $10^{-6}\ M_\odot$, and has a velocity at large distance from the star of several times 10^3 km/sec.

Problem 17.19. Estimate the nuclear-energy release per nova outburst, and the energy needed to eject the envelope. How does the energy release compare with the binding energy of the hot dwarf?

Problem 17.20. Carry out the derivation of (17.46), and discuss the approximations made in obtaining it.

The spectrum during the premaximum stage shows diffuse absorption lines against a strong continuum. There is little evidence of emission lines during this stage. The absorption features are Doppler-shifted toward the violet (indicating expansion velocities of about 10^3 km/sec), but there is little change in the spectral type until nearly maximum light. This suggests that the photosphere is expanding, but remains dense enough to maintain the absorption features. The diffuse absorption features imply photosphere temperatures of $1\text{–}2 \times 10^4$ K. These observations support the idea that the increase in luminosity is associated with the thermal energy content of the expanding gas. Denoting the radius of the white dwarf prior to the outburst by R_0, and the envelope's internal energy density at time t and radius r by $u(r, t)$, the time-rate of change of the luminosity is

$$\frac{dL}{dt} = -\frac{d}{dt}\int_{R_0}^{R(t)} u(r, t)\, 4\pi r^2\, dr$$
$$= -\frac{d}{dt}\int_0^{R(t)} 4\pi a T^4(r, t) r^2\, dr. \quad (17.47)$$

The last step is obtained by assuming that the radiation pressure dominates the gas pressure, and that $T(r, t)$ is a constant for $r < R_0$ (see Figure 17.19). The instantaneous edge of the advancing photosphere is $R(t)$. Equation (17.47) may be solved by the following approximations.

Using the Eddington approximation, we relate the temperature to the optical depth:

$$T^4 = \frac{1}{2} T_e^4\left(1 + \frac{3}{2}\tau\right)$$
$$= \frac{L(t)}{8\pi\sigma R^2}\left(1 + \frac{3}{2}\tau\right). \quad (17.48)$$

A simple approximation for the optical depth follows if $\kappa\rho$ is assumed constant throughout the expanding mass shell; taking $\kappa\rho = \kappa_0 \Delta M/(4\pi R^3/3)$, where ΔM is the mass that is ejected by the nova, we find that the optical depth is

$$\tau = \int_r^R \kappa\rho dr = \frac{3\kappa_0 \Delta M}{4\pi R^3}(R - r). \qquad (17.49)$$

Substituting (17.49) into (17.48) and the result into (17.47) yields an equation of the form

$$L(t) = -\frac{d}{dt}\frac{RLf(R)}{c}. \qquad (17.50)$$

Finally, assuming a constant expansion velocity $v = R/t$, we can solve (17.50) for the luminosity. The time to maximum light (the rise time) is easily shown to be

$$t_{max} = \left(\frac{3\kappa_0 \Delta M/16\pi vc}{1 + 2v/3c}\right)^{1/2}. \qquad (17.51)$$

Taking representative values for mass loss $\Delta M \approx 10^{-5} M_\odot$, expansion velocity $v \simeq 10^3$ km/sec, and $\kappa_0 = 0.4$, we find that the rise time is about 3.5 hr, which is typical of fast novae.

Near maximum light the envelope is believed to be ejected, forming a rapidly expanding circumstellar shell. At t_{max} the radius of the leading edge $R(t_{max}) \approx vt_{max}$, which is typically 10^{12} cm. Following maximum, the principal spectrum develops, consisting of broad symmetric emission bands with an absorption line at the violet edge. The former are a superposition of Doppler-shifted emission lines from the entire shell due primarily to H, CaII and NaI. The absorption line results from expanding gas lying between the observer and the white dwarf, and usually indicates a larger velocity than during the premaximum stage.

Observations suggest that several mass shells may be ejected following the initial outburst. The additional shells may be responsible for the acceleration of the initial shell following maximum light that is usually observed. There are also theoretical arguments indicating that shock waves formed at the surface of the white dwarf may contribute to the evolution of the photosphere. The late time-decay of the light curve is exponential in all novae.

Unlike supernovae, novae do not result in a restructuring of the entire star. Each outburst releases 10^{44}–10^{45} ergs as radiation, and about a tenth this amount as kinetic energy of the ejected shell. This is several orders of magnitude less than the binding energy of a hot white dwarf, whose structure is largely unaffected by each outburst. Normally an ultraviolet dwarf would cool in 10^5–10^6 years; however, the periodic accretion of H-rich matter may help to maintain the dwarf's high surface temperatures for an extended period of time.

Problem 17.21. Apply energy balance to a shell of thickness *dr* in the photosphere and derive (17.46).

The shells ejected by most novae are axially symmetric, which may indicate that nonradial pulsational modes play a role in the outburst. Finally, we note that some novae become unusually strong infrared emitters during the transition stage (Figure 17.18). The infrared excess is probably associated with the formation of dust grains in the expanding gas (see Chapter 19).

Appendix 1

CONSTANTS AND UNITS

Mathematical Constants

$\pi = 3.1416$
$e = 2.7183$
1 radian = 57.296 degrees
1 arc-sec = 1″ = 4.848×10^{-6} radians
$\log e = 0.4343 = (\ln 10)^{-1}$

Physical Constants

Speed of light	$c = 2.9979 \times 10^{10}$ cm/sec
Gravitational constant	$G = 6.670 \times 10^{-8}$ dynes cm^2 g^{-2}
Planck's constant	$h = 2\pi\hbar = 6.626 \times 10^{-27}$ erg sec
	$\hbar = 1.055 \times 10^{-27}$ erg sec
Electric charge	$e = 4.803 \times 10^{-10}$ esu
Mass of the electron	$m_e = 9.110 \times 10^{-28}$ g
Mass of the proton	$m_p = 1.673 \times 10^{-24}$ g
Mass of the pion	$m_\pi = 2.49 \times 10^{-25}$ g
Boltzmann's constant	$k = 1.381 \times 10^{-16}$ erg deg^{-1}
	$= 8.617 \times 10^{-5}$ eV deg^{-1}
Avogadro's number	$N_0 = 6.022 \times 10^{23}$ mole^{-1}
Bohr radius	$a_0 = \hbar^2/m_e e^2$
	$= 5.292 \times 10^{-9}$ cm
Bohr magneton	$\mu_B = he/4\pi m_e c = 9.274 \times 10^{-21}$ erg gauss^{-1}
Electron magnetic moment	$\mu_e = 1.001\ \mu_B$
Proton magnetic moment	$\mu_p = 1.521\ \mu_B$
Radiation constant	$a = 8\pi^5 k^4/15c^3h^3$
	$= 7.565 \times 10^{-15}$ erg cm^{-3} deg^{-4}
Stefan-Boltzmann constant	$\sigma = ac/4$
	$= 5.670 \times 10^{-5}$ erg cm^{-2} sec^{-1} deg^{-4}
Fine-structure constant	$\alpha = e^2/\hbar c = 1/137.04$
Classical electron radius	$e^2/m_e c^2 = 2.818 \times 10^{-13}$ cm
Compton wavelength of electron	$\lambda_e = \hbar/m_e c$
	$= 3.861 \times 10^{-11}$ cm
Thomson cross section	$\sigma_T = (8\pi/3)(e^2/m_e c^2)^2$
	$= 6.652 \times 10^{-24}$ cm^2
1 eV	$= 1.16 \times 10^4$ K
	$= 1.602 \times 10^{-12}$ ergs
1 Rydberg	$m_e e^4/2\hbar^2 = 13.606$ eV
	$m_p/m_e = 1836$

Astronomical Constants

Astronomical unit	A.U. $= 1.496 \times 10^{13}$ cm
Parsec	pc $= 3.086 \times 10^{18}$ cm
	$= 3.261$ light years
Solar mass	$M_\odot = 1.989 \times 10^{33}$ g
Solar radius	$R_\odot = 6.960 \times 10^{10}$ cm
Solar luminosity	$L_\odot = 3.862 \times 10^{33}$ erg/sec
Solar absolute magnitude	$M_{b,\odot} = +4.77$
Solar effective temperature	$T_\odot = 5{,}800$ K
Sidereal year	$= 3.156 \times 10^7$ sec

Appendix 2

ATOMIC MASS EXCESSES

Z	Element	A	M − A, Mev
0	*n*	1	8.07144
1	H	1	7.28899
	D	2	13.13591
	T	3	14.94995
	H	4	28.22000
		5	31.09000
2	He	3	14.93134
		4	2.42475
		5	11.45400
		6	17.59820
		7	26.03000
		8	32.00000
3	Li	5	11.67900
		6	14.08840
		7	14.90730
		8	20.94620
		9	24.96500
4	Be	6	18.37560
		7	15.76890
		8	4.94420
		9	11.35050
		10	12.60700
		11	20.18100
5	B	7	27.99000
		8	22.92310
		9	12.41860
		10	12.05220
		11	8.66768
		12	13.37020
		13	16.56160
6	C	9	28.99000
		10	15.65800
		11	10.64840
		12	0
		13	3.12460
		14	3.01982
		15	9.87320
7	N	12	17.36400
		13	5.34520
		14	2.86373
		15	0.10040
		16	5.68510
		17	7.87100
8	O	14	8.00800
		15	2.85990
		16	−4.73655
		17	−0.80770
		18	−0.78243
		19	3.33270
		20	3.79900
9	F	16	10.90400
		17	1.95190
		18	0.87240
		19	−1.48600
		20	−0.01190
		21	−0.04600
10	Ne	18	5.31930
		19	1.75200
		20	−7.04150
		21	−5.72990
		22	−8.02490
		23	−5.14830
		24	−5.94900

Z	Element	A	M − A, Mev
11	Na	20	8.28000
		21	−2.18500
		22	−5.18220
		23	−9.52830
		24	−8.41840
		25	−9.35600
		26	−7.69000
12	Mg	22	−0.14000
		23	−5.47240
		24	−13.93330
		25	−13.19070
		26	−16.21420
		27	−14.58260
		28	−15.02000
13	Al	24	0.1000
		25	−8.9310
		26	−12.2108
		27	−17.1961
		28	−16.8554
		29	−18.2180
		30	−17.1500
14	Si	26	−7.1320
		27	−12.3860
		28	−21.4899
		29	−21.8936
		30	−24.4394
		31	−22.9620
		32	−24.2000
15	P	28	−7.6600
		29	−16.9450
		30	−20.1970
		31	−24.4376
		32	−24.3027
		33	−26.3346
		34	−24.8300
16	S	30	−14.0900
		31	−18.9920
		32	−26.0127
		33	−26.5826
		34	−29.9335
		35	−28.8471
		36	−30.6550
		37	−27.0000
		38	−26.8000
17	Cl	32	−12.8100
		33	−21.0140
		34	−24.4510
		35	−29.0145
		36	−29.5196
		37	−31.7648
		38	−29.8030
		39	−29.8000
		40	−27.5000
18	Ar	34	−18.3940
		35	−23.0510
		36	−30.2316
		37	−30.9509
		38	−34.7182
		39	−33.2380
		40	−35.0383
		41	−33.0674
		42	−34.4200
19	K	36	−16.7300
		37	−24.8100
		38	−28.7860
		39	−33.8033
		40	−33.5333
		41	−35.5524
		42	−35.0180
		43	−36.5790
		44	−35.3600
		45	−36.6300
		46	−35.3400
		47	−36.2500
20	Ca	38	−21.6900
		39	−27.3000
		40	−34.8476
		41	−35.1400
		42	−38.5397
		43	−38.3959
		44	−41.4596
		45	−40.8085
		46	−43.1380
		47	−42.3470
		48	−44.2160
		49	−41.2880
21	Sc	40	−20.9000
		41	−28.6450
		42	−32.1410
		43	−36.1740
		44	−37.8130
		45	−41.0606
		46	−41.7557
		47	−44.3263
		48	−44.5050
		49	−46.5490
		50	−44.9600
22	Ti	42	−25.1230
		43	−29.3400
		44	−37.6580
		45	−39.0020
		46	−44.1226
		47	−44.9266
		48	−48.4831
		49	−48.5577
		50	−51.4307
		51	−49.7380
		52	−49.5400
23	V	46	−37.0600
		47	−42.0100
		48	−44.4700
		49	−47.9502
		50	−49.2158
		51	−52.1989
		52	−51.4360
		53	−52.1800
		54	−49.6300
24	Cr	48	−42.8130
		49	−45.3900
		50	−50.2490
		51	−51.4472
		52	−55.4107
		53	−55.2807
		54	−56.9305

Z	Element	A	M − A, Mev	Z	Element	A	M − A, Mev
		55	−55.1130			62	−66.7480
		56	−55.2900			63	−65.5160
25	Mn	50	−42.6480			64	−67.1060
		51	−48.2600			65	−65.1370
		52	−50.7020			66	−66.0550
		53	−54.6820	29	Cu	58	−51.6590
		54	−55.5520			59	−56.3590
		55	−57.7048			60	−58.3460
		56	−56.9038			61	−61.9840
		57	−57.4800			62	−62.8130
		58	−55.6500			63	−65.5831
26	Fe	52	−48.3280			64	−65.4276
		53	−50.6930			65	−67.2660
		54	−56.2455			66	−66.2550
		55	−57.4735			67	−67.2910
		56	−60.6054			68	−65.4100
		57	−60.1755	30	Zn	60	−54.1860
		58	−62.1465			61	−56.5800
		59	−60.6599			62	−61.1230
		60	−61.5110			63	−62.2170
		61	−59.1300			64	−66.0003
27	Co	54	−47.9940			65	−65.9170
		55	−54.0140			66	−68.8810
		56	−56.0310			67	−67.8630
		57	−59.3389			68	−69.9940
		58	−59.8380			69	−68.4250
		59	−62.2327			70	−69.5500
		60	−61.6513			71	−67.5200
		61	−62.9300			72	−68.1440
		62	−61.5280	31	Ga	63	−56.7200
		63	−61.9200			64	−58.9280
28	Ni	56	−53.8990			65	−62.6580
		57	−56.1040			66	−63.7060
		58	−60.2280			67	−66.8650
		59	−61.1587			68	−67.0740
		60	−64.4707			69	−69.3262
		61	−64.2200			70	−68.8970

SOURCE: D. Clayton, *Principles of Stellar Evolution and Nucleosynthesis* (New York: McGraw-Hill, 1968)

Bibliography

General

G. O. Abell, *Exploration of the Universe,* 2nd ed. (New York: Holt, Rinehart & Winston, 1975).

C. W. Allen, *Astrophysical Quantities* (London: Athlone Press, 1973).

F. H. Shu, *The Physical Universe* (Mill Valley, Calif.: University Science Books, 1982).

M. Harwit, *Astrophysical Concepts* (New York: Wiley, 1973).

F. Hoyle and J. Narlikar. *The Physics-Astronomy Frontier* (San Francisco: W. H. Freeman, 1980).

W. K. Rose, *Astrophysics* (New York: Holt, Rinehart & Winston, 1973).

A. Unsold, *The New Cosmos* (New York: Springer-Verlag, 1977).

VOLUME I

PART 1: INTRODUCTION

Chapter 1: An Overview of Stellar Structure and Evolution

I. S. Shklovskii, *Stars: Their Birth, Life and Death* (San Francisco: W. H. Freeman, 1978).

Chapter 2: Properties of Matter

H. B. Callen, *Thermodynamics* (New York: Wiley, 1961).

F. Reif, *Fundamentals of Statistical and Thermal Physics* (New York: McGraw-Hill, 1965).

G. H. Wannier, *Statistical Physics* (New York: Wiley, 1966).

Ya. B. Zeldovich and I. D. Novikov. *Relativistic Astrophysics* (Chicago: University of Chicago Press, 1971).

Chapter 3: Aspects of Observational Astronomy

G. O. Abell, *Exploration of the Universe,* 2nd ed. (New York: Holt, Rinehart & Winston, 1975).

E. E. Salpeter, "The Luminosity Function and Stellar Evolution." *Ap. J.,* **121** (1955): 161.

T. L. Swihart, *Astrophysics and Stellar Astronomy* (New York: Wiley, 1968).

R. Trumpler and W. Weaver, *Statistical Astronomy* (Berkeley, Calif.: University of California Press, 1953).

PART 2: STELLAR STRUCTURE

General

J. P. Cox and R. T. Guili, *Stellar Structure,* Vol. I (New York: Gordon and Breach, 1968).

H. Y. Chiu, *Stellar Physics* (Waltham, Mass.: Blaisdell, 1968).

L. Motz, *Astrophysics and Stellar Structure* (Waltham, Mass.: Ginn, 1970).

E. Novotny, *Introduction to Stellar Atmospheres and Interiors* (Oxford, England: Oxford University Press, 1973).

Chapter 4: Static Stellar Structure

S. Chandrasekhar, *An Introduction to the Study of Stellar Structure* (Chicago: University of Chicago Press, 1939).

M. Schwarzschild, *The Structure and Evolution of the Stars* (Princeton, N.J.: Princeton University Press, 1958).

H. Y. Chiu, *Stellar Physics* (Waltham, Mass.: Blaisdell, 1968).

Chapter 5: Radiation and Energy Transport

S. Chandrasekhar, *Radiative Transfer* (New York: Dover, 1960).

T. G. Cowling, "Magnetic Stars," in *Stellar Structure,* edited by L. H. Aller and D. B. McLaughlin (Chicago: University of Chicago Press, 1965).

L. Mestel, "Meridian Circulation in Stars," in *Stellar Structure,* edited by L. H. Aller and D. B. McLaughlin (Chicago: University of Chicago Press, 1965).

D. Mihalas, *Stellar Atmospheres,* 2nd ed. (San Francisco: W. H. Freeman, 1978).

E. A. Spiegel, "Convection in Stars. I: Basic Boussinesq Convection," *Ann. Rev. Ast. Ap.,* **9** (1971): 323.

———, "Convection in Stars. II: Special Effects," *Ann. Rev. Ast. Ap.,* **10** (1972): 197.

A. Unsold, *Physik der Sternatmosphären* (Berlin: Springer-Verlag, 1955).

Chapter 6: Atomic Properties of Matter

A. N. Cox, "Stellar Absorption Coefficients and Opacities," in *Stellar Structure,* edited by L. H. Aller and D. B. McLaughlin (Chicago: University of Chicago Press, 1965).

T. R. Carson, "Stellar Opacity," *Ann. Rev. Ast. Ap.,* **14** (1976): 95.

D. G. Hummer and G. Rybicki, "The Formation of Spectral Lines," *Ann. Rev. Ast. Ap.,* **9** (1971): 237.

J. T. Jeffries, *Spectral Line Formation* (Waltham, Mass.: Blaisdell, 1968).

PART 3: STELLAR EVOLUTION

General

D. D. Clayton, *Principles of Stellar Evolution and Nucleosynthesis* (New York: McGraw-Hill, 1968).

J. P. Cox and R. T. Guili, *Stellar Structure,* Vol. II (New York: Gordon and Breach, 1968).

H. Gursky, "Neutron Stars, Black Holes and Supernovae," in *Frontiers of Astrophysics,* edited by E. H. Avrett (Cambridge, Mass.: Harvard University Press, 1976).

I. Iben, Jr., "Normal Stellar Evolution," in *Stellar Evolution,* edited by H. Y. Chiu and A. Murriel (Cambridge, Mass.: MIT Press, 1972).

S. L. Shapiro and S. A. Teukolsky, *Black Holes, White Dwarfs and Neutron Stars* (New York: Wiley, 1983).

R. F. Stein, "Stellar Evolution: A Survey with Analytical Models," in *Stellar Evolution,* edited by R. F. Stein and A. G. W. Cameron (New York: Plenum Press, 1966).

Ya. B. Zeldovich and I. D. Novikov, *Relativistic Astrophysics* (Chicago: University of Chicago Press, 1971).

Chapter 7: Nuclear Energy Sources

D. D. Clayton, *Principles of Stellar Evolution and Nucleosynthesis* (New York: McGraw-Hill, 1968).

Chapter 8: Introduction to Stellar Evolution

R. J. Taylor, *The Stars: Their Structure and Evolution* (New York: Springer-Verlag, 1972).

Chapter 9: The Main Sequence

L. Motz, *Astrophysics and Stellar Structure* (Waltham, Mass.: Ginn, 1970).

Chapter 10: Evolution Away from the Main Sequence

I. Iben, Jr., "Post Main Sequence Evolution of Single Stars," *Ann. Rev. Ast. Ap.,* **12** (1974): 215.

L. Motz, *Astrophysics and Stellar Structure* (Waltham, Mass.: Ginn, 1970).

Chapter 11: Deviations from Quasistatic Evolution

N. Baker, "Stellar Stability and Stellar Pulsations," in *Stellar Evolution,* edited by H. Y. Chiu and A. Murriel (Cambridge, Mass.: MIT Press, 1972).

J. P. Cox, "Nonradial Oscillations of Stars: Theory and Observations," *Ann. Rev. Ast. Ap.,* **14** (1976): 247.

R. F. Christy, "Variable Stars—Realistic Models," in *Stellar Evolution,* edited by H. Y. Chiu and A. Murriel (Cambridge, Mass.: MIT Press, 1972).

K. J. Fricke and R. Kippenhahn, "Evolution of Rotating Stars," *Ann. Rev. Ast. Ap.,* **10** (1972): 45.

P. Ledoux, "Stellar Stability," in *Stellar Structure,* edited by L. H. Aller and D. B. McLaughlin (Chicago: University of Chicago Press, 1965).

Chapter 12: Final Stages of Stellar Evolution

W. D. Arnett, G. J. Hansen, J. W. Truran, and A. G. W. Cameron, eds., *Nucleosynthesis* (New York: Gordon and Breach, 1968).

D. D. Clayton, *Principles of Stellar Evolution and Nucleosynthesis* (New York: McGraw-Hill, 1968).

E. E. Salpeter, "Stellar Evolution Leading up to White Dwarfs and Neutron Stars," in *Relativity Theory and Stellar Structure, Lectures in Applied Mathematics,* Vol. 10, American Mathematical Society, 1967.

Chapter 13: Weak Interactions in Stellar Evolution

J. N. Bahcall and R. L. Sears, "Solar Neutrinos," *Ann. Rev. Ast. Ap.,* **10** (1972): 25.

Z. Barkat, "Neutrino Processes in Stellar Interiors," *Ann. Rev. Ast. Ap.,* **13** (1975): 45.

H. Y. Chiu, *Stellar Physics* (Waltham, Mass.: Blaisdell, 1968).

Chapter 14: Degenerate Stars

J. R. P. Angel, "Magnetic White Dwarfs," *Ann. Rev. Ast. Ap.,* **16** (1978): 487.

L. Mestel, "The Theory of White Dwarfs," in *Stellar Structure,* edited by L. H. Aller and D. B. McLaughlin (Chicago: University of Chicago Press, 1965).

J. P. Ostriker, "White Dwarfs," in *Stellar Evolution,* edited by H. Y. Chiu and A. Murriel (Cambridge, Mass.: MIT Press, 1972).

R. Sexl and H. Sexl, *White Dwarfs—Black Holes* (New York: Academic Press, 1979).

Chapter 15: Supernovae

G. E. Brown, H. A. Bethe, and G. Baym, "Supernovae," *Nucl. Phys.,* **A375** (1982): 481.

D. N. Schramm, *Supernovae* (Dordrecht, Holland: Reidel, 1977).

F. Zwicky, "Supernovae," in *Stellar Structure,* edited by L. H. Aller and D. B. McLaughlin (Chicago: University of Chicago Press, 1965).

Chapter 16: Compact Stellar and Relativistic Objects

G. Baym and C. Pethick, "Physics of Neutron Stars," *Ann. Rev. Ast. Ap.,* **17** (1979): 415.

R. Giaconi and R. Ruffini, *Physics and Astrophysics of Neutron Stars and Black Holes* (Amsterdam: North Holland, 1978).

H. Gursky and R. Ruffini, *Neutron Stars, Black Holes and Binary X-ray Sources* (Dordrecht, Holland: Reidel, 1975).

W. H. Press and K. S. Thorne, "Gravitational-Wave Astronomy," *Ann. Rev. Ast. Ap.,* **10** (1972): 335.

C. Misner, K. S. Thorne, and J. A. Wheeler, *Gravitation* (San Francisco: W. H. Freeman, 1973).

R. N. Manchester and J. H. Taylor, *Pulsars* (San Francisco: W. H. Freeman, 1977).

J. V. Narliker, *Introduction to Cosmology* (Boston: Jones and Bartlett, 1983).

M. Ruderman, "Pulsars: Structure and Dynamics," *Ann. Rev. Ast. Ap.,* **10** (1972): 427.

R. Sexl and H. Sexl, *White Dwarfs—Black Holes* (New York: Academic Press, 1979).

S. Weinberg, *Gravitation and Cosmology,* (New York: Wiley, 1972).

Ya. B. Zeldovich and I. D. Novikov, *Relativistic Astrophysics* (Chicago: University of Chicago Press, 1971).

Chapter 17: Close Binary Systems

B. Paczynski, "Close Binaries," in *Stellar Evolution,* edited by H. Y. Chiu and A. Murriel (Cambridge, Mass.: MIT Press, 1972).

J. S. Gallagher and S. Starrfield, "Theory and Observation of Classical Novae," *Ann. Rev. Ast. Ap.,* **16** (1978): 171.

W. K. Rose, "Novae," in *Stellar Evolution,* edited by H. Y. Chiu and A. Murriel (Cambridge, Mass.: MIT Press, 1972).

Index

Volume I ends on page 344. Volume II begins on page 345.

Volume I ends on page 344

Volume I ends on page 344

Volume I ends on page 344

Volume I ends on page 344

 Volume I ends on page 344

Volume I ends on page 344

Volume I ends on page 344

Volume I ends on page 344